# Emergence, Complexity and Computation

Volume 42

The Emergence, Complexity and Computation (ECC) series publishes new developments, advancements and selected topics in the fields of complexity, computation and emergence. The series focuses on all aspects of reality-based computation approaches from an interdisciplinary point of view especially from applied sciences, biology, physics, or chemistry. It presents new ideas and interdisciplinary insight on the mutual intersection of subareas of computation, complexity and emergence and its impact and limits to any computing based on physical limits (thermodynamic and quantum limits, Bremermann's limit, Seth Lloyd limits…) as well as algorithmic limits (Gödel's proof and its impact on calculation, algorithmic complexity, the Chaitin's Omega number and Kolmogorov complexity, non-traditional calculations like Turing machine process and its consequences,…) and limitations arising in artificial intelligence. The topics are (but not limited to) membrane computing, DNA computing, immune computing, quantum computing, swarm computing, analogic computing, chaos computing and computing on the edge of chaos, computational aspects of dynamics of complex systems (systems with self-organization, multiagent systems, cellular automata, artificial life,…), emergence of complex systems and its computational aspects, and agent based computation. The main aim of this series is to discuss the above mentioned topics from an interdisciplinary point of view and present new ideas coming from mutual intersection of classical as well as modern methods of computation. Within the scope of the series are monographs, lecture notes, selected contributions from specialized conferences and workshops, special contribution from international experts.

Indexed by zbMATH.

More information about this series at https://link.springer.com/bookseries/10624

Andrew Adamatzky
Editor

# Automata and Complexity

## Essays Presented to Eric Goles on the Occasion of His 70th Birthday

Springer

*Editor*
Andrew Adamatzky
Unconventional Computing Centre
University of the West of England
Bristol, UK

ISSN 2194-7287 ISSN 2194-7295 (electronic)
Emergence, Complexity and Computation
ISBN 978-3-030-92550-5 ISBN 978-3-030-92551-2 (eBook)
https://doi.org/10.1007/978-3-030-92551-2



This Springer imprint is published by the registered company Springer Nature Switzerland AG
The registered company address is: Gewerbestrasse 11, 6330 Cham, Switzerland

# Preface

Eric Goles is one of the world leaders in the field of automata and complexity. His made groundbreaking discovering theory and analysis of complex systems, particularly in the field of discrete systems dynamics such as neural networks, automata networks, majority networks, bootstrap percolation models, cellular automata, computational complexity theory, discrete mathematics and theoretical computer science. This book commemorates Eric Goles's achievements in science and engineering. The chapters are authored by world leaders in computer science, physics, mathematics and engineering.

The book will be a pleasure to explore for readers from all walks of life, from undergraduate students to university professors, from mathematicians, computers scientists and engineers to chemists and biologists.

Bristol, UK
July 2021

Andrew Adamatzky

# Contents

# Eric Goles

**Andrew Adamatzky**

Eric Goles, son of a theatre actress and a musician, was born in Antofagasta, in northern Chile, between the Pacific Ocean and the desert of Atacama. In 1970, he joined the University of Chile and graduated with a degree in mathematical engineering in 1975. He later went to the University of Grenoble, France, to carry out doctoral studies which culminated in 1980 with his thesis "Comportement oscillatoire d'une famille d'automates cellulaires non uniformes" where he proved the nowadays famous theorem that symmetric threshold automaton oscillates only with period one (fixed points) or two [12, 30, 57, 58, 62]. A nice review of this theorem can be found in [16]. In 1982, Eric joined the prestigious CNRS (French National Centre for Scientific Research) and moved to France. First to the Institute of Applied Mathematics, IMAG, at the University of Grenoble and later to the Laboratory on Network dynamics and Epistemology at the Polytechnic Institute in Paris. During this period, he carried out a state thesis in Mathematics, also in the field of automata networks [31]. After moving back to Chile to join the Engineering School of the Universidad de Chile, where he worked as a Professor till 2006. He then moved to the Faculty of Engineering and Sciences at the University Adolfo Ibáñez where he is still working. In 2004 he also founded the Institute of Complex Systems at Valparaiso, the first-ever Chilean research establishment devoted to complex systems. In 1993 he was honoured with the main scientific award of his country, the National Science Price. He has written more than two hundred articles and ten books. Further, he has trained a huge number of young scientists both in Chile and abroad. Eric excels in numerous fields like theoretical computer science, discrete mathematics, neural and Boolean networks, cellular automata and mathematical modelling in physics,

A. Adamatzky (✉)
Unconventional Computing Laboratory, University of the West of England,
Bristol BS16 1QY, UK
e-mail: andrew.adamatzky@uwe.ac.uk



A. Adamatzky (ed.), *Automata and Complexity*, Emergence, Complexity
and Computation 42, https://doi.org/10.1007/978-3-030-92551-2_1

biology, and social sciences. Here we provide just a few examples of his and his colleagues' outstanding research results.

**Sand Piles and Chip Firing Game**

The sand piles model, closely related to Spencer's chip firing game [75], was introduced as a tool to study self-organised criticality [2]. Eric analysed this model from complexity and computational universality points of view [38, 55, 56]. Results of the analysis were impressive. In 1992, Eric determines the lattice structure of the sand pile automata [33] and in Goles and Kiwi provided bounds for the transient time length of the sand pile modules, characterised the fixed points to which they converge and gave closed formulas for the sequential transient time [36, 37]. In the same year Bitar and Goles determined the two-cycle behaviour of the parallel chip firing game on trees and demonstrated that the chip firing game belongs to Wolfram class 4 automata [3]. In series of influential papers Eric and colleagues demonstrated that sand pile and chip firing game are universal computers. That is by representing logical truth by presence of a sand grain or a chip and logical false by absence of the grain/chip one route information as avalanches and implement logical gates via interaction of avalanches in an appropriate geometrical structure [23, 42, 43]. Other impressive results related to sand pile model, developed by Eric and colleagues, include estimation of computational complexity of sandpile avalanches [21], analysis of sandpile dynamical responses to reversal of the avalanche's source position [59], determination of non-polynomial (almost exponential) periods for the parallel chip firing game [65].

**Sakoda and Schelling Models**

In 1943 Sakoda developed a model of attitude-based social interaction of discrete agents, which was only published thirty years later in 1971 [73] at the same time as the similar model proposed by Schelling [74]. The models laid a foundation for studies of spatial mechanisms of social dynamics and economy. When Schelling was awarded a Nobel Prize in Economy "his models of segregation" was cited in an official documents as one reasons for the award [63]. In 2011 Goles et al. [17] studied exhaustively the behaviour of a generalized Schelling model in a two and three-dimensional grid with several neighbourhoods and they establish an energy operator associated with the model's dynamics. Other developments concerning the Schelling model can be seen in a combinatorial game developed by Eric and colleagues: a line (or cycle) graph with white and black tokens and an empty site where two players move alternately one of its colour's token to the hole trying to reach a connected configuration of its tokens [35]. Eric's interest in social models continued to develop precisely in the almost forgotten model of Sakoda through the characterization of the dynamics of all aptitude rules in one and two-dimensional grids [69]. More recently a generalisation of Schelling's model to other local functions in a two-dimensional grid has been done in [78].

**Communication Complexity**

This topic of computer science research attracted Eric's interest for a number of years. In 2008 Eric and colleagues proposed to define cellular automata via their communication protocols: "if we are able to give a protocol describing a cellular automaton, then we can understand its behaviour" [46]. In the same paper they proposed a hierarchy of complexity classes in cellular automata, based on their communication complexity of the automata. In [64] the same authors determine complexity equivalences between one round communication protocols and intrinsic cellular automata universality. Analysis of communicating protocols emerging from cellular automaton dynamics has been continued in [39] where authors developed a non-trivial communication protocol describing dynamics of elementary cellular automaton rule 218. In [28] Eric and colleagues have given protocols, and lower bounds of the same order, that together solve the one-round communication complexity of the prediction problem for nearly one-third of all elementary cellular automata, corresponding to the family of monotone rules. Most recent results of Eric's team have been about communication complexity of number-conserving cellular automata [28, 54].

**Computational Complexity and Universality**
These two topics usually go a pair in Eric and colleagues' works because the computational complexity of a system is typically estimated by embedding a relevant Boolean circuit in the system. In 1993 Eric and colleagues constructed a simulation of a Turing machine by cellular automata based on the equivalence between programmable machines and Turing machines [41]. They proved that for this class of cellular automata the associated limit language is regular. In paper [22] Gajardo and Goles present designs of Boolean circuits embedded in the space-time evolution of the two-dimensional three-state automaton. To enhance their proof of the universality of the automaton they also simulate a Turing machine in the reaction-diffusion automaton. Later, Goles and Montealegre analysed computational complexity of majority automata networks (the state of a vertex being the most represented in its neighbour). One of the first to study the relationship between the dynamics of the majority automata and its computational complexity was C. Moore proving in [6] that the majority automaton is P-Complete in three or more dimensions, leaving open the complexity characterization in a two-dimensional grid. Later, in [53] Eric and his colleagues characterize the complexity of the frozen majority (i.e., state 1 remains invariant): if the maximum degree of the network is 5 or more then the problem is P-complete, otherwise (maximum degree less than 5) the problem is in the class NC. Further, the same authors proved in [49] that the majority automaton in planar graphs is P-complete. To achieve that they used the two periodic behaviour of the majority automaton [12, 57, 62] as traffic lights to cross information. Further, by considering other iteration schemes they proved that the majority automaton iterated under a block sequential updating scheme is NP-Hard [48]. Result which is improved in [52] by proving that the problem is PSPACE-Complete. Previous results are obtained for the usual majority automata, i.e., the networks admit weights 0 or 1, say to sites are friends if they are connected (i.e., the weight in the incidence matrix is 1). In [50] they studied the more general case where the weights in the matrix may be $-1$, 0, 1. Other problems related to dynamics and computational complexity are related to the diffusion limited segregation studied recently in [4]. Their most recent results deal with the complexity boundaries of the graphs with polynomially growing treewidth [51].

**Neural and Boolean Automata Networks**
First Eric's works on these subjects are dated back to his thesis in 1980, where he characterizes the dynamical behaviour of disjunctive networks and also proved his very well know theorem about the periodicity of symmetric neural networks [12, 57, 58, 62] . But that was only the beginning, in [32], where he obtained bounds on the cycle and transient length for parallel iterations of antisymmetric sign functions. Later in Cosnard et al. [15] studied and derived formalism related to the threshold neural automata with memory: bounds on transient periods, and characterised reversibility versus the coupling coefficients. From 1985 Goles and colleagues introduced decreasing energy functions as a tool for analyses a huge class of neural automata networks [8, 29, 45, 62]. This tool has been successfully applied in studying various types of continuous state and discrete state neural networks [13, 14, 20, 61]. Other notable results in Boolean and neural networks include but not limited to the following: (1) necessary and sufficient conditions for the existence of fixed points in discrete neural networks and an upper bound for the number of fixed points [1], (2) exponential transient classes of symmetric neural networks for synchronous and sequential updating [44] (3) complexity of high-order neural networks [9], (4) characterisation of attractor space of neural networks over undirected graphs [11] (5) universality, via embedding Boolean circuits and simulating Turing machine, of discrete neural networks [10, 27]. One of Eric's recent works [11] characterizes completely the existence of cycles for symmetric network iterated both in parallel or block-sequentially. The authors define an index to the graphs of the automaton such that there exist cycles other than fixed points (period two or more) if and only if such index is non-negative.

**Discrete Ants**
Advances of Langton's ant model [66] have been done in several directions. Eric and collaborators developed a generation of Langton's ant model and analysed the complexity of the ants' behaviour on the graphs. They shown a high degree of unpredictability in general case, especially in the families of finite graphs where the period of the system growth exponentially with the size of the graph. They also shown that a prediction of the dynamics of the generalised ants on finite graphs is P-hard [25]. In paper [26] they constructed Boolean circuits with the trajectory of a single ant. They prove P-hardness of the ants system. Further studies dealt with detailed characterisaton of space-time dynamics of 64 ant's rules [24] and analyses of complex pattern formation and sensitivity of traces of two-dimensional ants [67].

**Tiling**
Eric and colleagues contribution to the tiling theory has been manifested in three key results. First, they demonstrated the existence of coding that allows for an efficient transformation of an arbitrary degrees of freedom tiling problem into a restricted four degrees of freedom problem (the tiling with rotation) [60]. Second, they provided solution to the following problem: to tile a rectangle or a torus with only vertical and horizontal bars of a given length such that the number of bars in every column and row equals to some given numbers [18]. Third, they established a bridge between tiling and folding. Namely, given a finite word coding vertical and horizontal folds

they provided a necessary and sufficient condition in order to tile the plane with a set of tiles constructed with copies of the unfold surface [34].

**Analysis of Natural Systems**
From the beginning, during his doctoral studies, Eric was interested in Boolean networks and its applications. Actually, he told me that one of the first articles that he read seriously was Stuart Kauffman's paper, Metabolic stability and epigenesis in randomly constructed genetic nets [64]. From that Eric and colleagues continue to apply Boolean networks in modelling and analysis of natural systems. The results included characterisation of two-dimensional Boolean dynamics in a grid such that each Boolean rule has only two inputs [19], representation of microbial interactions in a human microbiome as threshold Boolean networks [76], modelling the immune control of macrophages and the genetic control of the floral morphogenesis [40, 70], the analysis of cell cycle models [47] bacterium quorum sensing [72], a novel framework to study the influence of minimal cognitive mechanisms on the formation and evolution of languages [77], plants response to salt stress [71] and contagion phenomena in a two-dimensional grid [6].

**Decision Making and Social Science**
One of Eric's first works in this subject was to characterize the behaviour of a population to choose an opinion among several [61]. In paper [7] Eric and colleagues analyzed two simple dynamical models of decision that represent the school choice problem under two views: (1) individual expectations when deciding for a school, without major consideration of the social environment, and (2) focused on social expectations modelled by the neighbourhood preferences when deciding for a school. The computational experiments demonstrated that the social expectations model represents a more socially efficient situation that may help families to stay informed about accessibility, information, social capital, and improved school performances. Recently in [5] Eric and colleagues present a model of competing activist and political polarization and in [68] a model related to social crisis. Other key topics of Eric Goles and colleagues research include block invariance and reversibility of one-dimensional linear cellular automata, Lyapunov operators to study the convergence of extreme automata, properties of positive functions and the dynamics of associated automata networks, effects of firing memory in the dynamics of conjunctive networks, the complexity of asynchronous freezing cellular automata, on the robustness of update schedules in Boolean networks, naming game automata networks, complexity of the majority rule on planar graphs, learning gene regulatory networks using the bees algorithm, a sequential operator for filtering cycles in Boolean networks, prime number selection of cycles in a predator-prey model.

## References

1. Aracena J, Demongeot J, Goles E (2004) Positive and negative circuits in discrete neural networks. IEEE Trans Neural Netw 15(1):77–83
2. Bak P, Tang C, Wiesenfeld K (1987) Self-organized criticality: an explanation of the 1/f noise. Phys Rev Lett 59(4):381
3. Bitar J, Goles E (1992) Parallel chip firing games on graphs. Theor Comput Sci 92(2):291–300
4. Bitar N, Goles E, Montealegre P (2019) Computational complexity of restricted diffusion limited aggregation. arXiv:1904.10011
5. Lucas B, Pedro M, Eric G, Hans G (2020) Competing activists-political polarization. Physica A: Stat Mech Appl 545:123713
6. Böttcher L, Woolley-Meza O, Goles E, Helbing D, Herrmann HJ (2016) Connectivity disruption sparks explosive epidemic spreading. Phys Rev E 93(4):042315
7. Canals C, Goles E, Mascareño A, Rica S, Ruz GA (2018) School choice in a market environment: individual versus social expectations. Complexity
8. Goles EC (1985) Dynamics of positive automata networks. Theor Comput Sci 41:19–32
9. Goles EC, Matamala M (1994) Dynamical and complexity results for high order neural networks. Int J Neural Syst 5(3):241–252
10. Goles EC, Matamala M (1996) Symmetric discrete universal neural networks. Theor Comput Sci 168(2):405–416
11. Goles EC, Ruz GA (2015) Dynamics of neural networks over undirected graphs. Neural Netw 63:156–169
12. Goles EC (1980) Comportement oscillatoire d'une famille d'automates cellulaires non uniformes. PhD thesis, Grenoble Institute of Technology, France
13. Cosnard M, Goles EC (1995) A characterization of the existence of energies for neural networks. In: Fülöp Z, Gécseg F (eds) Proceedings of the automata, languages and programming, 22nd international colloquium, ICALP95, Szeged, Hungary, 10–14 July 1995. Lecture notes in computer science, vol 944. Springer, pp 570–580
14. Michel Cosnard and Eric Goles Ch (1997) Discrete state neural networks and energies. Neural Netw 10(2):327–334
15. Cosnard M, Moumida D, Goles E, de St Pierre T (1988) Dynamical behavior of a neural automaton with memory. Compl Syst 2(2):161–176
16. Delahaye JP (2018) Faut-il adopter l'avis de ces voisins. Pour la Sci 494
17. Domic NG, Goles E, Rica S (2011) Dynamics and complexity of the schelling segregation model. Phys Rev E 83(5):056111
18. Dürr C, Goles E, Rapaport I, Rémila E (2003) Tiling with bars under tomographic constraints. Theor Comput Sci 290(3):1317–1329
19. Fogelman-Soulie F, Goles-Chacc E, Weisbuch G (1982) Specific roles of the different boolean mappings in random networks. Bull Math Biol 44(5):715–730
20. Fogelman-Soulié F, Mejía C, Goles E, Aguilera SM (1989) Energy functions in neural networks with continuous local functions. Compl Syst 3(3)
21. Formenti E, Goles E, Martin B (2012) Computational complexity of avalanches in the kadanoff sandpile model. Fundamenta Informaticae 115(1):107–124
22. Gajardo A, Goles E (2001) Universal cellular automaton over a hexagonal tiling with 3 states. Int J Algebr Comput 11(03):335–354
23. Gajardo A, Goles E (2006) Crossing information in two-dimensional sandpiles. Theor Comput Sci 369(1–3):463–469
24. Gajardo A, Goles E (2004) Dynamics of a class of ants on a one-dimensional lattice. Theor Comput Sci 322(2):267–283
25. Gajardo A, Goles E, Moreira A (2001) Generalized langton's ant: dynamical behavior and complexity. In: Annual symposium on theoretical aspects of computer science. Springer, pp 259–270
26. Gajardo A, Moreira A, Goles E (2002) Complexity of langton's ant. Disc Appl Math 117(1–3):41–50

27. Goles E, Matamala M (1997) Reaction-diffusion automata: three states implies universality. Theory Comput Syst 30(3):223–229
28. Goles E, Moreira A, Rapaport I (2011) Communication complexity in number-conserving and monotone cellular automata. Theor Comput Sci 412(29):3616–3628
29. Goles E, Vichniac G (1990) Energy and attractors in parallel potts dynamics. J Phys A: Math Gen 23(7):1329
30. Goles E (1982) Fixed point behavior of threshold functions on a finite set. SIAM J Algebr Disc Methods 3(4):529–531
31. Goles E (1985) Comportement dynamique de reseaux d'automates. PhD thesis, Université de Grenoble, France
32. Goles E (1986) Antisymmetrical neural networks. Disc Appl Math 13(1):97–100
33. Goles E (1992) Sand pile automata. Ann de l'IHP Physique théorique 56:75–90
34. Goles E (2004) Folding and tiling. Theor Comput Sci 322(2):285–296
35. Goles E, Gómez L (2018) Combinatorial game associated to the one dimensional schelling's model of social segregation. Nat Comput 17(2):427–436
36. Goles E, Kiwi MA (1992) Dynamics of sand-piles games on graphs. In: Latin American symposium on theoretical informatics. Springer, pp 219–230
37. Goles E, Kiwi MA (1993) Games on line graphs and sand piles. Theor Comput Sci 115(2):321–349
38. Goles E, Latapy M, Magnien C, Morvan M, Phan HD (2004) Sandpile models and lattices: a comprehensive survey. Theoret Comput Sci 322(2):383–407
39. Goles E, Little C, Rapaport I (2008) Understanding a non-trivial cellular automaton by finding its simplest underlying communication protocol. In: International symposium on algorithms and computation. Springer, pp 592–604
40. Goles E, Lobos F, Ruz GA, Sené S (2020) Attractor landscapes in boolean networks with firing memory: a theoretical study applied to genetic networks. Nat Comput 19(2):295–319
41. Goles E, Maass A, Martinez S (1993) On the limit set of some universal cellular automata. Theor Comput Sci 110(1):53–78
42. Goles E, Margenstern M (1996) Sand pile as a universal computer. Int J Mod Phys C 7(02):113–122
43. Goles E, Margenstern M (1997) Universality of the chip-firing game. Theor Comput Sci 172(1–2):121–134
44. Goles E, Martinez S (1989) Exponential transient classes of symmetric neural networks for synchronous and sequential updating. Complex Syst 3(6):589–597
45. Goles E, Martţnez S (1991) Lyapunov functionals for automata networks defined by cyclically monotone functions. SIAM J Disc Math 4(2):200–206
46. Goles E, Meunier P-E, Rapaport I, Theyssier G (2009) Communications in cellular automata. In: Neary T, Woods D, Seda T, Murphy N (eds) Proceedings international workshop on the complexity of simple programs, Cork, Ireland, 6–7th Dec 2008, volume 1 of electronic proceedings in theoretical computer science. Open Publishing Association, pp 81–92
47. Goles E, Montalva M, Ruz GA (2013) Deconstruction and dynamical robustness of regulatory networks: application to the yeast cell cycle networks. Bull Math Biol 75(6):939–966
48. Goles E, Montealegre P (2014) Computational complexity of threshold automata networks under different updating schemes. Theor Comput Sci 559:3–19
49. Goles E, Montealegre P (2015) The complexity of the majority rule on planar graphs. Adv Appl Math 64:111–123
50. Goles E, Montealegre P, Perrot K, Theyssier G (2018) On the complexity of two-dimensional signed majority cellular automata. J Comput Syst Sci 91:1–32
51. Goles E, Montealegre P, Ríos-Wilson M, Theyssier G (2020) On the impact of treewidth in the computational complexity of freezing dynamics. arXiv:2005.11758
52. Goles E, Montealegre P, Salo V, Törmä I (2016) Pspace-completeness of majority automata networks. Theor Comput Sci 609:118–128
53. Goles E, Montealegre-Barba P, Todinca I (2013) The complexity of the bootstraping percolation and other problems. Theor Comput Sci 504:73–82

54. Goles E, Moreira A (2012) Number-conserving cellular automata and communication complexity: a numerical exploration beyond elementary CAS. J Cell Automata 7(3)
55. Goles E, Morvan M, Phan HD (2002) Lattice structure and convergence of a game of cards. Ann Comb 6(3–4):327–335
56. Goles E, Morvan M, Phan HD (2002) Sandpiles and order structure of integer partitions. Disc Appl Math 117(1-3):51–64
57. Goles E, Olivos J (1980) Periodic behaviour of generalized threshold functions. Disc Math 30(2):187–189
58. Goles E, Olivos J (1981) Comportement périodique des fonctions à seuil binaires et applications. Disc Appl Math 3(2):93–105
59. Goles E, Prisner E (2000) Source reversal and chip firing on graphs. Theor Comput Sci 233(1–2):287–295
60. Goles E, Rapaport I (1999) Tiling allowing rotations only. Theor Comput Sci 218(2):285–295
61. Goles E, Tchuenté M (1983) Iterative behaviour of generalized majority functions. Math Soc Sci 4(3):197–204
62. Goles-Chacc E, Fogelman-Soulié F, Pellegrin D (1985) Decreasing energy functions as a tool for studying threshold networks. Disc Appl Math 12(3):261–277
63. Hegselmann R, Schelling TC, Sakoda JM (2017) The intellectual, technical, and social history of a model. J Artif Soc Soc Simul 20(3)
64. Kauffman SA (1969) Metabolic stability and epigenesis in randomly constructed genetic nets. J Theor Biol 22(3):437–467
65. Kiwi MA, Ndoundam R, Tchuente M, Goles E (1994) No polynomial bound for the period of the parallel chip firing game on graphs. Theor Comput Sci 136(2):527–532
66. Langton CG (1986) Studying artificial life with cellular automata. Physica D: Nonlinear Phenom 22(1-3):120–149
67. Markus M, Schmick M, Goles E (2006) Tracks emerging by forcing langton's ant with binary sequences. Complexity 11(3):27–32
68. Mascareño A, Goles E, Ruz GA (2016) Crisis in complex social systems: a social theory view illustrated with the chilean case. Complexity 21(S2):13–23
69. Medina P, Goles E, Zarama R, Rica S (2017) Self-organized societies: on the sakoda model of social interactions. Complexity
70. Ruz GA, Goles E, Sené S (2018) Reconstruction of boolean regulatory models of flower development exploiting an evolution strategy. In: 2018 IEEE congress on evolutionary computation (CEC). IEEE, pp 1–8
71. Ruz GA, Timmermann T, Goles E (2016) Neutral space analysis of gene regulatory network models of salt stress response in arabidopsis using evolutionary computation. In: 2016 IEEE congress on evolutionary computation (CEC). IEEE, pp 4281–4288
72. Ruz GA, Zúñiga A, Goles E (2018) A boolean network model of bacterial quorum-sensing systems. Int J Data Mining Bioinf 21(2):123–144
73. Sakoda JM (1971) The checkerboard model of social interaction. J Math Sociol 1(1):119–132
74. Schelling TC (1971) Dynamic models of segregation. J Math Sociol 1(2):143–186 (1971)
75. Spencer J (1986) Balancing vectors in the max norm. Combinatorica 6(1):55–65
76. Travisany D, Goles E, Latorre M, Cortés M-P, Maass A (2020) Generation and robustness of boolean networks to model clostridium difficile infection. Nat Comput 19(1):111–134
77. Vera J, Goles E (2016) Automata networks for memory loss effects in the formation of linguistic conventions. Cogn Comput 8(3):462–466
78. Vieira AP, Goles E, Herrmann HJ (2020) Dynamics of extended schelling models. J Stat Mech: Theory Exp 2020(1):013212

# Seven Things I Know About Them

**Jacques Demongeot**

**Abstract** In this paper, we intend to present a series of 7 application examples inspired by the work of Eric Goles with his numerous collaborators, while remaining focused on the field of Boolean automata. This constitutes a sort of anthology showing the extent of the mathematical domain defined by the study of Boolean automata dynamics, showing the relevance of open paths and results obtained by Eric Goles and the high explanatory power of the models which arise from his work during 45 years.

**Keywords** Boolean automata · Automata gradient dynamics · Automata Hamiltonian dynamics · Updating schedule

## 1 Introduction

A discrete dynamical system has the same definition that the continuous ones. It involves a flow function *f* defined on ExT, where T is a discrete time space (in general $\mathbb{N}$) and E a discrete state space ($\{0, 1\}^n$ in the Boolean case and more generally a finite subset of $\mathbb{R}_+{}^n$), $f(x, t)$ representing for each state $x$ and time $t$, the state reached after time $t$ by the trajectory starting in state $x$ at time 0. We denote in general $f(x(0), t)$ *by* $x(t) = (x_i(t))_{i=1,\ldots,n}$, which permits to have a coherent notation for all the states of a trajectory. The set of such states is called the orbit of $x(0)$. Following [1], we can now define the discrete time derivative for the state vector $(x_i(t))_{i=1,\ldots,n}$ by:

$$\Delta x_i/\Delta t = (x_i(t + \Delta t) - x_i(t))/\Delta t,$$

which reduces to $x_i(t + 1) - x_i(t)$, if $\Delta t = 1$. By using the same formula, we can also define:

J. Demongeot (✉)
Faculty of Medicine, Laboratory AGEIS EA 7407, Team Tools for E-Gnosis Medical, University Grenoble Alpes, 38700 La Tronche, France
e-mail: Jacques.Demongeot@univ-grenoble-alpes.fr



A. Adamatzky (ed.), *Automata and Complexity*, Emergence, Complexity and Computation 42, https://doi.org/10.1007/978-3-030-92551-2_2

- the space derivative: $\Delta g(x)/\Delta i = (g(x_{i+\Delta i}) - g(x_i))/\Delta i = g(x_i + 1) - g(x_i)$, if $i$ is 1-dimensional and $\Delta i = 1$. If $i$ is $n$-dimensional, it is possible to partially derive in each dimension.
- the partial state derivative: $\Delta g(x)/\Delta x_i = [g(x_1,\ldots, x_i + \Delta x_i,\ldots, x_n) - g(x_1,\ldots, x_i,\ldots, x_n)]/\Delta x_i$.

A discrete automaton is defined by a transition function $F$:

$$\Delta x_i/\Delta t = (x_i(t + \Delta t) - x_i(t))/\Delta t = F_i(x(t)),$$

where $F_i$ depends only on coordinates $(x_j(t))_{j\in V(i)}$, V($i$) being a neighbourhood of $i$ in the space set (in general the Manhattan—or $L_1$—unit ball of $E \cap \mathbb{R}_+{}^n$ centred on $i$ and having a radius equal to 1), with conditions defining V on the boundary of E (e.g. periodic) and with constraints on the discrete velocity ensuring that flow remains in E.

We will first define a discrete analogous of a potential (or gradient) continuous dynamical system (called here potential automaton). A continuous potential differential equation on $\mathbb{R}^n$ is defined by: $\forall i = 1,\ldots,n$, $dx_i/dt = -\partial P/\partial x_i$, where $P$ is a real continuously differentiable function (e.g., a polynomial with real coefficients) on $\mathbb{R}^n$. In the same way, a potential automaton on the discrete state space E is defined by:

$$x_i(t + 1) = h(-\Delta P/\Delta x_i + x_i(t)), \tag{1}$$

where $P$ is a real function (e.g., a polynomial with real coefficients) on E and $h$ a function from $\mathbb{R}$ to E, with boundary conditions ensuring that the flow remains in E. For example, in the Boolean case, we will choose for $h$ the Heaviside function $H$: $H(s) = 1$, if $s > 0$, and $H(s) = 0$, if $s \leq 0$. In the integer case (E subset of $\mathbb{N}^n$), $h$ can be the identity, if $P$ has integer coefficients and if $\forall i = 1,\ldots,n$, $\Delta x_i \in \{-1, 0, 1\}$.

Provided by the above definitions, we will give now some examples of application of Boolean networks inspired by the work of Eric Goles and his numerous co-workers.

## 2 Seven Remarks About Boolean Automata Theory

### *2.1 Potential Boolean Automata (Inspired by Cosnard and Goles [2])*

**Proposition 1** In the Boolean case, let suppose that $A = 0$, $P(x) = {}^t xAx + Bx$, with $a_{ii} = 0$ and each sub-matrix on any subset $J$ of indices in $\{1, \ldots,n\}$ of $A$ is non positive and less than the linear operator -$B$ restrained on $J$. Then $P$ decreases on the trajectories of the potential automata defined by $x_i(t + 1) = H(-\Delta P/\Delta x_i + x_i(t))$

for any mode of implementation of the dynamics (sequential, block sequential and parallel). These Boolean automata constitute a Hopfield-like network whose weights are $w_{ii} = 1$ and $w_{ij} = -a_{ij} - a_{ji}$, $\forall j \neq i$, thresholds are the $b_i$'s, and stable fixed configurations correspond to the minima of $P$.

**Proof** It is easy to check that: $\Delta P/\Delta x_i = \Sigma_{j \neq i} (a_{ij} + a_{ji}) x_j + b_i$ and: $x_i (t+1) = H (-P/x_i + x_i(t)) = H (-[j \neq i (a_{ij} + a_{ji}) x_j(t) + b_i] + x_i(t)) = H (j\, w_{ij} x_j(t) - b_i)$

We can calculate for the block sequential iteration at any step of block $J$:

$$x_i(t+1) = H(-\Delta P/\Delta x_i + x_i(t)) = H(-[\sum\nolimits_{j \neq i}(a_{ij} + a_{ji})x_j(t) + b_i] + x_i(t))$$
$$= \mathrm{H}(\sum\nolimits_{j} w_{ij}x_j(t) - b_i)$$

$$P(x(t+1)) - P(x(t)) = \sum\nolimits_{(i,j)\in J\times J} a_{ij}\Delta x_i \Delta x_j + \sum\nolimits_{i\in J} b_i \Delta x_i \leq 0,$$

the result coming from the hypothesis on the sub-matrices $J$ of $W$ or from [2] ■

The interest of the Proposition 1 is to show that the Hopfield-like network defined by:

$$\forall i = 1, \ldots, n, x_i(t+1) = H\left(\sum\nolimits_j w_{ij}x_j(t) - b_i\right), \text{ with}$$
$$w_{ii} = 1, w_{ij} < 0, b_i \geq \mathrm{e} > 0,$$

has not only $P$ as Lyapunov function as proved in [2], but more it can be considered as a potential automaton with a potential equal to $P$, because the opposite of the gradient of $P$ is related to the velocity of the automaton, what is quite different in general for a system with simply a Lyapunov function (Fig. 1) [3].

Another example of potential Boolean automata is given by the n-switch often used in morphogenesis modelling [4, 5], for example in dorsal somites (Fig. 2) [6] or skin appendages [7, 8] models. It is easy to show that its Hopfield-like dynamics with all weights $w_{ij}$ equal to $-1$, except $w_{ii} = 1$ and all $b_i \geq \mathrm{e} > 0$, is gradient for any updating mode and its attractors are the 6 fixed points (10,000), (01,000), (00,100),

**Fig. 1** Potential automaton with $\Delta x$ = *-gradP* (on the left) and an automaton with a Lyapunov function decreasing on its trajectories (on the right) (after [3])

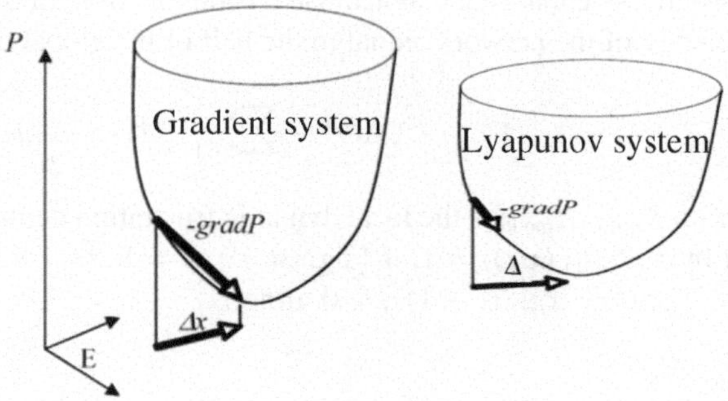

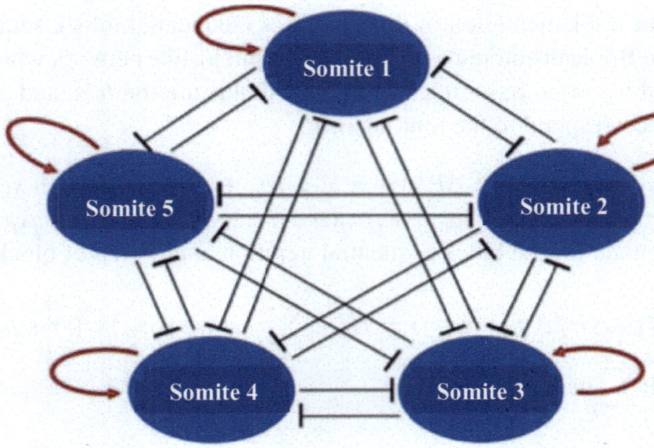


**Fig. 2** "Metatron" interaction graph of a 5-switch (after [4, 5])

(00,010) and (00,001) and (00,000). The equivalent continuous model [5] has the same property.

## 2.2 Hamiltonian Boolean Automata

**Proposition 2** Let us consider a deterministic Hopfield-like network of size $n$, which is a circuit sequentially or synchronously updated with constant absolute value $w$ for its non-zero interaction weights. Then, its dynamics is conservative, keeping constant on the trajectories the Hamiltonian function L defined by:

$$L(x(t)) = \sum_{i=1,n} \frac{(x_i(t) - x_i(t-1))^2}{2}$$

$$= \sum_{i=1,n} \frac{H\left(w_{i(i-1)modnxi-1}(t-1) - x_i(t-1)\right)^2}{2}$$

where $H$ denotes the classical Heaviside function. L($x(t)$) is the total discrete kinetic energy of the network, equal to the half of the global dynamic frustration:

$$F(x(t)) = \sum_{i=1-n} F_{i,(i-1)modn}(x(t)),$$

with $F_{i,\,(i-1)\,\mathrm{mod}n}$ is the local dynamic frustration defined between nodes $(i-1)$ and $i$ by:$F_{i,\,(i-1)}\,(x(t)) = 1$, if {sign($w_{i(i-1)}$) $= 1$, $x_i(t) \neq x_{i-1}(t-1)$} {sign($w_{i(i-1)}$) $= -1$, $x_i(t) = x_{i-1}(t-1)$}, $= 0$, if not.

| $F{=}H$ Global frustration | Configuration Nb | Attractor Nb | Attractor Length |
|---|---|---|---|
| 0 | 1 | 2 fixed points | 1 |
| 2 | $4 = C^{H-1}_{n-H+2}/2$ | 7 | 8 |
| 4 | $10 = C^{H-1}_{n-H+2}/2$ | 3 | 4 (symmetrized) |
| 4 | | 16 | 8 |
| 6 | $4 = C^{n-H-1}_{H+2}/2$ | 7 | 8 |
| 8 | 1 | 1 | 2 |

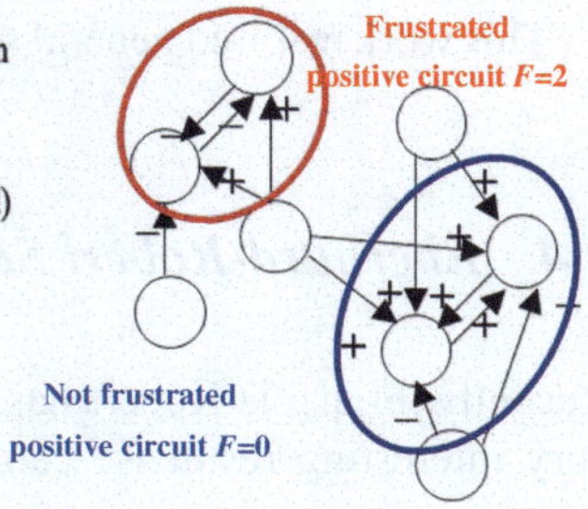


**Fig. 3** On the left: calculation of the global frustration $F$ for circuits of length 8 with only identities and negations as local transitions. On the right: "Arabidopsis" interaction graph with eight nodes and two circuits of length 2 (after [4, 5]), one frustrated (in red) and the other not frustrated (in blue)

Proposition 2 still holds if the network is a circuit whose transition functions are Boolean identity or negation [10], on which it is easy to calculate the global frustration $F$ and show that it characterizes attractors by remaining constant along them (Fig. 3).

A general program of characterizing potential and Hamiltonian Boolean automata could extend the energetic notion of dissipative (potential) and conservative (Hamiltonian) energy, from the Hopfield-like Boolean networks to the most general Boolean automata. This work would be in the direct continuation of the pioneering work of the two PhD students of François Robert, Eric Goles and Françoise Fogelman [11].

## 2.3 *Social Choice and Majority Rule*

In the spirit of the games theory of the seventies [12], Eric Goles and Maurice Tchuente considered in [13] a society of $n$ persons $\{P_1,\ldots,P_n\}$ having at time $t$ opinions $\{x_1(t),\ldots, x_n(t)\}$ with interaction coefficient $a_{ij} = a_{ji}$ between $P_i$ and $P_j$. Let $\{\theta_1,\ldots,\theta_p\}$ be the set of possible opinions which may be assumed by any person, with a local hierarchy $h_i$ adopted by each person $P_i$ (a reordering of opinion indices without ex-æquo, that is a permutation of $\{1,\ldots, p\}$). The dynamical behaviour of such a society depends on local majority rules, where, if $a(k)$ denotes the global weight of the opinion $\theta_k$, the change of opinion is made as follows:

$$x_i(t+1) = k, \text{ with}$$

$$k = \sup\left\{ i / \forall r = 1, \ldots, p, a(i) \sum\nolimits_{j/x_j(t)=k} a_{ij} \geq a(r) \sum\nolimits_{j/x_j(t)=r} a_{ij} \right\}$$

Then, the main result is the following.

**Proposition 3** In such a society the opinion of any member $P_i$, after a certain number of steps, either remains constant or oscillates between two values.

This work inspired general studies on social choices as those described in [14, 15].

## 2.4 Eberhard-Robert Scholia

Recently, François Robert has established in collaboration with André Eberhard a very interesting result we can call a "scholia", because it is an original explanatory comment opening a new domain of research concerning interactions between automata networks. The example treated in [16] concerns cross interactions between two Boolean automata of size 2, one with transition $F_0$ and the other with transition $G_0$, evolving in time with the following rules, where symbol $\neg$ denotes Boolean complementary (negation):

- Sequential dependence: $G_1(x_1, x_2) = G_0(F_0(x_1, x_2), x_2)$, $F_1(x_1, x_2) = F_0(x_1, G_1(x_1, x_2))$,..., $G_i(x_1, x_2) = G_{i\text{-}1}(F_{i\text{-}1}(x_1, x_2), x_2)$, $F_i(x_1, x_2) = F_{i\text{-}1}(x_1, G_i(x_1, x_2))$,...
- Parallel dependence: $G_1(x_1, x_2) = G_0(F_0(x_1, x_2), x_2)$, $F_1(x_1, x_2) = F_0(x_1, G_0(x_1, x_2))$,...,
  $G_i(x_1, x_2) = G_{i\text{-}1}(F_{i\text{-}1}(x_1, x_2), x_2)$, $F_i(x_1, x_2) = F_{i\text{-}1}(x_1, G_{i\text{-}1}(x_1, x_2))$,...
- Sequential opposition: $G_1(x_1, x_2) = G_0(\neg F_0(x_1, x_2), x_2)$, $F_1(x_1, x_2) = F_0(x_1, \neg G_1(x_1, x_2))$,..., $G_i(x_1, x_2) = G_{i\text{-}1}(\neg F_{i\text{-}1}(x_1, x_2), x_2)$, $F_i(x_1, x_2) = F_{i\text{-}1}(x_1, \neg G_i(x_1, x_2))$,...
- Parallel opposition: $G_1(x_1, x_2) = G_0(\neg F_0(x_1, x_2), x_2)$, $F_1(x_1, x_2) = F_0(x_1, \neg G_0(x_1, \ldots\ldots x_2))$,...,
  $G_i(x_1, x_2) = G_{i\text{-}1}(\neg F_{i\text{-}1}(x_1, x_2), x_2)$, $F_i(x_1, x_2) = F_{i\text{-}1}(x_1, \neg G_{i\text{-}1}(x_1, x_2))$,...

Then, the Eberhard-Robert scholia says:

**Proposition 4**

(i) Sequential and parallel dependence (resp. opposition) rules give same fixed points.
(ii) The transform by sequential (resp. parallel) opposition of $(\neg F, \neg G)$ is the complementary of the transform by sequential (resp. parallel) dependence of $(F, G)$.

Such a result can be applied to situations in which two groups of actors (political, ethnic, social, neural, genetic, etc.) or two age classes evolve by taking social choices with cross interactions. It could serve namely to revisit and interpret the asymptotic properties of complex Boolean biological networks as those studied in [17, 18].

## 2.5 Arabesques

A close friend of Eric Goles, René Thomas, invented the notion of arabesques systems, and studied their dynamics both in discrete and continuous framework [17, 19, 20]. The Boolean equations of such a system of size $n$ is given by:

$$\forall i = 1, \ldots, n, x_i = \mathrm{H}\big(w_{ii} X_i - w X_{j+1(mod n)} + w X_{i-1(mod n)}\big)$$

The weights verify: $0 \leq w_{ii} < w$. If $n = 2$ and $w_{22} = 0$, the arabesque is called regulon [17]. It is the smallest network containing one positive and one negative circuit (Fig. 4). The dynamics of the arabesque of size 2, with $w_{11} = w_{22} > 0$, has been used for representing for example the functioning of the hippocampus [19, 20]. The asymptotic behaviour of the arabesque of size 3 in sequential updating (Fig. 4) is represented by two unstable fixed points (000), (111), and a stable cycle of order 6 (010*, 0*11, 01*1, 001*, 0*01, 10*1, 101*, 1*00, 10*0, 110*, 1*10, 01*0), * denoting the node to update, with (011*, 0*10, 01*0), (00*1, 001*), (1*01, 0*01), (100*, 1*00) and (11*0, 110*) as attraction basin trajectories.

In parallel updating mode, (000) and (111) are unstable fixed point and there is a cycle of order 6 (010, 011, 001, 101, 100, 110). This cycle exists in the continuous analog (useful to interpret the mechanism of memory evocation [19]) and both in discrete parallel and in continuous case with $w_{ii} = 0$, the system is Hamiltonian, with conservation of the kinetic energy, independently of the number of tangent circuits in the arabesque. A deeper study of such intersecting circuits can be found in [20].

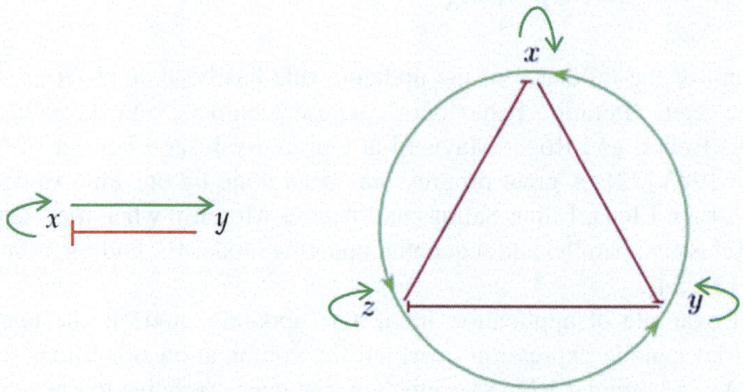


**Fig. 4** On the left, regulon of size 3. On the right, arabesque of size 3. The positive interactions (activations) are represented by green arrows, the negative (inhibitions) by red ones

| | |
|---|---|
| E. Goles | |
| F. Robert | |
| N. Gastinel | |
| J. Kuntzmann | |
| G. Valiron (also advisor of L. Schwartz, himself advisor of J.L. Lions) | |
| E. Borel | |
| G. Darboux (also advisor of H. Lebesgue) | |
| E. Borel | |
| M. Chasles | |
| S.D. Poisson | |
| J.L. Lagrange (also advisor of J. Fourier) | P.S. Laplace |
| L. Euler | J. Le Rond d'Alembert |
| Johann Bernoulli (also advisor of D. Bernoulli) | |
| Jacob Bernoulli | N. Eglinger |
| N. Malebranche | |
| G.W. Leibniz | |
| C. Huyghens | E. Weigel |
| F. van Schooten | J.J. Stampioen |
| J. Golius | M. Mersenne (also advisor of B. Pascal) |
| W. Snellius | T. Erpenius |
| R. Snellius | J.J. Scaliger |
| V. Naibod | A. Turnèbe |
| E. Reinhold | J. Toussain |
| J. Milich | G. Budé |
| D. Erasmus | |

**Fig. 5** Mathematical genealogy of Eric Goles. A mathematician under another is his PhD advisor

## 2.6 *Block Parallel Updating*

The problem of the influence of the updating rule has been emphasized since the start of the team “Iteration behaviours”, whose members were brought together by François Robert and Roger Maynard at University Joseph Fourier of Grenoble (IMAG) in 1983 [22]. A great progress has been done by our PhD students Julio Aracena, Adrien Elena, Lilian Salinas and Andrès Moreira, when they searched to escape the classical parallel and sequential updating modes for finding more realistic schedules [23, 24].

A good example of application for a new updating mode is the nuclear and mitochondrial genetic expression in which the combination of cellular repressors (microRNAs and circular RNAS), mitochondrial genes (relative to the cell respiration) and nuclear genes under the control of the chromatin clock, present a dynamical schedule close to a new updating mode called block-parallel in [22], for which the updating is sequential inside blocks, these blocks being updated parallelly with not necessary the same internal clock. The problems linked to the search for asymptotic behavior of such updated Boolean automata are very hard, but will be surely extensively studied in the future by the Goles' school.

## 2.7 *Discrete Convolution (to Regularize, We Convolve, L. Schwartz)*

Yann Le Cun (PhD student of a friend of Eris Goles, Maurice Milgram) has defined in 1983 at a Colloquium held at Les Houches (University J. Fourier of Grenoble) and co-organized by J. Demongeot and B. Lacolle from the IMAG team "Iteration behaviours", the notion of learning process [25] and then, he used for the deep learning [26, 27] several discrete convolution operators, useful in writing or image processing [29]. The Boolean automata $z(t)$ resulting from the discrete convolution between two Boolean automata $x(t)$ and $y(t)$ (called the kernel filter) can be defined as follows:

$$z_i(t) = \sum\nolimits_{i-k \in v(i)} x_{i-k}(t) y_k(t)$$

where the size of the neighbourhood $V(i)$ equals $p$ and the positive kernel filter of size $p$, $y(t)$, verifies: for any $t$, $k = 1,..., p$ $y_k(t) = 1$. If $x_i(t)$ has a regular derivative (in the sense of F. Robert [1]), that is, for any spatial derivative and any $t$, $\Delta x_i(t)/\Delta i \leq 1$, then $z_i(t)$ is also regular, because we have: $\Delta z_i(t)/\Delta i \leq {}_k\sum_{V(i)} \Delta x_{i-k}(t)/\Delta k \; y_k(t) \leq 1$.

# 3 Mathematical Genealogy of Eric Goles as a Conclusion

Eric Goles has a prestigious mathematical genealogy starting with François Robert and ending with the two most renowned philomaths (like him) of the Renaissance, Guillaume Budé and Desiderius Erasmus both born in 1467, 584 years before him. This genealogy is described in Fig. 5 (after [29]).

The first ancestor of Eric Goles intends to end this genealogy with the following tribute:

«Mon billet pour Éric», by François Robert.

«Éric, la scène se passe à Grenoble quelques années après ta soutenance de thèse. Retour d'Israël via Paris, et avant de rentrer à Santiago, tu débarques impromptu ce matin-là dans mon bureau de la Tour IRMA au Campus. Nous sommes contents de nous revoir, et tu commences à m'expliquer au tableau ce qui préoccupe ton esprit en ce moment. S'agissait-il du tas de sable qui s'effondre sur lui-même ? Je crois que oui. Et tu me montres la relation mathématique que tu as écrite au tableau: « Tu vois, cette expression, eh bien elle ne me plaît pas, elle n'est pas casher !» Devant mon air d'incompréhension, tu corriges immédiatement: « Disons qu'elle n'est pas très catholique !», ce qui n'a pas amélioré ma compréhension pour autant, mais j'ai alors saisi en un éclair le concept de mathématicien international que tu commençais à incarner alors.

J'ai par ailleurs été très sensible au témoignage chaleureux que tu as rendu récemment à Jacques Demongeot [30], dans lequel tu insistes beaucoup pour rendre effectivement sensible cette part d'éternité et de globalité de vos échanges dînatoires, mathématiques ou non, que ce soit au Bar national de Santiago ou ailleurs: une caractéristique importante qui vous situe bien. Et de plus, ce témoignage m'a redonné comme instantanément aussi la vive perception du compagnonnage artisanal qui animait les membres de notre groupe « Comportement d'Itérations». Quelle équipe, en effet, d'excellents chercheurs ! Je les cite par ordre d'entrée en scène: Michel Cosnard, Maurice Tchuente, toi Eric, Françoise Fogelman, Houcine Snoussi, Yves Robert, les premiers couteaux en quelque sorte, tous chercheurs C.N.R.S. (sauf Françoise, universitaire, et Houcine, boursier marocain) et une huitaine de jeunes en troisième cycle. Jacques Demongeot entretenait des relations suivies avec tout ce petit monde. Nous partagions chaque semaine notre conviction dans la dimension intemporelle de notre réalité mathématique en cours d'élaboration, assidûment méditée et travaillée, et c'était là que résidait le ciment du groupe.

En ce qui te concerne, ta façon très spécifique de « faire des maths» est maintenant largement connue: fougueux, jovial, parfois brouillon et fâché de l'être, tu déploies au tableau une puissante séduction mathématique qui jaillit de ta très forte conviction personnelle, manipulant un maelström de notions entrelacées... Tu convaincs, car tu es habité, et spécifiquement toi !

Il n'empêche: les deux théorèmes de Golès-Martinez sur les cycles de longueur deux dans les réseaux d'automates à seuils symétriques (il y a trente-cinq ans !) resteront pour moi le signal fort d'un accomplissement à venir, qui s'est grandement réalisé depuis. J'ai même vu récemment que tu avais contribué à faire, de la fourmi de Langton, la brique de base d'un calculateur universel, dans la ligne de ce que vous élaboriez à l'époque avec Maurice Tchuente et Yves Robert sur d'autres modèles.

Cher Éric, bon et heureux 70, et longue vie ! Avec toute mon amitié, François Robert.»

## References

1. Robert F (1986) Discrete iterations: a metric study. Springer, Berlin
2. Cosnard M, Goles E (1997) Discrete states neural networks and energies. Neural Netw 10:327–334
3. Demongeot J, Elena A, Weil G (2006) Potential-Hamiltonian decomposition of cellular automata. Application to degeneracy of genetic code and cyclic codes III. Comptes Rendus Biol 329, 953–962
4. Cinquin O, Demongeot J (2002) Positive and negative feedback : striking a balance between necessary antagonists. J Theor Biol 216:229–241
5. Cortez MJV, Rabajante JF, Tubay JM, Babierra AL (2017) From epigenetic landscape to phenotypic fitness landscape: evolutionary effect of pathogens on host traits. Infect Genet Evol 51:245–254
6. Goldbeter A, Pourquié O (2008) Modeling the segmentation clock as a network of coupled oscillations in the Notch, Wnt and FGF signaling pathways. J Theor Biol 252:574–585
7. Michon F, Forest L, Collomb E, Demongeot J, Dhouailly D (2008) BMP-2 and BMP-7 play antagonistic roles in feather induction. Development 135:2797–2805

8. Abbas L, Demongeot J, Glade N (2009) Synchrony in Reaction-diffusion models of morphogenesis: applications to curvature-dependent proliferation and zero-diffusion front waves. Phil Trans R Soc A 367:4829–4862
9. Demongeot J, Ben Amor H, Hazgui H, Waku J (2014) Stability, complexity and robustness in population dynamics. Acta Biotheor 62:243–284
10. Rachdi M, Waku J, Hazgui H, Demongeot J (2020) Entropy as robustness marker in genetic regulatory networks. Entropy 22:260
11. Goles E, Fogelman-Soulié F, Pellegrin D (1985) Decreasing energy functions as a tool for studying threshold networks. Disc Appl Math 12:261–277
12. Moulin H (1976) Extensions of two person zero sum games. J Math Anal Appl 55:490–508
13. Goles E, Tchuente M (1983) Iterative behaviour of generalized majority functions. Math Soc Sci 4:197–204
14. Demongeot J, Volpert V (2015) Dynamical system model of decision making and propagation. J Biol Syst 23:339–353
15. Banerjee M, Demongeot J, Volpert V (2016) Global regulation of individual decision making. Math Methods Appl Sci 39:4428–4436
16. Robert F (2020) Une aventure mathématique. Personal Commun
17. Demongeot J, Aracena J, Ben Lamine S, Meignen S, Tonnelier A, Thomas R (2001) Dynamical systems and biological regulations. In: Goles E, Martinez S (eds) COMPLEX SYSTEMS. Kluwer, Amsterdam, pp 105–151
18. Aracena J, Demongeot J, Goles E (2004) Mathematical modelling in genetic networks. IEEE Trans Neural Netw 15:77–83
19. Antonopoulos C, Basios V, Demongeot J, Nardone P, Thomas R (2013) Linear and nonlinear arabesques: a study of closed chains of negative 2-element circuits. Int J Bifurc Chaos 23:30033
20. Demongeot J, Elena A, Noual M, Sené S, Random boolean networks and attractors of their intersecting circuits. In: Barolli L, Xhafa F, Jeong HY (eds) IEEE AINA' 11. IEEE Proceedings, Piscataway, pp 483–487
21. Demongeot J, Khlaifi H, Istrate D, Mégret L, Taramasco C, Thomas R (2020) From conservative to dissipative non-linear differential systems. an application to the cardio-respiratory regulation. Disc Contin Dyn Syst 13:2121–2134
22. Robert F (1969) Blocs-H-matrices et convergence des méthodes itératives classiques par blocs. Linear Algebra Appl 2:223–265
23. Demongeot J, Elena A, Sené S (2008) Robustness in neural and genetic networks. Acta Biotheor 56:27–49
24. Aracena J, Goles E, Moreira A, Salinas L (2009) On the robustness of update schedules in Boolean networks. Biosystems 97:1–8
25. le Cun Y (1986) Learning processes in an asymmetric threshold network. In: Fogelman-Soulié F, Bienenstock E, Weisbuch G (eds) Disordered systems and biological organization, *Les Houches*, France, 1985. North Holland, Amsterdam, pp 233–240
26. le Cun Y, Bottou L, Bengio Y, Haffner P (1998) Gradient-based learning applied to document recognition. Proc IEEE Soc 86:2278–2324
27. Kavukcuoglu K, Sermanet P, Boureau Y, Gregor K, Mathieu M, Le Cun Y (2010) Learning convolutional feature hierarchies for visual recognition. In: Lafferty JD, Williams CKI, Shawe-Taylor J, Zemel RS, Culotta A (eds) Advances in neural information processing systems (NIPS), vol 23. Curran Associates, Red Hook, pp 1090—1098
28. Demongeot J, Bezy-Wendling J, Mattes J, Haigron P, Glade N, Coatrieux JL (2003) Multiscale modeling and imaging: the challenges of biocomplexity. Proc IEEE Soc 91:1723–1737
29. Genealogy (2020). https://www.genealogy.math.ndsu.nodak.edu/id.php?id=160281. Accessed 20 May 20
30. Goles E (2020) Tribute to Prof. Jacques Demongeot: conversations at the *National Bar*. Nat Comput 19:3–4

# Distortion in Automorphisms of Expansive Systems

**Nicolas Bitar, Sebastian Donoso, and Alejandro Maass**

**Abstract** In this work we study the role distortion plays on automorphisms groups of expansive dynamical systems. We begin by generalizing results from subshifts, linking distortion and non-expansivity, to arbitrary expansive systems, and explore the subset of symmetrically distorted automorphisms. Due to the generalization, we are able to determine that expansive automorphisms can never be distorted.

## 1 Introduction

In symbolic dynamics and group theory, distortion generally refers to an object that grows or moves sub-linearly. In particular, we say that a cellular automaton or an endomorphism acting on a symbolic space is *range distorted* if the local radius of the iterated applications grows at the aforementioned rate. Examples of this behavior can be constructed from Turing machines when viewed as endomorphisms or automorphisms of a symbolic space. Indeed, Guillon and Salo showed in [7] that any aperiodic Turing machine is range distorted. Also, using the so-called conveyor belt technique one can show that on any sofic shift one can define a non-trivial range distorted automorphism.

Analogously, an element of a finitely generated group is said to be *distorted* if its minimal expression on the generating set grows sub-linearly with successive iterations of the element. The two mentioned concepts are related: a group distorted

---

All authors thank ANID Grant BASAL FB210005 and ACE210010.

---

N. Bitar · S. Donoso · A. Maass (✉)
Department of Mathematical Engineering and Center for Mathematical Modeling,
Universidad de Chile, IRL-CNRS 2807 Santiago, Chile
e-mail: amaass@dim.uchile.cl

N. Bitar
e-mail: nbitar@dim.uchile.cl

S. Donoso
e-mail: sdonoso@dim.uchile.cl

 
A. Adamatzky (ed.), *Automata and Complexity*, Emergence, Complexity
and Computation 42, https://doi.org/10.1007/978-3-030-92551-2_3

automorphism is always range distorted (here we are considering the group of automorphisms). This prompts the fundamental question: does the converse hold?

To address this question, several notions of discrete Lyapunov exponents have been introduced. These objects are closely related to range distortion due to the fact that they quantify the average rate at which any kind of endomorphism on a symbolic space moves information. One recent example are the exponents introduced by Cyr, Franks and Kra in [6], that quantify the rate at which information is moved asymptotically. The novelty of these exponents is that they relate distortion to geometrical properties of the space-time defined by endomorphisms.

In this article, we will show that the phenomenon of distortion is not exclusive to the realm of symbolic systems. This is achieved through the use of M. Boyle and D. Lind's work on expansive dynamical systems [2]. We generalize both the notions of local radius and asymptotic Lyapunov exponents to endomorphisms of arbitrary expansive dynamical systems. Furthermore, the connections between distortion and geometry are preserved. We exemplify this generalization by considering automorphisms of the $n$-torus.

There is a second question we want to address in this article. Since Boyle and Lind introduced the notions of expansive and non-expansive directions for the study of directional dynamics of an action, there has been one persistent question: which sets can occur as sets of non-expansive directions?

In [2] they showed that this set is closed and, if the domain is infinite, non-empty. Furthermore, they showed that any closed set of directions, with two or more elements is the set of non-expansive directions for some action. Later, Hochman showed in [9] that for every direction, there exists an automorphism of a subshift such that its unique non-expansive direction is the selected one, effectively solving the realization problem. Nevertheless, the subshift built to achieve this result lacks of many natural dynamical properties one would like to get, as transitivity or minimality. This motives the author to ask the following, still open, question: Does any closed non-empty set of directions arise as the set of non-expansive directions of a $\mathbb{Z}^2$-action that is transitive or minimal?

We begin by introducing the necessary concepts from the field of symbolic dynamics from the theory of expansive dynamical systems. We then proceed to generalize the concept of radius to the context of expansive systems, introducing the concept of distorted automorphism. Next, we introduce alternative notions of distortion through the generalization of discrete Lyapunov exponent to the realm of expansive systems. This allows us to establish a connection between non-expansive directions and the asymptotic behavior of an automorphism. In addition, we establish that no expansive automorphism can be distorted. We continue by addressing the question of group distortion in relation to range distortion by studying the set of distorted automorphisms with distorted inverse. Finally, we look at examples of distorted automorphisms, first in the context of subshifts through the use of Turing Machines and then in a non-symbolic example through the study of automorphisms of the torus.

## 2 Definitions

### 2.1 Symbolic Spaces

Let $\Sigma$ be a finite set, which we will henceforth call the alphabet. A *full-shift* is the dynamical system given by the space $\Sigma^{\mathbb{Z}}$, endowed with the product topology, and the shift function $\sigma\colon \Sigma^{\mathbb{Z}} \to \Sigma^{\mathbb{Z}}$ given by $\sigma(x)_i = x_{i+1}$, for all $x = (x_i)_{i\in\mathbb{Z}}$ and $i \in \mathbb{Z}$. A $\sigma$-invariant and closed subset $X$ of a full-shift $\Sigma^{\mathbb{Z}}$ is called a subshift. Usually we denote a subshift by $(X, \sigma)$ or $X$ indistinctly.

Given a configuration $x \in X$, for $i < j$ we denote the finite word composed by the symbols in $x$ from index $i$ up to $j$, $x_i x_{i+1} \dots x_{j-1} x_j$, by $x_{[i,j]}$.

The set of finite words appearing in points or configurations of $\Sigma^{\mathbb{Z}}$ is denoted by $\Sigma^*$. Define the language of a subshift $X \subseteq \Sigma^{\mathbb{Z}}$, as the set of all finite words that appear on configurations in $X$,

$$\mathcal{L}(X) = \{w \in \Sigma^* : w = x_{[n,m]}, \text{ for some integers } n \le m \text{ and } x \in X\}$$

**Definition 1** A subshift $X \subseteq \Sigma^{\mathbb{Z}}$ is said to be a shift of finite type (SFT), if there exists a finite set of (forbidden) words $\mathcal{F} \subseteq \Sigma^*$ such that $x \in X$ if and only if no word in $\mathcal{F}$ appears as a subword of $x$.

A subshift $X \subseteq \Sigma^{\mathbb{Z}}$ is said to be mixing if there exists $N \in \mathbb{N}$ such that for every pair of words $u, v \in \mathcal{L}(X)$ and $n \ge N$ there exists $w \in \mathcal{L}(X)$ such that $|w| \ge n$ and $uwv \in \mathcal{L}(X)$.

### 2.2 Endomorphisms of a Shift Space

A map $f\colon X \to Y$ between two subshifts $X$ and $Y$ is called a *morphism* if it is continuous and shift commuting, that is, $f \circ \sigma = \sigma \circ f$. One says the function is an *endomorphism* if $X = Y$ and an *automorphism* if it is also bijective. We will denote the set of all automorphisms of a subshift $X$ by $\mathrm{Aut}(X)$.

Due to the Curtis–Hedlund–Lyndon Theorem [8] we know that every endomorphism $\phi\colon X \to X$, with $X \subseteq \Sigma^{\mathbb{Z}}$, is determined by a local function $\Phi\colon \Sigma^{2N+1} \to \Sigma$ such that $\phi(x)_i = \Phi(x_{[-N+i,N+i]})$. In other words, for two configurations $x, y \in X$,

$$x_{[-N,N]} = y_{[-N,N]} \implies \phi(x)_0 = \phi(y)_0.$$

The minimum $N \in \mathbb{N}$ such that the previous property is satisfied is called the *range* of $\phi$, and is denoted by $\mathrm{range}(\phi)$.

We can further classify automorphisms according to how they act on specific configurations.

**Definition 2** Let $(X, \sigma)$ be a subshift and $\phi \in \text{Aut}(X)$. Given a configuration $x \in X$, we say $\phi$ is *weakly periodic* on $x$ if there exists $p \in \mathbb{N}$ and $q \in \mathbb{Z}$ such that $\phi^p(x) = \sigma^q(x)$.

If $\phi$ is not weakly periodic for any configuration, then we say it is *aperiodic*.

## 2.3 Algebraic Distortion

Let us briefly introduce the classical notion of distortion in group theory.

**Definition 3** Let $G$ be a finitely generated group and $S \subseteq G$ a symmetric generating set. Given $g \in G$, we define the length of $g$ with respect to $S$, $\ell_S(g)$, as the smallest non-negative integer $n$ such that $g$ can be written as a product of $n$ elements of $S$. We write,

$$\ell_S(g) = n.$$

By convention, we use that $\ell_S(e) = 0$.

We note that the function $\ell_S$ depends on the generating set $S$ only up to a multiplicative constant.

**Lemma 1** ([4], Lemma 2.4) *If $S_1$ and $S_2$ are two generating sets of $G$, then there exists a constant $c \geq 1$ such that*

$$\frac{1}{c}\ell_{S_2}(g) \leq \ell_{S_1}(g) \leq c\ell_{S_2}(g), \ \forall g \in G.$$

**Definition 4** Let $G$ be a finitely generated group and $S$ a symmetric generating set. The translation length of an element $g \in G$ is defined as the limit:

$$\|g\|_S := \lim_{n\to\infty} \frac{\ell_S(g^n)}{n}.$$

We say $g$ is a distorted element if $\|g\|_S = 0$.

**Remark 1** It is important to note that due to Lemma 1, the property of being distorted is independent of the generating set.

## 2.4 Expansive Dynamical Systems

Let $(X, \rho)$ be a compact metric space with metric $\rho$, which we assume to be infinite. A $\mathbb{Z}^d$-action $\Psi$ on $X$ is a homomorphism from the additive group $\mathbb{Z}^d$ to the group $\text{Homeo}(X)$ (homeomorphisms of $X$ with composition). Given a subset $F \subseteq \mathbb{R}^d$ we define:

$$\rho_{\Psi}^{F}(x, y) = \sup\{\rho(\Psi^n(x), \Psi^n(y)) : n \in F \cap \mathbb{Z}^d\},$$

where $\Psi^n$ is the element in Homeo($X$) associated to $n \in \mathbb{Z}^d$. If $F \cap \mathbb{Z}^d = \emptyset$, we write $\rho_{\Psi}^{F}(x, y) = 0$.

In this context, an *automorphism* $\phi \colon X \to X$ *for the action of* $\Psi$ is a bi-continuous function such that it commutes with the $\mathbb{Z}^d$-action, that is, $\phi \circ \Psi^n = \Psi^n \circ \phi$, $\forall n \in \mathbb{Z}^d$. The group of all automorphisms for the action of $\Psi$ will be denoted by Aut($X$, $\Psi$). In the sequel, if there is no ambiguity on the action $\Psi$, we will just speak about automorphisms of $X$.

**Definition 5** A $\mathbb{Z}^d$-action $\Psi$ on $X$ is *expansive* if there exists $c > 0$ such that

$$\rho_{\Psi}^{\mathbb{R}^d}(x, y) \le c \implies x = y.$$

In such a case, $c$ is called the expansivity constant of $\Psi$.

When $d = 1$, we say $\Psi$ is *positively expansive* if there exists $c > 0$ such that

$$\rho_{\Psi}^{\mathbb{R}_{0,+}}(x, y) \le c \implies x = y.$$

For a subset $F \subseteq \mathbb{R}^d$ and $v \in \mathbb{R}^d$, we define

$$\mathrm{dist}(v, F) = \inf\{\|v - w\| : w \in F\},$$

where $\| \cdot \|$ denotes the Euclidean norm on $\mathbb{R}^d$. For $t > 0$ we define the *thickening of $F$ by $t$* as $F^t = \{v \in \mathbb{R}^d : \mathrm{dist}(v, F) \le t\}$.

**Definition 6** Let $\Psi$ be a $\mathbb{Z}^d$-action on $X$ and $F \subseteq \mathbb{R}^d$. Then, $F$ is expansive for $\Psi$ if there exists $\varepsilon > 0$ and $t > 0$ such that

$$\rho_{\Psi}^{F^t}(x, y) \le \varepsilon \implies x = y.$$

If $F$ does not satisfy this condition, it is said to be non-expansive.

When a $\mathbb{Z}^d$ action $\Psi$ is expansive, the following lemma allows us to consider a uniform $\varepsilon$ in Definition 6.

**Lemma 2** ([2], Lemma 2.3) *Let $\Psi$ be an expansive $\mathbb{Z}^d$-action on $X$, with expansivity constant $c$. Then, for each expansive subset $F \subseteq \mathbb{R}^d$ for $\Psi$ there exists $s > 0$ such that*

$$\rho_{\Psi}^{F^s}(x, y) \le c \implies x = y.$$

# 3 Generalizing the Range

We are interested in defining a notion of distortion for an automorphism on an arbitrary expansive system. With this in mind, we present the following general lemma.

**Lemma 3** *Let $\Psi$ be an expansive $\mathbb{Z}^d$-action on $X$, with expansivity constant $c$. Then, for all $\varepsilon > 0$ there exists $M \in \mathbb{N}$ such that $\forall x, y \in X$,*

$$\rho_{\Psi}^{B_\infty(0,M)}(x, y) \leq c \implies \rho(x, y) \leq \varepsilon,$$

*where $B_\infty(0, M) = \{v \in \mathbb{R}^d : ||v||_\infty = \max_{1\leq i\leq d} |v_i| \leq M\}$.*

***Proof*** We proceed by contradiction. Let $\varepsilon$ be such that for all $m \in \mathbb{N}$, there are $x_m, y_m \in X$ such that $\rho_{\Psi}^{B_\infty(0,m)}(x_m, y_m) \leq c$ and $\rho(x_m, y_m) > \varepsilon$. Since $X$ is compact, we have a subsequence $(m_i)_{i\in\mathbb{N}}$ such that the sequences $(x_{m_i})_{i\in\mathbb{N}}$ and $(y_{m_i})_{i\in\mathbb{N}}$ converge to $\bar{x}$ and $\bar{y}$ respectively.

Let us consider $\eta > 0$, $n \in \mathbb{Z}^d$ and $I \in \mathbb{N}$ such that $\forall i \geq I, m_i \geq \max\{\|n\|_\infty, m_I\}$

$$\rho(\Psi^n x_{m_i}, \Psi^n \bar{x}) \leq \frac{\eta}{2} \text{ and } \rho(\Psi^n y_{m_i}, \Psi^n \bar{y}) \leq \frac{\eta}{2}.$$

Then,

$$\begin{aligned}\rho(\Psi^n \bar{x}, \Psi^n \bar{y}) &\leq \rho(\Psi^n x_{m_i}, \Psi^n \bar{x}) + \rho(\Psi^n x_{m_i}, \Psi^n y_{m_i}) + \rho(\Psi^n y_{m_i}, \Psi^n \bar{y})\\ &\leq \eta + c.\end{aligned}$$

By taking $\eta \to 0$ ($m_i$ is always greater than $\|n\|_\infty$) we obtain

$$\rho(\Psi^n \bar{x}, \Psi^n \bar{y}) \leq c, \ \forall n \in \mathbb{Z}^d.$$

Thus, using that $\Psi$ is expansive with constant $c$, we get that $\bar{x} = \bar{y}$. Therefore, for a sufficiently large $i$

$$\begin{aligned}\rho(x_{m_i}, y_{m_i}) &\leq \rho(x_{m_i}, \bar{x}) + \rho(y_{m_i}, \bar{y})\\ &\leq \varepsilon,\end{aligned}$$

which is a contradiction.

## *3.1 Automorphisms on Expansive Systems*

From this point onward, we will work in the following context. Let $T : X \to X$ be a homeomorphism on the compact metric space $(X, \rho)$. In this setting we write

Aut$(X, T)$ for the group of automorphisms of $X$ commuting with $T$. Observe that $T$ defines a $\mathbb{Z}$-action on $X$, so we can use the previously defined notations. Recall that given a subset $F \subseteq \mathbb{R}$

$$\rho_T^F(x, y) = \sup\{\rho(T^n x, T^n y) : n \in F \cap \mathbb{Z}\},$$

and that the system $(X, T)$ is expansive if there exists a constant $c > 0$ such that

$$\rho_T^{\mathbb{R}}(x, y) \leq c \implies x = y.$$

**Definition 7** Let $(X, T)$ be an expansive system of constant $c > 0$ and $\phi \in$ Aut $(X, T)$. We call *range* of $\phi$ to the minimum $M \in \mathbb{N}$ such that $\forall x, y \in X$,

$$\rho_T^{[-M,M]}(x, y) \leq c \implies \rho(\phi(x), \phi(y)) \leq c,$$

and we denote it by range$(\phi)$.

It is clear that if $(X, T)$ is a subshift, with $T$ being the shift map, the previous definition coincides with the usual notion of the radius of an automorphism, as this space is expansive of constant $c = 1/2$.

In the general case where $(X, T)$ is an expansive system of constant $c > 0$, the existence of $M$ is ensured by the following. Since $\phi \in$ Aut$(X, T)$ is continuous over the compact set $X$, then it is uniformly continuous. Therefore, we have that there exists $\delta > 0$ such that for all $x, y \in X$:

$$\rho(x, y) \leq \delta \implies \rho(\phi(x), \phi(y)) \leq c.$$

By applying Lemma 3, we have that there exists $M \in \mathbb{N}$ such that for all $x, y \in X$:

$$\rho_T^{[-M,M]}(x, y) \leq c \implies \rho(x, y) \leq \delta,$$

and therefore,

$$\rho_T^{[-M,M]}(x, y) \leq c \implies \rho(\phi(x), \phi(y)) \leq c.$$

Now consider $\phi \in$ Aut$(X, T)$. With both $\phi$ and $T$ we can construct a $\mathbb{Z}^2$-action $\Psi$ defined by: if $n = (n_1, n_2) \in \mathbb{Z}^2$ then $\Psi^n = \phi^{n_2} \circ T^{n_1}$. We also denote the dynamical system defined by the action $\Psi$ by $(X, T, \phi)$.

**Definition 8** Let $E, F \subseteq \mathbb{R}^2$, and $\Psi$ an expansive $\mathbb{Z}^2$-action on $X$. We say $E$ codifies $F$ if for all $v \in \mathbb{R}^2$,

$$\rho_\Psi^{E+v}(x, y) \leq c \implies \rho_\Psi^{F+v}(x, y) \leq c.$$

Using this terminology, the range of $\phi \in \mathrm{Aut}(X, T)$ can be understood as the minimum $M \in \mathbb{N}$ such that $[-M, M] \times \{0\}$ codifies $\{(0, 1)\}$ for the $\mathbb{Z}^2$-action $(X, \Psi)$ induced by $T$ and $\phi$ as defined above.

**Lemma 4** *Let $(X, T)$ be an expansive system of constant $c > 0$, and consider $\phi, \psi \in \mathrm{Aut}(X, T)$. Then,*

$$\mathrm{range}(\phi \circ \psi) \leq \mathrm{range}(\phi) + \mathrm{range}(\psi).$$

*In particular, the sequence $(\mathrm{range}(\phi^n))_{n\in\mathbb{N}}$ is subadditive.*

***Proof*** Let $M = \mathrm{range}(\phi)$ and $N = \mathrm{range}(\psi)$. If we have $\rho_T^{[-(M+N), M+N]}(x, y) \leq c$, then

$$\forall t \in [-(M+N), M+N] : \ \rho(T^t x, T^t y) \leq c.$$

If we fix $m \in [-M, M]$ and define $\bar{x} = T^m x,\ \bar{y} = T^m y$, from the previous inequality we obtain that:

$$\forall n \in [-N, N] : \ \rho(T^n \bar{x}, T^n \bar{y}) \leq c.$$

By definition of range, this means that $\rho(\psi(\bar{x}), \psi(\bar{y})) \leq c$. Since this is possible for any $m \in [-M, M]$, we have:

$$\forall m \in [-M, M] : \ \rho(T^m \psi(x), T^m \psi(y)) \leq c,$$

which implies that $\rho(\phi \circ \psi(x), \phi \circ \psi(y)) \leq c$, and therefore, $\mathrm{range}(\phi \circ \psi) \leq N + M$.

Since $(\mathrm{range}(\phi^n))_{n\in\mathbb{N}}$ is a subadditive sequence, by Fekete's Lemma, we have that the following definition makes sense.

**Definition 9** Let $(X, T)$ be an expansive system of constant $c > 0$. The *asymptotic range* of $\phi \in \mathrm{Aut}(X, T)$ is defined by

$$\mathrm{range}_\infty(\phi) := \lim_{n\to\infty} \frac{\mathrm{range}(\phi^n)}{n}.$$

If $\mathrm{range}_\infty(\phi) = 0$ we say $\phi$ is *range distorted*, and denote the set of all range distorted automorphisms in $\mathrm{Aut}(X, T)$ by $RD(X, T)$.

The following simple proposition can be deduced from definition and previous lemma.

**Proposition 1** *Let $\phi, \psi \in \mathrm{Aut}(X, T)$. We have,*

1. $\mathrm{range}_\infty(\psi \circ \phi \circ \psi^{-1}) = \mathrm{range}_\infty(\phi)$,
2. $\mathrm{range}_\infty(\phi^p) = p \cdot \mathrm{range}_\infty(\phi)$ *for* $p \in \mathbb{N}$,
3. *if $\psi$ and $\phi$ commute, then* $\mathrm{range}_\infty(\psi \circ \phi) \leq \mathrm{range}_\infty(\psi) + \mathrm{range}_\infty(\phi)$.

## 4 Alternative Notion of Distortion

The definition of asymptotic range defined in previous section concerns the average evolution of the symmetric window with which an automorphism of an expansive system $(X, T)$ is computed. To complement this analysis, we introduce an alternative notion of distortion through the use of the Lyapunov exponents presented by Cyr et al. in [6]. These exponents serve to study the average speed at which information asymptotically propagates through the automorphism. This notion will later be shown to be very important because of their connection to some kind of geometry associated to the automorphism.

In what follows we fix an expansive dynamical system $(X, T)$ of expansive constant $c > 0$. Recall $(X, \rho)$ is a compact metric space and $T$ is a homeomorphism of $X$.

**Lemma 5** *Let $(X, T)$ be an expansive system with expansivity constant $c > 0$ and $\phi \in \text{Aut}(X, T)$. Then,*

$$\rho_T^{[0,+\infty)}(x, y) \leq c \implies \rho_T^{[\text{range}(\phi),+\infty)}(\phi(x), \phi(y)) \leq c.$$

*That is,* $[0, +\infty) \times \{0\}$ *codifies* $[\text{range}(\phi), +\infty) \times \{1\}$ *in* $(X, T, \phi)$.

***Proof*** We begin by simplifying the notation by writing $r = \text{range}(\phi)$. Let $x, y \in X$ be such that $\rho_T^{[0,+\infty)}(x, y) \leq c$. Then, in particular, we have that

$$\rho_T^{[-(r+k),r+k]}(T^{r+k}x, T^{r+k}y) \leq c, \quad \forall k \geq 0.$$

By the definition of range, we have that

$$\rho(\phi(T^{r+k}x), \phi(T^{r+k}y)) \leq c, \quad \forall k \geq 0,$$

which, by taking supremum over $k$, can be re-written as

$$\rho_T^{[r,+\infty)}(\phi(x), \phi(y)) \leq c.$$

This finishes the proof. □

We recall from the previous section that we can apply the notion of coding to the $\mathbb{Z}^2$-action $\Psi$ defined by the expansive action $T$ and an automorphism $\phi \in \text{Aut}(X, T)$. We consider the following sets:

$$C^-(\phi) = \{k \in \mathbb{Z} : (-\infty, 0] \times \{0\} \text{ codifies } (-\infty, k] \times \{1\}\},$$

$$C^+(\phi) = \{k \in \mathbb{Z} : [0, \infty) \times \{0\} \text{ codifies } [k, \infty) \times \{1\}\}.$$

Due to Lemma 5, both sets are non-empty. This allows us to define the quantities:

$$W^-(n,\phi) = \sup C^-(\phi^n),$$

$$W^+(n,\phi) = \inf C^+(\phi^n).$$

By definition, we have that for $n \geq 1$, $W^\pm(n,\phi) = W^\pm(1,\phi^n)$. In addition,

**Lemma 6** *Let $(X,T)$ be an expansive system and consider $\phi, \psi \in \mathrm{Aut}(X,T)$. Then,*

$$W^+(n,\phi\psi) \leq W^+(n,\phi) + W^+(n,\psi) \text{ and } W^-(n,\phi\psi) \geq W^-(n,\phi) + W^-(n,\psi).$$

*In particular, the sequences $(W^+(n,\phi))_{n\in\mathbb{N}}$ and $(-W^-(n,\phi))_{n\in\mathbb{N}}$ are subadditive.*

Again, Fekete's Lemma allows us to make the following definition:

**Definition 10** ([6], *Definition 3.12*) Let $(X,T)$ be an expansive system. Given $\phi \in \mathrm{Aut}(X,T)$, we define the exponents of Cyr, Franks and Kra by:

$$\alpha^-(\phi) = \lim_{n\to\infty} \frac{W^-(n,\phi)}{n},$$

$$\alpha^+(\phi) = \lim_{n\to\infty} \frac{W^+(n,\phi)}{n}.$$

**Definition 11** We say an automorphism $\phi \in \mathrm{Aut}(X,T)$ of an expansive system $(X,T)$ is *Lyapunov distorted* if $\alpha^\pm(\phi) = 0$. We denote the set of all Lyapunov distorted automorphisms in $\mathrm{Aut}(X,T)$ by

$$AD(X,T) = \{\phi \in \mathrm{Aut}(X,T) : \alpha^\pm(\phi) = 0\}.$$

These exponents satisfy some very useful properties.

**Proposition 2** *Let $(X,T)$ be an expansive system and consider $\phi \in \mathrm{Aut}(X,T)$. We have the following properties:*

1. *For all $k \in \mathbb{Z}$, $\alpha^\pm(T^k\phi) = \alpha^\pm(\phi) + k$.*
2. *For all $m \in \mathbb{N}$, $\alpha^\pm(\phi^m) = m\alpha^\pm(\phi)$.*
3. *If $\psi \in \mathrm{Aut}(X,T)$ commutes with $\phi$, then:*

$$\alpha^+(\phi\psi) \leq \alpha^+(\phi) + \alpha^+(\psi) \text{ and } \alpha^-(\phi\psi) \geq \alpha^-(\phi) + \alpha^-(\psi).$$

4. *$\alpha^+(\phi) + \alpha^+(\phi^{-1}) \geq 0$ and $\alpha^-(\phi) + \alpha^-(\phi^{-1}) \leq 0$.*
5. *If $X$ is an infinite subshift, then $\alpha^-(\phi) \leq \alpha^+(\phi)$.*

***Proof*** The first property follows directly from the fact that $W^\pm(n, T^k\phi) = nk + W^\pm(n,\phi)$. The second one comes from:

$$\lim_{n\to\infty} \frac{W^\pm(nm,\phi)}{n} = m \cdot \lim_{n\to\infty} \frac{W^+(nm,\phi)}{nm}.$$

Next, for property 3, we see that if $\phi, \psi \in \mathrm{Aut}(X, T)$ commute, then

$$W^+(n, \phi\psi) = W^+(1, (\phi\psi)^n) = W^+(1, \phi^n\psi^n) \le W^+(n, \phi) + W^+(n, \psi).$$

Property 4 follows from property 3 and the fact that $\alpha^{\pm}(\mathrm{id}) = 0$.

The last property was proved in Proposition 3.15 of [6]. □

Using the following lemma we can see that, in the case of subshifts, Lyapunov distortion is weaker than range distortion. Given $\phi \in \mathrm{Aut}(X, T)$ we denote the interval $[-W^+(n, \phi), -W^-(n, \phi)]$ by $I(n, \phi)$.

**Lemma 7** *Let $(X, T)$ be an expansive system and consider $\phi \in \mathrm{Aut}(X, T)$. If $J$ is an interval that $\phi^n$-codes $\{0\}$, then $I(n, \phi) \subseteq J$.*

***Proof*** Let $J = [a, b]$ be an interval that $\phi^n$-codes $\{0\}$. Then, $(-\infty, 0]$ must $\phi^n$-code $(-\infty, -b]$ and $[0, \infty)$ must $\phi^n$-code $[-a, \infty)$. We conclude by using the definition of $I(n, \phi)$.

□

**Lemma 8** *Let $(X, T)$ be an expansive system and consider $\phi \in \mathrm{Aut}(X, T)$. Then,* $\mathrm{range}_\infty(\phi) \ge \max\left\{\alpha^+(\phi), -\alpha^-(\phi)\right\}$.

***Proof*** The result follows from previous lemma by noting that the interval $[-\mathrm{range}(\phi^n), \mathrm{range}(\phi^n)]$ $\phi^n$-codes $\{0\}$. □

**Proposition 3** *Let $(X, \sigma)$ be an infinite subshift. Then, $RD(X, \sigma) \subseteq AD(X, \sigma)$.*

***Proof*** For $\phi \in \mathrm{Aut}(X, \sigma)$, due to (5) on Proposition 2 and Lemma 8, we know that

$$\mathrm{range}_\infty(\phi) \ge \alpha^+(\phi) \ge \alpha^-(\phi) \ge -\mathrm{range}_\infty(\phi),$$

which concludes the proof. □

We can see that in the context of SFTs, the two notions are in fact equivalent. To see this, we first need an auxiliary result.

**Lemma 9** ([6], Lemma 3.21) *Let $(X, \sigma)$ be an SFT and $\phi \in \mathrm{Aut}(X, \sigma)$. Then, there is a constant $C(\phi)$ such that*

$$\frac{|I(n, \phi)| - 1}{2} \le \mathrm{range}(\phi^n) \le |I(n, \phi)| + C(\phi).$$

*If $X$ is a full-shift we can take $C(\phi) = 0$.*

**Theorem 1** *Let $(X, \sigma)$ be an SFT. Then, $AD(X, \sigma) = RD(X, \sigma)$.*

***Proof*** By diving by $n$ and taking limit on the expression given by Lemma 9, we conclude. □

## 5 Geometry and Distortion

In [6], Cyr, Franks and Kra showed that there is a connection between discrete Lyapunov exponents and the geometry of the $\mathbb{Z}^2$-system $(X, \sigma, \phi)$, where $\phi \in \text{Aut}(X, \sigma)$ and $X$ is a subshift. This connection was first explored by Hochman in [9] through the notion of prediction shapes. We generalize these result to the context of expansive systems $(X, T)$. Finally, in the context of subshifts, we connect the newly introduced direction exponents to the standard ones. We relate the fact of having these exponents equal to zero to having non-expansive directions.

The following theorems connect the geometry of the space-time of the automorphisms with its asymptotic behavior.

**Theorem 2** *Let $(X, T)$ be an expansive system and consider $\phi \in \text{Aut}(X, T)$. Then, the lines defined by $x = \alpha^+(\phi)y$ and $x = \alpha^-(\phi)y$ are not expansive.*

The proof of this fact is very technical and can be retraced step by step from [6].

***Proof*** Let $c > 0$ be the expansive constant of $(X, T)$. We will only look at the case where there exists a constant $D > 0$ such that

$$0 \leq W^+(k, \phi) - k\alpha^+(\phi) < D, \quad \forall k \geq 0.$$

It is possible to see that due to the definitions at play, if we have two elements $x, y \in X$, such that

$$\rho_T^{[0,\infty)}(x, y) \leq c,$$

then for all $j \geq 0$ and $i \geq D + \alpha^+(\phi)j$,

$$\rho(T^i\phi^j x, T^i\phi^j y) \leq c.$$

We see that this can be interpreted as $x$ and $y$ coinciding on the lower half space of the line $i = D + \alpha^+(\phi)j$.

For every $n$, because of the definition of $W^+(n, \phi)$, we have $x_n, y_n \in X$ such that

$$\rho_T^{[0,\infty)}(x_n, y_n) \leq c,$$

but, $\rho(T^{W^+(n)-1}\phi^n x_n, T^{W^+(n)-1}\phi^n y_n) > c$. By defining $\hat{x}_n := T^{W^+(n)}\phi^n x_n$ and $\hat{y}_n := T^{W^+(n)}\phi^n y_n$, we can see that,

$$\rho(T^{-1}\hat{x}_n, T^{-1}\hat{y}_n) > c,$$

and for all $j \geq -n$ and $i \geq D + \alpha^+(\phi)j$,

$$\rho(T^i\phi^j\hat{x}_n, T^i\phi^j\hat{y}_n) \leq c.$$

Next, we use the compactness of $X$ to find a convergent subsequences, that converge to $\hat{x} = \lim \hat{x}_n$ and $\hat{y} = \lim \hat{y}_n$.

Let us have an arbitrary $\varepsilon > 0$. Then, due to the continuity of $T$ and $\phi$, for sufficiently large $n$,

$$\begin{aligned} c &< \rho(T^{-1}\hat{x}_n, T^{-1}\hat{y}_n), \\ &\leq \rho(T^{-1}\hat{x}_n, T^{-1}\hat{x}) + \rho(T^{-1}\hat{x}, T^{-1}\hat{y}) + \rho(T^{-1}\hat{y}, T^{-1}\hat{y}_n), \\ &\leq 2\varepsilon + \rho(T^{-1}\hat{x}, T^{-1}\hat{y}). \end{aligned}$$

Therefore, $c < \rho(T^{-1}\hat{x}, T^{-1}\hat{y})$. Analogously, for an arbitrary $\varepsilon > 0$, sufficiently large $n$ and $(i, j)$ such that $i > D + \alpha^+(\phi)j$,

$$\begin{aligned} \rho(T^i\phi^j\hat{x}, T^i\phi^j\hat{y}) &\leq \rho(T^i\phi^j\hat{x}_n, T^i\phi^j\hat{x}) + \rho(T^i\phi^j\hat{x}, T^i\phi^j\hat{y}) + \rho(T^i\phi^j\hat{y}, T^i\phi^j\hat{y}_n), \\ &\leq 2\varepsilon + c. \end{aligned}$$

In other words, for $i > D + \alpha^+(\phi)j$,

$$\rho(T^i\phi^j\hat{x}, T^i\phi^j\hat{y}) \leq c.$$

This proves that the line defined by $x = \alpha^+ y$ is not expansive. □

**Theorem 3** *Let $(X, T)$ be an expansive dynamical system, $\phi \in \mathrm{Aut}(X, T)$ and $L$ a line in $\mathbb{R}^2$ given by $x = my$. If $m > \max\{\alpha^+(\phi), -\alpha^-(\phi^{-1})\}$ or $m < \min\{\alpha^-(\phi), -\alpha^+(\phi^{-1})\}$, then $L$ is expansive.*

***Proof*** Let us first show that if $m > \alpha^+(\phi)$, then $L$ is left-expansive.

We take $x, y \in X$ such that:

$$\rho(T^n\phi^k(x), T^n\phi^k(y)) \leq c, \quad \forall (n, k) \in \mathbb{Z}^2 \text{ such that } n > mk.$$

Because $m > \alpha^+(\phi)$, the vector defined by $(\alpha^+(\phi), 1)$ is not parallel to $L$.

By Definition 10, for sufficiently large $n$, the vector $(W^+(n), n)$ is not parallel to $L$.

Next, let us have $(u_0, v_0)$, an arbitrary point to the left of $L$ (that is, $u_0 < mv_0$). There exists $n_0 > 0$ such that if $u_1 = u_0 - W^+(n_0)$ and $v_1 = v_0 - n_0$, then $(u_1, v_1)$ is to the right of $L$. Therefore, the line given by $\{(t, v_1) : u_1 \leq t\}$ is to the right of $L$ and codifies $(u_0, v_0)$ by definition of $W^+(n_0)$. This shows that $L$ is left expansive.

Analogously, if $m < \alpha^-(\phi)$ then $L$ is right expansive.

Lastly, we can see that the transformation $r(x, y) = (x, -y)$ allows us to move between the $\mathbb{Z}^2$-systems $(X, T, \phi)$ and $(X, T, \phi^{-1})$. Consequently, $L$ is right-expansive (left) on the first system if and only if $r(L)$ is left-expansive (right) one the second one. This fact concludes the proof □

By combining these results, we arrive at the fundamental connection between distortion and non-expansive subspaces.

**Corollary 1** *Let* $(X, T)$ *be an expansive system and consider* $\phi \in \text{Aut}(X, T)$. *Then,* $\phi, \phi^{-1} \in AD(X, T)$ *if and only if* $x = 0$ *is the only non-expansive direction of* $\phi$.

A first consequence of this connection is the fact that expansive automorphisms can not be Lyapunov distorted.

**Theorem 4** *Let* $(X, T)$ *be an expansive system of constant* $c > 0$ *and consider an expansive automorphism* $\phi \in \text{Aut}(X, T)$ *of constant* $\delta > 0$. *Then,* $\phi \notin AD(X)$.

***Proof*** Let us call $\Psi$ the joint $\mathbb{Z}^2$-action of $T$ and $\phi$. Due to Lemma 3 on the expansive system $(X, T)$, we know that there exists $M \in \mathbb{N}$ such that:

$$\forall x, y \in X : \; \rho_T^{[-M,M]}(x, y) \leq c \implies \rho(x, y) \leq \delta.$$

Now, let us see that $L_0$, defined by $x = 0$, is an expansive direction. If we have $x, y \in X$ such that

$$\rho_\Psi^{L_0^M}(x, y) \leq c,$$

in particular we have that,

$$\forall n \in \mathbb{Z} : \; \rho_T^{[-M,M]}(\phi^n(x), \phi^n(y)) \leq c.$$

This implies that,

$$\forall n \in \mathbb{Z} : \; \rho(\phi^n(x), \phi^n(y)) \leq \delta.$$

Given that $\phi$ is expansive, this means that $x = y$. We conclude by using Theorem 2. □

## 6 Symmetric Distortion Subset

Let $(X, T)$ be an expansive system. We denote the subset of range distorted automorphisms of $(X, T)$ with a range distorted inverse by:

$$\mathfrak{D}(X, T) = \{\phi \in \text{Aut}(X, T) : \text{range}_\infty(\phi) = \text{range}_\infty(\phi^{-1}) = 0\},$$

and the subgroup of distorted elements in $\text{Aut}(X, T)$ (in the algebraic sense) by $GD(X, T)$.

**Proposition 4** $GD(X, T) \subseteq \mathfrak{D}(X, T)$.

***Proof*** Let $\phi$ be a distorted element of $\text{Aut}(X, T)$. This means that there exists a finitely generated subgroup $G$ of $\text{Aut}(X, T)$ such that $\|\phi\|_S = 0$, for a symmetric generating set $S$. Then, by Lemma 4,

$$\text{range}(\phi^n) \leq \ell_S(\phi^n) \cdot \max_{s \in S}\{\text{range}(s)\}.$$

Dividing the expression by $n$ and taking limit, $\phi$ is range distorted. We conclude by noting that if $\phi$ is group distorted, its inverse also is. □

For the next Lemma, we recall that a map $f: X \to X$ is said to be equicontinuous if for all $\varepsilon > 0$ there exists $\delta > 0$ such that

$$\rho(x, y) \leq \delta \implies \rho(f^n(x), f^n(y)) \leq \varepsilon, \ \forall n \in \mathbb{Z}.$$

**Lemma 10** *Let $\phi \in \mathrm{Aut}(X, T)$ be an equicontinuous automorphism of the expansive system $(X, T)$. Then $\phi \in \mathfrak{D}(X, T)$.*

***Proof*** Because $\phi$ is equicontinuous, for all $\varepsilon > 0$ there exists a $\delta > 0$ such that

$$\rho(x, y) \leq \delta \implies \rho(\phi^n(x), \phi^n(y)) \leq \varepsilon, \ \forall n \in \mathbb{Z}.$$

By picking $\varepsilon = c$, Lemma 3 tells us that there exists $M > 0$ such that

$$\rho_T^{[-M,M]}(x, y) \leq c \implies \rho(x, y) \leq \delta.$$

This implies that,

$$\rho_T^{[-M,M]}(x, y) \leq c \implies \rho(\phi^n(x), \phi^n(y)) \leq c, \ \forall n \in \mathbb{Z},$$

that is, $\mathrm{range}(\phi^n) \leq M$ for all $n \in \mathbb{Z}$. We conclude that $\mathrm{range}_\infty(\phi) = \mathrm{range}_\infty(\phi^{-1}) = 0$. □

**Remark 2** As a consequence of the Arzelà–Ascoli Theorem, any compact subgroup $K$ of $\mathrm{Aut}(X, T)$ satisfies $K \subseteq \mathfrak{D}(X, T)$.

We are interested in understanding the structure of $\mathfrak{D}(X, T)$. In the general setting, this set is not a subgroup of $\mathrm{Aut}(X, T)$, as is shown in Remark 3 below, where we show an automorphism which is not distorted, but is a composition of two distorted automorphisms through an example due to Schmieding [12].

**Remark 3** Let $X = \{0, 1, 2\}^{\mathbb{Z}}$ be the full-shift on 3 symbols. Let $\phi_1$ be the marker automorphism that permutes 000111 with 002111, and $\phi_2$ the marker automorphism that permutes 000111 with 002111 (marker automorphisms are presented in great detail in [3]). If we define $\phi = \phi_2 \circ \phi_1$, it is possible to see $\phi$ is not distorted even though both $\phi_1$ and $\phi_2$ are symmetrically distorted.

The next lemma follows directly from Proposition 1.

**Lemma 11** *Let $(X, T)$ be an expansive system and consider $\phi, \psi \in \mathfrak{D}(X, T)$. We have the following properties:*

1. *If $[\phi, \psi] = \mathrm{id}$, then $\phi \circ \psi \in \mathfrak{D}(X, T)$, where $[\phi, \psi] = \phi\psi\phi^{-1}\psi^{-1}$.*
2. *For all $\varphi \in \mathrm{Aut}(X, T)$, $\varphi \circ \phi \circ \varphi^{-1} \in \mathfrak{D}(X, T)$.*

*3.* $\phi^p \in \mathfrak{D}(X, T)$, *for all* $p \in \mathbb{N}$.

**Proposition 5** *Let* $(X, \sigma)$ *be a mixing SFT. Then,* $\mathfrak{D}(X, \sigma)$ *contains an isomorphic copy of every finite group.*

***Proof*** This result follows from the fact that every finite order automorphism is equicontinuous and the Kim and Roush Theorem [10], that states that the automorphism group of a mixing SFT contains an isomorphic copy of the automorphisms group of any $n$-full shift for $n \geq 2$. □

Finally, let us generalize the fact that the subgroup generated by the action $T$ has a trivial intersection with $\mathfrak{D}(X, T)$.

**Lemma 12** *Let* $(X, T)$ *be an expansive system of expansive constant* $c > 0$*. If* $X$ *is infinite, then* $\text{range}_\infty(T) = 1$.

To prove this result we make use of a following result by Schwartzman about infinite systems.

**Theorem 5** ([2], Theorem 3.9) *Let* $T$ *be a homeomorphism of an infinite compact metric space* $(X, \rho)$ *and* $\delta > 0$*. Then, there exists two distinct* $x, y \in X$ *such that* $\rho(T^n x, T^n y) \leq \delta$ *for all* $n \geq 0$.

***Proof*** (of Lemma 12) It is evident that $\text{range}(T^n) \leq n$. To obtain the other bound, by applying Theorem 5 to $T^{-1}$, we obtain two distinct points $x, y \in X$ such that $\rho_T^{(-\infty,0]}(x, y) \leq c$. Given that $T$ is expansive and the points are different, we choose the smallest $m > 0$ such that $\rho(T^m x, T^m y) > c$.

For $n \in \mathbb{N}$, we define $\bar{x} = T^{m-n} x$ and $\bar{y} = T^{m-n} y$. Then, $\rho_T^{[-n+1,n-1]}(\bar{x}, \bar{y}) \leq c$, with $\rho(T^n \bar{x}, T^n \bar{y}) > c$. This means that $\text{range}(T^n) > n - 1$, and as a consequence $\text{range}(T^n) = n$. □

**Corollary 2** *Let* $(X, T)$ *be an infinite expansive system. Then we have* $\mathfrak{D}(X, T) \cap \langle T \rangle = \{\text{id}\}$.

## 7 Turing Machines as Dynamical Systems

Let us briefly look at a particular type of automorphisms in the subshift case. Specifically, we will take a look at automorphisms coming from Turing machines to better understand how the asymptotic range is related to the rate at which information is transmitted. These examples also serve to establish the existence of non-trivial (that is, of infinite order) range distorted automorphisms on a broad class of SFTs.

We assume some knowledge about the basics of complexity theory and Turing Machines (TM), for a complete reference see [13]. In the context of dynamical systems, there are several ways of representing a Turing Machine as a dynamical system. In this section we will use a model where the head moves presented in [11].

We will denote the set of states of the TM by $Q$, the alphabet by $A$ and $\delta\colon Q \times A \to Q \times A \times \{-1, 0, 1\}$ its transition function.

For $n \geq 0$ we define the subshift,

$$X_n = \{x \in (Q \cup A)^{\mathbb{Z}} : |\{i \in \mathbb{Z} : x_i \in Q\}| \leq n\}.$$

It is possible to show that $X_n$ is a sofic subshift. That is, can be obtained as a factor of a subshift of finite type. We do not need to be more precise in this article.

**Definition 12** We define a moving head Turing Machine (TMH) as $g \in \mathrm{End}(X_1)$, where the head (given by the coordinate $x_i \in Q$) points to the site in its right and $g$ executes the machine given by the transition function $\delta\colon Q \times A \to Q \times A \times \{-1, 0, 1\}$. It is easy to see that for all TMH $g$, $\mathrm{range}(g) = 2$. See next figure for an illustration of this definition.

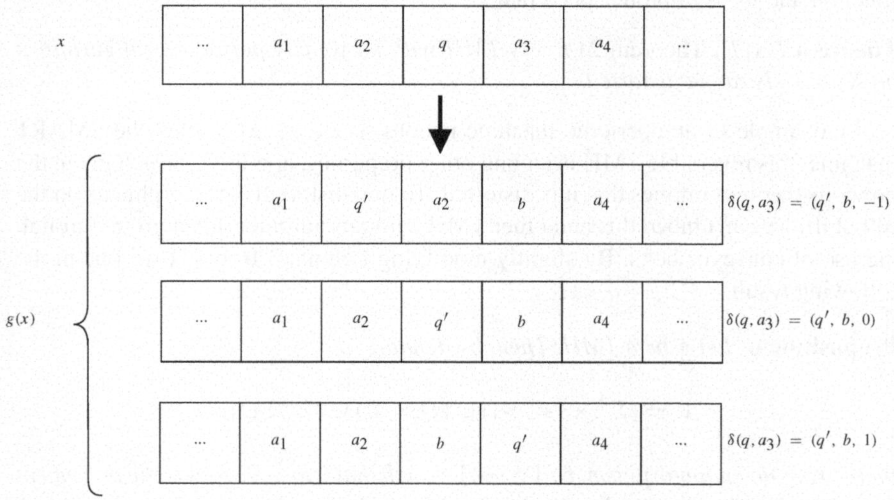


**Definition 13** The position function $\mathfrak{p}\colon X_1 \to \mathbb{Z} \cup \{\infty\}$ of a TMH $g$ is defined by $\mathfrak{p}(x) = n$ if $x_n \in Q$ and $\mathfrak{p}(x) = \infty$ on the other case.

The furtherest site the machine visits up to time $t$ by the machine on configuration $x \in X_1$ is:

$$s_t(x) := \max\{|\mathfrak{p}(g^s(x))| : 0 \leq s \leq t\}.$$

Then, we define the movement function of the machine at time $t$ as:

$$m(t) = \max_{x \in X_1} s_t(x).$$

It is clear that $\mathrm{range}(g^t) = m(t)$. To carefully examine the possible growth rates of $m(t)$, we introduce the following asymptotic growth notation.

**Definition 14** Let $\kappa, \eta \colon \mathbb{N} \to \mathbb{N} \setminus \{0\}$. We write $\kappa(n) = O(\eta(n))$ if there exists a constant $K$ such that $\kappa(n) \leq K\eta(n)$ for all sufficiently large $n$. Similarly, we write $\kappa(n) = \Omega(\eta(n))$ if $\eta(n) = O(\kappa(n))$. Furthermore, we write $\kappa(n) = \Theta(\eta(n))$ if $\kappa(n) = O(\eta(n))$ and $\kappa(n) = \Omega(\eta(n))$.

There exists a trichotomy with respect to the velocity that machines can have,

**Theorem 6** ([7], Theorem 1) *Let $g$ be a TMH with movement function $m$. Then, exactly one of the following holds:*

- *$m$ is bounded,*
- *$m(t) = \Omega(\log(t))$ and $m(t) = O(t/\log(t))$,*
- *$m(t) = \Theta(t)$.*

In addition, it is possible to establish a connection between the periodicity of the function and its asymptotic speed rate.

**Theorem 7** ([7], Theorem 2) *Every TMH with no weakly periodic configurations on $X_1 \setminus X_0$ is range distorted.*

An example of an aperiodic machine is constructed in [5]. Called the SMART machine, this reversible TMH is among other properties, aperiodic, which given the previous theorem implies that it is distorted. To find distorted automorphisms on the full-shift, we can embed this and other TMH's into its automorphism group through the use of conveyor belts. By slightly modifying Lemma 3 from [7] we obtain the following result:

**Proposition 6** *Let $g$ be a TMH. Then, by defining*

$$\Gamma = (\Sigma^2 \times \{<, >\}) \cup (Q \times \Sigma) \cup (\Sigma \times Q),$$

*there exists an endomorphism $f \colon \Gamma^{\mathbb{Z}} \to \Gamma^{\mathbb{Z}}$ such that if $m \colon \mathbb{N} \to \mathbb{N}$ is the movement function of $g$, then* $\mathrm{range}(f^t) \leq m(t)$ *for all $t \in \mathbb{N}$. Furthermore, $f$ is reversible if and only if $g$ is.*

Because the conveyor belt method allows us to see every reversible TMH within the automorphism group of a full-shift, we can conclude the following:

**Theorem 8** *Let $(X, \sigma)$ be a mixing SFT. Then, the set of reversible TMH is contained in* $\mathrm{Aut}(X, \sigma)$. *In particular, it contains an infinite order distorted automorphism.*

The proof of this fact follows directly from the previous proposition and the Kim-Roush Theorem for automorphisms groups of mixing shifts of finite type [10].

## 8 Non-shift Examples

Let us look at an example of the presented results for an expansive system that is not a subshift. Let $\mathbb{T}^n = \mathbb{R}^n/\mathbb{Z}^n$ be the n-dimensional torus. We endow this space with a metric induced by the 2-norm on $\mathbb{R}^n$:

$$\rho(x, y) = \inf_{k \in \mathbb{Z}^n} \|x - y - k\|_2.$$

To find expansive homeomorphisms on this space, we use the following result about automorphisms of the torus.

**Proposition 7** *Let $T_A$ be an automorphism of the n-torus, with $n \geq 2$ and A its corresponding matrix on $GL(n, \mathbb{Z})$ over $\mathbb{R}^2$. Then, the following statements are equivalent:*

1. *$T_A$ is expansive,*
2. *$A \in GL(n, \mathbb{Z})$ is expansive in $\mathbb{R}^n$,*
3. *A has no eigenvalue of modulus 1.*

A matrix $A \in \mathrm{GL}(n, \mathbb{Z})$ is said to be expansive if there exists a constant $c > 1$ such that $\|Ax\| \geq c\|x\|$ for all $x \in \mathbb{R}^n$.

We note that $n$ must be grater or equal than two, due to the fact that there are no expansive automorphisms on the 1-torus. The proof of this fact and of the proposition is outlined in [14]. By following its procedure, we can obtain the following lemma.

**Lemma 13** *Let $T_A$ be an expansive automorphism of the n-torus, where $n \geq 2$. Then, if we define $L'(A) = \max\{\|A\|, \|A^{-1}\|\}$, the expansive constant for the automorphism is given by*

$$c = \min\left\{\frac{1}{2L'(A)}, \frac{1}{4}\right\}.$$

***Proof*** Due to the previous Proposition, we know that if $T_A$ is expansive, $A$ is expansive. This in turn means that the set $\{\|A^m x\| : m \in \mathbb{Z}\}$ is unbounded.

Because $T_A$ is linear, we only have to prove the following: for $x \in \mathbb{T}^n$ such that $x \neq 0$, then there exists $m \in \mathbb{Z}$ such that $\rho(T_A^m x, 0) > c$. We do this in two cases. If $\|x\| > c$, it is evident that:

$$\rho(T_A^0 x, 0) = \|x\|_2 > c.$$

If $\|x\| \leq c$, due to the aforementioned set being unbounded we can define:

$$k = \inf\{|m| : \|A^m x\| > c,\ m \in \mathbb{Z}\}.$$

Let us suppose without loss of generality that $\|A^k x\| > c$. Then we have that

$$c < \|A^k x\| \leq \|A\| \|A^{k-1} x\| \leq L'(A) c \leq \frac{1}{2},$$

which means that $A^k x \in (-1, 1)^n$. Finally, $\rho(T_A^k x, 0) = \|A^k x\| > c$. □

Let us see how everything works by taking the matrix

$$A = \begin{bmatrix} 1 & 1 \\ 2 & 1 \end{bmatrix}.$$

For simplicity's sake we will use the same notation for $A$ and $T_A$.

Its eigenvalues are $\lambda_1 = 1 + \sqrt{2}$ and $\lambda_2 = 1 - \sqrt{2}$, which due to Proposition 7 means that it defines an expansive homeomorphism. It is possible to see that its expansive constant is in fact $c = \frac{1}{4}$.

Furthermore, by examining at the matrices that commute with $A$, we find that

$$\text{Aut}(\mathbb{T}^2, A) = \left\{ \begin{bmatrix} a & b \\ 2b & a \end{bmatrix} : a^2 \neq 2b^2, \ a, b \in \mathbb{Z} \right\}.$$

It is possible to observe that a matrix such as

$$M = \begin{bmatrix} 0 & 1 \\ 2 & 0 \end{bmatrix} \in \text{Aut}(\mathbb{T}^2, A)$$

satisfies range$(M) = 1$, due to the fact that for $x, y \in \mathbb{T}^2$

$$\rho(Ax, Ay) \leq \frac{1}{4} \implies \rho(Mx, My) \leq \frac{1}{4}.$$

Finally, we notice that the eigenvalues of matrices in $\text{Aut}(\mathbb{T}^2, A)$ are given by $\lambda_1 = a + \sqrt{2}b$ and $\lambda_2 = a - \sqrt{2}b$. By Theorem 4, we have that

$$AD(\mathbb{T}^2, A) = \{I, -I\},$$

where $I$ is the identity matrix.

## 9 Conclusion

Throughout this work we have seen and developed the connections between distortion and non-expansivity. First, by generalizing the concept of range to general expansive systems, we have seen the aforementioned connection is not exclusive to subshifts. This ultimately led to the fact that expansive automorphisms cannot be distorted. What the obtained results suggest, is that a non-expansive direction is one in which the rate at which information propagates with sub-linear speed. Also, this generalization allows us to use concepts and tools from the study of automorphisms on symbolic systems to general expansive ones.

Nevertheless, the greatest question concerning distortion on automorphism groups remains open: is every range distorted automorphism group distorted? Even though there seems to be a direct path for solving this question, constructing a Turing machine-like automorphism that is range distorted but not group distorted, it is not clear how the construction of this automorphism can be achieved. It is possible that the study of the group generated by symmetrically distorted automorphisms, $\langle \mathrm{d}(X) \rangle$, can shed some light on this mystery. It can also be possible to answer the question by studying the group distorted Turing machines on the group of reversible Turing machines presented by Barbieri, Kari and Salo in [1]. There also remains to see if it is possible to have an automorphism with a unique non-expansive direction of irrational slope over a domain which is transitive or minimal. A possible approach consists on codifying a subshift suspension in a way that the non-expansive directions of the suspension are preserved.

## References


1. Barbieri S, Kari J, Salo V (2016) The group of reversible Turing machines. In: International workshop on cellular automata and discrete complex systems. Springer, Berlin, pp 49–62
2. Boyle M, Lind D (1997) Expansive subdynamics. Trans Amer Math Soc 349(1):55–102
3. Boyle M, Lind D, Rudolph D (1988) The automorphism group of a shift of finite type. Trans Amer Math Soc 306(1):71–114
4. Calegari D, Freedman MH (2006) Distortion in transformation groups. Geom Topol 10(1):267–293
5. Julien C, Nicolas O, Rodrigo T-A (2017) A small minimal aperiodic reversible Turing machine. J Comput Syst Sci 84:288–301
6. Cyr V, Franks J, Kra B (2019) The spacetime of a shift endomorphism. Trans Amer Math Soc 371(1):461–488
7. Guillon P, Salo V (2017) Distortion in one-head machines and cellular automata. In: International workshop on cellular automata and discrete complex systems. Springer, Berlin, pp 120–138
8. Hedlund GA (1969) Endomorphisms and automorphisms of the shift dynamical system. Theory Comput Syst 3(4):320–375
9. Hochman M (2011) Non-expansive directions for $\mathbb{Z}^2$ actions. Ergodic Theory Dyn Syst 31(1):91–112
10. Kim KH, Roush FW et al (1990) On the automorphism groups of subshifts. Pure Math Appl 1(4):203–230
11. Kurka P (1997) On topological dynamics of Turing machines. Theor Comput Sci 174(1–2):203–216
12. Schmieding S (2019) Automorphisms of the shift: Lyapunov exponents, entropy, and the dimension representation. Ergodic Theory Dyn Syst, pp 1–19
13. Sipser M (1996) Introduction to the theory of computation. ACM Sigact News 27(1):27–29
14. Walters P (2000) An introduction to ergodic theory, vol 79. Springer Science & Business Media, Berlin

# Periods in the Q2R, X2R and Kawasaki-Q2R Cellular Automata

**Lidia Stocker and Hans J. Herrmann**

**Abstract** We study global and local periods of the cellular automaton Q2R, the equivalent model on a triangular grid (X2R) and a Kawasaki-Q2R update. The first two show similar results in both the global and local periods. We find a critical energy $E_{cp} = -0.8700 \pm 0.0006$ for Q2R and $E_{cp} = -0.88408 \pm 0.0004$ for X2R. In the Kawasaki-Q2R automaton dynamics finite global periods are present exclusively at very low energies and no critical energy is found. However, in all three cellular automata there is an evident formation of clusters of finite cycles. Ergodicity is violated.

## 1 Introduction

Cellular automata are defined by a set of deterministic rules where the dynamics of a configuration depends on the neighbourhood of each cell [1, 8]. A famous example is Q2R, which has been studied under different points of view. Vichniac first proposed this model and suggested its potential for simulating magnetism [7]. He also raised the issue about its ergodicity. Later, Pomeau argued how Q2R could be seen as a simulation of the Ising model [5]. Herrmann et al. investigated the ergodicity, concluding that at low energies the system is locked into dynamical clusters with finite periodicity [2]. Also more recently, new results about Q2R have been obtained: Urbina and Rica considered a master equation for reversible and conservative discrete systems [6]. Montava-Medel et al. presented a phase space classification of their topological space [4].

Here, we aim to improve and further investigate the results of Herrmann et al. [2], studying three cellular automata. We observe that only limited regions of the phase space of Q2R are visited. We compare the results of Q2R to the same cellular

---

L. Stocker
Institute for Theoretical Physics, ETH Zurich, 8093 Zurich, Switzerland
e-mail: stockerl@ethz.ch

H. J. Herrmann (✉)
PMMH, ESPCI, 75005 Paris, France
e-mail: hans.herrmann@espci.fr

 
A. Adamatzky (ed.), *Automata and Complexity*, Emergence, Complexity and Computation 42, https://doi.org/10.1007/978-3-030-92551-2_4

automaton on a triangular, instead of a square, lattice. We name it, inspired by the classification discussed in [7], the X2R automaton. We also consider a Kawasaki-Q2R dynamics by imposing additionally conservation of the magnetisation. The three cellular automata studied are characterised by discrete space, time and spin variables $\pm 1$. They are completely deterministic and reversible. Their energy is conserved in time and the change of a spin is determined by the configuration of its nearest neighbours, conferring a local character to the dynamics.

The structure of our contribution is the following: we first discuss the methods used in Sect. 2, the theoretical framework and the implementations. Next, in Sects. 3–5, we treat the three cellular automata separately identifying their differences and interpreting the results. Finally, in Sect. 6, we summarize the latter points by giving a broader overview.

## 2 Methods

In this section, we introduce the methods and notation used, as well as the main theoretical concepts. Our work focuses on binary 2D systems with $L \times L$ sites and periodic boundary conditions. Each site $i$ is singly occupied by a spin pointing up or down: $s_i = \pm 1$. In Q2R and Kawasaki-Q2R we work on a *square* lattice, thus each site has four nearest neighbours, notated by $\langle\rangle$. The X2R automaton is defined on a *triangular* lattice leading to six nearest neighbours per site. We define the normalised Ising energy of these systems as

$$E = -\frac{2}{kL^2} \sum_{\langle i,j \rangle} s_i s_j \tag{1}$$

where $k$ is the number of nearest neighbours in the grid. With this normalisation a comparison between square and triangular grid is achievable.

In Q2R and X2R a spin is *flipped* when the Ising energy is conserved. This occurs if the number of its nearest neighbours with spin pointing up is equal the number of nearest neighbours with spin down (two or three respectively). In Kawasaki-Q2R, a neighbouring pair $\langle i, i' \rangle$ *exchanges* its spin $s_i \leftrightarrow s_{i'}$ if they point in opposite directions and, again, energy is conserved. In other words, they exchange their spin if the number of spin-up nearest neighbour of the lattice site $i$, excluding $i'$, is equal the number of spin-up nearest neighbour of $i'$ excluding $i$. Conventionally, during time evolution, flips or exchanges of single spins should not depend on the dynamics of other sites during the same time step. For this reason, we performed a partition of the grid into different sublattices. During a time step each site is updated once. The choice of partitions is based on the condition that no nearest neighbour of a site should belong to the same sublattice of the site itself. This is because the nearest neighbour spins determine if a flip, respectively an exchange occurs. The partition of the grid is different for each of the cellular automata studied and will be discussed in the next sections.

According to the definition of the Ising energy, Eq. (1), we have $E = -1$ if $s_i = -1\ \forall i$ on both the square and triangular grid. This convention differs from the one used in [2] by a factor two. $E$ is easily computable, constant under the dynamics of our cellular automata and is thus chosen as the reference parameter to characterise a configuration. To get a configuration with a given energy, we first set all $s_i = -1$ and then randomly flip single sites until we reach the desired value. Our main interest is the dynamics of these configurations, in particular, their global and local periodicity as discussed in [2]. We first study the *global* periods treating systems of different size $32 \leq L \leq 6400$. A configuration has global period $t_g$ if it comes back to its initial state after $t_g$ time steps. As the cellular automata are completely deterministic and reversible, this occurs after maximal $t_{max} = 2^{L \times L}$ steps. Because $t_{max}$ increases exponentially with $L$, the computational time required to determine the global period of a lattice can be very large, even for system with relatively small size. We are interested in short global periods and therefore define any $t > 1000$ as infinity. We determine the fraction $p(E)$ of configurations at a fixed energy $E$ having a global period $t_g > t_{g,max} = 1000$. We also calculate the average period among the configurations with $t_g \leq 1000$ for the energies with $p(E) < 0.9$. The data with higher fractions are not considered, because of the too-small number of statistical samples.

After having studied the global properties, we examine the period of each single lattice site (local periods) in systems with $L = 1280$. In [2], it was was shown that Q2R is not ergodic: clusters of periodic cycles tend to appear in a "sea" of static sites. Two sites belong to the same clusters if they are nearest neighbour, their spins change in time and the local period of one is a multiple of the other. To calculate the local periods we evolve the system during $2 \cdot t_{max}$ time steps storing the configurations each time. The time evolution of each single lattice site is extracted from these data. We are then able to determine its period $t$ by starting with $t = 1$. If this guess results to be wrong, we increase $t$ by one and check if the periodicity of the site is two. In the opposite case, $t$ is again increased until having found the correct value or reached $t_{max}$. We observe the fraction of sites with a "finite" period for different $t_{max} = 32, 80, 160, 320$ and calculate the average local period. For the calculation of the average local periods, sites with $t > t_{max}$ are defined to have local period $t = 0$.

Finally, we analyse the cluster size distribution of local periods of different configurations. This was done with an implementation of the Hoshen-Kopelman algorithm, which determines clusters and their size [3]. We also study the cluster period distribution. The period of a cluster is defined as the maximal local period among the sites belonging to it. As for each cellular automaton different neighbours are involved, the cluster identification method must be adapted respectively. The value of the local period of two neighbouring sites is a multiple of both periods. Consequently, the algorithm is the following: we first go through the configuration and label the sites having their local period $t > 1$. We then join neighbouring labelled sites with the Hoshen-Kopelman algorithm. During this process, we also store the maximal local period within each cluster.

All global averages are calculated over $n = 1000$ different configurations. Local averages are performed over $n = 100$ configurations. It is not possible to determine reliable error bars for the mean periods (local and global) because of the large fluctuations. The data do not seem to be self-averaging.

## 3 Q2R: The Square Lattice Cellular Automaton

Q2R is defined on the square lattice, and a single site flips if two of its nearest neighbour point up and the other two point down. We split the system into two sublattices as defined by a checker-board. We first update the "white" sites and then the "black" ones. It is easy to prove that no spin does belong to the same sublattice of any of its nearest neighbours, fulfilling the requirement just explained.

To determine the global period, we implement the following algorithm: we first store the initial configuration and magnetisation $M_0$ of the configuration. At any time step $i$ with $M_i = M_0$, the initial configuration is compared to the one at time $i$. If the two are equal, the global period $t = i$ is found. Otherwise, the time evolution continues, stopping at $t_{max} = 1000$ and labelling configurations as infinite for which no smaller period is found.

Our outcome for local and global periods coincides with the results of [2], see Fig. 1. With higher energies, the probability that a configuration has a finite period decreases. Correspondingly, the average period of the finite configurations steadily increases.

The local results also reproduce the previous work, as shown in Fig. 2. A central point is the presence of a critical energy $E_{cp}$ for the local average periods. With larger $t_{max}$, we get a higher mean local period at the critical point, which diverges for larger $t_{max}$. Thus, we perform a finite size analysis. $E_{cp}$ is then determined as follows:

$$E_{cp}(t_{max}) = E_{cp}(1 - bt_{max}^{x}) \tag{2}$$

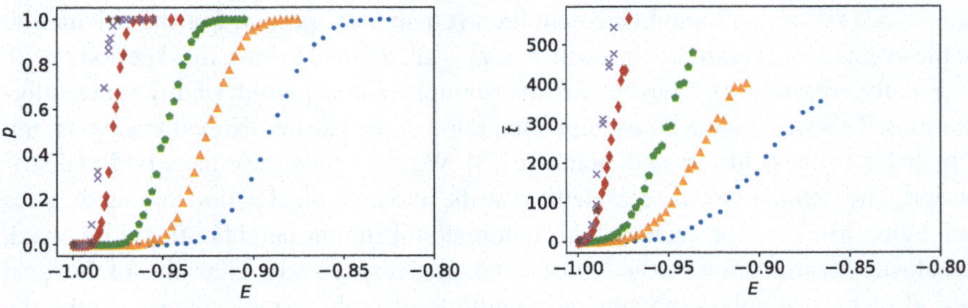


**Fig. 1** Fraction of configurations (left) $p(E)$ with global period $t > 1000$ and average period (right) as a function of their energy $E$ for Q2R. L = •32, ▲64, •128, ♦640, ×1280, ★6400

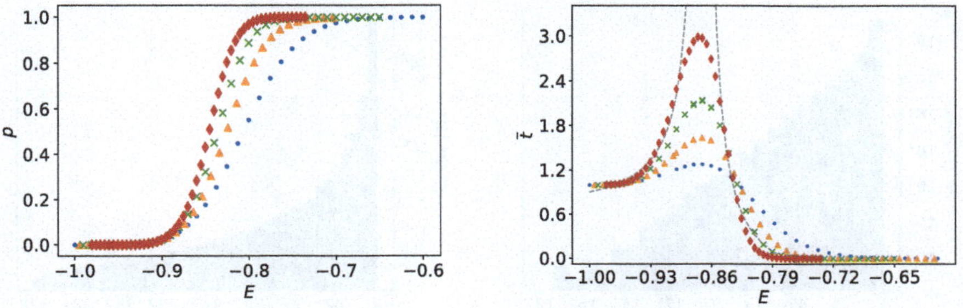


**Fig. 2** Fraction of sites with local finite period (left) and average local period (right) in systems of $L = 1280$ for Q2R. $t_{max}$ = •32, ▲80, ×160, ♦320. There is a critical energy at $E_{cp} = -0.8700 \pm 0.0006$

where $E_{cp}(t_{max})$ is the energy with maximal local period for $t_{max} = 32, 80, 160, 320$ and $b$ a constant. A numerical fit using Eq. (2) leads to $E_{cp} = -0.8700 \pm 0.0006$ and $x = 0.34 \pm 0.09$. Next, we calculate the $z$ exponent in the relation $t_{max}(L) = L^z$. Defining $z = -\gamma/x$ it is possible to determine the critical exponent $\gamma$. This exponent characterizes the mean local period $\bar{t} \propto |E - E_{cp}|^{-\gamma}$. According to our definition we have $\bar{t} = 1$ for $E \ll E_{cp}$ and $\bar{t} = 0$ for $E \gg E_{cp}$. In order to adjust to these two limiting cases, we add one constant for $E < E_{cp}$ and another for $E > E_{cp}$. In fact, we make a fit

$$\bar{t} = c|E - E_{cp}|^{-\gamma} + a_{\pm} \tag{3}$$

where $a_{\pm}$ denotes a constant which is different below and above the critical energy. We obtain $\gamma = 1.2 \pm 0.3$, which agrees well with the fit to the $t_{max} = 320$ curve. The fit is represented in the plot by the dashed grey line.

At the critical energy $E_{cp}$, we determine the cluster size distribution, shown in Fig. 3. Among 1000 simulations we find a maximal cluster size $S_{max} = 179$. We observe that, even at the critical energy, the clusters are rather small (in our grid there are $1280^2$ sites!). Nevertheless, at the critical energy the cluster size distribution seems to follow a power-law $n_S \propto S^{-a}$, where $n_S$ is the number of clusters with size $S$. With a numerical fit we calculate $a = 1.112 \pm 0.004$. The maximal local period observed is $t_{max} = 320$ (the maximal possible value according to our choice). In this case, the number $n_t$ of clusters found with period $t$ behaves like $n_t = t^{-b}$ with $b = 1.35 \pm 0.03$. In Fig. 4. we show some local period configurations, observing that clusters have a typical rectangular shape. This is not surprising, as the flip function stabilizes rectangular spin up clusters in a spin down environment (or vice-versa).

## 4 X2R: Q2R on Triangular Lattice

X2R is like Q2R because except that it is defined on the triangular- instead of a square lattice. In this case, the spin of a single site is, again, flipped if energy is conserved.

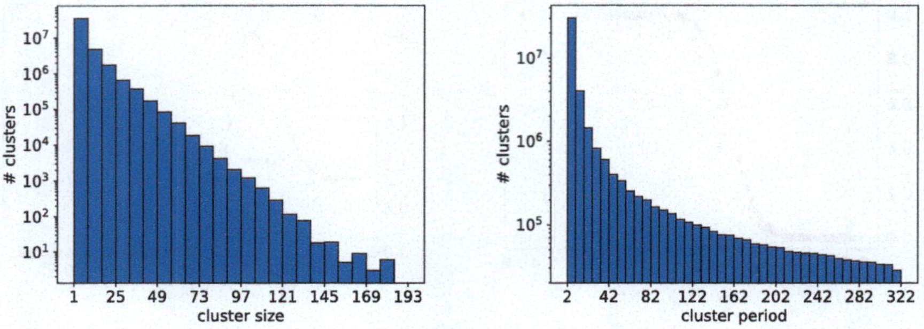


**Fig. 3** Log plot of the cluster size distribution (left) and log-log plot of the cluster period distribution (right) of Q2R at the critical energy $E_{cp} \simeq 0.874$ for 100 configurations of size $L = 1280$

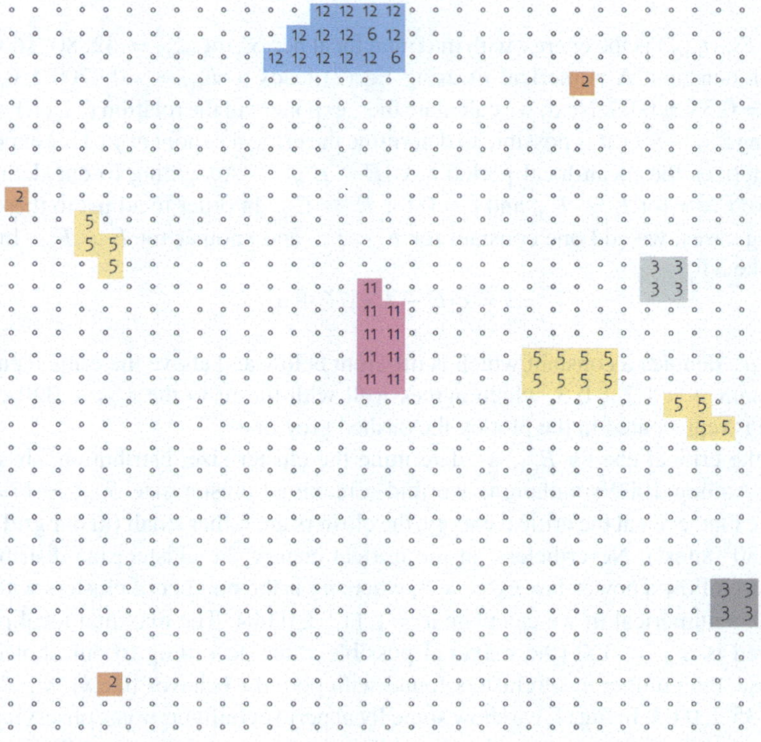

**Fig. 4** Local periods in a configuration with $L = 32$ at the critical energy $E_{cp} \simeq -0.87$ for Q2R. o denotes period 1, colours distinguish local periods. We observe clusters of rectangular shape

This occurs if the spin of exactly three of its six nearest neighbours point up. The system is divided into three sublattices for the same reasons argued before and as sketched in Fig. 5. Possible global and local periods are established and calculated in the same way as for the square lattice. We modify the Hoshen-Kopelman algorithm, accounting as nearest neighbour each of the six adjacent sites.

For the global periods, we obtain a very similar behaviour as for Q2R (Fig. 6). We observe that the curves representing the fraction of configurations with an infinite period are shifted to the right. This can be justified, as in this case a flip occurs if exactly three, not two, nearest neighbours have $s = 1$. This arrangement occurs with a lower probability than in the case of four nearest neighbours. Also, in this case, the mean global period increases monotonously.

In the analysis of the local periods, we observe a similarity with Q2R, see Fig. 7. We perform, as described for the Q2R automaton case, a finite size analysis. In the first step, for the relation of Eq. 2 we find $E_{cp} = -0.8408 \pm 0.0004$ and $x = 0.92 \pm 0.04$. The critical point is slightly higher than for Q2R. This is again probably due to large number of nearest neighbours. Also in this case, we have a phase transition charac-

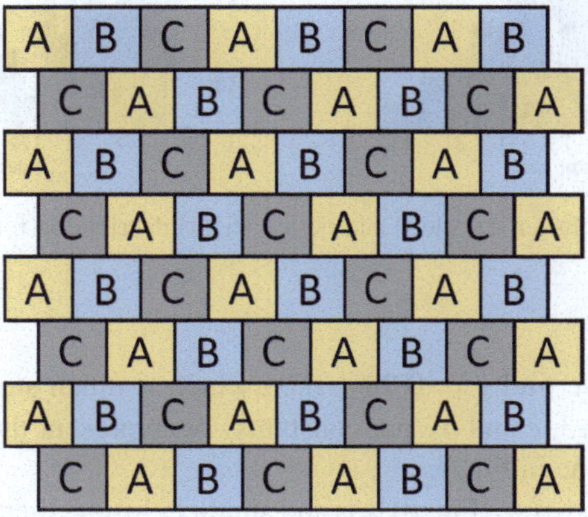


**Fig. 5** Division in sublattices $A$, $B$, $C$ required for a consistent update of X2R on the triangular grid

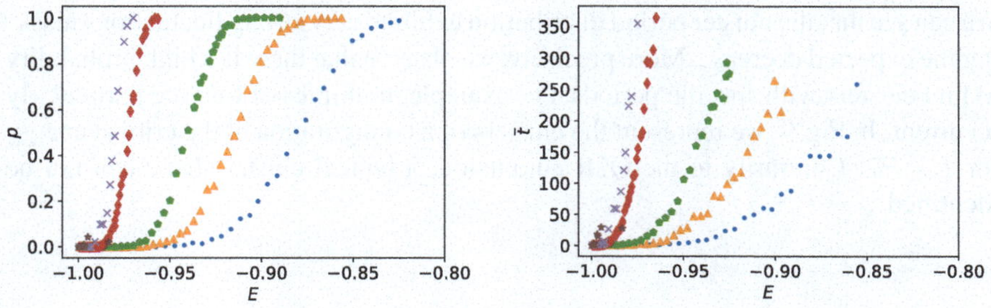


**Fig. 6** Fraction of configurations (left) $p(E)$ with global period $t > 1000$ and average period (right) as a function of their energy $E$ for X2R. L = •32, ▲64, ⬢128, ♦640, ×1280, ⋆6400

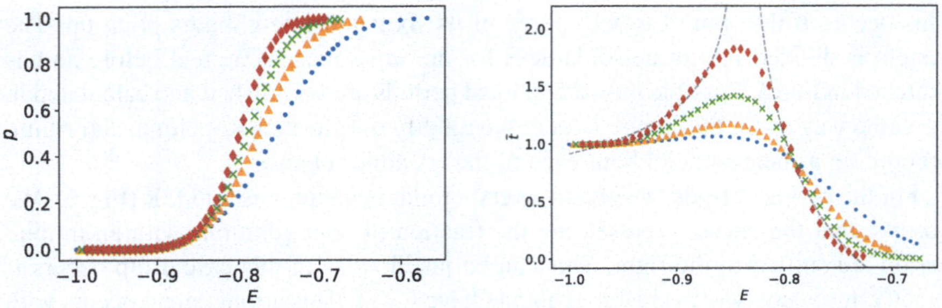


**Fig. 7** Fraction of sites with local finite period (left), and average local period (right) in systems of $L = 1280$ for X2R. $t_{max}$ = •32, ▲80, ×160, ♦320. A critical energy at $E_{cp} = -0.8408 \pm 0.0004$ is observed

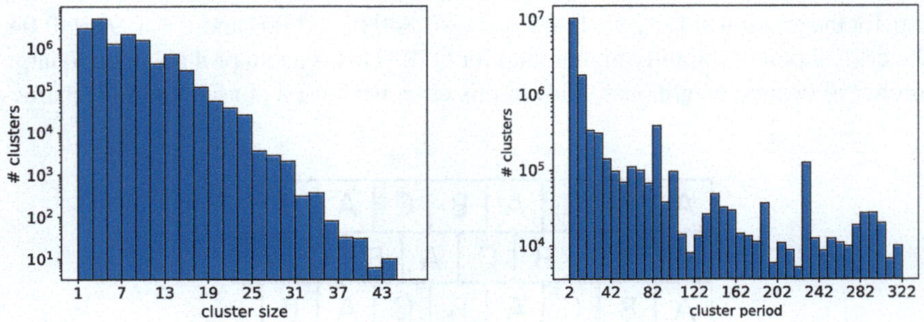


**Fig. 8** Log plot of the cluster size (left) and period (right) distribution for 100 configurations at $E_{cp} \simeq -0.848$ for X2R

terised by Eq. (3). We find $\gamma = 0.28 \pm 0.05$ which is much smaller than for Q2R. As described in Sect. 3, $c$ and $a_{\pm}$ are constants. We represent the scaling behaviour by the dashed grey line in Fig. 7.

For X2R, we observe (see Fig. 8) less and smaller clusters than for Q2R. The maximal cluster size, again among 1000 simulations, is only $S_{max} = 44$. Their number decays as $n_S \propto S^{-a}$, although with more fluctuations as compared to Q2R. We find $a = 1.35 \pm 0.05$. The maximal local period is also chosen in this case $t_{max} = 320$. We can see that the cluster period distribution exhibits very strong fluctuations masking the expected decrease. More precisely, we observe that there is a high probability to find clusters with specific periods. For example, multiples of four are particularly recurrent. In Fig. 9, we represent the clusters of a configuration at the critical energy for $L = 32$. Contrarily to the Q2R automaton, a typical cluster shape can not be identified.

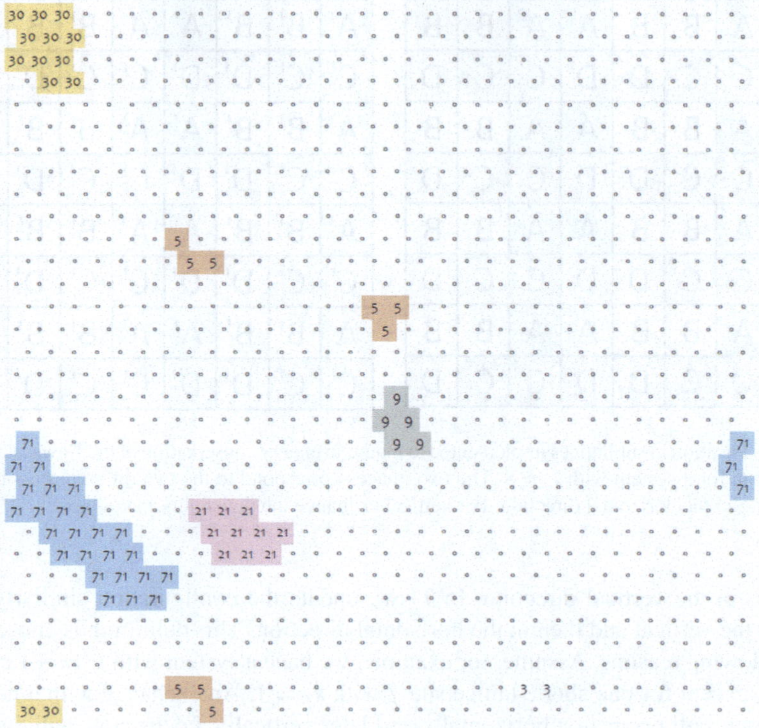

**Fig. 9** Example of local periods for X2R. The system is of size $L = 32$ at the critical energy $E_{cp} \simeq -0.848$. No apparent typical cluster shape can be identified

## 5 Kawasaki-Q2R with Energy Conservation

The dynamics of this cellular automaton consists of an exchange between neighbouring sites. If the two spins $s_i$, $s_i'$ of a neighbouring pair are anti-parallel, the spin of both of them can flip. This occurs if the two involved sites have an equal number of "spin up" nearest neighbour, excluding the other site of the pair. As we deal with pairs, we have to distinguish between horizontal and vertical direction. Additionally to the energy, we have another conserved quantity: the magnetisation $M$.

Any pair of spins has six nearest neighbours, and for this reason, the partition into sublattices is more challenging. We treat, for both vertical and horizontal direction, the two possible combinations of pairing for each site (right and left nearest neighbour, upper and lower nearest neighbour respectively). Each of the two different combinations has to be split into four sublattices, as shown in Fig. 10 for the horizontal exchanges. The vertical case is equivalent. We have two time steps $T$ and $T'$. In $T$ we treat each couple of neighbours of the first sublattice in the horizontal direction, performing exchanges in the sequence A, B, C, D. The same procedure is repeated for the second pairing combination (A', B', C', D'). Then, we anagously implement

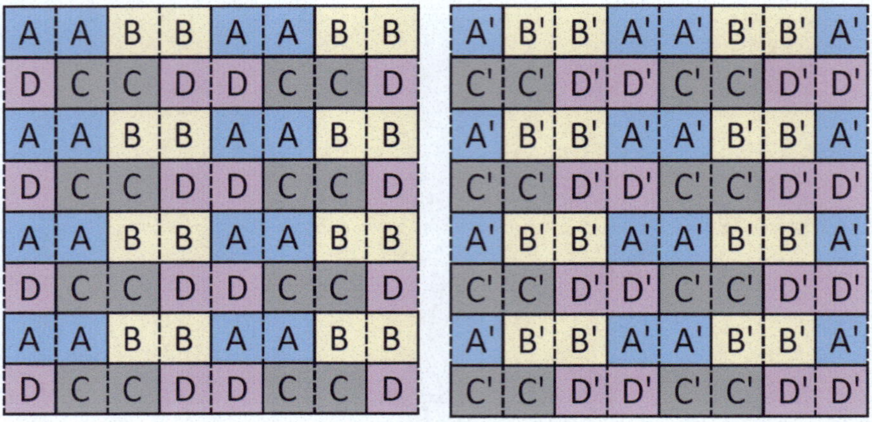


**Fig. 10** Partition in sublattices implemented to maintain energy conservation of the Kawasaki-Q2R automaton for a system with $L = 8$. The two images correspond to the two different possibilities of pairing in the horizontal direction. For vertical exchange the division is analogous

updates in the vertical direction. In $T'$ we update the configuration similarly, but first in the vertical and then in the horizontal direction. This definition is chosen for the following reasons. Assume, for example, we have a system with $s_i = -1$ everywhere, except for one single lattice site $j$ with $s_j = 1$. An exchange with a nearest neighbour will occur first horizontally and later vertically. Focus now, without loss of generality, on the case (this depends on the initial $j$, see sublattice partition) when $s_j = 1$ first exchanges with its right neighbour, and then with its vertical neighbour. The dynamics repeats the next steps continuously. Thus due to periodic boundary conditions, the system would come back to its initial configuration after $t = 4L$ time steps. We want to establish global periods for $t \leq 1000$ and compare the average to the value for different sizes (up to 6400). For these reasons, we introduce the previously defined even and odd time steps and implement the sequence $TT'TT'$. In fact, with this definition, most of the sites will have $t = 2$. This behaviour can be seen as some oscillating blinkers. It is also important to note that, again in the case of a singular spin-up site $s_j$, after a single time step, the cell with $s'_j = 1$ will be its fifth nearest neighbour. This is a consequence of the definition of our time step. We have to modify the Hoshen-Kopelman algorithm previously used by taking in account the fifth nearest neighbour belonging to the same cluster for consistency. This effect can be observed in Fig. 11, where we show the local periods for a random configuration with $L = 32$ and $E \simeq -0.97$.

For the calculation of the global periods, we can not rely on the algorithm presented for $Q2R$, as the latter is based on a comparison with the magnetisation $M_i$ at each time step. In the Kawasaki-Q2R automaton $M$ is constant. We therefore implemented a straightforward algorithm comparing the configuration after each time step with the initial one. This process requires more computational time.

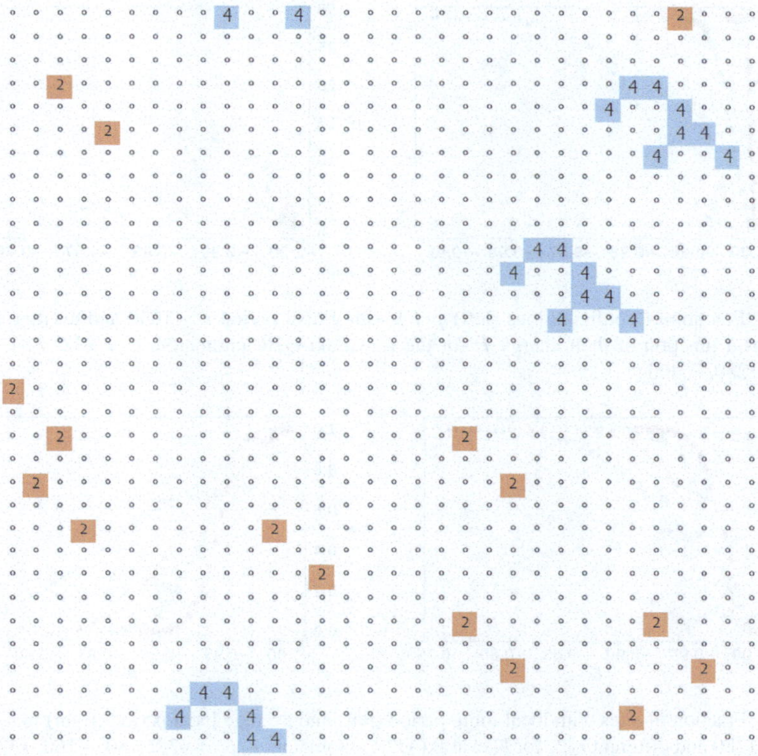

**Fig. 11** Local periods for a configuration at energy $E \simeq -0.97$ and size $L = 32$. ∘ indicates a site of period 1, orange clusters are of period 2 and blue of period 4

The behaviour for the Kawasaki automaton dynamics is quite different from Q2R and X2R. Configurations with finite periods are only present at low energies (see Fig. 12). Additionally, finite configurations have a lower average global period and this value does not systematically increase with energy but exhibits strong fluctuations. We can observe and conclude that either a configuration has a short period, or we will observe an important motion, especially for large systems.

Analysing the local periods, we observe additional differences. Also in this case, for all $t_{max}$ between 32 and 320, we have an increase in the fraction of lattice sites, but with an infinite local period in a smaller energy range compared to the previously studied cellular automata (see Fig. 13). For $t_{max} \leq 160$ the mean local period steadily decreases: No critical point can be detected. As shown in Fig. 13, however, for $t_{max} = 320$ we observe, after a short decrease, an unexpected jump at $E = 0.9669 \pm 0.0003$. With a simulation for $t_{max} = 640$, we obtain essentially the same values as for $t_{max} = 320$. Also a different number generator leads to the same results. Observing the cluster period distribution for the energy at the jump (see Fig. 14), we can extract a possible explanation. In fact, most of the clusters have either low $2 \leq t < 8$ or

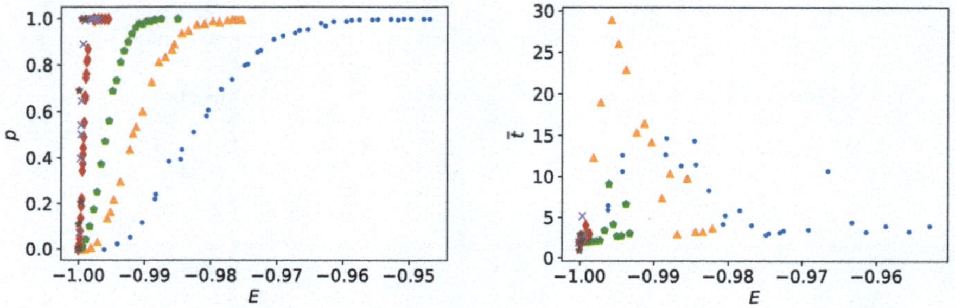


**Fig. 12** Fraction of configurations (left) $p(E)$ with global period $t > 1000$ and average period (right) as a function of their energy $E$ for the Kawasaki-Q2R automaton. L = •32, ▲64, ⬟128, ♦640, ×1280, ⋆6400

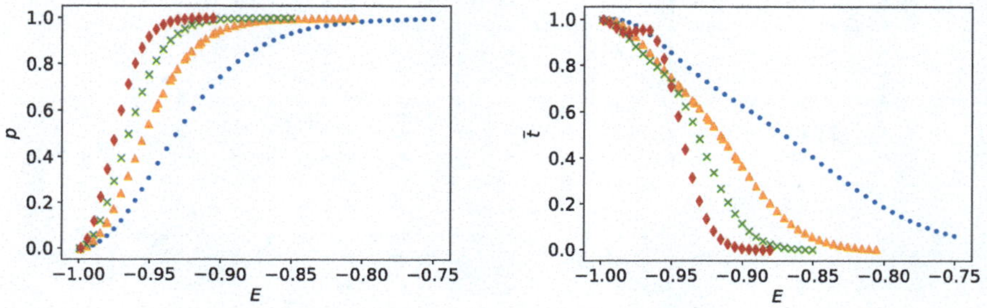


**Fig. 13** Fraction of sites with local finite period (left) and average local period (right) in systems of $L = 1280$ and different $t_{max}$ for Kawasaki-Q2R automaton. $t_{max}$ = •32, ▲80, ×160, ♦320. We do not observe any critical point opposed to the previous cellular automata

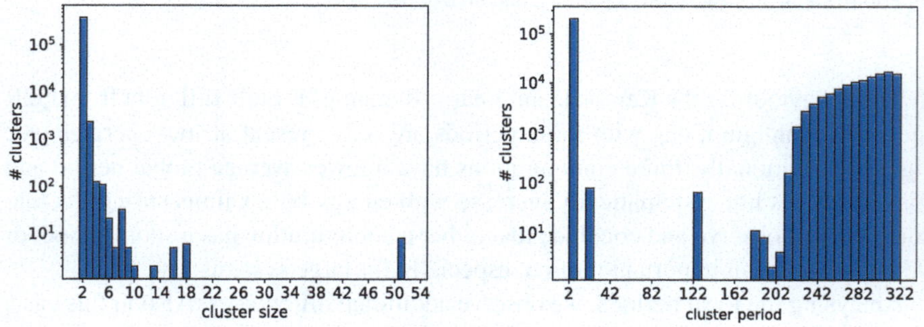


**Fig. 14** Cluster size (left) and cluster period (right) distribution of Kawasaki-Q2R at the jump energy $E \simeq -0.97$ on 100 systems of size $L = 1280$

large $t > 128$ period. We have a possible interpretation of this substantial difference encountered between $t_{max} \leq 160$ and $t_{max} = 320$. Contrarily to Q2R and X2R, in the Kawasaki-Q2R automaton, we have three different possibilities for the number of spin-up nearest neighbour to flip a pair of spins. An increase in energy could then lead to configurations in which less pairs of spins undergo an exchange: Less sites will have an infinite period, resulting in an average value closer to one.

## 6 Discussion and Conclusion

The similarities in the definition of Q2R and X2R lead to analogous behaviour. In both cellular automata, we observe a comparable evolution of the finite global period as a function of the energy. Additionally, we have a critical energy characterized by a divergence of mean local periods. Although the probability of finding a configuration with global period $t \leq 1000$ steadily decreases for higher energies, we verify (as observed by Herrmann et al. [2]) that dynamical sites arrange in clusters and have a limited periodicity. These clusters lie in a "sea" of sites with fixed spin. Configurations in both Q2R and X2R at low energy are thus clearly not ergodic. The Kawasaki-Q2R automaton exhibits very different results. In this case, the probability of finding an infinite global configuration increases more rapidly and reaches its maximum for very low energies. Also in this case, we can observe the formation of clusters with finite periods, but for smaller energies compared to Q2R and X2R. No critical point is observed and the mean local period steadily decreases with the energy. We can not provide an explication for the disappearance of the maximum in the mean local period, although we suspect that the larger number of possibilities under which an exchange takes place could play an important role. The unexpected jump at $t_{max} = 320$ can be partially justified by the cluster period distribution: this value results to be either less than 10, or larger than 200 at the jump energy $E_j \simeq -0.97$. The configurations are not ergodic, instead there are dynamical sites arranging in clusters with finite periodicity in all Q2R, X2R and Kawasaki-Q2R.

## References

1. Herrmann HJ (1989) Cellular Automata. In: Proto AN (ed), Nonlinear phenomena in complex systems. North-Holland Delta Series, pp 151–199
2. Herrmann HJ, Carmesin H-O, Stauffer D (1987) J Phys A 20:4939
3. Hoshen J, Kopelman R (1976) Phys Rev B 14:3438
4. Montalva-Medel M, Rica S, Urbina F (2020) Chaos Solitons Fractals 133:109618
5. Pomeau Y (1984) J Phys A 17:L415
6. Urbina F, Rica S (2016) Phys Rev E 94:6
7. Vichniac GY (1984) Physica D 10:12
8. Wolfram S (1983) Rev Mod Phys 55:601

# The Mirage of Universality in Cellular Automata

**Guillaume Theyssier**

**Abstract** This note is a survey of examples and results about cellular automata with the purpose of recalling that there is no 'universal' way of being computationally universal. In particular, we show how some cellular automata can embed efficient but bounded computation, while others can embed unbounded computations but not efficiently. We also study two variants of Boolean circuit embedding, transient versus repeatable simulations, and underline their differences. Finally we show how strong forms of universality can be hidden inside some seemingly simple cellular automata according to some classical dynamical parameters.

## 1 Eric, the Collector

The present note responds to an invitation to contribute to a book at the occasion of Eric Goles 70th birthday. Before diving into the scientific content, I should say a few words about Eric and the motivation behind this note.

Anyone knowing Eric certainly had the pleasure to listen to some of his colorful anecdotes (I certainly did). He owns a large collection, large enough to adapt to a wide variety of listeners and circumstances. The collection is in fact twice as large, because each anecdote, usually told to an international audience, is doubled with a more confidential Chilean version full of slang words. Eric's pleasure of telling stories is obvious, he has generously shared his collection, but nobody has listened to the same sequence of anecdotes and we all end up with a different global picture, much like the adventurous readers of the antinovel of Cortázar.

The collection of models and systems studied by Eric in its numerous scientific publications is equally striking. It abounds in small examples that are carefully analyzed and shown to capture important phenomena. It connects different points of view and different communities of researchers. It seems to never end up in the exact same theoretical framework and invites us to think about details that make a difference. In short, there is an anti-Bourbakist quality to it.

G. Theyssier (✉)
CNRS, Université Aix-Marseille, Marseille, France
e-mail: guillaume.theyssier@cnrs.fr

 
A. Adamatzky (ed.), *Automata and Complexity*, Emergence, Complexity and Computation 42, https://doi.org/10.1007/978-3-030-92551-2_5

At the heart of this scientific collection (at least from what I can tell from my collaboration with Eric), there is the question of the computational universality of small dynamical systems, and how it manifests itself in the complexity of various associated decision problems. Computational universality of dynamical systems is a topic that might seem boring to the classical computer scientist (after all Turing showed the existence of a universal machine in the 1930s) and not serious for the dynamical systems community (this is not real maths[1]). Part of the problem is that this kind of research is endangered by what I would call the *mirage of universality*: the illusion that there must be universal consequences to the fact of being "computationally universal" independently of the precise definition used, and that such a statement, even given without technical details, gives information by itself. Pursuing this mirage, one is tempted to put forward vague theorem statements and hide the concrete mathematical result in the proofs (or sketch of). To make an analogy, no paper in computational complexity would use theorem statements like "Problem X is hard" and then, hidden in the proof details, unveil the definition of "hard". On the contrary, computational complexity theory has been extremely fruitful by putting forward a vast "zoo" of precisely defined complexity classes, often with a corresponding notion of reduction.

Of course, there is a lot to say and a lot as already been said about the mathematical formalization of computational universality in dynamical systems, but my intention here is clearly not to start a comprehensive survey on the topic [7, 21]. Instead, I would like to invite the reader to a quick tour of examples and properties breaking this mirage of universality. Most of them were encountered or established during my collaboration with Eric, and I hope this note can give a clue about the richness of Eric's scientific collection.

**Content of the note**: To simplify exposition, I chose to restrict to (classical) cellular automata and tackle three main topics in three separate sections. Each topic shows examples of "computationally universal" cellular automata that, in some sense, do not behave as expected, or pair of examples that behave differently with respect to some parameter:

- efficient versus unbounded computations: how some cellular automata are able to embed one type of computations but not the other;
- transient versus repeatable circuit simulations: about the existence of (at least) two fundamentally different ways to simulate Boolean circuits in cellular automata, and their consequences;
- hidden universality: how cellular automata might seem 'simple' according to some parameter despite being actually universal.

Before starting, some standard definition are given below to set up our framework.

[1] It should be noted however that a growing trend in symbolic dynamics has shown the importance of computability considerations. Some of these results were even published in real math journals.

## 2 Standard Definitions and Notations

For any finite set $Q$ (the alphabet) and positive integer $d$ (the dimension), we consider the space of configurations $Q^{\mathbb{Z}^d}$, i.e. the set of maps giving a state from $Q$ to each position in the lattice $\mathbb{Z}^d$. The state of configuration $c \in Q^{\mathbb{Z}^d}$ at position $z \in \mathbb{Z}^d$ will be denoted either $c(z)$ or $c_z$.

For $n \in \mathbb{N}$, let $\mathcal{B}(n)$ be the set of positions of $\mathbb{Z}^d$ of maximum norm at most $n$:

$$\mathcal{B}(n) = \{z \in \mathbb{Z}^d : \|z\|_\infty \leq n\}.$$

Then for any $u \in Q^{\mathcal{B}(n)}$, we define the *cylinder set* $[u]$ centered on cell 0 by:

$$[u] = \{c \in Q^{\mathbb{Z}^d} : \forall z \in \mathcal{B}(n), c_z = u_z\}.$$

These cylinder sets can be chosen as a base of open sets of the space $Q^{\mathbb{Z}^d}$ endowing it with a compact topology [17]. Equivalently, the same topology can be defined by the Cantor distance:

$$\delta(c, c') = 2^{-\min\{\|z\|_\infty : c_z \neq c'_z\}}.$$

A cellular automaton of dimension $d$ and state set $Q$ is a map $F$ acting continuously on configurations and translation invariant way. Equivalently (Curtis-Lyndon-Heldund theorem [14]), it can be defined locally by a neighborhood $V$ (a finite subset of $\mathbb{Z}^d$) and a local transition map $f : Q^V \to Q$ as follows:

$$\forall z \in \mathbb{Z}^d, \quad F(c)_z = f\left(c\big|_{z+V}\right)$$

where $c\big|_{z+V}$ denotes the map $z' \in V \mapsto c_{z+z'}$.

The *radius* of $F$ is the smallest integer $r$ such that $V \subseteq \mathcal{B}(r)$ where $V$ is some neighborhood for which there is a local map $f_V : Q^V \to Q$ defining $F$ as above. $F$ induces an action on finite patterns as follows. For any $n \in \mathbb{N}$ and any $u \in Q^{\mathcal{B}(n+r)}$, $F(u)$ is the finite pattern $v \in Q^{\mathcal{B}(n)}$ obtained by application of $f$ on $u$ at each position from $\mathcal{B}(n)$, *i.e.* such that

$$\forall c \in [u], F(c) \in [v].$$

We are now going to define a notion of universality for cellular automata. We choose this one for two reasons: first it is one of the strongest form of universality and will serve us as a benchmark in the following, and second, it is intrinsic to the model of cellular automata and make no reference to other models of computation (for more details, see [5, 6, 22]).

This notion, called *intrinsic universality*, is based on a notion of (intrinsic) simulation that is defined through two ingredients [5, 6].

The first ingredient is a notion of cell-wise simulation that works by restriction to a sub alphabet and then projection onto the target alphabet. To be more precise let $F$ and $G$ be cellular automata of dimension $d$. We denote by $F \triangleleft G$ the fact that $F$ is obtained from $G$ by cell-wise restriction and projection, formally: $\exists \pi : Q \subseteq Q_G \rightarrow Q_F$ surjective such that for all $c \in Q^{\mathbb{Z}^d}$

$$\overline{\pi} \circ G(c) = F \circ \overline{\pi}(c)$$

where $\overline{\pi} : Q^{\mathbb{Z}^d} \rightarrow Q_F^{\mathbb{Z}^d}$ is the cell-wise application of $\pi$. In the language of dynamical systems, $(F, Q_F^{\mathbb{Z}^d})$ is a factor of $(G, Q^{\mathbb{Z}^d})$ which is a sub-system of $(G, Q_G^{\mathbb{Z}^d})$.

Now we add the second ingredient, *rescaling*, that allows to turn a cell-wise simulation into a simulation that works by blocks: blocks of cell of the first CA are simulated by blocks of cell of the second CA. Given a rectangular shape $\mathbf{m} = (m_1, \ldots, m_d)$ and some alphabet $Q$ we define the bloc recoding map $B_{\mathbf{m}}$ from $Q^{\mathbb{Z}^d}$ to $\left(Q^{m_1 m_2 \cdots m_d}\right)^{\mathbb{Z}^d}$ by:

$$\begin{aligned} B_{\mathbf{m}}(x)(z_1, \ldots, z_d) = \big(&x(m_1 z_1, \ldots, m_d z_d), \ldots, \\ &x(m_1 z_1 + m_1 - 1, m_2 z_2, \ldots, m_d z_d), \\ &x(m_1 z_1, m_2 z_2 + 1, m_3 z_3, \ldots, m_d z_d), \ldots \\ &x(m_1 z_1 + m_1 - 1, \ldots, m_d z_d + m_d - 1)\big) \end{aligned}$$

It is a bijection that recodes any configurations by blocks of shape $\mathbf{m}$. Now if $t$ is a positive integer and $\mathbf{z} \in \mathbb{Z}^d$, we define the rescaling of $F$ of parameters $\mathbf{m}$ and $t$ as the CA $F^{<\mathbf{m},t>} = B_{\mathbf{m}} \circ F^t \circ B_{\mathbf{m}}^{-1}$.

We finally say that $G$ *simulates* $F$, denoted $F \leq G$, if there are parameters $\mathbf{m}$, $t$, $\mathbf{m}'$ and $t'$ such that $F^{<\mathbf{m},t>} \triangleleft G^{<\mathbf{m}',t'>}$. We also say that $G$ *strongly simulates* $F$ if there are parameters $\mathbf{m}$ and $t$ such that $F \triangleleft G^{<\mathbf{m},t>}$. Then, a CA $G$ is *intrinsically universal* if for any CA $F$ we have $F \leq G$. It can be shown that an intrinsically universal CA can in fact *strongly simulate* any CA [6].

Finally, we assume the reader is familiar with basic notions and results of computability and complexity theory. We will use the following standard classes of decision problems:

- P is the set of problems which can be solve be a deterministic Turing machine in polynomial time;
- NLOGSPACE is the set of problems which can be solve be a non-deterministic Turing machine in logarithmic space;
- $\Sigma_1^0$ is the set of recursively enumerable problems (which contains the halting problem).

Without explicit mention and when speaking about P-completeness we consider LOGSPACE reductions. When speaking about $\Sigma_1^0$-completeness we usually consider many-one reductions.

## 3 Efficient Versus Unbounded Computations

It is well-known that, besides the reference model of Turing machines, there are other ones that fundamentally differ because they either only allow efficient but bounded computation (like Boolean circuits) or unbounded but slow computations (like Minsky machines) [19]. We would like to illustrate this aspect in the framework of cellular automata in a precise manner. To simplify, we restrict to dimension 1 in this section. We first define two classical problems associated to any CA which will serve as canonical indicators for both aspects mentioned above: efficiency and unboundedness of computations.

The first one is about short-term predictability within a bounded time range and provides a fine-grained complexity measurement within class P.

**Definition 1** Let $F$ be any CA of radius $r$ and alphabet $Q$. The prediction problem $\text{PRED}_F$ is defined as follows:

- input: $t > 0$ and $u \in Q^{\mathcal{B}(rt)}$
- output: $F^t(u) \in Q$.

The second one asks for a prediction about an unbounded future and provides a coarse-grained complexity measure allowed to cross the decidable barrier. It could be refined in many ways as in the definition of universality for dynamical symbolic systems from [7]. We prefer to keep it simple for the clarity of exposition. We say a configuration $c \in Q^{\mathbb{Z}}$ is eventually bi-periodic if it is eventually periodic to the left and eventually periodic to the right, said differently if it is of the form ${}^{\infty}u_L \cdot u \cdot u_R^{\infty}$ where $u_L$, $u$ and $u_R$ are finite words.

**Definition 2** Let $F$ be any CA of radius $r$ and dimension 1. The reachability problem $\text{UBPRED}_F$ is defined as follows:

- input: an eventually bi-periodic configuration $c = {}^{\infty}u_L \cdot u \cdot u_R^{\infty}$ and a state $q$.
- output: decide whether there is $t \in \mathbb{N}$ such that $F^t(c)_0 = q$.

One of the well-know results of computational universality in cellular automata is about elementary rule 110 given by the local rule $\delta : \{0, 1\}^3 \to \{0, 1\}$ with

$$\delta(x, y, z) = (1 - xyz) \cdot \max(y, z).$$

It is interesting to note that the first proof of computational universality of this cellular automaton due to Cook [3] was enough to prove undecidability of $\text{UBPRED}_\delta$ but did not give information about problem $\text{PRED}_\delta$. It is only later, by a strong improvement in one step the the reduction, that $\text{PRED}_\delta$ was proven to be P-complete [20]. The purpose of this section is precisely to make clear that there is generally no implication in either direction between the P-hardness of PRED and the undecidability of UBPRED.

**Definition 3** A CA $F$ is a *freezing CA* if, for some (partial) order $\leq$ on states, the state of any cell can only decrease, *i.e.*

$$F(c)_z \leq c_z$$

for any configuration $c$ and any cell $z$.

The definition above was introduced in [9] in studied more in depth in [23]. Similar cellular automata corresponding to bounded changes or bounded communications were also considered in the literature with the point of view language recognizers [2, 18, 24]. Under the hypothesis that NLOGSPACE $\neq$ P, the following results show examples of cellular automata that can embed arbitrary unbounded computation, but not in an efficient way.

**Theorem 1** (Sect. 4.3 of [23]) *For any freezing CA $F$ of dimension 1, the problem* $\text{PRED}_F$ *is in* NLOGSPACE. *There exists a 1D freezing CA $F$ such that* $\text{UBPRED}_F$ *is $\Sigma_1^0$-complete.*

We are now going to build an example with the opposite computation embedding properties: as hard as it can be in the short term (it can embed efficiently bounded computations), but decidable in the long term (it can not embed unbounded computations). It is inspired from [23, Example 7] and consists in a simulation of some P-complete cellular automaton inside finite zones, with some head controlling the simulation and forced to move back and forth inside the zone and shrink it by one cell at each bounce on a boundary. The simulation is such that one step of the simulated cellular automaton is done at each pass so that there is only a quadratic slowdown (see Fig. 1).

Let $F$ be any 1D CA on alphabet $Q$ with radius 1 and local map $\delta : Q^3 \to Q$. We define $\mathcal{Z}_F$ on alphabet $R = \{b, b_+, e\} \cup Q'$ with $Q' = Q \times Q \times \{\leftarrow, \rightarrow, l, r\}$ and radius 1 as follows:

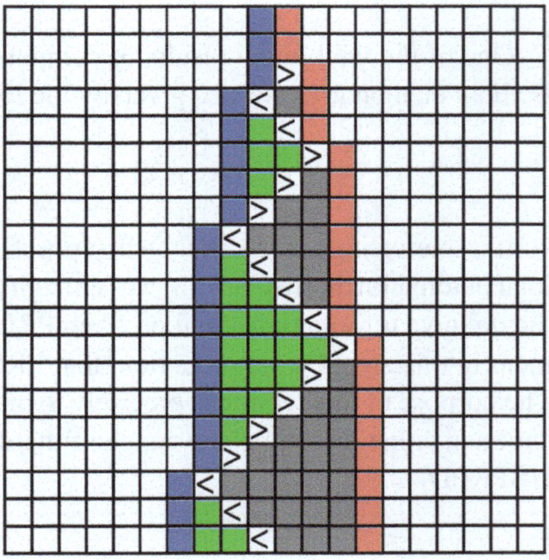

**Fig. 1** The shrinking zone trick behind the construction of $\mathcal{Z}_F$ (time goes from bottom to top)

- $e$, the *error state*, is a spreading state: any cell with $e$ in its neighborhood turns into state $e$; a configuration $c$ is *valid* if $e$ never appears in its orbit;
- $b$, the blank state, never changes except in presence of the error state; $b_+$ becomes $b$ except in presence of the error state; a maximal connect component of cells in state $Q'$ is a *working zone*;
- in a working zone, patterns of the form $(x, y, r)(x', y', l)$, or $(x', y', l)(x, y, r)$, or $(x, y, z)(x', y', z')$ with $\{z, z'\} \subseteq \{\leftarrow, \rightarrow\}$, or $(x, y, r)(x', y', z)$ or $(x, y, z)$ $(x', y', l)$ with $z \in \{\leftarrow, \rightarrow\}$, are forbidden and generate an $e$ state when detected; therefore in a valid configuration and in each working zone there is at most one occurrence of a state of the form $(x, y, \{\leftarrow, \rightarrow\})$ called the *head*;
- a cell without forbidden pattern (from previous item) and without head in its neighborhood doesn't change its state;
- the movements and actions of the heads are as follows:
  - inside a working zone, the head in state $\leftarrow$ moves left, the head in state $\rightarrow$ moves right; the local map $\delta$ is only applied the head moves left to right; precisely we have the following transitions:

$$
\begin{aligned}
(x, y, l),\ (x', y', \leftarrow),\ (x'', y'', r) &\mapsto (x', y', r)\\
(x, y, l),\ (x', y', l),\ (x'', y'', \leftarrow) &\mapsto (x', y', \leftarrow)\\
(x, y, l),\ (x', y', \rightarrow),\ (x'', y'', r) &\mapsto (x', y', l)\\
(x, y, \rightarrow),\ (x', y', r),\ (x'', y'', r) &\mapsto (\delta(y, x', x''), x', \rightarrow)
\end{aligned}
$$

  - when a boundary of the working zone is reached, the head bounces, changes of direction and the working zone get shrinked by one cell; precisely we have the following transitions:

$$
\begin{aligned}
b,\ (x, y, l),\ (x', y', \leftarrow) &\mapsto (x, y, \leftarrow)\\
b,\ (x, y, \leftarrow),\ (x', y', r) &\mapsto (x, y, \rightarrow)\\
b,\ (x, y, \rightarrow),\ (x', y', r) &\mapsto (y, x, l)\\
b,\ (x, y, l),\ (x', y', \rightarrow) &\mapsto b_+\\
(x, y, \rightarrow),\ (x', y', r),\ b &\mapsto (x', y', \rightarrow)\\
(x, y, l),\ (x', y', \rightarrow),\ b &\mapsto (x', y', \leftarrow)\\
(x, y, l),\ (x', y', \leftarrow),\ b &\mapsto (x', y', r)\\
(x, y, \leftarrow),\ (x', y', r),\ b &\mapsto b_+
\end{aligned}
$$

    *(note the swap between $x$ and $y$ in the third transition above to initialize the sequential application of $\delta$)*
  - finally the head disappears in a working zone of size 1, precisely:

$$
b', (x, y, z), b'' \mapsto (x, y, r)
$$

    for any $\{b', b''\} \subseteq \{b, b_+\}$.

**Theorem 2** *For any $F$, the problem* $\text{UBPRED}_{\mathcal{Z}_F}$ *is decidable in polynomial time. If $F$ is chosen so that* $\text{PRED}_F$ *is* P*-complete, then* $\text{PRED}_{\mathcal{Z}_F}$ *is* P*-complete.*

***Proof*** For the first part of the Theorem, let us consider an eventually bi-periodic configuration $c = {}^{\infty}u_L \cdot u \cdot u_R^{\infty}$. There are four cases:

- $c$ is not a valid configuration, which means that it contains a working zone with a forbidden pattern. Since the forbidden pattern are locally detectable, such a forbidden pattern must be detected inside the finite word $u_L u_L u_L u u_r u_R u_R$ (considering the worst case where $u_L$ or $u_R$ is of size 1). Therefore, in time $t$ which is linear in the sizes of $u_L$, $u$ and $u_R$ we have $\mathcal{Z}_F^t(c)_0 = e$;
- $c$ is a valid configuration and position 0 belong to a finite working zone in $c$. Since the left and right boundary of this zone must belong either to $u$, or $u_L$ or $u_R$, the zone is of linear size and it gets completely shrinked in quadratic time, meaning that the state of position 0 will no longer change after a quadratic time;
- $c$ is a valid configuration and position 0 belongs to an infinite zone in $c$. In this case, the position of the head must belong to either $u$, $u_L$ or $u_R$ and the same for the eventual (unique) boundary of the zone. Therefore, after a linear time in the worst case, cell 0 will never change again (for instance, the head comes from the right, bounces to the left boundary, crosses once more position 0, but never comes back again);
- $c$ is a valid configuration but position 0 does not belong to some working zone, then for any $t \geq 1$ we have $\mathcal{Z}_F^t(c)_0 = b$.

We deduce that that after a quadratic time the state of cell 0 does not change any more, so it is sufficient to simulate $\mathcal{Z}_F$ on $c$ for this number of states to solve problem $\text{UBPRED}_{\mathcal{Z}_F}$.

For the second part of the Theorem, see [23, Lemma 1 and Proposition 5]. □

## 4 Transient Versus Repeatable Circuit Simulation

In this section we focus on dimension 2 and simulation of Boolean circuit and logical gates by cellular automata. Showing how a cellular automaton can embed Boolean circuits is one of the common methods used to claim its Turing universality (see for instance [1, 8, 21]).

We are now going to describe two modes of simulation of a set of logical gates by a cellular automaton, which were formalized in [11]. The basic simulation mechanism behind both simulation modes uses square blocks concatenated in a grid-like fashion. Each such square block represents a part of a concrete Boolean circuit (either a node or wire). The definition doesn't require any specific way of representing information inside the blocks, just that the family of blocks use coherent representation of information so that the Boolean logic works when assembling them. More concretely, they communicate information with their four neighbors (north, east, south, west) in such a way that each one implements a Boolean function with at most 2 inputs and at most 2 outputs.

In the sequel all considered blocks will compute one of the following maps (we represent them using type $\{0,1\}^4 \to \{0,1\}^4$ in order to make explicit the position of inputs and outputs among the neighbors in the order north, east, south, west):

$$\begin{aligned}
\text{AND}(x,*,*,y) &= \big(0, \min(x,y), 0, 0\big)\\
\text{OR}(x,*,*,y) &= \big(0, \max(x,y), 0, 0\big)\\
\text{CROSS}(x,*,*,y) &= \big(0, y, x, 0\big)\\
\text{NOP}(*,*,*,*) &= \big(0,0,0,0\big)\\
\text{FORK}(*,*,*,x) &= \big(0, x, x, 0\big)\\
\text{WIRE}_{i,o}\big(c \in \{0,1\}^4\big) &= k \in \{0,\dots,3\} \mapsto \begin{cases} c(i) & \text{if } k = o\\ 0 & \text{else}\end{cases}
\end{aligned}$$

for any $i \neq o \in \{0,\dots,3\}$. Note that any function $f$ above is such that

$$f(0,0,0,0) = (0,0,0,0).$$

We denote by $Img(f)$ the set of 4-uple that can be obtained as an image of $f$. The $\text{WIRE}_{i,o}$ functions are just all the possible ways to read a bit on one side and transmit it to another side. Together with the NOP and FORK function they represent the basic planar wiring toolkit denoted $W$ in the sequel. The NOP gate is special in that one considers it has 4 inputs and 4 outputs.

The two circuit simulation modes share the same block representation of circuit and information, but they differ in their requirement about the dynamical evolution of blocks. In the first mode, called *transient mode*, the gates can be used only once and nothing is granted concerning their evolution afterwards. The second mode, called *repeatable mode*, asks for each gate to go back to some acceptable state each time they are used so that they can be used again. Both modes require the simulation to work in constant time.

Let $\mathcal{G} \subseteq \{\text{AND}, \text{OR}, \text{CROSS}\}$ be a set of gates. Let $F$ be a CA with states set $Q$ and $N > 0$ be an integer. Consider a set $V \subseteq Q^{N\times N}$ of patterns, the valid blocks, each of which as a type $f_u$ where $f \in \mathcal{G} \cup W$ and $u \in Img(f)$ and such that, for any $f_u$, there is some block of $V$ of type $f_u$. If a block $B \in V$ has type $f_{(a,b,c,d)}$ for some $f$, we say it has *north value* $a$, *east value* $b$, *south value* $c$ and *west value* $d$. Finally let $\Delta > 0$ be some constant. A configuration is *valid* if it is a concatenation of valid blocks where output sides of a block must face input sides of its corresponding neighbors. Given a block $B \in V$ of type $f_u$ in a valid configuration, we say that it *makes the correct transition* if it becomes a block of type $f_v$ after $\Delta$ steps where $v = f(n,e,s,w)$ is the output of $f$ on the input read from surrounding blocks, precisely: the block at the north of $B$ has south value $n$, the block at the east of $B$ has west value $e$, etc.

**Transient simulation**. We say that $F$ *simulates the set of gates* $\mathcal{G}$ *in transient mode* with delay $\Delta$ and valid blocks $V$ if for any valid configuration $c$, the configuration $F^{\Delta}(c)$ is valid and for any $f \in \mathcal{G} \cup W$, any block of type $f_{(0,0,0,0)}$ in $c$ makes the correct transition.

**Repeatable simulation**. The simulation is *repeatable* if any block in any valid configuration makes the correct transition.

Before stating some theorems, let us define a decision problem associated to any 2D cellular automaton that will serve as a benchmark to separate the two kinds of circuit simulation above.

**Definition 4** Let $F$ be any 2D CA of radius $r$ and alphabet $Q$, and $\phi$ a non-decreasing function such that $1 \leq \phi(n) \leq 2^{O(n)}$. The prediction problem $\mathrm{CYCLE}_F^{\phi}$ is defined as follows:

- input: a periodic configuration $c$ of period $n \times n$
- output: is the length of the temporal cycle reached from $c$ strictly greater than $\phi(n)$?

The first mode of simulation (the repeatable mode) is the strongest one, and is actually equivalent to intrinsic universality even if we use only monotone gates. In this case, although the definition does not explicitly provide crossing gates, it is always possible to realize a dynamical crossing and build arbitrary reusable bloc elements leading to intrinsic universality.

**Theorem 3** ([11]) *A 2D CA F is intrinsically universal if and only if it can simulate a AND, OR circuitry in a repeatable way. In this case* $\mathrm{PRED}_F$ *is P-complete,* $\mathrm{UBPRED}_F$ *is* $\Sigma_1^0$*-complete and* $\mathrm{CYCLE}_F^{\phi}$ *is* PSPACE*-complete for some* $\phi$.

To illustrate the difference between the two modes we shall use the symmetric signed majority cellular automaton: it is essentially a majority rule where the state of each neighboring cell can be inverted or not before evaluating majority, this being done according to a local invariant sign vector and in a symmetric way: if cell $z$ inverts the value of its neighbor $z'$, then $z'$ will also invert the value of $z$. We use the von Neumann neighborhood $V = \{(0,0), (0,1), (1,0), (0,-1), (-1,0)\}$. The symmetric signed majority cellular automaton $F_1$ is defined over state set $Q = \{-1, 1\}^6$. To simplify notation, we will see each state $q \in Q$ as a pair $(I(q), S(q))$ where $I(q) \in \{0, 1\}$ represent the *inner state* and $S(q) \in \{-1, 1\}^V$ is a *sign vector* associating a sign to each neighbor of the von Neumann neighborhood. For any configuration $c \in Q^{\mathbb{Z}^2}$, any cell $z$ and any cell $z' \in z + V$ we define the symmetric weight $w_{zz'} \in \{-1, 1\}$ as $w_{zz'} = (S(c_z)(z' - z))(S(c_{z'})(z - z'))$. We note that $w_{zz'} = w_{z'z}$, hence the name symmetric. $F_1$ is then defined as follows.

$$F_1(c)_z = (\alpha, S(c_z))$$

where

$$\alpha = \begin{cases} 1 & \text{if } \sum_{z' \in z+V} w_{zz'} I(c_{z'}) > 0, \\ -1 & \text{otherwise.} \end{cases}$$

The following theorem shows that transient simulations are strictly weaker than repeatable simulations, $F_1$ being an example capable of the former, but not the latter.

**Theorem 4** ([11]) *If a 2D CA $F$ can simulate a $\{AND, OR, CROSS\}$ circuitry in transient mode, then its associated problem* $\text{PRED}_F$ *is $P$-complete. $F_1$ defined above can simulate a $\{AND, OR, CROSS\}$ circuitry in transient mode. However, $F_1$ is not intrinsically universal and such that the problem* $\text{CYCLE}_{F_1}^{\phi}$ *is in* P *if $\phi \equiv 1$ and trivial else.*

## 5 Hidden Universality

When proving that some cellular automaton is computationally universal, it can be acceptable to avoid a precise definition of universality if the construction makes it clear enough. However, a precise definition seems necessary when ones want to show that some cellular automaton is **not** universal. To avoid formalism, one could be tempted to use a shorter path: prove that, according to some well-chosen parameter, the considered cellular automaton is too simple to be universal. The intuition is that a universal cellular automaton should have roughly the highest complexity for the parameter. This approach can be made precise and yield some proof tools of non-universality in some contexts [12]. The purpose of this section is to recall that things can get counter-intuitive and such a parameter must be chosen carefully.

### *5.1 Hidden Minsky Machines Simulation*

The *limit set* of a cellular automaton $F$ is the nonempty closed subset

$$\Omega_F = \bigcap_{t \in \mathbb{N}} F^t(X).$$

It represents the set of configurations that may appear arbitrarily far in the evolution and the restriction of $F$ to $\Omega_F$ is often considered as the asymptotic dynamics of $F$. The limit language is the set of finite patterns that occur in some configuration of $\Omega_F$. It is not difficult to see that the limit language is always co-recursively enumerable. However, there are known examples of non-recursive ones [15]. The attentive reader of [4] has probably spotted the affirmation that universal cellular automata have a non-recursive limit set. Depending on the definition of universality, this affirmation can be false. The following theorem shows that arbitrary Minsky machine simulations can be realize while maintaining a simple limit set (see [10] for the precise definition of simulation).

**Theorem 5** ([10]) *For any Minsky machine M there exists a CA of dimension* 1 *that simulates M but whose limit language is regular.*

## 5.2 *Hidden Intrinsic Universality*

Following Theorem 5, one can go one step further and hide intrinsic universality behind a simple limit set (at the price of a complexity increase from regular to NLOGSPACE). The main trick of the next theorem is inspired from [16]: adding to a given cellular automaton $F$ on alphabet $Q$, a firing squad component (see Fig. 2) that is able to fill-in the limit set restricted to the $Q$ component, and therefore make it simple independently of $F$.

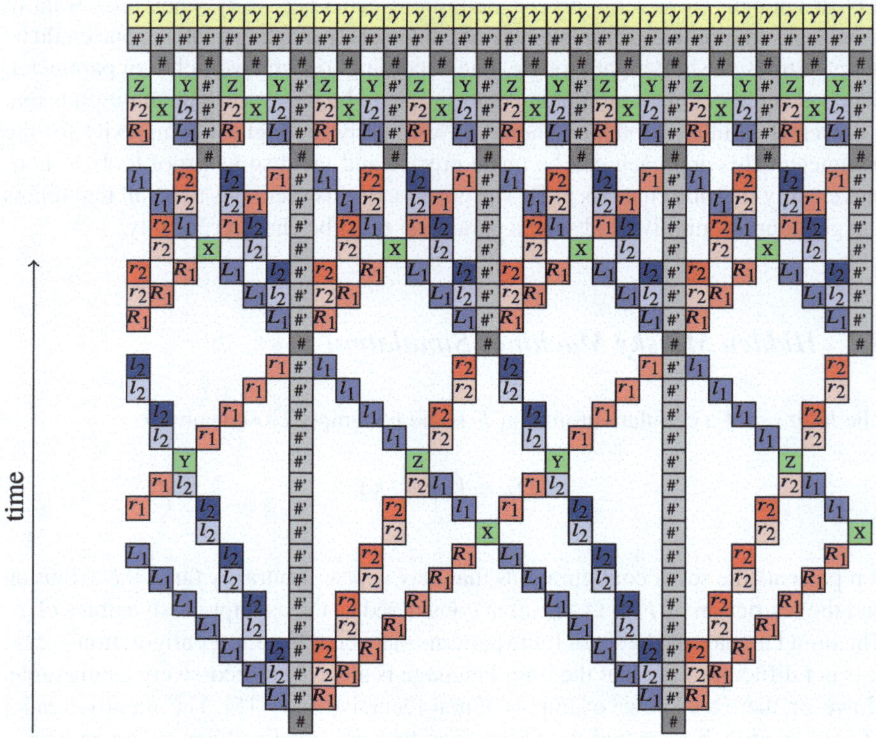

**Fig. 2** J. Kari's firing squad trick: a synchronous apparition of $\gamma$ can be triggered arbitrarily far in time, thus allowing to complete the limit set on some component of states to the full-shift

**Theorem 6** ([13]) *There exists an intrinsically universal CA whose limit language is* NLOGSPACE.

Given $n \in \mathbb{N}$, the *column factor* of width $n$ of $F$, $\Sigma_n(F)$, is the set of columns that can appear in space-time diagrams of $F$:

$$\Sigma_n(F) = \left\{(u_t)_{t\in\mathbb{N}} : u_t \in Q^n, u_t = F^t(c)_{[1,n]}, c \in Q^{\mathbb{Z}^d}\right\}.$$

To $\Sigma_n(F)$ we associate its language of finite patterns $L(\Sigma_n(F))$ defined as the set of words $u_t \cdots u_{t+k}$ for some $(u_t)_{t\in\mathbb{N}} \in \Sigma_n(F)$ and $t, k \in \mathbb{N}$.

The approach of [7] to define universality for general dynamical systems translates into the following in our settings. To $F$ we associate the model checking problem:

- **input**: $n$ and a regular language $L_n$ over alphabet $Q^n$,
- **question**: decide whether $L_n$ intersects $L(\Sigma_n(F))$.

$F$ is *BDK-universal* if its associated model checking problem is r.e.-complete. Like for limit sets, column factors can be filled up and thus simplified by increasing the alphabet starting from an arbitrarily complex cellular automaton.

**Theorem 7** ([13]) *There exists an intrinsically universal CA $F$ such that $L(\Sigma_n(F))$ are regular languages computable from $n$. In particular, such $F$ is not BDK-universal.*

## References


1. Banks ER (1970) Universality in cellular automata. In: Eleventh annual symposium on switching and automata theory, Santa Monica, CA. IEEE
2. Carton O, Guillon B, Reiter F (2018) Counter machines and distributed automata—a story about exchanging space and time. In: Cellular Automata and discrete complex systems—Proceedings of the 24th IFIP WG 1.5 international workshop, AUTOMATA 2018, Ghent, Belgium, 20–22 June 2018, pp 13–28
3. Cook M (2004) Universality in elementary cellular automata. Compl Syst 15:1–40
4. Čulik K II, Pachl J, Yu S (1989) On the limit sets of cellular automata. SIAM J Comput 18(4):831–842 August
5. Delorme M, Mazoyer J, Ollinger N, Theyssier G (2011) Bulking II: classifications of cellular automata. Theor Comput Sci 412:3881–3905
6. Delorme M, Mazoyer J, Ollinger N, Theyssier G (2010) Bulking I: an abstract theory of bulking. oai:hal.archives-ouvertes.fr:hal-00451732, Jan 2010
7. Delvenne J-C, Kurka P, Blondel VD (2006) Decidability and universality in symbolic dynamical systems. Fundam Inform 74(4):463–490
8. Durand B, Róka Z (1999) Cellular automata: a parallel model, volume 460 of mathematics and its applications, chapter The game of life:universality revisited. Kluwer Academic Publishers, pp 51–74
9. Goles E, Ollinger N, Theyssier G (2015) Introducing freezing cellular automata. In: Kari J, Törmä I, Szabados M (eds) Exploratory papers of cellular automata and discrete complex systems (AUTOMATA 2015), pp 65–73
10. Goles E, Maass A, Martinez S (1993) On the limit set of some universal cellular automata. Theor Comput Sci 110:53–78

11. Goles E, Montealegre P, Perrot K, Theyssier G (2018) On the complexity of two-dimensional signed majority cellular automata. J Comput Syst Sci 91:1–32
12. Goles EC, Meunier P-E, Rapaport I, Theyssier G (2011) Communication complexity and intrinsic universality in cellular automata. Theor Comput Sci 412(1-2):2–21
13. Guillon P, Meunier P-E, Theyssier G (2010) Clandestine simulations in cellular automata. In: Kari J (ed) Proceedings of the second symposium on cellular automata "Journées Automates Cellulaires", JAC 2010, Turku, Finland, 15–17 Dec 2010. Turku Center for Computer Science, pp 133–144
14. Hedlund GA (1969) Endomorphisms and automorphisms of the shift dynamical systems. Math Syst Theory 3(4):320–375
15. Hurd LP (1990) Nonrecursive cellular automata invariant sets. Compl Sys 4:131–138
16. Kari J (1994) Rice's theorem for the limit sets of cellular automata. Theor Comput Sci 127:229–254
17. Kůrka P (2003) Topological and symbolic dynamics. Société Mathématique de France
18. Kutrib M, Malcher A (2010) Cellular automata with sparse communication. Theor Comput Sci 411(38–39):3516–3526
19. Minsky M (1967) Computation: finite and infinite machines. Prentice Hall, Englewoods Cliffs
20. Neary T, Woods D (2006) P-completeness of cellular automaton rule 110. In: International colloquium on automata languages and programming (ICALP), volume 4051 of LNCS. Springer, pp 132–143
21. Ollinger N (2008) Universalities in cellular automata a (short) survey. In: Durand B (ed) First symposium on cellular automata "Journées Automates CellulairesâŁž (JAC 2008), Uzès, France, 21–25 Apr 2008. Proceedings. MCCME Publishing House, Moscow, pp 102–118
22. Ollinger N (2008) Universalities in cellular automata a (short) survey. In: JAC, pp 102–118
23. Ollinger N, Theyssier G (2019) Freezing, bounded-change and convergent cellular automata. CoRR. arxiv:1908.06751. https://doi.org/10.46298/dmtcs.5734. https://dmtcs.episciences.org/9004
24. Vollmar R (1981) On cellular automata with a finite number of state changes. In: Knödel W, Schneider HJ (eds), Parallel Processes and Related Automata/Parallele Prozesse und damit zusammenhängende Automaten, volume 3 of Computing Supplementum. Springer Vienna, pp 181–191

# Connectivity and Connected Components in the Number-in-Hand Computation Model

**Antonio Lizama and Ivan Rapaport**

**Abstract** We study the multiparty communication model where players are the nodes of a graph $G$ and each of these nodes knows his/her own identifier together with the identifiers of his/her neighbors. The nodes simultaneously send a unique message to a referee who must give the output (which is a fiunction of $G$). In this paper we prove that counting the numbers of connected components of $G$ and deciding the connectivity of $G$ are equivalent problems (in terms of the size of the messages).

## 1 Introduction

In the *number-in-hand* multiparty communication model there are $k$ players. Each of these $k$ players receives an $n$-bit input string $x_i$ and they all need to collaborate in order to compute some function $f(x_1, \ldots, x_k)$. Despite its simplicity, the case $k > 2$ started to be studied very recently [1–9].

There are different communication modes for the *number-in-hand* model. In this paper we focus on the *simultaneous message* communication mode, in which all players simultaneously send a unique message to a referee. The referee collects the messages and computes the function $f$. The computational power of both the players and the referee is unlimited. When designing a protocol for a function $f$, the goal is to minimize the size of the longest message generated by the protocol. This minimum, usually depending on $n$, is called the *message size complexity* of $f$. Typical questions in communication complexity consist in designing protocols with small messages, and proving lower bounds on the size of such messages.

Several authors considered the case where the data distributed among the players is a graph [1, 3, 4, 8, 9]. Informally, each player knows a set of edges of the graph and together they must compute some function that depends on the graph. Again we

A. Lizama
Universidad Adolfo Ibáñez, Santiago, Chile
e-mail: antonio.lizama.o@uai.cl

I. Rapaport (✉)
DIM-CMM (UMI 2807 CNRS), Universidad de Chile, Santiago, Chile
e-mail: rapaport@dim.uchile.cl

 
A. Adamatzky (ed.), *Automata and Complexity*, Emergence, Complexity and Computation 42, https://doi.org/10.1007/978-3-030-92551-2_6

can observe two different settings. In one of them, the edges are distributed among the players in an adversarial way [1, 9]. In this work, following [1, 3], we consider the setting where each player corresponds to a node of the graph, and thus each player knows the identifier of this node together with the identifiers of its neighbors, represented as an $n$-bit vector (in the vector $x_i$ of the $i$-th node, the bit number $j$ is set to 1 if and only if the $i$-th and the $j$-th nodes are adjacent).

Formally, an $n$-node network is represented by a graph $G = (V, E)$, where $V = \{v_1, \ldots, v_n\}$ and $E = \{e_1, \ldots, e_m\}$. It is assumed that the identifier of each node $v_i \in V$ is $ID(v_i) = i$. The entry of $v_i$ is $x_i \in \{0, 1\}^n$, where $x_i(j) = 1$ if and only if $v_i v_j \in E$.

The goal is to compute a function $f(x_1, \ldots, x_n)$. In order to achieve this, every processors $v_i$ sends a message $out_i$ to an external entity, called the *referee*. With these messages, the *referee* should be able to decide the value of function $f$. This process is called a *protocol*. The internal computing power of each node and the *referee* is unbounded.

It is clear that, if there is no restriction on the size of the messages that each node sends to the *referee*, then it is enough for each one to send the entire neighborhood. With those messages, each of size $O(n)$, computing the value of $f(x_1, \ldots, x_n)$ becomes trivial. The challenge is to find non-trivial lower bounds on $C(f)$, the message size complexity of $f$.

In problem CONNECTED-COMPONENTS the goal is to compute the number of connected components of the graph. On the other hand, in problem CONNECTIVITY the goal is to decide whether the graph is connected. It is obvious that $C(\text{CONNECTIVITY}) \leq C(\text{CONNECTED} - \text{COMPONENTS})$. The goal of this paper is to show that $C(\text{CONNECTED} - \text{COMPONENTS}) = O(\text{CONNECTIVITY})$.

## 2 The Reduction

In this section we show that, if there is a protocol $\mathcal{P}$ that solves CONNECTIVITY, then there is another protocol $\mathcal{P}'$, with complexity of the same order, that solves CONNECTED-COMPONENTS. The following theorem shows the existence of such reduction.

**Theorem 1** $C(\text{CONNECTED} - \text{COMPONENTS}) = O(\text{CONNECTIVITY})$.

***Proof*** We are going to prove that, if CONNECTIVITY can be solved with messages of size $\ell(n)$, then CONNECTED-COMPONENTS can be solved using messages of size $2\ell(n + 2)$.

Let $G = (V, E)$ be a graph and $\mathcal{P}$ a protocol that decides the problem CONNECTIVITY using $\ell(n)$ bits. That is, the maximum length of the messages is $\ell(n)$. Consider the following family of graphs $\mathcal{F}$, which is built from $G$, as follows.

$$\mathcal{F} := \{G'_A = (V', E') \mid A \subseteq V,\ V' = V \cup \{a, b\},\ E' = E \cup \{au, bw \mid u \in A, w \in V \backslash A\}\}.$$

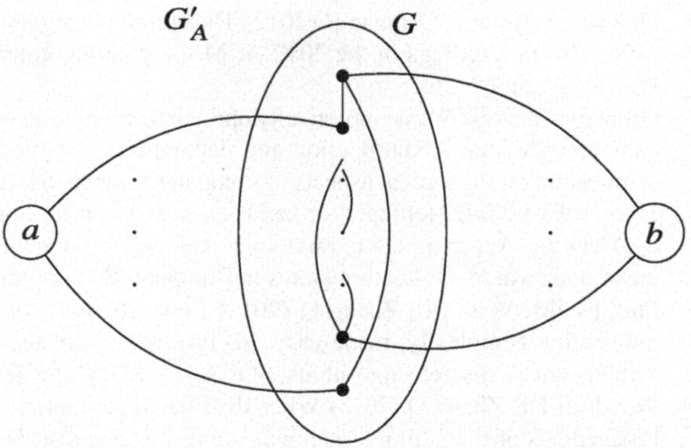


**Fig. 1** Graph that results from the reduction

In words, we add two new vertices $a$ and $b$, and connect every $v \in V$ to either $a$ ar $b$ according to the set $A$. Thus, each graph $G'_A$ in $\mathcal{F}$ is defined by fixing the set $A \subseteq V$ (see Fig. 1).

We define now the new protocol $\mathcal{P}'$ that computes the number of connected components in $G$. The new protocol $\mathcal{P}'$ is as follows. Each node $v \in V$ sends the message $m_v = m_v^1 m_v^2$, where $m_v^1$ corresponds to the message it would have sent in $\mathcal{P}$ if $v \in A$, while $m_v^2$ is the message it would have sent if $v \in V \backslash A$.

It follows that, in $\mathcal{P}'$, the message length sent by each node is $2f(n+2)$. With this information the *referee* is able to infer whether each graph $G'_A \in \mathcal{F}$ is connected. Indeed, as previously mentioned, a graph $G'_A \in \mathcal{F}$ is determined uniquely by the choice of $A \subseteq V$. Then, the *referee* must consider all possible subsets $A$ in the graph $G$. With this sequence of messages, and thanks to the existence of protocol $\mathcal{P}$, it is possible to determine whether $G'_A$ is connected or not for every $A$.

Analyzing the answer on all the graphs of the family $\mathcal{F}$, it is possible to determine the number of connected components of the original graph $G$. For this, it is enough to note that, if $G$ is a graph with $k$ connected components $V_1, \ldots, V_k$, then $G'_A$ is not connected if and only if $A = \{\cup_{i \in I} V_i \mid I \subseteq \{1, \ldots, k\}\}$. □

## References


1. Ahn KJ, Guha S, McGregor A (2012) Analyzing graph structure via linear measurements. In: Proceedings of the 23rd annual ACM-SIAM symposium on discrete algorithms, SODA '12, pp 459–467
2. Ahn KJ, Guha S, McGregor A (2012) Graph sketches: sparsification, spanners, and subgraphs. In: Proceedings of the 31st symposium on principles of database systems, PODS '12, pp 5–14
3. Becker F, Matamala M, Nisse N, Rapaport I, Suchan K, Todinca I (2011) Adding a referee to an interconnection network: What can(not) be computed in one round. In: IPDPS. IEEE, pp 508–514
4. Becker F, Montealegre P, Rapaport I, Todinca I (2014) The simultaneous number-in-hand communication model for networks: private coins, public coins and determinism. In: International colloquium on structural information and communication complexity, pp 83–95. Springer

5. Drucker A, Kuhn F, Oshman R (2012) The communication complexity of distributed task allocation. In: Proceedings of the 2012 ACM symposium on principles of distributed computing, PODC '12, pp 67–76
6. Gronemeier A (2009) Asymptotically optimal lower bounds on the NIH-multi-party information complexity of the AND-function and disjointness. In: Proceedings of the 26th international symposium on theoretical aspects of computer science, STACS '09, pp 505–516
7. Jaymar TS (2009) Hellinger strikes back: a note on the multi-party information complexity of AND. In: Approximation, randomization, and combinatorial optimization. Algorithms and techniques, vol 5687. Lecture Notes in Computer Science, pp 562–573
8. Phillips JM, Verbin E, Zhang Q (2012) Lower bounds for number-in-hand multiparty communication complexity, made easy. In: Proceedings of the twenty-third annual ACM-SIAM symposium on discrete algorithms, SODA '12. SIAM, pp 486–501
9. Woodruff DP, Zhang Q (2013) When distributed computation is communication expensive. In: Proceedings of the 27th international symposium on distributed computing, vol 8205. Lecture Notes in Computer Science, DISC '13, pp 16–30

# Contagion Dynamics in Complex Networks

**Lucas Böttcher**

**Abstract** Models of spreading processes in networks can help to provide insights into phenomena such as epidemic outbreaks, opinion formation, and failure propagation. Many contagion models are based on the idea that spreading occurs from an "infected" source to "non-infected" components, which may recover. In the case of social opinion formation, external factors such as media influence have to be taken into account, and in some cases multiple infectious sources are necessary to sustain spreading (*complex contagion*). In this chapter, I provide a brief overview of common models of contagious processes and show that many of them can be treated as special cases of a general contagion model. Interestingly, despite its general formulation, the stationary behavior of the model is characterized by only three distinct classes. As an application of the general contagion model, I discuss its ability to describe activist-voter interactions in election campaigns.

## 1 Introduction

Mathematical models of epidemic processes [1–3] contributed to a better understanding of disease outbreaks and their control. Traditional spreading models have been complemented with results from the study of complex networks [4, 5] to also account for the influence of different interaction networks. One possible classification of spreading processes is based on the number of contacts to "infected" sources, which are necessary to sustain spreading. In the case of a so-called *simple contagion* (i.e., contact process [6, 7]) such as a "flu-like" epidemic, a disease may spread from *one* single infected source to non-infected components. On the other hand, *complex contagions* describe phenomena such as the diffusion of innovations [8, 9], political mobilization [10, 11], viral marketing [12], and coordination games [13], where individuals may change their opinion/state if they are connected to *multiple* "infected" sources [14, 15]. In addition to the direct transmission of certain attributes

L. Böttcher (✉)
Frankfurt School of Finance and Management, Adickesallee 32-34, 60322 Frankfurt am Main, Germany
e-mail: l.boettcher@fs.de

 
A. Adamatzky (ed.), *Automata and Complexity*, Emergence, Complexity and Computation 42, https://doi.org/10.1007/978-3-030-92551-2_7

from one individual/component to another, external factors such as media influence and spontaneous failure have to be also taken into account [16, 17].

In the first part of this chapter, I provide a brief overview of some common models of contagious processes and show that they are specific cases of a general contagion model, which exhibits hysteresis effects [16, 18] and limit cycles [16, 19]. Despite its general formulation, the mean-field critical behavior of the model is restricted to only three possible regimes: (a) uncorrelated spontaneous transition dynamics, (b) contact process dynamics, and (c) cusp catastrophes. Cusp catastrophes [17, 20, 21] can entail abrupt transitions and hysteresis effects. Such phenomena may complicate the control of networked systems, because small variations in the system's control parameters (e.g., transmission rates) may cause abrupt transitions from a seemingly well-functioning state to global malfunction [5, 17, 20, 22, 23]. In the second part of this chapter, I outline the application of general contagion dynamics in the context of activist-voter interactions during election campaigns [24, 25].

## 2 General Contagion Model

A simple epidemic process may be described by so-called *susceptible-infected-suscpetible* (SIS) dynamics [1]. In statistical physics, this model is better known under the names contact process or Schlögl's first model [6, 7]. In SIS dynamics, infected nodes in a network can transmit a disease to their susceptible neighbors with rate $r$. In addition to induced transitions, susceptible nodes may change their state spontaneously with rate $p$. Infected nodes can then recover with rate $q$. The corresponding transitions on a square lattice are shown in Fig. 1. Some characteristic features of SIS dynamics such as epidemic threshold effects can be captured with a mean-field approach:

$$\frac{\mathrm{d}i(t)}{\mathrm{d}t} = \langle k\rangle r\, i(t)(1-i(t)) + p(1-i(t)) - qi(t)\,, \tag{1}$$

where $\langle k\rangle$ is the mean degree of the underlying network structure. A generalization of the SIS model that describes complex contagions with two sources is Schlögl's second model, also known as quadratic contact process [26]. Instead of Eq. (1), the corresponding mean-field approximation is

$$\frac{\mathrm{d}i(t)}{\mathrm{d}t} = \langle k\rangle r\, i(t)^2(1-i(t)) + p(1-i(t)) - qi(t)\,. \tag{2}$$

Complex contagions that require more than two contacts to "infected" sources can be described with threshold rules. In the following description of such general contagion dynamics, the components of a network are regarded as either active or inactive and change their state according to three fundamental processes: (i) an active node becomes spontaneously inactive in a time interval $\mathrm{d}t$ with probability $p\,\mathrm{d}t$, (ii) if

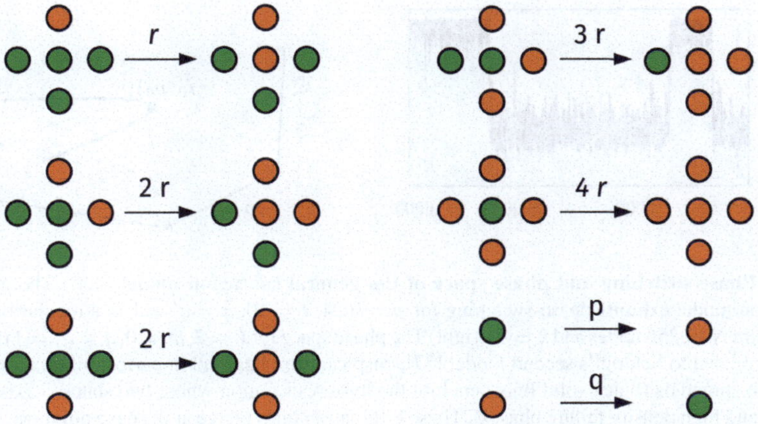


**Fig. 1 Simple epidemic on a square lattice.** Possible transitions for SIS dynamics on a square lattice (spatial rotation-invariance of the reactions is assumed). Orange nodes are infected and green nodes are susceptible

fewer than or equal to $m$ nearest neighbors of an active node are active, this node becomes inactive due to external causes with probability $r\,\mathrm{d}t$, and (iii) inactive states spontaneously become active with probability $q\,\mathrm{d}t$ (if inactive due to process (i)) or probability $q'\,\mathrm{d}t$ (if inactive due to process (ii)). The threshold $m$ determines if the neighborhood of a node is able to induce a transition from an active to an inactive state. A low value of $m$ describes the case where a large number of inactive neighbors is required to sustain the spread of failure, an innovation or opinion. The resulting dynamics of processes (i–iii) is particularly rich owing to the coexistence of limit cycles, random phase switching, and hysteresis [16, 18]. Before further discussing these phenomena, I will give an overview of the corresponding mean-field equations.

Let $a(t) \in [0, 1]$ and $z(t) = 1 - a(t)$ denote the total fractions of *inactive* and *active* nodes, respectively. The fraction of inactive nodes is the sum of the fractions of nodes $u_{\mathrm{spon}}(t)$ and $u_{\mathrm{ind}}(t)$ that failed spontaneously and in an induced manner (i.e., $a(t) = u_{\mathrm{spon}}(t) + u_{\mathrm{ind}}(t)$). The total fraction of inactive nodes in the stationary state is $a_{\mathrm{st}}$. For the derivation of the mean-field rate equations, I assume perfect mixing and first concentrate on the spontaneous dynamics:

$$\frac{du_{\mathrm{spon}}(t)}{dt} = p\,(1 - a(t)) - q u_{\mathrm{spon}}(t)\,, \tag{3}$$

where the first term accounts for spontaneous "failure" with rate $p$ and the second term describes recovery of spontaneously failed nodes with rate $q$.

The probability that a node of degree $k$ is located in a neighborhood where the number of active neighbors is smaller than or equal to $m$ is $E_k = \sum_{j=0}^{m} \binom{k}{k-j} a^{k-j}(1 - a)^j$. Consequently, the time evolution of nodes that become inactive in an induced way is described by:

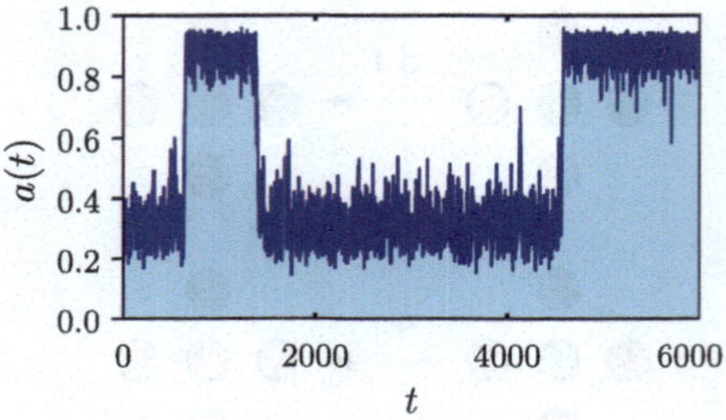


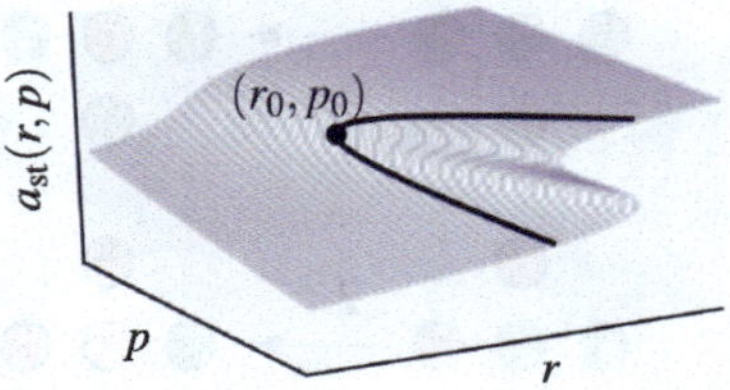


**Fig. 2 Phase-switching and phase space of the general contagion model.** (left) The general contagion model exhibits phase-switching for $p = 0.24$, $r = 10$, $q = q' = 1$ in a regular random graph with $N = 256$ nodes and $k = 4$. (right) The phase space for $k = 2, m = 0, q = q' = 1$ (mean-field) is related to Schlögl's second model [27], cusp catastrophes, and imperfect bifurcations [17, 19]. Two spinodals (black solid lines) enclose the hysteresis region where two states coexist (low density and high density failure phases). These bifurcation lines merge at the cusp point $(r_0, p_0) = (27/8, 1/8)$

$$\frac{du_{\text{ind}}(t)}{dt} = r \sum_k f_k E_k \left(1 - a(t)\right) - q' u_{\text{ind}}(t)\,, \tag{4}$$

where $f_k$ is the degree distribution. The first term describes transitions of nodes from an active to an inactive state with rate $r$ if their neighborhoods contain a sufficient number of inactive nodes, and the second term accounts for recovery of these nodes with rate $q'$. An interesting feature of the general contagion model is the possibility of phase-switching between two stationary states as illustrated in Fig. 2 (left) for a regular random graph with $N = 256$ nodes. The switching phenomenon is a consequence of the occurrence of two fixed points that are stable in the thermodynamic limit. However, fluctuations in finite networks allow the dynamics to stochastically switch between the two coexisting states. These unpredictable and potentially unintended switching effects are restricted to a bifurcation region as shown in Fig. 2 (right). Two bifurcation lines enclose this region and merge at the *cusp point*, where [17]

$$a_0(k, m) = \frac{k - 1 - m}{k + 1}\,, \tag{5}$$

$$r_0(k, m) = \frac{1}{S(a_0) + S'(a_0)(1 - a_0)}\,, \tag{6}$$

$$p_0(k, m) = \frac{S'(a_0)a_0 - S(a_0)}{S(a_0) + S'(a_0)(1 - a_0)}\,, \tag{7}$$

with $S(a) = (1 - a)E_k$. At the cusp point, the (mean-field) fraction of failed nodes scales as $a(r) \propto r^{1/3}$ and $a(p) \propto p^{1/3}$ [17]. Hysteresis effects within the bifurcation region lead to a path dependence of the dynamics—following a closed trajectory in the phase space that crosses this region may lead to a final state that differs substantially from the initial one. The smaller the hysteresis region, the less likely it is for the

**Table 1** Examples of models and processes that are related to the classes (a–c)

| (a) $m = k$ | (b) $m = k - 1$ | (c) $m < k - 1$ |
|---|---|---|
| Exogenous factors influencing adoption of innovations [28]* | Schlögl I [27, 30] | Schlögl II [27, 30]* |
| Social response to exogenous factors [29]† | Contact process [6, 7, 31]* | Quadratic contact process [26]* |
| | SIS model [1, 4]* | General contact process [35]* |
| | Reggeon field theory [32]* | Behavioral adoption [36]† |
| | Directed percolation [33]* | Threshold models of complex contagions [9, 12, 14, 15, 37–39]*† and coordination games [13]† |
| | Bass model [28, 34]* | |

*Exact mean-field correspondence
†Phenomenological correspondence

system to end up in this uncontrollable situation. Interestingly, the Euclidean lattice is characterized by an extremely narrow metastable domain compared to random networks [16].

Despite the general formulation of processes (i–iii) and the resulting rich dynamics resulting, the stationary behavior of the general contagion model is characterized by only three distinct classes. For a regular network with degree $k$ the model's critical behavior can be classified as follows: (a) a parameter space characterized by purely spontaneous dynamics if $m = k$, (b) contact process, SIS or Schlögel's first model's dynamics [6, 7] for $m = k - 1$, and (c) cusp catastrophes [19, 21] if $m < k - 1$ [17]. In particular, the occurrence of cusp catastrophes (see inset in Fig. 2 (right)) implies the possibility of abrupt transitions for all $m < k - 1$. In other words, if the contact to two or more (inactive) neighbors is necessary to sustain spreading, the phase space always corresponds to a cusp catastrophe [17]. More general degree distributions $f_k$ can be treated as a weighted sum over regular networks with different degrees [17]. Table 1 summarizes some common spreading models and their relation to classes (a–c).

## 3 Campaign Dynamics

One possible application of the general contagion model is to model activist-voter interactions in election campaigns with two competing parties [24, 25]. Political activists $A^+$ and $B^+$ respectively target persuadable nodes $B^0$ and $A^0$ and their $k$ neighbors (see Fig. 3 (left)). During a persuasion attempt, activists try convince these $k + 1$ nodes. Let $a^0$ and $b^0 = 1 - a^0$ be the fractions of persuadable nodes in states $A^0$ and $B^0$, respectively. Since $\dot{b}^0 = -\dot{a}^0$, the dynamics of $b^0$ is determined by the time evolution of $a^0$ and vice versa:

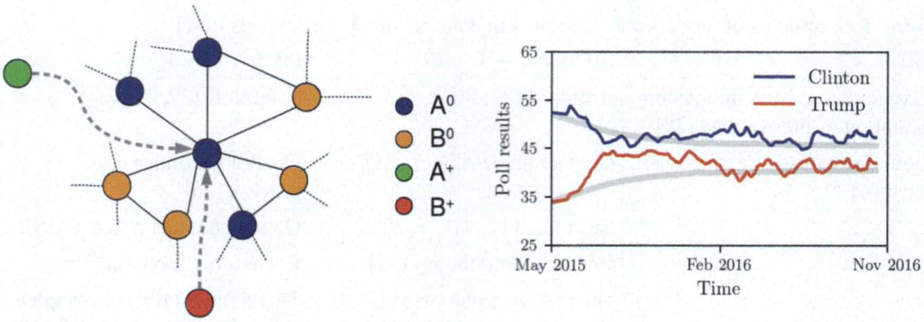


**Fig. 3 Activist-voter interactions and campaign dynamics.** The left panel shows an illustration of a campaign model where nodes and their corresponding edges are represented by colored circles and black lines respectively. Activists $A^+$ and $B^+$ can target neighborhoods with a sufficiently large number of persuadable voters $A^0$ and $B^0$ (see Eq. (8)). Blue and red curves in the right panel represent the poll results for Hillary Clinton and Donald Trump during the 2016 US presidential election campaign. We compare the poll results with simulations (grey lines). See Ref. [24], for further details

$$\begin{aligned}\dot{a^0}(t) = {} & \underbrace{\rho_A a^+ \frac{\sum_{j=\lceil \tau_A (k+1)\rceil}^{k+1} j \binom{k+1}{j} b^0(t)^j a^0(t)^{k+1-j}}{\sum_{j=\lceil \tau_A (k+1)\rceil}^{k+1} \binom{k+1}{j} b^0(t)^j a^0(t)^{k+1-j}}}_{\text{gain: } f_{B^0 \to A^0}(t)} \\ & - \underbrace{\rho_B b^+ \frac{\sum_{j=\lceil \tau_B (k+1)\rceil}^{k} j \binom{k+1}{j} a^0(t)^j b^0(t)^{k+1-j}}{\sum_{j=\lceil \tau_B (k+1)\rceil}^{k+1} \binom{k+1}{j} a^0(t)^j b^0(t)^{k+1-j}}}_{\text{loss: } f_{A^0 \to B^0}(t)},\end{aligned} \tag{8}$$

Note the similarity of Eq. (8) to the threshold model of Eq. (4). The first term in Eq. (8) describes the gain (loss) of "voters" and the second the loss (gain) of "voters" for party $A$ ($B$). The prefactor $\rho_A$ ($\rho_B$) is the probability that a voter changes the opinion as result of an activist persuasion attempt. Moreover, the proportion of $A$ ($B$) activists in the population is $a^+$ ($b^+$). The two fractions represent the expected numbers of voters that are contacted by corresponding activists. According to the lowest index $\lceil \tau_A (k+1)\rceil$ ($\lceil \tau_B (k+1)\rceil$) in the sums, the parameter $\tau_A$ ($\tau_B$) constrains the set of nodes with the neighborhood the activists will visit. One can interpret $\tau_A$ and $\tau_B$ as different technological advantages that result from different *microtargeting* approaches [25]. According to Eq. (8), an activist $A^+$ ($B^+$) will only visit neighborhoods with at least $\lceil \tau_A (k+1)\rceil$ ($\lceil \tau_B (k+1)\rceil$) nodes in state $B^0$ ($A^0$). An application of Eq. (8) to the 2016 US presidential election campaign is shown in Fig. 3 (right).

## 4 Conclusions and Outlook

In this chapter, I briefly described different models for simple and complex contagion phenomena and outlined that they can be treated as a special case of a more general (threshold-based) formulation of contagion dynamics. The stationary behavior of the general contagion model is characterized by only three distinct classes: (a) purely spontaneous dynamics, (b) contact process dynamics, and (c) cusp catastrophes. I discussed the application of the general contagion model in the context of activist-voter dynamics during election campaigns, where a threshold can be seen as a model for different "technological" advantages of campaign groups. In addition to the characterization of the stationary behavior of the general contagion model, future studies may focus on corresponding ageing (i.e., universal dynamical) effects [40, 41]. Furthermore, it would be interesting to connect these models to the research on societal polarization phenomena [42–44].

**Acknowledgements** The author acknowledges financial support from the Swiss National Fund grant *Multispecies interacting stochastic systems in biology*.

## References

1. Keeling MJ, Rohani P (2008) Modeling infectious diseases in humans and animals. Princeton University Press, Princeton, New Jersey
2. Hethcote HW (2000) SIAM Rev 42(4):599
3. Kermack WO, McKendrick AG (1927) Proc R Soc Ser A 115(772):700
4. Pastor-Satorras R, Castellano C, Van Mieghem P, Vespignani A (2015) Rev Mod Phys 87(3):925
5. Böttcher L, Woolley-Meza O, Goles E, Helbing D, Herrmann HJ (2016) Phys Rev E 93(4):042315
6. Marro J, Dickman R (2005) Nonequilibrium phase transitions in lattice models. Cambridge University Press, Cambridge, United Kingdom
7. Henkel M, Hinrichsen H, Lübeck S (2008) Non-equilibrium phase transitions volume I: Absorbing phase transitions. Springer
8. Coleman J, Katz E, Menzel H (1957) Sociometry 20(4):253
9. Rogers EM (2010) Diffusion of innovations. Simon and Schuster
10. Chwe MSY (1999) Am J Soc 105(1):128
11. Böttcher L, Woolley-Meza O, Brockmann D (2017) PloS One 12(5):e0178062
12. Leskovec J, Adamic LA, Huberman BA (2007) ACM Trans Web (TWEB) 1(1):5
13. Easley D, Kleinberg J (2010) Networks, crowds, and markets: Reasoning about a highly connected world. Cambridge University Press, Cambridge, United Kingdom
14. Granovetter M (1978) Am J Soc, pp 1420–1443
15. Centola D, Macy M (2007) Am J Soc 113(3):702
16. Böttcher L, Luković M, Nagler J, Havlin S, Herrmann H (2017) Sci Rep 7:41729
17. Böttcher L, Nagler J, Herrmann HJ (2017) Phys Rev Lett 118:088301
18. Majdandzic A, Podobnik B, Buldyrev SV, Kenett DY, Havlin S, Stanley HE (2014) Nat Phys 10:34
19. Strogatz SH (2014) Nonlinear dynamics and chaos: with applications to physics, biology, chemistry, and engineering. Westview Press, Boulder, Colorado
20. Ludwig D, Jones DD, Holling CS et al (1978) J Anim Ecol 47(1):315
21. Zeeman EC (1979) Catastrophe theory. Springer, Berlin/Heidelberg, Germany

22. Helbing D (2013) Nature 497(7447):51
23. Böttcher L, Woolley-Meza O, Araújo NA, Herrmann HJ, Helbing D (2015) Sci Rep 5:16571
24. Böttcher L, Herrmann HJ, Gersbach H (2018) PloS One 13(3):e0193199
25. Hoferer M, Böttcher L, Herrmann HJ, Gersbach H (2020) Physica A 538:122795
26. Durrett R (1999) SIAM Rev 41(4):677
27. Grassberger P (1982) Z Phys B 47(365–374)
28. Ghanbarnejad F, Gerlach M, Miotto JM, Altmann EG (2014) J R Soc Interface 11(101):20141044
29. Crane R, Sornette D (2008) Proc Natl Acad Sci 105(41):15649
30. Schlögl F (1972) Zeitschrift für Physik 253(2):147
31. Harris TE (1974) Contact interactions on a lattice. Ann Probab 2(6):969–988
32. Grassberger P, De La Torre A (1979) Ann Phys 122(2):373
33. Cardy JL, Sugar R (1980) J Phys A 13:L423
34. Bass FM (1969) Manag Sci 15(5):215
35. Tomé T, De Oliveira MJ (2015) Stochastic dynamics and irreversibility. Springer, Berlin/Heidelberg, Germany
36. Centola D (2010) Science 329(5996):1194
37. Watts DJ (2002) Proc Natl Acad Sci 99(9):5766
38. López-Pintado D (2008) Game Econ Behav 62:573
39. Gleeson JP (2013) Phys Rev X 3(2):021004
40. Böttcher L, Herrmann HJ, Henkel M (2018) J Phys A 51(12):125003
41. Richter P, Henkel M, Böttcher L (2020) Aging and equilibration in bistable contagion dynamics. Phys Rev E 102(4):042308. https://doi.org/10.1103/PhysRevE.102.042308
42. Xu S, Böttcher L, Chou T (2020) Diversity in biology: definitions, quantification and models. Phys Biol 17(3):031001
43. Böttcher L, Montealegre P, Goles E, Gersbach H (2019) Physica A, p 123713
44. Böttcher L, Gersbach H (2020) The great divide: drivers of polarization in the US public. EPJ Data Sci 9(1):1–13. https://doi.org/10.1140/epjds/s13688-020-00249-4

# On Birth, Death and Symmetry: Some Principles of Complex Ecological Systems

**Pablo A. Marquet**

**Abstract** For quite apparent reasons, much of the phenomenology of what we call life can be described by a birth and death process. Less apparent, however, is how symmetry comes into play. In this chapter we briefly summarize some of the finding that come about by putting together birth and death processes in the context of a symmetric system, or one where its components have identical per capita rates of birth and death, being in practical term identical. We will illustrate this process of birth and death in symmetric systems using a one dimensional diffusion model to account for the proportional abundance ecological entities. We show that the first principles of birth death and symmetry are fundamental to our understanding of ecological dynamics and the emergence of patterns in ecological systems. They represent first principles, that can be useful to generating theory, and their integration, in ecology.

## 1 Introduction

Upon his return from Spain in 1921 Jorge Luis Borges publishes his first book, entitled Fervor de Buenos Aires, a collection of poems that, he will later state, contained the basic ideas that he repeated ever after. Indeed, in the prologue to his Complete Works [2] he says that the man that wrote that first book in 1923 and the man that is now correcting it for the compilation were essentially the same men and adds..." we both disbelieve in the failure and success of literary schools and their dogmas; we are both devotee of Schopenhauer, of Stevenson and of Whitman. For me, Fervor de Buenos Aires prefigures everything I would do later." Life, Borges's included,

P. A. Marquet (✉)
Departamento de Ecología, Pontificia Universidad Católica de Chile, Alameda Bernado O'Higgins 340, 8331150 Santiago, Chile

Instituto de Ecología y Biodiversidad (IEB), Las Palmeras 3425, Santiago, Chile

The Santa Fe Institute, Santa Fe, NM 87501, USA

Instituto de Sistemas Complejos de Valparaíso (ISCV), Artillería 470, Cerro Artillería, Valparaíso, Chile
e-mail: pmarquet@bio.puc.cl



A. Adamatzky (ed.), *Automata and Complexity*, Emergence, Complexity and Computation 42, https://doi.org/10.1007/978-3-030-92551-2_8

is the quintessential phenomenon, the quintessential question and mistery. Borges captures all these features in his poem La Recoleta, which alludes to the Paris style neighborhood in Buenos Aires famous for the Recoleta Cemetery. He says:

...only life exists.
Space and time are forms of it;
they are magical instruments of the soul, (1)
and when life is extinguished,
space, time and death will be extinguished with it...

How it comes that Borges literary life may have been prefigured in his first book? thus making him to be eternally returning to the same concepts in an endless recurrence of the same theme like Golberg's variations or an ergodic system that fills the whole phase space of narrative with the same basic themes? Was this what made Borges different? Is all life the same recurrent dynamics of the same, albeit different, themes?

Part of the answers to these questions are provided by Borges himself. Indeed Borges warns us at the beginning of his book, in a sort of prologue or introduction, saying:

TO WHOMEVER WOULD READ
If the pages of this book consent to a happy
verse, forgive me the reader the discourtesy of having (2)
usurped I, previously. Our nothingness differs little;
it is trivial and fortuitous that
you were the reader of these exercises, and I the writer.

While poem (1) deals with the issue of birth and death, poem (2) deals with ergodicity in the sense of the equiprobability of visiting all possible available state to a system (being the poet or the reader for example) and also the fundamental symmetry and equality of all possible trajectories. These two poems are in fact poetic theorems that capture three fundamental first principles of biology, that is birth, death and symmetry, which I will elaborate on in the rest of this contribution.

## 2 On Symmetry

Before talking about symmetry and birth and death processes it is important to provide a more precise definition of our subject of study, and the special stance from where we will try to analyze it. In general terms, life is a different state of matter, to which physicist usually refer to as activated matter, but I rather use the word adaptive matter [15], as it interacts with its environment in a fast and adaptive way, establishing a "dialog" where both parties (the being and its environment) modify each other through natural selection and niche construction respectively. Further, the nature of the biological game is to transform the environment, in terms on the materials and energy it contains, into copies of itself, the faster you can do that, the higher your fitness. It is important to bear in mind that what we call the environment, it is usually understood as composed of a biotic interaction and a abiotic components associated to physical and chemical processes with which whatever the being or focal system we consider, interacts. For a long part of its history, ecology have tried to understand the relative importance of biotic and abiotic components and their interactions in affecting species and populations [1, 21].

One of the most earlier realizations regarding interactions among living entities (e.g., biotic interactions) was that entities too alike, in terms of habits and resource use where unlikely to coexist in the long-run, as shown by Lotka and Volterra [10, 23], because even minute differences in growth rate would amplify and one species would always win at the expenses of the other, and since species are different in many subtle ways the difference is axiomatic, sort to speak (see Hardin 1970). This result became to be known and the "competitive exclusion principle", which is a mathematical truth that provides a standars against which to compare reality [5]. The fascinating fact is that when you compare it to reality one finds that coexistence is the rule! eventhough we known than more than 99% of all species that have existed are extinct. Ever since this principle was born (sensu [5]), scientists have been pondering the question of what are the processes that foster coexistence. Modern coexistence theory (e.g., [3]) recognize several biotic and abiotic mechanisms that reduce fitness differences among species and/or foster differences in species resource use through, for example, niche partitioning (by eating different food items or in different places or times). Among fitness equalizing mechanisms are tradeoffs between dispersal capacity and competitive ability, which is usually associated with a fugitive species or one that can move quickly and exploit the resources found in areas where the competitive dominant species have not yet reached (see the recent stochastic model of Tejo et al. [20] for an example), this differences or asymmetry in the scale at which organisms perceive and interact with their environment is paradoxically the source of fitness symmetry that allows them to coexist, by retarding competitive exclusion [7]. Thus, species become in practice identical in terms of their percapita rates of birth and death and thus a symmetric world emerges and the fitness landscape becomes flat and, as Borges stated in Poem (2), there is a fundamental symmetry between entities performing different roles; between writer and reader and, we can add, among species within trophic levels. Organisms satisfying this symmetry would according to Van Valen [21] constitute a "homogeneous" group of entities inhabiting the same adaptive zone, and showing a constant extinction rate no matter the age or duration of this homogeneous group in the fossil record, implying that ([21], p. 16) "The effective environment of the members of any homogenous group of organisms deteriorates at a stochastically constant rate." This implies that, irrespective of geological duration, young and old taxa go extinct at the same rate, implying that the environment deteriorates in a similar way for every species within the homogeneous group and that, irrespective of longevity, they have to keep adapting to stay in the same place. This evidence gave raise to the now famous Red Queen hypothesis, which ([18], p. 612) implies that "...all taxa are running on a treadmill powered by an environment which deteriorates at a stochastically constant rate. The result is that an ancient taxon is no better adapted than a younger one; it has just been running in place longer." The name Red Queen is taken from a character in Alice's adventures in wonderland, a 1865 novel by Lewis Carroll -pen name of british mathematician Charles Dogson-, which has to keep running in order to stay in the same place. Interestingly, the reason for this constancy, is heterogeneity; taxa differ in many traits correlated with extinction, which seems to be present in any adaptive zone, such as variations in size, density, and distribution (see Van Valen 1978 cited in McCune [18]), and, the same as with the fitness equalizing mechanisms fostering coexistence, differences among species level traits render them similar regarding to extinction rates, irrespective of duration. So, paradoxically, it seems that symmetry with regard to fitness among species, requires asymmetry or differentiation in other traits that may affect fitness.

## *2.1 Symmetric Models*

One of the most fundamental models in ecology was developed in order to understand changes in species richness in a community composed by S species. The model is relatively simple and coarse, as it includes processes that tend to decrease the number of species (i.e. extinction) and those that tend to increase it (i.e. colonization and speciation). This stochastic model was inspired by empirical regularities associated to the observed increase in the number of species found in islands, as the area of the island increases and to the "Equilibrium model of island biogeography" [11, 12], which tries to understand the driver of species richness in insular or insular-like habitats. MacArthur and Wilson's stochastic model for the number of species found on a focal island, corresponds to a birth-death process whose time evolution follows the master equation:

$$\frac{dP_s(t)}{dt} = P_{s-1}(t)\lambda_{s-1} + P_{s+1}(t)\mu_{s+1} - P_s(t)[\lambda_s + \mu_s], \qquad (3)$$

for $s = 0, 1, \ldots, S$, where $P_s$ is the probability of observing $s$ species in a focal island, $S$ is the *pool* of species, $\lambda_s$ is the birth (i.e. colonization) rate associated to the transition from $s$ to $s + 1$ species and $\mu_s$ is the death (i.e. extinction) rate associated to the transition between $s$ to $s - 1$ species.

It is quite puzzling that MacArthur and Wilson, did not provide a formal solution for this equation, which would have entailed solving for the invariant density of the stochastic process, which would have been the Species Richness Distribution (SRD, see Marquet et al. [17]). The invariant distribution was obtained by Goel and Richter-Dyn [4]. To solve Eq. (3) one needs to specify an initial condition $P_0$ and some boundary conditions, which in this case can be deduced by noticing that when there are no species present in the island, $\mu_0 = 0$ and $\lambda_0 \neq 0$, and when the island is saturated with species, that is, the number of species is equal to the number of species in the pool $K$, then $\mu_K \neq 0$ and $\lambda_K = 0$. Thus, the stochastic process associated with the number of species is confined between the two reflecting states 0 and $K$. Now the task is to define the functional form of the extinction and colonization rates. Goel and Richter-Dyn [4] considered two scenarios for these rates. The first scenario reasons that because the area and hence the amount of resources present on the island are fixed, as the number of species in the island increases, the average population size of any given species decreases, hence the probability of extinction of a species can be hypothesized to monotonically increase with the number of species present in the island, hence $\mu = \mu s$. Similarly, it is reasonable to assume that the probability for a new species to become present in the island depends upon the species already present there, because the more species that become established on the island the lower the chances that a new immigrant individual will belong to a new species, hence the immigration rate should decrease monotonically as the number of species established on the island increases $\lambda_s = \lambda(K - \lambda)$, and $\lambda_K = 0$ when the number of species present on the island is equal to the number of species in the pool ($K$). Under these assumption the invariant distribution of the process is a binomial distribution.

The symmetric model of MacArthur and Wilson [11] was very influential and generated several applications and further developments (e.g., [9]) and provided the cornerstone of the Island Biogeography Theory. One of those developments, exposed one of its major weaknesses. I am referring to The Unified Neutral Theory of Biodiversity and Biogeography (UNTB) by Hubbell [6]. This theory basically takes MacArthur and Wilson's theory to the level of individuals and proposes that all individual are equal (or become equal as we discussed above) in their per capita rate of birth and death. Then, we can write the individual level equivalent of Eq. (3) (see [22])

$$\frac{dP_{n,k}(t)}{dt} = P_{n-1,k}(t)b_{n-1,k} + P_{n+1,k}(t)d_{n+1,k} - P_{n,k}(t)[b_{n,k} + d_{n,k}], \tag{4}$$

where $n$ corresponds to the number of individuals and $b$ is the birth rate and $d$ the death rate. If this are assumed density independent (i.e., if $b_n = nb$ and $d_n = dn$) then one obtains an equilibrium distribution that is the called the Fisher's Log Series ([8, 22]). Unfortunately, Eqs. (3) and (4) cannot be true at the same time, since given that each species is now represented by a given number of individuals, the transition between number of species are no longer markovian [17]. There are two solutions to this problem. The first or the weak one, it to make the number of species to become an observable on the abundance distribution, which corresponds to the state defining the system. If this is done we get a binomial distribution for the species richness. A more radical solution implies abandoning master equations and move to diffusion approximations. This entails to work on proportional abundances and richness or frequencies instead of numbers of individuals or species. As pointed out in Marquet et al. [16] this is the approximation taken by population geneticists back in the 30 s (e.g., [24]) where it was shown that the equilibrium frequency of a gene in a local population follows a Beta distribution.

## 2.2 Diffusion Approximation

Under this approximation we envision ecological systems as open system whose boundaries are defined by the observer. The system could be, for example, a 50 ha plot in a tropical forest or a $1\text{m}^2$ plot in the intertidal. What is important is to realize that once the observer defines the spatial scale of the system, it defines a boundary or an inside and an outside, where the focal system is embedded. We call this observer defined scale the focal community that is embedded into a bath or environment with which it interacts. The focal community dynamics is driven by birth and death processes and by immigration from the outside. Indeed the spatial scale of analysis is to some extent dictated by which is the dominant process adding new entities to the focal community; immigrations of individuals from species not yet found in the focal community but somewhere else in the bath, or new species arising through speciation within the focal community. If the later is the dominant process, then the spatial scale is likely to be large, since all species in the potential pool are already present and the only way a new species can arrive would be through speciation. Similarly, the processes that remove individuals and species from the focal community include death and emigration towards the bath or environment.

To model the dynamics of this focal community we used the diffusion approximation of birth and death processes independent of a focal community size $J$ (see details in Marquet et al. [16]). By community size we mean the total number of individuals regardless of species identity.

Let $N_J(t)$ denote the number of living individuals of a given species within a focal community of size $J$, at time $t \geq 0$ (so that $N_J(t)$ is less or equal to $J$ for all $t$). This is assumed to be a birth and death process, with transition matrix $P(t) = (P_{n,m}(t);\ n, m = 0, \ldots, J)$ ($n$ and $m$ denotes the number of individuals). For a small time increment $h$, this matrix satisfies as $h \to 0$ for $n \geq 0$

$$P_{n,n+1}(h) = B_J(n)h + o(h), \ \text{for} n \geq 0, \tag{5}$$
$$P_{n,n-1}(h) = D_J(n)h + o(h), \ \text{for} n \geq 1, \tag{6}$$
$$P_{n,n}(h) = 1 - (B_J(n) + D_J(n))h, \ \text{for} n \geq 0, \tag{7}$$
$$P_{n,m}(0) = \delta_{n,m}, \tag{8}$$

where $B_J(n)$ and $D_J(n)$ are the birth and death rates, respectively, $D_J(0) = 0$, $B_J(0) > 0$, $\delta_{n,m}$ is the customary Kronecker delta, and $o(h)$ denotes the Landau-symbol, which satisfies $\lim_{h\to 0} \frac{o(h)}{h} = 0$. Here, in addition, we assume that these rates are decomposed as follows

$$B_J(n) = b_J(n) + c_J(n) \tag{9}$$
$$D_J(n) = d_J(n) + c_J(n). \tag{10}$$

The terms $b_J$ and $d_J$ represent birth and death rates in the focal community, respectively, which will be asymptotically independent of $J$, while $c_J$ takes into account the variations on the above rates due to the interaction between the focal system and the environment wherein it is embedded, proper to an open system approach, and which as we will see later are associated to the noise term of the corresponding stochastic differential equation. Since we are interested in proportions $n/J$, we introduce the variable $x = n/J$, which takes values in $\{0, 1/J, 2/J, \ldots, 1\}$, and analyze the behavior of the system as the size of the population grows indefinitely: $J \to \infty$. At this stage it is important to state meaningful hypotheses for the previous rates for large $J$, as all changes of scales in the dynamics of the open system are driven by this community size.

We first assume that $b_j$ and $d_J$ will lead, respectively, to the $J$-invariant (or endogenous) birth and death rates of the focal system, that satisfy

$$\lim_{J\to\infty} b_J(xJ) = b(x); \ \lim_{J\to\infty} d_J(xJ) = d(x), \ (x \in [0, 1]). \tag{11}$$

On the contrary, the rate $c_J$, should vary significantly with $J$, however, we require that it satisfies

$$\lim_{J\to\infty} \frac{c_J(xJ)}{J} = c(x), \ (x \in [0, 1]). \tag{12}$$

We can now define the stochastic process $Z_J = (Z_J(t) = N(tJ)/J; \ t \geq 0)$ that we call the stochastic proportional abundance. This family of processes has a limit $Z = (Z(t); \ t \geq 0)$ as $J \to \infty$, that corresponds to a diffusion process satisfying the stochastic differential equation [16]

$$dZ(t) = (b(Z(t)) - d(Z(t)))dt + \sqrt{2c(Z(t))}dW(t), \tag{13}$$

where $W(t)$ denotes a Brownian motion.

It is worth noticing that the process $Z_J = (Z_J(t); \ t \geq 0)$ converges in distribution towards a diffusion process $Z = (Z(t); \ t \geq 0)$ as proven in Rebolledo [19], and so, any continuous functional $F(Z_J)$ of the trajectory of $Z_J$ converges in distribution to $F(Z)$.

Correspondingly, the Fokker-Planck equation associated with the probability density $\rho_t(x)$ of $Z(t)$, is given by

$$\frac{\partial}{\partial t}\rho_t(x) = \frac{\partial^2}{\partial x^2}(c(x)\rho_t(x)) - \frac{\partial}{\partial x}([b(x) - d(x)]\rho_t(x)), \tag{14}$$

with the additional condition that $\int_{\rho t}(x)dx = 1$. The stationary solution $\rho_\infty$ is determined as the solution to the equation

$$\frac{\partial^2}{\partial x^2}(c(x)\rho_\infty(x)) - \frac{\partial}{\partial x}([b(x) - d(x)]\rho_\infty(x)) = 0 \tag{15}$$

In order to find the stationary distribution we need to make a hypothesis for each of the rates $b(x)$, $d(x)$ and $c(x)$, the simplest ones are that

$$b(x) = b_0 + b_1 x \tag{16}$$
$$d(x) = d_0 + d_1 x \tag{17}$$
$$c(x) = \gamma x(1 - x), \tag{18}$$

where $b_i$, $d_i$, $(i = 0, 1)$, and $\gamma$ are positive constants. Under these hypotheses the stationary solution takes the form of a typical Beta distribution

$$\rho_\infty(x) = \frac{\Gamma(\alpha + \beta)}{\Gamma(\alpha)\Gamma(\beta)} x^{\alpha-1}(1 - x)^{\beta-1}. \tag{19}$$

which under the assumption of neutrality becomes [16]

$$\rho_\infty(x) = \frac{1}{B(\alpha, \alpha(S - 1))} x^{\alpha-1}(1 - x)^{\alpha(S-1)-1}, \tag{20}$$

where $B(\alpha, \alpha(S - 1)) = \int_0^1 x^{\alpha-1}(1 - x)^{\alpha(S-1)-1}$ (normalization constant) and $\alpha = \frac{m}{S(1-m)\lambda}$.

Thus, we have shown that diffusion approximations predict an equilibrium proportional abundance of species that also corresponds to the equilibrium distribution describing the frequencies of genes within local populations, showing that both proceses are deeply connected. Further, as shown in Marquet et al. [16] Eq. (20) has a strong empirical support, and does not have the problem of incompatibility shown for the master equation approach and thus can be applied to any level of organization from genes to individuals to species.

## 3 Concluding Remarks

The Beta distribution arise only if birth and death rates are additive, implying that entities are equal and independent. We know species are different, but what we have tried to show here is that to assume that they are, provides a good enough approximation since difference is associated to symmetry. To put it in a different way, different entities can be think of variations on same theme, as it is suggested by the existence of scaling or power law within ecological systems (e.g. [13]). The same as Borges was prefigured in his first poetry book, life may haven prefigured in its origin, everything we have been seen after that, are just variations and expansions on the same theme.

We showed that the first principles of birth death and symmetry are fundamental to our understanding of ecological dynamics and the emergence of patterns in ecology. They represent first principles, that can be useful to generating ecological theory using the language of mathematics [14]. But more fundamentally, their generality across levels of organization provide evidence that the dynamics of different levels may be more similar that we thought. Many biological phenomena remain to be understood using simple concepts such as birth, death and symmetry and the power of open systems analysis.

**Acknowledgements** The author acknowledges support from project FONDECYT 1200925. I would like to thanks my dear friend Rolando Rebolledo for introducing me to open system approaches, and teaching me the little math I can master about them. Evandro Ferrada, Mauricio Tejo and Cristobal Quiñinao have accompany me in this stochastic trip and I thank them for their patience, companionship and passion for understanding complex biological systems. Finally, my deepest appreciation to Eric Goles, a dear friend and partner in several scientific initiatives, whom I dedicate this essay in his 70th birthday. I tried to blend a bit of literature into my discussion as a token of appreciation to Eric, whose life has moved between science, arts, and the humanities.

## References

1. Barnosky AD (2001) Distinguishing the effects of the Red Queen and Court Jester on Miocene mammal evolution in the northern Rocky Mountains. J Vertebr Paleontol 21(1):172–185
2. Borges JL (1974) Jorge Luis Borges. Obras Completas I. Emecé Editores, Buenos Aires
3. Chesson P (2000) Mechanisms of maintenance of species diversity. Ann Rev Ecol Syst 31(1):343–366
4. Goel NS, Richter-Dyn N (1974) Stochastic models in biology. Elsevier
5. Hardin G (1960) The competitive exclusion principle. Science 131(3409):1292–1297
6. Hubbell SP (2001) The unified neutral theory of biodiversity and biogeography (MPB-32), vol 32. Princeton University Press
7. Hubbell SP (2005) Neutral theory in community ecology and the hypothesis of functional equivalence. Funct Ecol 19(1):166–172
8. Kendall DG (1948) On some modes of population growth leading to RA Fisher's logarithmic series distribution. Biometrika 35(1/2):6–15
9. Losos JB, Ricklefs RE (eds) (2009) The theory of island biogeography revisited. Princeton University Press
10. Lotka AJ (1920) Analytical note on certain rhythmic relations in organic systems. Proc Natl Acad Sci 6(7):410–415
11. MacArthur RH, Wilson EO (1963) An equilibrium theory of insular zoogeography. Evolution, pp 373–387
12. MacArthur RH, Wilson EO (1967) The theory of island biogeography, vol 1. Princeton University Press
13. Marquet PA, Quiñones RA, Abades S, Labra F, Tognelli M, Arim M, Rivadeneira M (2005) Scaling and power-laws in ecological systems. J Exp Biol 208(9):1749–1769
14. Marquet PA, Allen AP, Brown JH, Dunne JA, Enquist BJ, Gillooly JF, Gowaty PA, Green JL, Harte J, Hubbell SP, O'Dwyer J, Okie JG, Ostling A, Ritchie M, Storch D, West GB (2014) On theory in ecology. BioScience 64:701–710
15. Marquet PA (2017) Integrating macroecology through a statistical mechanics of adaptive matter. Proc Natl Acad Sci 114(40):10523–10525
16. Marquet PA, Espinoza G, Abades SR, Ganz A, Rebolledo R (2017) On the proportional abundance of species: integrating population genetics and community ecology. Sci Rep 7(1):1–10
17. Marquet PA, Tejo M, Rebolledo R (2020). In: Dobson A, Holt RD, Tilman D (eds) What is the species richness distribution? Princeton University Press, pp 177–188
18. McCune AR (1982) On the fallacy of constant extinction rates. Evolution, pp 610–614
19. Rebolledo R (1979).La méthode des martingales appliquée à l'étude de la convergence en loi de processus. Société Mathématique de France 62

20. Tejo M, Quiñinao C, Rebolledo R, Marquet PA (2021) Coexistence, dispersal and spatial structure in metacommunities: a stochastic model approach. Theor Ecol. https://doi.org/10.1007/s12080-020-00496-1
21. Van Valen L (1973) A new evolutionary law. Evol Theory 1:1–30
22. Volkov I, Banavar JR, Hubbell SP, Maritan A (2003) Neutral theory and relative species abundance in ecology. Nature 424(6952):1035–1037
23. Volterra V (1926) Fluctuations in the abundance of a species considered mathematically. Nature 118:558–560. https://doi.org/10.1038/118558a0
24. Wright S (1931) Evolution in Mendelian populations. Genetics 16(2):97–159

# A Spectral Outlook on the Elementary Cellular Automata with Cyclic Configurations and Block-Sequential Asynchronous Updates

**P. P. Balbi, G. S. Etchebehere, and E. L. P. Ruivo**

**Abstract** Spectral analysis has been used previously in the literature to analyse the space-time diagrams of the elementary cellular automata in the frequency domain, as it allows for a distinct perspective on the dynamics and limit behaviour of the rules. Asynchronous cellular automata are variants of cellular automata whose cells have their states updated at different time steps, either stochastically or deterministically. Here, by relying on the latter – the block-sequential update scheme – the entire elementary rule space is computationally probed, over cyclic configurations, according to their discrete Fourier spectra. The rule space is accounted for in its compact description, which became possible after we introduced a characterisation of dynamically equivalent rules under asynchronous updates. Since the number of possible update schemes depends on the configuration size, a reasoned choice had to be made in order to define an appropriate set of updates, which led us to three distinct families, each one with ten updates. Analysis of the spectra obtained was then carried out by means of a proposed measure of asynchronism of the updates, and by grouping the rules into similarity classes, according to the Fourier spectra entailed by each update.

---

P. P. Balbi (✉)
Universidade Presbiteriana Mackenzie, Faculdade de Computação e Informática (FCI) and Programa de Pós-Graduação em Engenharia Elétrica e Computação (PPGEEC), Rua da Consolação, 930, Consolação, 01302-907 São, Paulo, SP, Brazil
e-mail: pedrob@mackenzie.br

G. S. Etchebehere
Universidade Presbiteriana Mackenzie—PPGEEC, Rua da Consolação, 930, Consolação, 01302-907, São Paulo, SP, Brazil

E. L. P. Ruivo
Universidade Presbiteriana Mackenzie—FCI, Rua da Consolação, 930, Consolação, 01302-907, São Paulo, SP, Brazil
e-mail: eurico.ruivo@mackenzie.br

 
A. Adamatzky (ed.), *Automata and Complexity*, Emergence, Complexity and Computation 42, https://doi.org/10.1007/978-3-030-92551-2_9

# 1 Introduction

Cellular automata (CAs) are fully discrete dynamical systems capable of universal computation, based on the concept that complexity may emerge from local actions, even simple ones. Structurally, they consist of a regular lattice, of arbitrary dimension and size, with predefined boundary conditions, and made up of finite state machines (the cells), each one with an identical pattern of connection with its neighbours. The dynamics of cellular automata derives from the action of a state transition rule, which locally defines the state of each cell according to the state of its neighbourhood, usually in a synchronous fashion [6, 14]. Although such a characterisation is the standard, any of its aspects may be modified, such as the update scheme, changing to asynchronous, stochastically or deterministically [4].

Here we rely on the deterministic asynchronism variant, by which the lattice is partitioned in cell blocks and each one is assigned a priority of being updated, with the relative orders remaining fixed throughout the space-time (or temporal, for short) evolution. This kind of asynchronism, known as *block-sequential* [1], has had an increasing attention in the literature, both as a theoretical object [2, 3, 7, 12, 13] and as new framework for modelling [17].

The study of the dynamical behaviour of cellular automata has been traditionally carried out by analysing aspects of the temporal evolution entailed by the rules over finite or infinite initial configurations. However, the possibility of analysing the rules in the frequency domain allows for a completely new perspective on the rules' dynamics. For this, the *Discrete Fourier Transform* provides the natural way for the transposition between the temporal to the frequency domains. Curiously, not much has happened along this line since the first developments in [16]; among them we should mention [8, 10, 11], all of them with standard, synchronous updates.

Most significantly for present purposes is the work discussed in [11], where the authors computationally analysed all elementary cellular automata rules by their Fourier spectra and managed to prove the existence of a partition of the space into classes of spectral equivalence, obtained via refinement of the partition of the dynamical equivalence classes. What we do in the present work is a first extension to the latter, this time analysing the same space with block-sequential updates. But due to the computational impossibility of an exhaustive analysis out of all possible asynchronous update schemes, three special families of update were defined, in a total of 30 update schemes. This led us to the characterisation of the elementary space in various spectral similarity classes, according to the distinct updates employed, whose intrinsic level of asynchrony we tried to define by means of a measure of asynchrony degree herein proposed. The determination of which rules in a similarity class are also equivalent in terms of their spectra is yet to be properly investigated, beyond present purposes.

In what follows, the next two sections provide the formal definitions of all concepts needed in the presentation, initially the basic definitions, and then others, related to the notions we propose here, namely, the asynchrony degree and a precise characterisation of the notion of dynamical equivalence under block-sequential updates. The two subsequent sections discuss the methodological aspects of the work, firstly those related to spectral analysis of cellular automata, and then, to the definition of the asynchrony families of update employed. The following sections, respectively, discuss aspects of the similarity classes obtained and provide some concluding remarks.

## 2 Basic Definitions

### 2.1 Cellular Automata

In general terms, a (synchronous) *cellular automaton* is a quadruple $(S, N, f, d)$, where $S = \{0, 1, \ldots, k-1\}$, $k \in \mathbb{Z}_+$ is the set of states; $N = (\overrightarrow{x_1}, \overrightarrow{x_2}, \ldots, \overrightarrow{x_n})$, with $\overrightarrow{x_i} \in \mathbb{Z}^d$ and $n \in \mathbb{N}$, is the neighbourhood vector; $f : S^n \rightarrow S$ is the local transition function (or local rule); and $d \in \mathbb{Z}_+$ is the dimension [6].

For one-dimensional cellular automata ($d = 1$), at each time step the value of each cell is updated according to the cell's values within a neighbourhood radius $r \in \mathbb{N}$, with $2r + 1$ cells, according to the local rule $f : S^{2r+1} \rightarrow S$.

The elementary cellular automata (ECAs) are those that make up the space comprised by the binary ($S = \{0, 1\}$) one-dimensional local rules $f : \{0, 1\}^3 \rightarrow \{0, 1\}$. The rules in this space can be described as a binary sequence $(s_7 s_6 s_5 s_4 s_3 s_2 s_1 s_0)$, where $s_0$ is the new state of cell $x_i$ given by neighbourhood 000, $s_1$ by 001, and so on, up to $s_7$, which is given by neighbourhood 111. The elementary space has a total of $2^{2^3} = 256$ distinct rules, each one being referred to by its unique rule Wolfram, which is the decimal number corresponding to the binary sequence $(s_7 s_6 s_5 s_4 s_3 s_2 s_1 s_0)$ [15].

### 2.2 Cyclic Configurations

A *configuration* $c$ is a function $c : \mathbb{Z} \longrightarrow \{0, 1\}$. For convenience, for any $i \in \mathbb{Z}$ we denote $c(i)$ by $c_i$. In this context, $i$ is a *cell* and $c_i \in \{0, 1\}$ is its *state*. If there is $L \in \mathbb{Z}_+$, $L \geq |N|$, such that $c_i = c_{i+L}$ for any cell $i$, $c$ is said to be a *cyclic* configuration of length $L$. In that case, the configuration may be regarded as a vector $c = (c_1, c_2, \ldots, c_L) \in S^L$, as $c_i = c_j$ whenever $i \equiv j \bmod L$.

A one-dimensional radius-$r$ CA with local rule $f$ induces a *global rule* $F : S^{\mathbb{Z}} \longrightarrow S^{\mathbb{Z}}$ over the set of configurations given by

$$(F(c))_i = f(c_{i-\lceil r \rceil}, \ldots, c_i, \ldots, c_{i+\lfloor r \rfloor})$$

### 2.3 Asynchronous Cellular Automata with Block-Sequential Updates

For present purposes, an *asynchronous cellular automaton* over configurations of length $L$ is a quintuple $(S, N, f, d, \sigma_L)$, in which $(S, N, f, d)$ is a cellular automaton, and $\sigma_L : \{1, 2, \ldots, L\} \longrightarrow \{1, 2, \ldots, m\}$ is a surjective function for some $1 \leq m \leq L$, which is is the *update schedule* of the asynchronous CA.

The update schedule basically tells about the update priority of each cell. Therefore, a synchronous CA would have $\sigma(i) = 1$ for all $1 \leq i \leq L$, since every cell would be updated simultaneously. On the other hand, for $\sigma(i) = i$, for all $1 \leq i \leq L$ (the *sequential* update schedule), cell 1 would be updated first, then cell 2, cell 3 and so on. In general, since the cell positions in the configuration can be grouped in arbitrary blocks, the kind of asynchronism of interest here is usually referred to as *block-sequential* [1].

More precisely, given an update schedule $\sigma_L$, such that $\sigma(\{1, 2, \ldots, L\}) = \{1, 2, \ldots, m\}$, and $c$ a configuration of length $L$, let $I_k = \{i \in \{1, \ldots, L\} : \sigma(i) = k\}$ for all $1 \leq k \leq m$ and define

$$\left(F_{(\sigma,k)}(c)\right)_i = \begin{cases} (F(c))_i\,, & \text{if } i \in I_k \\ c_i\,, & \text{otherwise} \end{cases}.$$

The *asynchronous global rule* defined by $\sigma$ is given by

$$F_\sigma = F_{(\sigma,m)} \circ \cdots \circ F_{(\sigma,1)}.$$

For convenience, we define $F_{(\sigma,1,k)} = F_{(\sigma,k)} \circ \cdots \circ F_{(\sigma,1)}$ for all $1 \leq k \leq m$.

An example of the changes that the deterministic asynchrony causes in the evolution of cellular automata is depicted in Fig. 1, where the updating sequence consists in alternating the update of odd and even cell positions.

## 3 Specific Conceptual Proposals

### 3.1 Average Asynchrony Degree

Given an update schedule $\sigma$ over configurations of length $L$, we introduce the *asynchrony degree* of cell $i$, $1 \leq i \leq L$, denoted by $d_{(\sigma,i)}$, as the number of times, minus 1, the local function $f$ appears in the expression of $(F_\sigma(c))_i$, for any configuration $c$. The quantity is taken "$-1$" because $f$ appears exactly once in the *synchronous* update, in which case $d_{(\sigma,i)} = 0$.

For instance, if $\sigma_8 : [1, 8] \longrightarrow [1, 8]$ is given by $\sigma(i) = i$ and $f$ is a radius-1 local rule, for cell 3 in a configuration $c$, we have

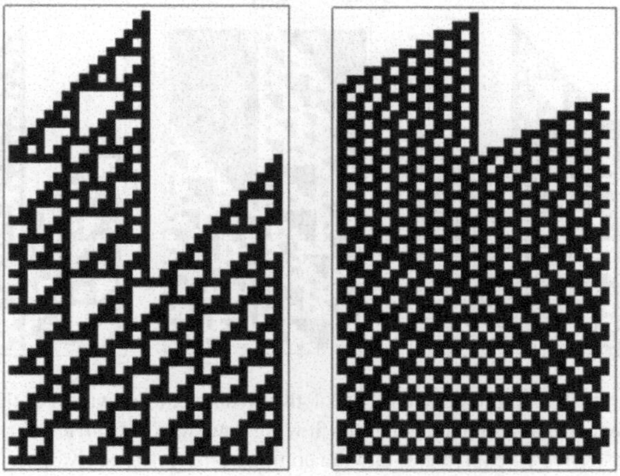

**Fig. 1** Temporal evolution of the elementary rule 110 under synchronous update (on the left) and under asynchronous blocks with alternating odd and even cell positions (on the right), with time flowing downwards

$$\left(F_{\sigma_8}(c)\right)_3 = f(f(f(c_8, c_1, c_2), c_2, c_3), c_3, c_4).$$

Therefore, $d_{(\sigma_8,3)} = 3 - 1 = 2$.

Now, given an update schedule $\sigma$ over configurations of length $L$, we define the *average asynchrony degree* of $\sigma$, denoted by $d_\sigma$, by

$$d_\sigma = \frac{1}{L} \cdot \sum_{i=1}^{L} d_{(\sigma,i)} \cdot$$

For instance, the average asynchrony of $\sigma_8$ in the example above is

$$d_{\sigma_8} = \frac{\sum_{i=1}^{8} d_{\sigma_8,i}}{8} = \frac{0+1+2+3+4+5+6+8}{8} = \frac{29}{8} = 3.625.$$

### 3.2 Dynamical Equivalence for Asynchronous Updates

It is well known that symmetries in the state transition table of synchronous cellular automata lead to the definition of rule classes of dynamical equivalence [9, 14]. In the case of the elementary space, for instance, in order to analyse the dynamics of its rules it suffices to look at only 88 rules, instead of the 256 of the entire space. Figure 2 illustrates the matter for ECA 110, under synchronous update.

However, it turns out that the notion of dynamical equivalence classes is more restricted in the context of block-sequential asynchronous CAs; this is what we discuss and formalise below.

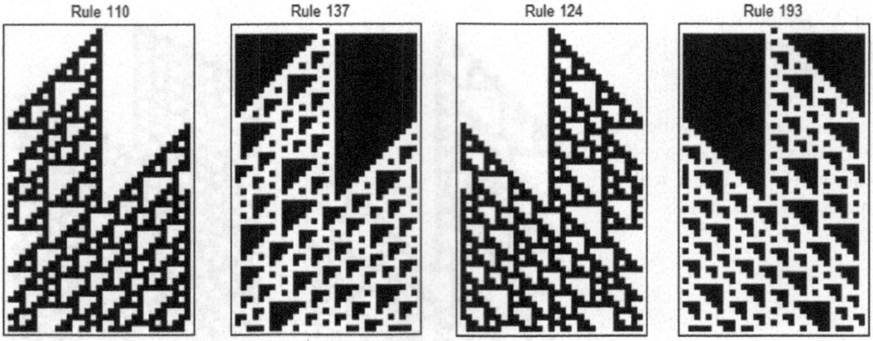


**Fig. 2** Temporal evolutions of the dynamical equivalence class of elementary rule 110 under synchronous update, with time flowing downwards; from left to right: the original rule; its conjugate; its reflected equivalent; and the conjugate-reflected equivalent

### 3.2.1 Conjugacy

Given an elementary cellular automaton local rule $f$, the *conjugate* of $f$ is the local rule $\overline{f}$ such that $f(x_1, x_2, x_3) = \overline{\overline{f}(\overline{x_1}, \overline{x_2}, \overline{x_3})}$, where $\overline{b} = 1 - b$, for any binary value $b$. Analogously, given a configuration $c$ with length $L$, the *conjugate* of $c$ is the configuration $\overline{c}$ given by $\overline{c}_i = \overline{c_i}, \forall 1 \le i \le L$. Notice that $\overline{(\overline{c})} = c$.

For synchronous update, the image of a configuration $c$ by $F$ is exactly the conjugate of the image of $\overline{c}$ by $\overline{F}$, that is

$$F(c) = \overline{(\overline{F}(\overline{c}))}. \tag{1}$$

**Proposition 1** *Given an elementary cellular automaton local rule $f$, a configuration $c$ and an update schedule $\sigma$, both of length $L$, $F_\sigma(c) = \overline{(\overline{F}_\sigma(\overline{c}))}$ holds.*

***Proof*** Let $c_i^{\sigma,k}$ be the state of cell $\left(F_{(\sigma,1,k)}(c)\right)_i$.

For all $i \in I_1$, we have $(c_{i-1}, c_i, c_{i+1}) = \overline{(\overline{c}_{i-1}, \overline{c}_i, \overline{c}_{i+1})}$, from the definition of $\overline{c}$. Since no cells were updated, $(c_{i-1}^{\sigma,1}, c_i^{\sigma,1}, c_{i+1}^{\sigma,1}) = (c_{i-1}, c_i, c_{i+1})$ and $(\overline{c}_{i-1}^{\sigma,1}, \overline{c}_i^{\sigma,1}, \overline{c}_{i+1}^{\sigma,1}) = (\overline{c}_{i-1}, \overline{c}_i, \overline{c}_{i+1})$. Hence

$$(c_{i-1}^{\sigma,1}, c_i^{\sigma,1}, c_{i+1}^{\sigma,1}) = \overline{(\overline{c}_{i-1}^{\sigma,1}, \overline{c}_i^{\sigma,1}, \overline{c}_{i+1}^{\sigma,1})} \tag{2}$$

and

$$f(c_{i-1}^{\sigma,1}, c_i^{\sigma,1}, c_{i+1}^{\sigma,1}) = \overline{(\overline{f}(\overline{c}_{\phi(i-1)}^{\sigma,1}, \overline{c}_{\phi(i)}^{\sigma,1}, \overline{c}_{\phi(i+1)}^{\sigma,1}))}. \tag{3}$$

Therefore,

$$F_{(\sigma,1,1)}(c) = \overline{(\overline{F}_{(\sigma,1,1)}(\overline{c}))}. \tag{4}$$

Now, suppose that for all $1 \leq j \leq k$ we have $F_{(\sigma,1,k)}(c) = \overline{(\overline{F}_{(\sigma,1,k)}(\overline{c}))}$. We will show that $F_{(\sigma,1,k+1)}(c) = \overline{(\overline{F}_{(\sigma,1,k+1)}(\overline{c}))}$.

Let $i \in I_{k+1}$. If $\sigma(i-1) \geq (k+1)$, then $c_{i-1}^{\sigma,(k+1)} = c_{i-1} = \overline{\overline{c}_{i-1}} = \overline{\overline{c}_{i-1}^{\sigma,(k+1)}}$, since such cells have not been updated and $c = \overline{(\overline{c})}$. On the other hand, if $\sigma(i-1) < (k+1)$, by the induction hypothesis $F_{(\sigma,1,k)}(c) = \overline{(\overline{F}_{(\sigma,1,k)}(\overline{c}))}$ we also have $c_{i-1}^{\sigma,(k+1)} = \overline{\overline{c}_{i-1}^{\sigma,(k+1)}}$. Analogously, $c_{i+1}^{\sigma,(k+1)} = \overline{\overline{c}_{i+1}^{\sigma,(k+1)}}$.

Hence,

$$(c_{i-1}^{\sigma,(k+1)}, c_{i}^{\sigma,(k+1)}, c_{i+1}^{\sigma,(k+1)}) = \overline{(\overline{c}_{(i-1)}^{\sigma,(k+1)}, \overline{c}_{i}^{\sigma,(k+1)}, \overline{c}_{i+1}^{\sigma,(k+1)})} \tag{5}$$

and

$$f(c_{i-1}^{\sigma,(k+1)}, c_{i}^{\sigma,(k+1)}, c_{i+1}^{\sigma,(k+1)}) = \overline{(\overline{f}(\overline{c}_{i-1}^{\sigma,(k+1)}, \overline{c}_{i}^{\sigma,(k+1)}, \overline{c}_{i+1}^{\sigma,(k+1)}))}. \tag{6}$$

Therefore,

$$F_{(\sigma,1,(k+1))}(c) = \overline{(\overline{F}_{(\sigma,1,(k+1))}(\overline{c}))}. \tag{7}$$

By induction on $k$, $F_\sigma(c) = \overline{(\overline{F}_\sigma(\overline{c}))}$.

Proposition 1 allows us to define the notion of a conjugate rule for asynchronous updates and shows that such a rule is the same one for the synchronous case. More precisely:

**Definition 1** Given an elementary cellular automaton local rule $f$, the *conjugate* of $f$ is the rule $\overline{f}$ such that, for any update schedule $s$ and any configuration $c$, $F_\sigma(c) = \overline{(\overline{F}_\sigma(\overline{c}))}$.

Now, regarding the sensibility of a rule and of its conjugate to update schedules, we have the following result:

**Corollary 1** *Given an elementary cellular automaton local rule $f$, $\sigma_1$ and $\sigma_2$ two update schedules, and a configuration $c$, such that $F_{\sigma_1}(c) \neq F_{\sigma_2}(c)$, there is a configuration $\tilde{c}$ such that $\overline{F}_{\sigma_1}(\tilde{c}) \neq \overline{F}_{\sigma_2}(\tilde{c})$. Moreover, $\tilde{c} = \overline{c}$.*

***Proof*** Take $\tilde{c} = \overline{c}$. By Proposition 1, $F_{\sigma_1}(c) = \overline{(\overline{F}_{\sigma_1}(\overline{c}))}$ and $F_{\sigma_2}(c) = \overline{(\overline{F}_{\sigma_2}(\overline{c}))}$. Since by hypothesis $F_{\sigma_1}(c) \neq F_{\sigma_2}(c)$, we have

$$\overline{(\overline{F}_{\sigma_1}(\overline{c}))} \neq \overline{(\overline{F}_{\sigma_2}(\overline{c}))} \Rightarrow \overline{F}_{\sigma_1}(\overline{c}) \neq \overline{F}_{\sigma_2}(\overline{c}). \tag{8}$$

**Corollary 2** *Both $f$ and $\overline{f}$ have the same number of different dynamics.*

### 3.2.2 Reflection

Given $f$ an elementary cellular automaton local rule, the *reflection* of $f$ is the local rule $f'$ such that $f(x_1, x_2, x_3) = f'(x_3, x_2, x_1)$.

Analogously, given a configuration $c$ with length $L$, the *reflection* of $c$ is the configuration $c'$ given by $c'_i = c_{L-i+1}, \forall 1 \le i \le L$.

Given $L \in \mathbb{Z}_+$, let $\phi_L : 1, 2, \ldots, L \longrightarrow 1, 2, \ldots, L$ be given by $\phi_L(i) = L - i + 1$. With such a notation, we may write $c'_i = c_{\phi_L(i)}$. Notice that $\phi_L^2 = Id$, hence $(c')' = c$.

For synchronous update, the image of a configuration $c$ by $F$ is exactly the reflection of the image of $c'$ by $F'$, that is

$$F(c) = (F'(c'))'. \tag{9}$$

In order to generalise the notion of a reflected rule for asynchronous updates, we must first define the *reflection of an update schedule*. Given $\sigma$ and update schedule of length $L$, define $\sigma'$ *the reflection of* $\sigma$ as $\sigma'(i) = \sigma(\phi_L(i))$.

**Proposition 2** *Given an elementary cellular automaton local rule $f$, a configuration $c$ and an update schedule $s$, both of length L, we have $F_\sigma(c) = (F'_{\sigma'}(c'))'$.*

***Proof*** Let $I_j = \{i \in \mathbb{N} : 1 \le i \le L, \sigma(i) = j\}$ and $I'_j = \{i \in \mathbb{N} : 1 \le i \le L, \sigma'(i) = j\}$, for $1 \le j \le L$. That is, $I_j$ is the set of indices who are updated at timestep $j$ in $\sigma$ and $I'_j$ is the analogous for $\sigma'$. Notice that $I'_j = \phi_L(I_j)$.

For all $i \in I_1$, we have $(c_{i-1}, c_i, c_{i+1}) = (c'_{\phi(i-1)}, c'_{\phi(i)}, c'_{\phi(i+1)})'$, from the definition of $c'$. Since no cells were updated, $(c^{\sigma,1}_{i-1}, c^{\sigma,1}_i, c^{\sigma,1}_{i+1}) = (c_{i-1}, c_i, c_{i+1})$ and $(c'^{\sigma',1}_{\phi(i-1)}, c'^{\sigma',1}_{\phi(i)}, c'^{\sigma',1}_{\phi(i+1)}) = (c'_{\phi(i-1)}, c'_{\phi(i)}, c'_{\phi(i+1)})$. Hence

$$(c^{\sigma,1}_{i-1}, c^{\sigma,1}_i, c^{\sigma,1}_{i+1}) = (c'^{\sigma',1}_{\phi(i-1)}, c'^{\sigma',1}_{\phi(i)}, c'^{\sigma',1}_{\phi(i+1)})' \tag{10}$$

and

$$f(c^{\sigma,1}_{i-1}, c^{\sigma,1}_i, c^{\sigma,1}_{i+1}) = f'(c'^{\sigma',1}_{\phi(i-1)}, c'^{\sigma',1}_{\phi(i)}, c'^{\sigma',1}_{\phi(i+1)}). \tag{11}$$

Therefore,

$$F_{(\sigma,1,1)}(c) = (F'_{(\sigma',1,1)}(c'))'. \tag{12}$$

Now, suppose that for all $1 \le j \le k$ we have $F_{(\sigma,1,k)}(c) = (F'_{(\sigma',1,k)}(c'))'$. We will show that $F_{(\sigma,1,k+1)}(c) = (F'_{(\sigma',1,k+1)}(c'))'$.

Let $i \in I_{k+1}$. If $\sigma(i-1) \ge (k+1)$, then $\sigma'(\phi_L(i-1)) \ge (k+1)$ and $c^{\sigma,(k+1)}_{i-1} = c_{i-1} = c'_{\phi(i-1)} = c'^{\sigma',(k+1)}_{\phi_L(i-1)}$, since such cells have not been updated and $c = (c')'$. On the other hand, if $\sigma(i-1) < (k+1), \sigma'(\phi_L(i-1)) < (k+1)$ and by the induction hypothesis $F_{\sigma,k}(c) = (F'_{\sigma',k}(c'))'$, we also have $c^{\sigma,(k+1)}_{i-1} = c'^{\sigma',(k+1)}_{\phi_L(i-1)}$. Analogously, $c^{\sigma,(k+1)}_{i+1} = c'^{\sigma',(k+1)}_{\phi_L(i+1)}$.

Hence,

$$(c^{\sigma,(k+1)}_{i-1}, c^{\sigma,(k+1)}_i, c^{\sigma,(k+1)}_{i+1}) = (c'^{\sigma',(k+1)}_{\phi_L(i-1)}, c'^{\sigma',(k+1)}_{\phi_L(i)}, c'^{\sigma',(k+1)}_{\phi_L(i+1)})' \tag{13}$$

and

$$f(c_{i-1}^{\sigma,(k+1)}, c_i^{\sigma,(k+1)}, c_{i+1}^{\sigma,(k+1)}) = f'(c_{\phi(i-1)}^{\prime\sigma',(k+1)}, c_{\phi(i)}^{\prime\sigma',(k+1)}, c_{\phi(i+1)}^{\prime\sigma',(k+1)}). \tag{14}$$

Therefore,

$$F_{(\sigma,1,(k+1))}(c) = (F'_{(\sigma',1,(k+1))}(c'))'. \tag{15}$$

By induction on $k$, $F_\sigma(c) = (F'_{\sigma'}(c'))'$.

Proposition 2 allows us to define the notion of a reflected rule for asynchronous updates and shows that such a rule is the same one for the synchronous case. More precisely:

**Definition 2** Given an elementary cellular automaton local rule $f$, the *reflection* of $f$ is the rule $f'$ such that, for any update schedule $s$ and any configuration $c$, $F_\sigma(c) = (F'_{\sigma'}(c'))'$.

Now, regarding the sensibility of a rule and of its reflection to update schedules, the following result holds:

**Corollary 3** *Given an elementary cellular automaton local rule $f$, $\sigma_1$ and $\sigma_2$ two update schedules, and a configuration $c$, such that $F_{\sigma_1}(c) \neq F_{\sigma_2}(c)$, there is a configuration $\tilde{c}$ such that $F'_{\sigma'_1}(\tilde{c}) \neq F'_{\sigma'_2}(\tilde{c})$. Moreover, $\tilde{c} = c'$.*

***Proof*** Take $\tilde{c} = c'$. By Proposition 2, $F_{\sigma_1}(c) = (F'_{\sigma'_1}(c'))'$ and $F_{\sigma_2}(c) = (F'_{\sigma'_2}(c'))'$. Since by hypothesis $F_{\sigma_1}(c) \neq F_{\sigma_2}(c)$, we have

$$(F'_{\sigma'_1}(c'))' \neq (F'_{\sigma'_2}(c'))' \Rightarrow F'_{\sigma'_1}(c') \neq F'_{\sigma'_2}(c'). \tag{16}$$

**Corollary 4** *Both $f$ and $f'$ have the same number of different dynamics.*

### 3.2.3 Reflected Conjugacy (or Conjugate Reflection)

As a consequence of Corollaries 2 and 4, the *reflected conjugate* (or the *conjugate reflection*) of $f$, $\overline{f'}$ (or $(\overline{f})'$) has the same number of different dynamics as $f$.

Therefore, when studying the different dynamics of a rule over *all* possible update schedules, one may consider a single representative of each dynamical equivalence class in the space at issue, instead of all the rules in the space, which, for the elementary space would mean the 88 rules already mentioned, usually taking the representative rule as the one with the smallest number in the class.

Nevertheless, when not all updates are taken into account – which is the present case, since specific families of updates will be used – only conjugacy always holds, meaning the necessity of analysing 136 elementary rules, out of the 256 that make up the space.

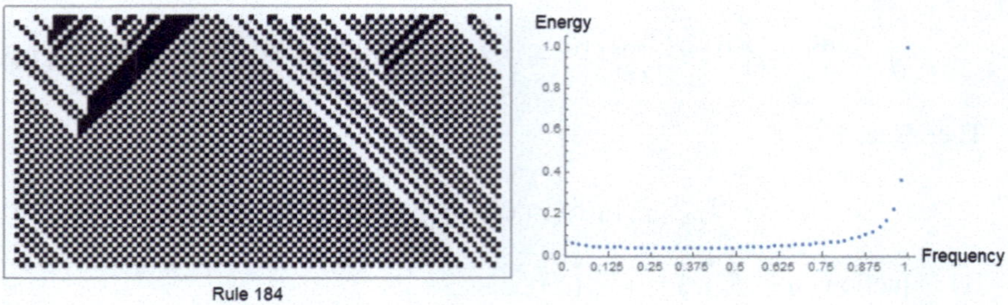


**Fig. 3** Temporal evolution of the elementary rule 184 (on the left) and the rule's corresponding discrete Fourier spectrum, worked out over a sample of initial configurations with 40% of 1s

# 4 Spectral Analysis of Cellular Automata

The Fourier spectra of binary, one-dimentional cellular automata is meant to provide information about the limit configuration of the CAs' temporal evolutions. Informally, it refers to an energy type quantity associated to the frequencies of state changes—all binary block sizes considered – present in the limit configurations. As such, a homogenous configuration has the minimum value of the quantity (the lowest possible frequency of bit changes), whereas the one with alternating 1s and 0s has the maximum value (the highest possible frequency of bit changes); intermediate situations account for the frequency of state changes of all blocks of bits present in the configuration at issue.

A nice example relating the temporal evolution of a rule (ECA 184, in the case) and the rule's spectrum is displayed in Fig. 3 [11]. The plotted spectrum (on the right) displays the level of bit variations due to all blocks of bits present in the limit configuration obtained out of the temporal evolution (on the left). ECA 184 is a well known number conserving rule, whose dynamical behaviour has been perfectly characterised. Since the temporal evolution shown derives from an initial configuration with 60% of 0s, it converges to a shifting configuration where the 1s and 0s are arranged in alternation, and the 0s in excess get organised in blocks of at least two 0s. So, the limit configuration has only these two features, therefore explaining the quickly rising levels of the highest frequencies, including the peak at the maximum possible frequency (due to the preponderance of alternating bits in the configuration), the small increase in the levels of very small frequencies (due to the blocks of 0s), and the overall flatness of the plot, due to the absence of other kinds of binary blocks.

The technique of analysis employed as the basis for the present work is the one described in [11]. This consists in calculating the discrete Fourier transforms for the set of limit configurations obtained out of the temporal evolution of a rule over a set of randomly generated initial configurations; the limit configurations have been approximated by disregarding a number of the initial time steps of each temporal evolution, a sufficiently large time that would allow reaching dynamical stability, and by taking the configuration of the lattice after that time.

The discrete Fourier transform (DFT) is calculated over the set of all (approximated) limit configurations according to the procedure described in the sequence. Given a complex vector $u = (u_1, \ldots, u_n)$, the DFT of $u$, denoted $DFT(u)$, corresponds to the complex vector $v = (v_1, \ldots, v_n)$ given by:

$$v_k = \frac{1}{n} \sum_{j=1}^{n} u_j e^{2\pi i (j-1)(k-1)/n}.$$

As such, the $DFT(u)$ refers to the absolute value, calculated coordinate by coordinate. In other words, if $DFT(u) = (v_1, \ldots, v_n)$, then $|DFT(u)| = (|v_1|, \ldots, |v_n|)$.

Given a one-dimensional cellular automaton local rule, the Fourier spectrum associated to it can be calculated as follows [11]:

- Generation of a set $C = \{c_1, \ldots, c_p\}$ of $p$ random initial configurations of length $L$, each one obtained by randomly setting the state value of each cell independently;
- Application of the local rule $f$ to each initial configuration $c_i \in \mathbb{C}$ a number $t \in \mathbb{N}$ of times, gathering a set $C' = \{c_1^{'}, \ldots, c_p^{'}\}$ of $p$ final configurations, with $c_i^{'} = F(t)(c_i), \forall i \in \{1, \ldots, p\}$;
- Calculation of the DFT for each $c_i^{'} \in \mathbb{C}'$, gathering the set $F(C') = \{F(c_1^{'}), \ldots, F(c_p^{'})\}$ of the final conditions DFTs;
- Definition of the Fourier spectrum (or simply spectrum) $f_c$ for rule $f$ as the arithmetic average of the module of the transforms $F(c_1^{'}), \ldots, F(c_p^{'})$ taken coordinate by coordinate; that is, $f_c = \frac{1}{p} \sum_{c' \in C'} |F|(c')$.
- Normalisation of the spectrum with values between 0.0 and 1.0.

Under synchronous update, the Fourier spectra associated to dynamically equivalent rules are identical, since reflection and/or conjugacy do not affect the dynamics. In [11], the spectral analysis of the entire elementary space under synchronous update has been carried out, which led to a partition of the space into 59 classes of spectral equivalence, which represents a refined partition of the 88 classes of dynamically equivalent rules. Those 59 classes have then been precisely explained in terms of dynamical equivalence among the rules in the class, as well as by the fact that some rules eventually become dynamically equivalent, after some initial time steps of their temporal evolution. The issue of similarity among Fourier spectra of different rules is depicted in Fig. 4, concerning elementary rules under synchronous updates; the figure illustrates the fact that even rules which are not dynamically equivalent may share the same Fourier spectrum.

In order to probe the reasons for the existence of distinct spectral classes to have similar (possibly the same) spectra, in [11] it was necessary to group the spectra of all ECAs according to their similarity, according to some distance measure between them; this was achieved by relying on the Euclidian distance $d(f_s, g_s)$ between the spectra $f_s$ and $g_s$ vectors of the local rules $f$ and $g$. Here we also follow a rule clustering procedure.

In order to balance reliability and accuracy of the generated spectra and fair computational time to obtain and analyse them, choices had to be made for the computational parameters involved. As such, all parameter values have been taken from [11, 14], but finally established in face of the computational demands of the present case, which are heavier than those related to the synchronous case. More precisely, the following are the quantities involved and their values:

– The number of time steps to assume stability convergence of the temporal evolution: 200 iterations.
– The lattice size (and consequent length of the configurations): 1024 cells.
– The random sample size of the initial configurations: 40. This is a much smaller sample than 1000, as used in [11], a decision derived from the heavier computational effort to generate each spectrum, let alone the fact that the number of spectra herein was substantially larger than the one in [11], that is, 4080 *versus* 88, respectively, because of 88 rules and 1 update scheme in the synchronous case, and 136 rules and 30 updates presently. Moreover, although the spectra formed with 1000 initial configurations is cleaner, the general shape is far from being lost, as Fig. 5 makes it evident.

In the next section we introduce and discuss the three families of asynchronous updates we defined for the spectral analyses of the 136 rules that represent each class of dynamical equivalence under asynchronous updates of the elementary space.

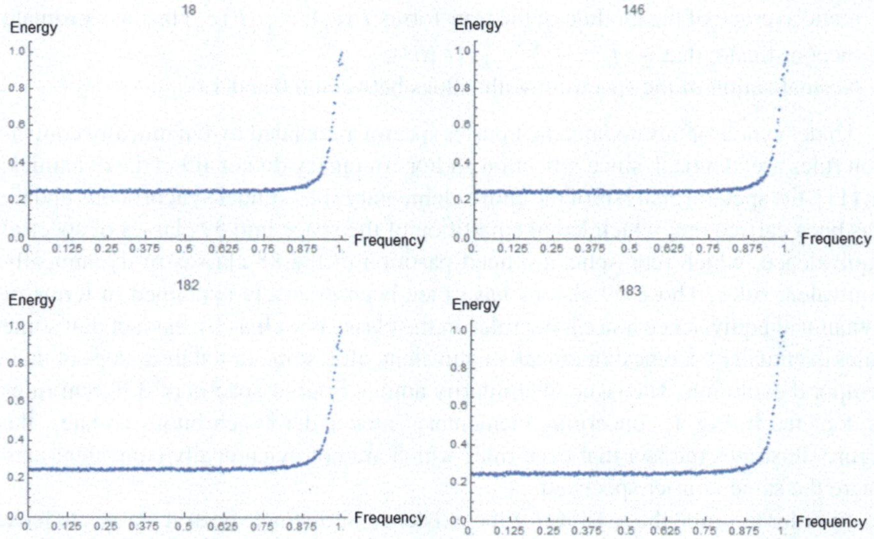


**Fig. 4** In spite of the rule pairs {18, 183} and {146, 182} defining distinct dynamical equivalence classes, they display the same spectra [11]

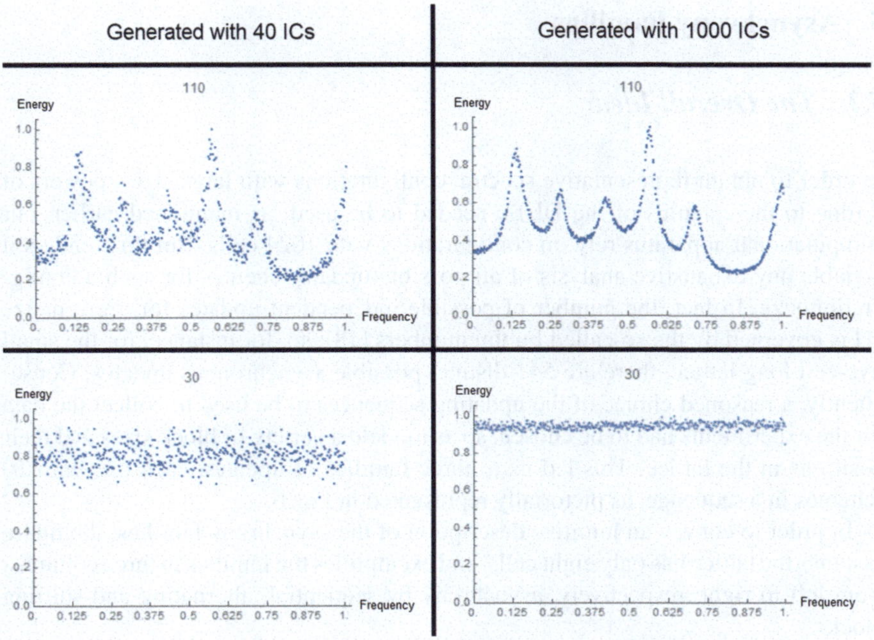


**Fig. 5** The spectra on the left, formed with 40 initial configurations (ICs), is less clean than those on the right (generated with 1000 ICs), but the general shapes are preserved. The spectra at the top refer to ECA 110, and those at the bottom, to ECA 30

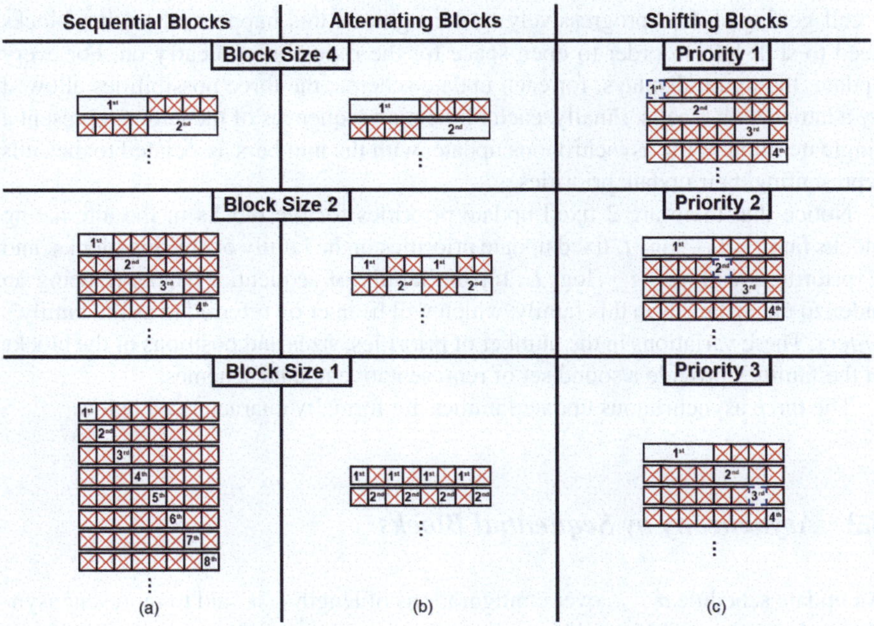


**Fig. 6** Pictorial representation of the families of asynchronous update schemes employed

# 5 Asynchrony Families

## 5.1 The Overall Idea

In order to obtain representative spectra, configurations with large sizes, powers of 2 (due to the calculus of the DFT), needed to be used; as mentioned earlier, our computational apparatus rely on configurations with 1024 cells. But this renders it inviable any exhaustive analysis of all possible update schemes for such a configuration size. In fact, the number of possible independent updates for a given size [1] is governed by the so-called Fubini numbers [18]; so, for instance, for the small five-cell long lattice, there are 541 distinct possible asynchronous updates. Consequently, a reasoned choice of the updating sequences to be used to collect the data for the experiments had to be chosen, so as to yield a variety of block sizes and their positions in the lattice. This led us to three families of updates, each one with 10 schemes in a sequence, as pictorially represented in Fig. 6.

In order to convey an intuitive description of the asynchrony families, the figure assumes the lattice has only eight cells, and exemplifies the families in three columns, from left to right, respectively, asynchrony by sequential, alternating and shifting blocks.

The update schemes for sequential and alternating blocks rely on cell blocks of equal size, in the figure's example, 4, 2 and 1 cells. In contrast, asynchrony with shifting blocks is defined by blocks with fixed sizes, in the case, two blocks with 1 cell and the others with 2 and 4, with the position of the left-most block with 1 cell getting shifted progressively to the right; as this happens, the other blocks need to shift left, in order to open space for the right shifts to carry on. For every update, the figure displays, for each update scheme, the three possibilities allowed by a lattice with 8 cells. Finally, each of the nine sequences of the lattice represent a single iteration of the asynchronous update, with the numbers associated to the cells representing their update priorities.

Notice that there are 2 fixed update priorities for the blocks in the alternating blocks family; $1 + \log_2 L$ fixed update priorities in the family of shifting blocks; and $2^i$ priorities, with $1 \leq i \leq \log_2 L$, for the family of sequential blocks ($i$ being an index to the instances in this family, which will be later on referred to as the family's *order*). These variations in the number of priorities, sizes and positions of the blocks in the families provide a sound set of representative update schemes.

The three asynchronous update families are formally characterised below.

## 5.2 Asynchrony by Sequential Blocks

An update schedule $\sigma_{(L,m)}$ over configurations of length $L$ is said to represent asynchrony by *sequential blocks* if:

- $L = 2^M$ for some $M \in \mathbb{N}$; and
- There is $m \in \mathbb{N}$, $m \leq M$ such that $\sigma(i) = \lceil \frac{i}{2^m} \rceil$, where $2^m$ is the *number of cells in each block*.

For instance, with $L = 8$ and $m = 2$, we have
$\sigma_{(8,2)}(i) = \lceil \frac{i}{4} \rceil = \begin{cases} 1, & \text{if } i \in \{1, 2, 3, 4\} \\ 2, & \text{if } i \in \{5, 6, 7, 8\} \end{cases}$.

## 5.3 Asynchrony by Alternating Blocks

Firstly, given $j, j_1, j_2 \in \mathbb{Z}$, we denote the inequality $j_1 \leq j \leq j_2$ simply by $j \in [j_1, j_2]$. Now, an update schedule $\sigma_{(L,m)}$ over configurations of length $L$ is said to represent asynchrony by *alternating blocks* if:

- $L = 2^M$ for some $M \in \mathbb{N}$; and
- There is $m \in \mathbb{N}$, $M$ divisible by $2m$, such that
$\sigma(i) = \begin{cases} 1, & \text{if } i \in [1 + 2mk, m + 2mk] \text{ for some } k \in \mathbb{Z}, k \geq 0 \\ 2, & \text{otherwise} \end{cases}$,
with $m$ representing the *length of each alternating block* in the update schedule.

For instance, with $L = 8$ and $m = 2$, we have
$\sigma_{(8,2)}(i) = \begin{cases} 1, & \text{if } i \in [1 + 4k, 2 + 4k] \text{ for some } k \in \mathbb{Z}, k \geq 0 \\ 2, & \text{otherwise} \end{cases} = \begin{cases} 1, & \text{if } i \in [1, 2] \cup [5, 6] \\ 2, & \text{if } i \in [3, 4] \cup [7, 8] \end{cases}$.

## 5.4 Asynchrony by Shifting Blocks

In order to ease the notation, consider the partition $\Pi_M = \{B_{(M,1)}, B_{(M,2)}, \ldots, B_{(M,M)}\}$ of $[1, 2^M - 1]$, with its blocks given by:

$$\begin{cases} B_{(M,1)} = & [1, 2^{M-1}] \\ B_{(M,j)} = & \left[1 + \sum_{l=M+1-j}^{M-1} 2^l, \sum_{l=M-j}^{M-1} 2^l\right], \text{ if } 2 \leq j \leq M-1 \\ B_{(M,M)} = & \{2^M - 1\} \end{cases}.$$

For instance, with $M = 3$, the partition becomes $\Pi_M = \{B_{(3,1)}, B_{(3,2)}, B_{(3,3)}\} = \{\{1, 2, 3, 4\}, \{5, 6\}, \{7\}\}$ of $[1, 2^3 - 1] = [1, 7]$

An update schedule $\sigma_{(L,l)}$ over configurations of length $L$ is said to represent asynchrony by *shifting blocks* if:

- $L = 2^M$ for some $M \in \mathbb{N}$; and

- There is $l \in \mathbb{Z}$, $1 \leq l \leq M$ such that $\sigma(i) = \begin{cases} j, & \text{if } i \in B_{(M,j)} \text{ with } j < l \\ j+1, & \text{if } (i-1) \in B_{(M,j)} \text{ with } j \geq l \\ l, & \text{otherwise} \end{cases}$, with $l$ representing the *position of the shift* in the update schedule

For instance, with $L = 8$ and $l = 2$, we have

$$\sigma_{(8,2)}(i) = \begin{cases} j, & \text{if } i \in B_{(3,j)} \text{ with } j < 2 \\ j+1, & \text{if } (i-1) \in B_{(3,j)} \text{ with } j \geq 2 \\ 2, & \text{otherwise} \end{cases}.$$

Notice that, in this case,

$$\begin{cases} 1,2,3,4 \in \left[1, 2^2\right] = B_{(3,1)} \Longrightarrow \sigma(1) = \sigma(2) = \sigma(3) = \sigma(4) = 1, \\ (6-1), (7-1) \in \left[2^2+1, 2^1+2^2\right] = B_{(3,2)} \Longrightarrow \sigma(5) = \sigma(6) = 2+1 = 3, \\ (8-1) \in \{2^3-1\} = B_{(3,3)} \Longrightarrow \sigma(7) = 3+1 = 4. \end{cases}$$

Therefore, we have $\sigma_{(8,2)}(i) = \begin{cases} 1, & \text{if } i \in \{1,2,3,4\} \\ 2, & \text{if } i = 5 \\ 3, & \text{if } i \in \{6,7\} \\ 4, & \text{if } i = 8 \end{cases}$.

# 6 Results: Spectral Similarity Classes

## 6.1 Size and Asynchronous Degree of the Spectral Classes

Since the lattice size we used to obtain the results to be shown below is made up of 1024 cells, by following the rationale exemplified above this implies that 10 updates are considered for each asynchrony family, concerned with block sizes with $2^i$ cells, $0 \leq i \leq 9$, for the update schemes with sequential and alternating blocks, and 10 distinct positions of the shifting 1-cell block. In order to make it easier to present our results, we refer to each one of the 10 update schemes of an asynchrony family by its *order*, a number from 1 to 10.

For the families of sequential and alternating blocks the meaning of the order is that $2^{10-\text{order}}$ yields the size of the cell blocks that are updated (or not) at any time. So, order 1 means the largest possible block size (512 cells), whereas order 10 refers to the smallest, i.e., blocks made of a single cell. So, the smallest the order of an update, the largest its associated blocks, and therefore, the less asynchronous it is.

Due to its different nature, the interpretation of the update's order for the family of shifting blocks needs to be reappraised. In this case, order means the update priority of the 1-cell block that progressively shifts to the right. Putting another way, $2^{10-\text{order}}$ yields the size of the cell block which is on the right-hand side of the shifting 1-cell block.

In order to group the Fourier spectra into similarity classes, various clustering procedures were attempted, with the acceptation criterion for a successful result having been that two or more spectra would be considered members of a same class, only if their spectra would differ by at most 0.1 energy units, all frequencies considered. The first one of them relied on automatically grouping the rules, iteratively, according to their relative distance, measured in terms of their Euclidean distance. However, the result was not satisfactory, because of various false positives and false negatives. Subsequently, various attempts were made with the off-the-shelf *Mathematica* software (by means of its native FindCluster function), which automatically groups data according to various clustering algorithms and distances; after attempting various parameterisations, the Jarvis-Patrick similarity clustering algorithm [5] was chosen, with neighbourhood radius of 0.1932, as it yielded the more sensible results, characterised in terms of the smallest number of false positives and false negatives. But since the result up to that point still had clear flaws, a subsequent procedure was employed, in which the classes were visually analysed and the spectra manually rearranged.

Table 1 displays the number of spectral similarity classes for each asynchrony order of each family of asynchrony, for the 136 dynamical representatives of the elementary space; naturally, larger number of similarity classes implies that a given family can lead to higher diversity of spectra. For comparison, also included in the table is the number of spectral equivalence classes of the elementary cellular automata under synchronous update, obtained in [11], mentioned earlier. Notice that, while the classes for the synchronous case are actually of an *equivalence* nature, the ones for the asynchronous cases carry a weaker notion, in that the classes only reflect spectral *similarity*. Although the determination of which rules in a similarity class are also equivalent in terms of their spectra is not addressed herein, it is reasonable to expect that the number of equivalent asynchronous classes should increase from the current number of similarity classes. Furthermore, it is clear that asynchrony brings much more spectral diversity to the elementary space than synchrony.

Figure 7 displays the number of similarity classes in the families, according to the orders of the updates that defined them. As the asynchrony order increases, the number of similarity classes of the families of sequential and alternating blocks have a general tendency to increase, until reaching a maximum at order 8, from which they decline. Notice that this behaviour is more intense for the family of sequential blocks, and that, at order 10, the number of similarity classes in this family is smaller than the number of equivalence classes in the synchronous case. In contrast, the number of similarity classes in the family of shifting blocks is kept stable at every asynchrony order. So, for the families considered, altering the blocks' positions, causes smaller changes in the number of similarity classes than altering their sizes.

Looking at the average asynchrony degree (defined earlier) associated to every update scheme in each family, brings a new perspective to the analysis. As shown in Fig. 8, the asynchrony degree of the families of sequential and alternating blocks become progressively higher, the larger the order becomes, as their constituting blocks become progressively smaller. But although the asynchronous degrees of both families are practically the same up to order 8, the updates with sequential

**Table 1** Number of similarity classes of the elementary space, according to the order of each asynchronous update scheme, and its number of spectral equivalence classes under synchronous update

| Asynchrony order | Sequential blocks | Alternating blocks | Shifting blocks | Synchronous update |
|---|---|---|---|---|
| 1 | 62 | 62 | 71 | . |
| 2 | 62 | 62 | 68 | . |
| 3 | 64 | 62 | 68 | . |
| 4 | 69 | 64 | 67 | . |
| 5 | 74 | 64 | 68 | 59 |
| 6 | 81 | 65 | 65 | . |
| 7 | 91 | 65 | 67 | . |
| 8 | 106 | 75 | 64 | . |
| 9 | 94 | 68 | 64 | . |
| 10 | 45 | 62 | 69 | . |

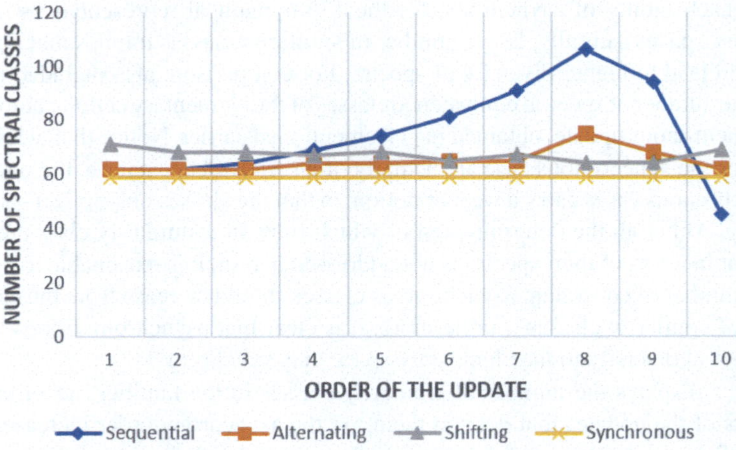


**Fig. 7** Number of similarity classes of the asynchrony scheme orders for each asynchrony family type

blocks become incomparably higher compared to the ones with alternating blocks at orders 9 and 10; this is due to the fact that only a single block is updated each time in the family of sequential blocks, while the family of alternating blocks causes groups of blocks to be updated each time, in parallel, more akin to the synchronous update. All in all, it seems reasonable in the present context to associate an intrinsic asynchrony degree to each family, averaged over all the 10 orders, as a property of the family, in which case the sequential blocks family can be regarded as the most asynchronous of the families, followed by that of alternating blocks, and finally, the

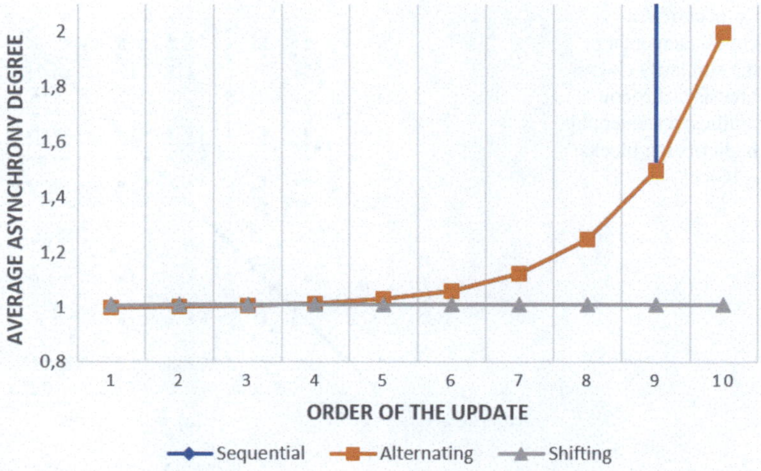


**Fig. 8** Average asynchrony degree of the three asynchrony families, for each order

shifting blocks family; in doing so, the following values of asynchrony degree of the families are obtained, respectively, 52.2, 1.2 and 1.0.

## 6.2 *Rule Co-occurrences in the Spectral Classes*

The similarity classes have also been probed in terms of their sizes and compositions. As such, we looked at the number of times a rule co-occurred with each other in a similarity class, for each asynchrony family, therefore leading to three $136 \times 136$ co-occurrence matrices.

Figure 9 displays the values in the matrices in gray scale, white relating to no co-occurrence, black representing 10 co-occurrences, and gray-levels for the values in between; naturally, 10 is the maximum possible value because of the 10 existing orders within a family. The following observations can be made:

- Family of sequential blocks (Fig. 9a): More rule pairs with a single co-occurrence comparatively to the other matrices of individual families, and moderate numbers of rule pairs with co-occurrences between 8 and 10.
- Family of alternating blocks (Fig. 9b): Smaller number of rule pairs with co-occurrences between 8 and 10, and the highest number of rule pairs with co-occurrences between 4 and 6, among all the single-family matrices.
- Family of shifting blocks (Fig. 9c): Almost every rule pair features co-occurrences from 8 to 10, and there is an almost zero number of the other co-occurrence values.

The figure clearly indicates the existence of co-occurring rules in all the matrices, even though with varied frequencies. It also reveals a relation between the diversity of co-occurrence values in each individual family and the variation of the number

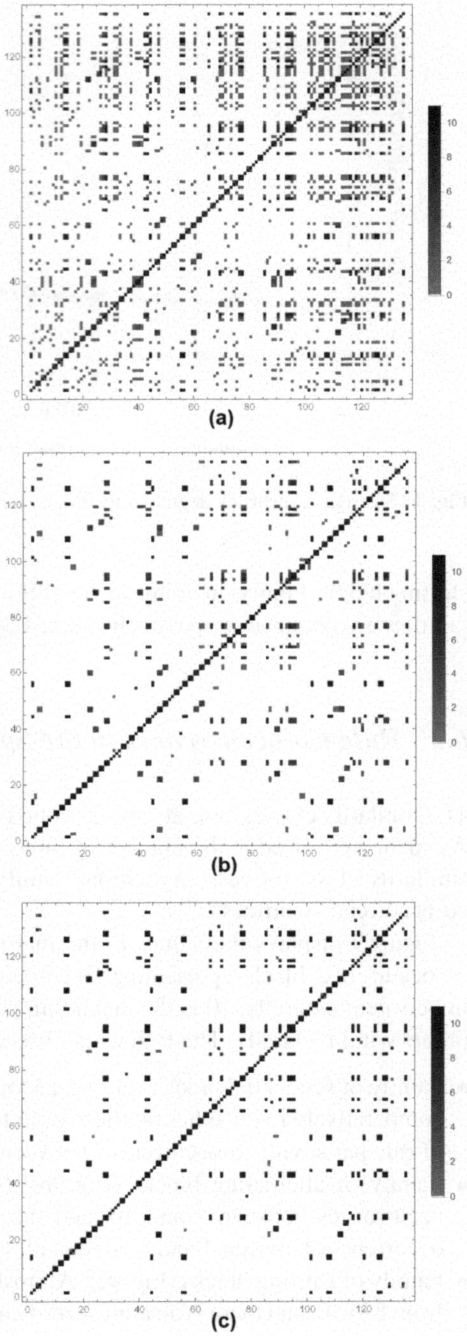


**Fig. 9** Co-occurrence matrices of all elementary rules in the similarity classes, for the three asynchronous update families: **a** sequential blocks; **b** alternating blocks; **c** shifting blocks

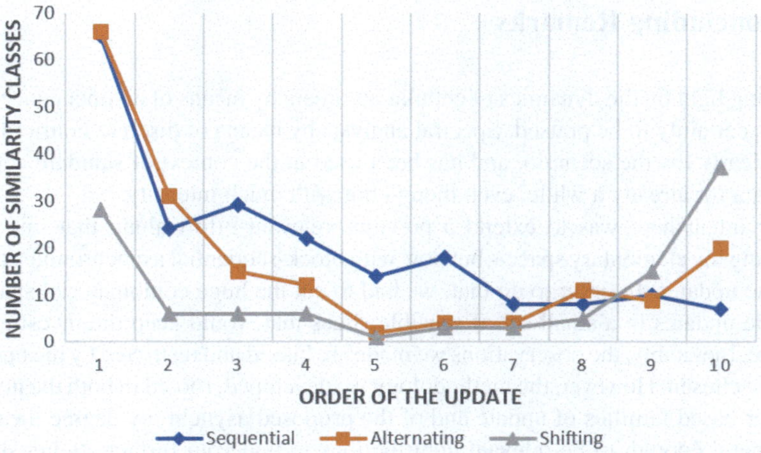


**Fig. 10** Number of similarity classes in each family, for every one of its update schemes

of similarity classes in each one, across their 10 different orders. So, it is clear that the diversity of co-occurrence values is the largest in the sequential blocks family, followed by the alternating blocks family, and then, by the shifting blocks family, and this ordering is the same as the variation of the number of similarity classes in each family, which are, respectively, 61, 13 and 7; in other words, as the number of classes within a family changes significantly, this strongly affects the co-occurrence of rules. Naturally, this is in tune with the fact that the intrinsic average asynchrony degree of each family decreases in the same order, that is, the families of sequential, alternating and shifting blocks.

From yet another perspective of analysis, Fig. 10 exhibits the number of similarity classes for all possible orders of each asynchrony family. Clearly, there is a predominance of similarity classes in just a single order for all families, but more intensely for the families of sequential and alternating blocks. This means that the classes do not possess a significant stability across the different orders, therefore having more chance to be confined to a single order. The plot also shows that the shifting blocks family leads to more stability because of its smaller intrinsic asynchrony degree among the families, therefore explaining its associated smaller number of classes for order 1 and the highest for order 10. As for the intermediate order values, the correlation between the number of classes and the intrinsic asynchrony degree of each family is not as clear as in the extremes of the plot, but some coherence is still noticeable for the sequential blocks family, in that the number of classes progressively decreases as the orders increase (i.e., asynchrony increases), as well as for the three families jointly considered, in that the number of their classes for orders from 3 to 7 maintain the expected ordering correspondingly to their asynchronous degrees, with the sequential blocks family at the top and that of shifting blocks at the bottom.

## 7 Concluding Remarks

Shedding light on the dynamics of cellular automata by means of distinct conceptual tools is certainly to be praised. Spectral analysis by means of discrete Fourier transform clearly fits the scenario, and has been used in the context of standard cellular automata for already a while, even though not with much intensity.

Our intent here was to extend a previous relevant effort along that line – on analysing the elementary space – but now with block-sequential asynchronous, deterministic updates. In order to do that, we had to cut the huge combinatorial space of possible updates in a small set of sensible slices that would keep the investigation feasible. Inevitably, the observations we made are biased and restricted by the updates we have chosen. However, the methodology we developed, rooted in both the notions of order based families of update and of the proposed asynchrony degree measure, are general enough to be relevant contributions to underpin further studies on the subject. In fact, although our emphasis was on the elementary space, these conceptual constructs are totally applicable to any other one-dimensional cellular automata space.

Although we defined asynchrony degree as an average quantity related to the number of applications of the local function to the positions of a configuration, variants are certainly possible, such as the total number, the maximum number, the mode, etc. Other contexts of usage of the concept might well suggest one of the alternatives as the most appropriate.

As expected, the presence of asynchrony was shown to induce changes in the spectra of the elementary cellular automata; but since the dynamical possibilities in the space is limited, the rules could be grouped into similarity classes, according to the specific update strategy employed. Based on them, it was possible to show, although not strictly, a generally positive correlation between the number of similarity classes of a family of updates and its asynchrony degree. This also explains the significantly smaller number of equivalent classes previously obtained in the literature for the synchronous elementary space.

Although the computational experiments have been carried out with a small number of initial configurations, as discussed, we understand that the qualitative conclusions drawn would remain basically the same should a larger sample of initial configurations had been used. Notwithstanding, with more initial configurations we would expect some numerical change in the distribution of the number of the similarity classes in each family.

Finally, it was observed that the larger the asynchrony degree of an update family, the higher the tendency for the similarity classes in the family to be made up of a single rule. Consequently, block-sequential asynchrony can lead to more richness of dynamics in the elementary cellular automata and, very likely, also in other spaces. It is tempting to try to correlate this spectral richness with the dynamical analysis of the rules, under their temporal domain.

**Acknowledgements** We thank the Brazilian agencies CAPES (Coordenação de Aperfeiçoamento de Pessoal de Nível Superior) and CNPq (Conselho Nacional de Desenvolvimento Científico e

Tecnológico), the former for a postgraduate fellowship grant and for the projects STIC-AmSud (CoDANet) no. 88881.197456/2018-01 and Mackenzie-PrInt no. 88887.310281/2018-00, and the latter for the research grant PQ 305199/2019-6.

## References

1. Aracena J, Fanchon E, Montalva M, Noual M (2011) Combinatorics on update digraphs in boolean networks. Discret Appl Math 159:401–409
2. Bersini H, Detours V (1994) Asynchrony induces stability in cellular automata based models. In: Artificial Life V. Cambridge, pp 382–387
3. Dennunzio A, Formenti A, Manzoni L, Mauri G, Porreca AE (2017) Computational complexity of finite asynchronous cellular automata. Theor Comput Sci Amsterdam 664:131–143
4. Fatès N (2014) A guided tour of asynchronous cellular automata. J Cell Autom Philadelphia 9(5–6):387–416
5. Jarvis RA, Patrick EA (1973) Clustering using a similarity measure based on shared near neighbors. IEEE Trans Comput 100(11):1025–1034
6. Kari J (2005) Theory of cellular automata: a survey. Theor Comput Sci Amsterdam 334(1–3):3–33
7. Goles E, Montalva-Medel M, Mortveit H, Ramirez-Flandes S (2015) Block invariance in elementary cellular automata. J Cell Autom 10(1–2):119–135
8. Li W (1987) Power spectra of regular languages and cellular automata. Complex Syst Champaign 1(1):107–130
9. Li W, Packard N (1990) The structure of the elementary cellular automata rule space. Complex Syst Champaign 4(3):281–297
10. Ninagawa S (2008) Power spectral analysis of elementary cellular automata. Complex Syst Champaign 17(4):399–411. http://www.mathematica-journal.com/2014/08/representing-families-of-cellular-automata-rules. Cited 5 Oct 2018
11. Ruivo ELP, de Oliveira PPB (2013) A spectral portrait of the elementary cellular automata rule space. Irreducibility Comput Equiv Berlin 2:211–235
12. Ruivo ELP, Montalva-Medel M, de Oliveira PPB, Perrot K (2018) Characterisation of the elementary cellular automata in terms of their maximum sensitivity to all possible asynchronous updates. Chaos, Solitons Fractals Amsterdam 113, 209–220
13. Schonfisch B, de Roos A (1999) Synchronous and asynchronous updating in cellular automata. BioSystems Amsterdam 51(3):123–143
14. Wolfram S (2002) A new kind of science. Wolfram Media, Champaign
15. Wolfram S (1994) Cellular automata and complexity: collected papers. Basic Books, New York
16. Wolfram S (2018) Computation theory of cellular automata. Commun Math Phys 96(1):15–57. http://mathworld.wolfram.com/Rule90.html. Cited 23 Mar 2018
17. Takada Y, Isokawa T, Peper F, Matsui N (2006) Construction universality in purely asynchronous cellular automata. J Comput Syst Sci Amsterdam 72(8):1368–1385
18. The On-line Encyclopedia of Integer Sequences—OEIS (2020) Fubini numbers: number of preferential arrangements of n labeled elements. https://oeis.org/A000670

# Sandpile Toppling on Penrose Tilings: Identity and Isotropic Dynamics

**Jérémy Fersula, Camille Noûs, and Kévin Perrot**

**Abstract** We present experiments of sandpiles on grids (square, triangular, hexagonal) and Penrose tilings. The challenging part is to program such simulator; and our javacript code is available online, ready to play! We first present some identity elements of the sandpile group on these aperiodic structures, and then study the stabilization of the maximum stable configuration plus the identity, which lets a surprising circular shape appear. Roundness measurements reveal that the shapes are not approaching perfect circles, though they are close to be. We compare numerically this almost isotropic dynamical phenomenon on various tilings.

## Preamble

The experiments presented in this paper were conducted using *JS-Sandpile*, a javascript sandpile simulator we developed. The code is available at https://github.com/huacayacauh/JS-Sandpile, and it is ready for the reader to play at https://huacayacauh.github.io/JS-Sandpile/. Color codes of all our pictures are a gradation, from *almost white* = 0 grain, to *almost black* = $deg(v) - 1$ grains (for any vertex $v$), *i.e.* the darker it is, the closer to the stability threshold. Stable vertices have greyscale colors, unstable vertices have flashy colors. Table 1 presents color codes.

---

J. Fersula
Université publique, Marseille, France
e-mail: jeremy.fersula@etu.univ-amu.fr

C. Noûs
Cogitamus laboratory, Marseille, France
e-mail: camille.nous@cogitamus.fr

K. Perrot (✉)
Université publique, Marseille, France
e-mail: kevin.perrot@lis-lab.fr

 
A. Adamatzky (ed.), *Automata and Complexity*, Emergence, Complexity and Computation 42, https://doi.org/10.1007/978-3-030-92551-2_10

**Table 1** Color codes of the pictures in this article, according to the degree of the vertices (all our tilings have the same degree for all its tiles/vertices)

| $deg(v)$ | 0 grain | 1 grain | 2 grains | 3 grains | 4 grains | 5 grains | 6 grains | 7 grains |
|---|---|---|---|---|---|---|---|---|
| 3 | | | | | | | | |
| 4 | | | | | | | | |
| 6 | | | | | | | | |

## 1 Introduction

It all started in 1987 with the work of Bak, Tang and Wiesenfled [2]. Sandpiles have initially been introduced as number conserving cellular automata on the two dimensional square grid, defined by the local toppling of sand grains to neighboring cells, with statistics on chain reactions presenting scale invariance typical of phase transition phenomena in physics, the so called *self-organized criticality* [3, 34]. Soon after, Dhar realized that sandpiles have a beautiful algebraic structure, which generalizes to graphs and digraphs [10]. Since then, sandpiles have raised great interests for their simple local definitions exhibiting complex global behaviors. Following the work of Goles [23], numerous researches have been conducted on one-dimensional models under sequential update mode [24, 25, 32, 50], parallel update mode [13], and some variants such as symmetric [17, 21, 49, 51] or Kadanoff rule [48].

On the two-dimensional side, the identity of the sandpile group on square grids retains its mysteries, though the relaxation of hourglasses (toppling from a single site) begin to reveal its structure through involved partial differential equation developments [35–37, 44, 45].

The sandpile model on graphs is general enough to embed arbitrary computation, as attested by its Turing-universality [27]. In order to analyse the apparent complexity of the model in a formal framework, particular interest has been raised on the computational complexity of predicting the dynamics of sandpiles. Despite strong efforts started with Moore and Nilsson in [42] and important contributions from Goles, the computational complexity of the problem on two dimensional grids remains open between NC and P. The one-dimensional model is in NC [40, 42], as well as the Kadanoff rule [16, 19] and more general variants [20]. It is P-complete from dimension three [42], and even allows for some undecidable problems [7]. The (im)possibility of building crossing gates in two dimensions is studied in [22, 43], and similar issues on closely related threshold automata such as the majority rule have been fruitful [26, 28–31, 41]. See [18] for a survey of the results on lattices.

Achieving isotropy in a cellular automaton is not a trivial matter, as cells evolve on intrinsically anisotropic supports (grids). The authors of [9] manage to build parabolas and circles in a two-dimensional cellular automaton with $5^{13}$ states. Though optimizing the number of state was not an objective of their work, it reflects the difficulty of the task. Approximative (and simpler) approaches to build circles (isotropic diffusion) include: probabilistic methods [39, 54, 55, 58], using a continuous state space [38], or a large neighborhood to alleviate the anisotropy [14, 56, 59].

We give the definition of the sandpile model on general graphs in Sect. 2, and present its algebraic structure at the heart of the experiments conducted in this article. We also define the tilings considered as supports for the sandpile dynamics: square, triangular and hexagonal grids, Penrose tilings (kite-dart and rhombus) obtained by substitution, and Penrose tilings (rhombus) obtained by the cut and project (multigrid) method. Identity elements of the sandpile group on Penrose tilings are exposed in Sect. 3. Our main contribution is in Sect. 4, where we study numerically the isotropy observed during the stabilization process from the maximum stable configuration plus the identity.

## 2 Model

Bak, Tang and Wiesenfeld first defined in [2] the sandpile model on two-dimensional grids with von Neumann neighborhood. We present in Sect. 2.1 a general definition on any undirected multi graph, in Sect. 2.2 the algebraic structure first revealed by Dhar in [10], and in Sect. 2.3 we introduce various grids and more generally tilings on which the sandpile model can be studied.

### *2.1 Sandpiles on Graphs*

Given a (connected) finite undirected multi graph $G = (V, E)$ with a distinguished *sink* $s \in V$, let $\tilde{V} = V \setminus \{s\}$. A *configuration* $c : \tilde{V} \to \mathbb{N}$ assigns a number of sand grains to each non-sink vertex of $G$. The basic local evolution rule is the *toppling* at some vertex $v \in \tilde{V}$, which may occur when $c(v) \geq deg v$, and consists in vertex $v$ giving as many grains as it has edges to each of its neighbors. When $c(v) \geq deg v$ we say that vertex $v$ is *unstable*, and a configuration with no unstable vertex is called *stable*. Formally, let $\Delta = D - A$ be the *graph Laplacian* of $G$, *i.e.* with $D$ the degree matrix having $deg v$ on the diagonal and $A$ the adjacency matrix of $G$, and let $\tilde{\Delta}$ be the *reduced graph Laplacian* obtained from $\Delta$ by deleting the row and column corresponding to the sink $s$. With $\tilde{\Delta}_v$ the row of $\tilde{\Delta}$ corresponding to vertex $v \in \tilde{V}$, performing a toppling at $v$ corresponds to going from $c$ to $c' = c - \tilde{\Delta}_v$, which we denote $c \overset{v}{\to} c'$, or simply $c \to c'$, and $\to^*$ its reflexive-transitive closure.

As such the system is non-deterministic on configuration space $\mathbb{N}^{\tilde{V}}$, but it is straightforward to notice that any configuration $c$ converges (because of the absorbing sink vertex $s$) to a unique (because topplings commute) stable configuration, denoted $c^\circ$. This is the so called *abelian property* of sandpiles, or *fundamental theorem* of sandpiles. The vector $x$ such that $c - \tilde{\Delta} \cdot x = c^\circ$ is called the *shot vector* or *odometer function* of configuration $c$, it records how many times each vertex has toppled during the *stabilization* of $c$. The fundamental theorem of sandpiles furthermore states that the odometer function of any configuration is unique.

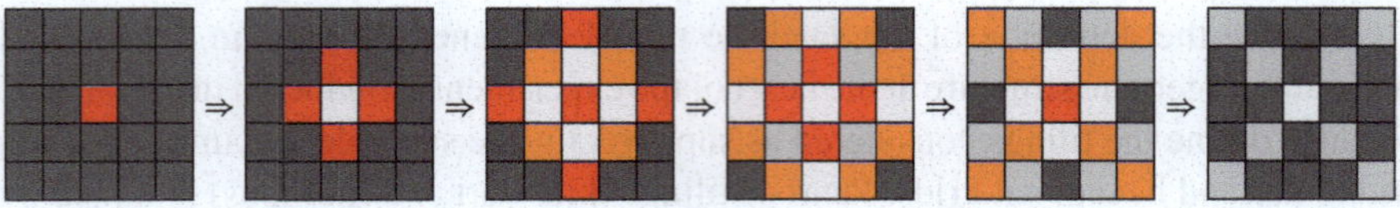

**Fig. 1** Five steps of the sandpile model on a square grid of size $5 \times 5$, corresponding to the graph with one vertex per square and north-east-south-west adjacencies (all vertices have degree 4). The sink $s$ is not pictured, squares/vertices on the border have one edge connected to $s$, and squares/vertices on the corners have two edges connected to $s$

For the purpose of simulations that take some importance in the present work, we define the deterministic *synchronous* dynamics as the evolution toppling synchronously all unstable vertices, at each step. Formally, we define $c \Rightarrow c'$ with

$$c' = c - \sum_{\substack{v \in \tilde{V} \\ c(v) \geq deg v}} \tilde{\Delta}_v.$$

An example is given on Fig. 1. For the sake of simplicity, in the following *graph* will stand for *finite undirected multi graph*.

## 2.2 Abelian Group Structure

Given a graph $G = (V, E)$ with sink $s$, let $C = \mathbb{N}^{\tilde{V}}$ denote its set of configurations. The set of *stable configurations* $C_{\texttt{stab}}$ is naturally equipped with the *operation* $\oplus$ defined as

$$c \oplus c' = (c + c')^{\circ}$$

where $c + c'$ is the componentwise addition of two configurations, *i.e.* $(c + c')(v) = c(v) + c'(v)$ for all $v \in \tilde{V}$. It is straightforward to notice that $(C_{\texttt{stab}}, \oplus)$ is a *commutative monoid* (closure, associativity, identity, commutativity), the identity being the configuration $z$ such that $z(v) = 0$ for all $v \in \tilde{V}$.

Here comes the magics of sandpiles. The set of *recurrent configurations*

$$\begin{aligned} C_{\texttt{rec}} =& \{c \in C_{\texttt{stab}} \mid \forall c' \in C : \exists c'' \in C : c' \oplus c'' = c\} \\ =& \{c \in C_{\texttt{stab}} \mid \forall c' \in C_{\texttt{stab}} : \exists c'' \in C_{\texttt{stab}} : c' \oplus c'' = c\} \end{aligned}$$

corresponds to the intersection of ideals of $C_{\texttt{stab}}$, *i.e.*

$$C_{\texttt{rec}} = \bigcap_{\substack{I \subseteq C_{\texttt{stab}} \\ I \text{ ideal of } C_{\texttt{stab}}}} I$$

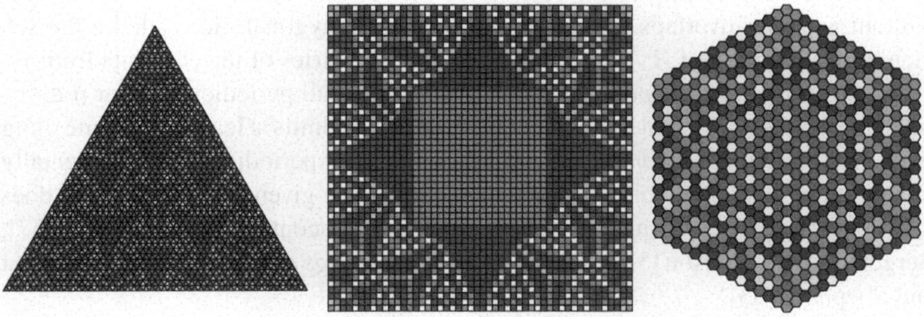

**Fig. 2** Identity elements of the sandpile group on graphs corresponding to the triangular grid of side length 50 (left, $deg(v) = 3$ for any vertex $v$), the square grid of side length 50 (center, $deg(v) = 4$ for any vertex $v$), the hexagonal grid of side length 14 (right, $deg(v) = 6$ for any vertex $v$). All at the same scale

with $I$ (right) *ideal* of $C_{\mathtt{stab}}$ if and only if $I \oplus C_{\mathtt{stab}} \subseteq C_{\mathtt{stab}}$ where $I \oplus C_{\mathtt{stab}} = \{i \oplus c \mid i \in I \text{ and } c \in C_{\mathtt{stab}}\}$. Moreover, it is a classical result of algebra that the intersection of all ideals of a *commutative semigroup* (closure, associativity, commutativity) gives an *abelian group* (closure, associativity, identity, inverse, commutativity). So $(C_{\mathtt{rec}}, \oplus)$ is an abelian group, called the *sandpile group* on graph $G$ with sink $s$. Now remark that the configuration $z$ containing no grain is (except on very restricted cases) not an element of $C_{\mathtt{rec}}$, hence the[1] identity element $e \in C_{\mathtt{rec}}$ of the sandpile group is *a priori* not obvious to construct, and it turns out that few is know about its structure on numerous interesting cases, as presented on Fig. 2. It can be proven that

$$e = (2m - (2m)^{\circ})^{\circ} \tag{1}$$

with $m$ the *maximum stable* configuration defined as $m(v) = deg(v) - 1$ for all $v \in \tilde{V}$, and $(2m)(v) = 2\,c(v)$ for all $v \in \tilde{V}$. Indeed, $2m - (2m)^{\circ}$ contains at least $deg(v) - 1$ grains at each vertex $v \in \tilde{V}$ hence its stabilization is recurrent, and furthermore it corresponds to subtracting two configurations from the same equivalence class according to relation $\leftrightarrow^*$ (the symmetric closure of $\rightarrow^*$), therefore to a configuration in the class of the identity (see [11] for details). The equality follows since the identity element of $(C_{\mathtt{rec}}, \oplus)$ is unique.

## 2.3 Tilings

We will concentrate on finite tilings, but still introduce general definitions. An *infinite tiling* $\mathcal{T}$ by $\tau$ is a covering of $\mathbb{R}^2$ by finitely many polygonal *tiles* and their images by isometry (translation, rotation, flip), *i.e.* copies of the tiles from $\tau$ cover the plane

[1] The identity element of the sandpile group is unique.

without gaps nor overlaps. Let $\tau$ be a finite set of polygonal tiles called a *tile set*, then $\mathcal{T}$ is a partition of $\mathbb{R}^2$ into countably many isometries of the elements from $\tau$.

An infinite tiling $\mathcal{T}$ is *periodic* when it has a non-null periodicity vector $\mathbf{p} \in \mathbb{R}^2$, such that $\mathcal{T} + \mathbf{p} = \mathcal{T}$. A tile set $\tau$ is *aperiodic* when it admits at least one infinite tiling (we say that $\tau$ *tiles the plane*), and none are periodic. Aperiodicity is fundamentally related to the uncomputability of the *domino problem*: given a (finite) tile set, does it tile the plane? Let us quickly mention the seminal contributions of Wang [57], Berger [5] and Robinson [52], along with the book *Tilings and Patterns* by Grünbaum and Shephard [33].

In order to consider sandpiles on tilings, we explain now how to construct finite undirected multi graphs with a distinguished sink. A *finite tiling* is simply a subset of some infinite tiling $\mathcal{T}$. Remark that this requires a finite tiling to be *extensible* into an infinite tiling, which will be the case for all our finite tilings. Indeed, given some tile set $\tau$ we will generate arbitrarily large finite tilings, which implies the existence of an infinite tiling by compactness of the set of infinite tilings by $\tau$. Two tiles are *adjacent* in $\mathcal{T}$ when they share an edge. All the tilings we consider will see their adjacent tiles share full sides, i.e. no tile will share a partial side or more than one side with another tile. These are called *edge-to-edge* tilings. Given a finite tiling, each tile corresponds to a vertex, plus an additional sink vertex corresponding to the *outside face*. The tile to tile adjacencies are given by the edge-to-edge connections, and the number of edges connecting a tile to the sink is equal to its number of sides connected to the outside face, so that the degree of every tile is equal to its number of sides. Tiles connected to the sink are said to be on the *border* of the Tiling. Note that the sandpile dynamics is given by this underlying graph, but tiles furthermore have coordinates in $\mathbb{R}^2$.

We now present the finite tilings considered in this article. The size of tiles and the position of coordinate $(0, 0)$ will be important for the observations presented in Sect. 4. Grids are illustrated on Fig. 2.

**Square grids** They correspond to the original two-dimensional sandpile model by Bak, Tang and Wiesenfled [2]. Given size $n$, it basically consists in a $n \times n$ square grid with adjacencies given by von Neumann neighborhood (north, east, south, west). Tiles on the borders have one edge connected to the sink, and tiles on the corners have two edges connected to the sink. Each tile is a square of side length 1, and the finite tiling is centered with coordinate $(0, 0)$ in the middle of the grid: if $n$ is even then coordinate $(0, 0)$ corresponds to four tiles corners; otherwise coordinate $(0, 0)$ corresponds to the center of a tile.

**Triangular grids** Given size $n$, it is made of equilateral triangles of side length 1, arranged up and down to form an equilateral triangle of side length $n$, where the three outer sides of the tiling are made of $n$ triangular tiles. Tiles on the border have one edge connected to the sink, and tiles on the corners have two edges connected to the sink. The finite tiling is centered with coordinate $(0, 0)$ at the barycenter of its three corners.

**Hexagonal grids** Given size $n$, it is made of regular hexagons of side length 1, arranged to form an hexagonal grid (orientation is *flat*) with six sides each made of $n$ tiles. Tiles on the border have two edges connected to the sink, and tiles on corners

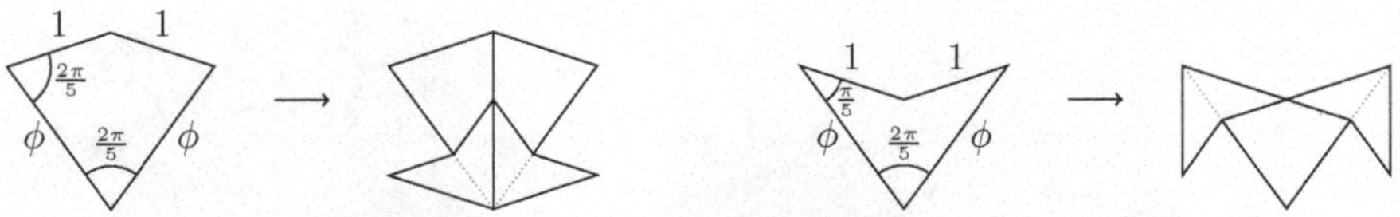


**Fig. 3** Substitutions of P2 kite (left) and dart (right) tiles

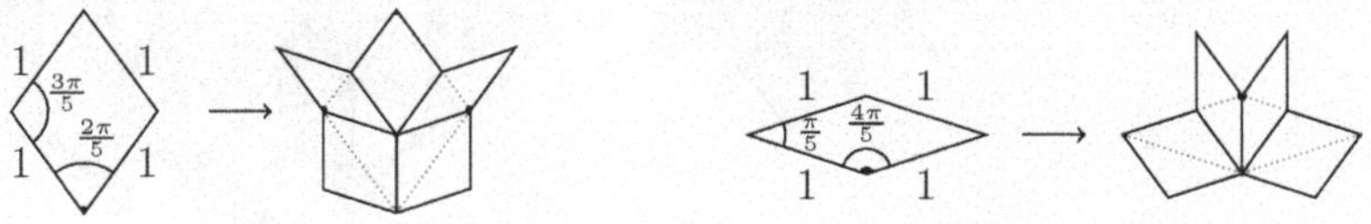


**Fig. 4** Substitutions of P3 fat (left) and thin (right) rhombi tiles. Due to the symmetry of P3 tiles, we highlight the origin point of each tile

**Fig. 5** Three iterations of the susbtitution from a *P2 Sun*, with identity elements of the sandpile group pictured

have three edges connected to the sink. The finite tiling is centered with coordinate (0, 0) at the center of the central hexagon.

**Penrose tilings** Penrose developed in [46, 47] a series of elegant aperiodic tile sets. We consider *P2* (*kite-dart*), and *P3* (*rhombus*) (tilings by P2 and P3 are *mutually locally derivable*, see [1]). Penrose tilings may be obtained by *substitution*[2] as described on Fig. 3 for P2, and Fig. 4 for P3. After substituting, we rescale all tiles up by the substitution factor $\phi = \frac{1+\sqrt{5}}{2}$, so that the tiles of the final tiling have the exact same size as the base tiles (kite, dart, fat, thin). Figure 5 presents three iterations of the substitution from a *P2 Sun*, Fig. 6 presents three iterations of the substitution from a *P2 Star*, and Fig. 7 presents three iterations of the substitution from a *P3 Sun*.

[2] Note that in order to enforce aperiodicity in P2 and P3, matching constraints should be added on tile edges, for example via notches, but the finite tiling generation methods we employ do not require such considerations.

**Fig. 6** Three iterations of the susbtitution from a *P2 Star*, with identity elements of the sandpile group pictured

**Fig. 7** Three iterations of the susbtitution from a *P3 Sun*, with identity elements of the sandpile group pictured

Coordinate $(0, 0)$ is at the center of the initial Suns and Stars, and remains at the symmetry center of tilings obtained by subsequent substitutions.

Penrose P3 tilings may alternatively be generated by *cut and project* method: consider the 5-dimensional plane $E$ spaned by vectors $(\cos \frac{2k\pi}{5})_{0\leq k<5}$ and $(\sin \frac{2k\pi}{5})_{0\leq k<5}$, passing through the point $(\frac{1}{5}, \frac{1}{5}, \frac{1}{5}, \frac{1}{5}, \frac{1}{5})$. The orthogonal projection of the 5-dimensional grid lines (1-simplices) contained in $E + [0, 1]^5$ onto $E$ gives a tiling by P3 tiles. We implement it via the dual *multigrid* method of de Bruijn [6], where one considers a 2-dimensional space and five line families (our *pentagrid*) given by the intersections of $E$ with the five 5-dimensional hyperplanes $G_i = \{x \in \mathbb{R}^5 \mid x \cdot e_k \in \mathbb{Z}\}$ for $0 \leq k < 5$, where $e_k$ is the $k$-th unit vector of $\mathbb{R}^5$. Pentagrid lines of a family are *indexed* by the value of $x \cdot e_k$, and we denote $\ell_i^k$ the line from family $0 \leq k < 5$ of index $i \in \mathbb{Z}$. Note that with this setting, no more than two pentagrid lines intersect at a given position. The 2-dimensional space is divided into polygonal *cells* delimited by pentagrid lines. Each cell $p$ is labeled by a 5-tuple of integers $(p_k)_{0\leq k<5}$ such

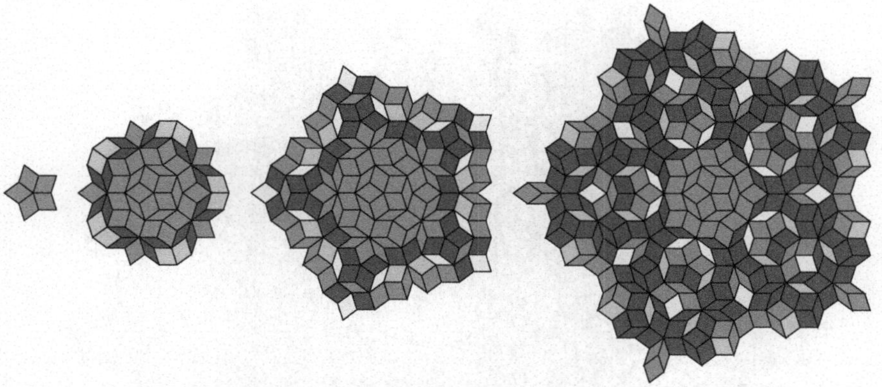

**Fig. 8** P3 cut and project tilings of sizes 1, 2, 3 and 4, with identity elements of the sandpile group pictured

that cell $p$ lies in between lines $\ell^k_{p_k}$ and $\ell^k_{p_k+1}$. To each cell $p$ corresponds a point (a tile's bound coordinate) in the 2-dimensional space, at $\sum_{0 \leq k < 5} p_k(\cos \frac{2k\pi}{5}, \sin \frac{2k\pi}{5})$. It follows that to each intersecting pair of pentagrid lines corresponds a tile, whose two edge orientations are given by the two line families. Details can be found in [15]. We bound this process to a finite *P3 cut and project tiling of size n* by considering only tiles corresponding to points of intersection lying in between $\ell^i_{-n}$ and $\ell^i_n$ for all families $0 \leq i < 5$. The coordinate $(0, 0)$ of this tiling is given by the cell labeled $(0, 0, 0, 0, 0)$, and is therefore a tile bound (shared by multiple tiles). Rhombus tiles have side length 1 as on Fig. 4. Figure 8 presents some P3 cut and project tilings.

## 3 Sandpile Identity on Penrose Tilings

We were curious to see what the identity element of the sandpile group would look like on (finite) Penrose tilings, and it seems that no particular structure appears.[3] Examples are presented on Figs. 9, 10 and 11, *JS-Sandpile* computes them from Formula 1.

We nevertheless discovered an interesting phenomenon on P3 cut and project tilings, where the identity elements seem to display some stability. Indeed, identity elements on successive sizes have a somewhat large central part of the configuration in common. This obeservation is presented on Fig. 12. We may therefore conjecture a convergence of the sandpile identity on P3 cut and project tilings: as the size $n$ increases, a larger part of the identity is fixed (remains the same for all sizes $n' \geq n$). Let us remark that this phenomenon does not seem to take place on P2 nor P3 tilings obtained by subsitution.

[3] Well, this is a bit disappointing, but we think that it is worth showing that it does not appear to be a fruitful research direction, or maybe a more insightful reader would encounter something out there....

**Fig. 9** Identity element of the sandpile group on the tiling obtained after 7 iterations of the substitution from a P2 Sun (6710 tiles)

## 4 Isotropic Dynamics

On square, triangular, hexagonal grids and Penrose tilings, a very interesting phenomenon appears[4] during the stabilization process

$$(m + e)^{\circ} = m$$

where $m$ is the maximum stable configuration ($m(v) = deg(v) - 1$ for all tile $v$) and $e$ is the identity element of the sandpile group. Indeed, during the last phase of the stabilization process leading back to $m$ (which is a uniform configuration in these cases since the number of neighbors is identical for all tiles), one can see the configuration $m$ reappear from the outside (near the border) towards the center, outside a

[4] This also takes place on other tilings, outside the scope of the present work.

**Fig. 10** Identity element of the sandpile group on the tiling obtained after 6 iterations of the substitution from a P3 Sun (5415 tiles)

shrinking circular shape, in a process step by step covering the whole configuration with tiles containing $deg(v) - 1$ grains. The first part of the stabilization process is quite involved. Two illustrations are given on Figs. 13 and 14.

## 4.1 Roundness

In order to measure this phenomenon, we introduce the *roundness* as follows. First, we partition a configuration into two parts: the outside part with all tiles having $deg(v) - 1$ grains connected to the border, and the inner part. Given a tiling $G = (V, E)$ with sink $s$ and a configuration $c : \tilde{V} \to \mathbb{N}$, let the *maximum stable components*, $\texttt{MSC}(c) = \{V_1, V_2, \ldots, V_k\}$, be the connected components of tiles (from $\tilde{V}$) having $deg(v) - 1$ grains. Then the *outer tiles* is the set

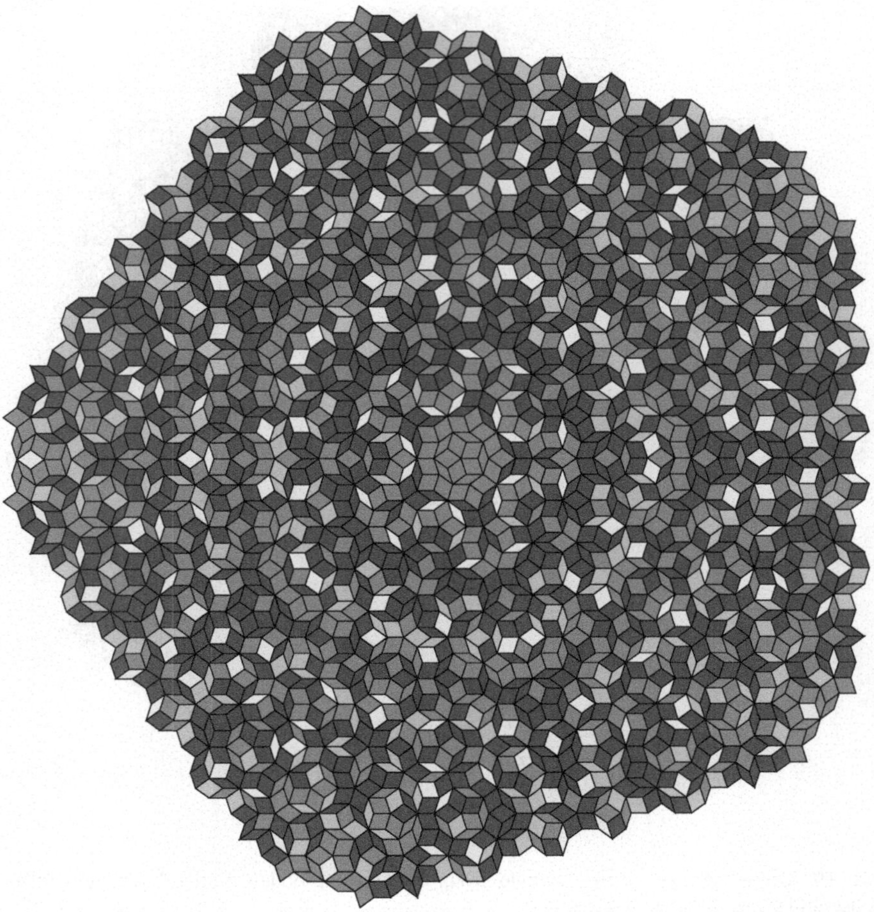

**Fig. 11** Identity element of the sandpile group on a P3 cut and project tiling of size 10 (2440 tiles)

$$\texttt{outer}(c) = \bigcup_{\substack{V_i \in \texttt{MSC}(c) \\ \exists v \in V_i : \{v,s\} \in E}} V_i$$

and the *inner tiles* is the set $\texttt{inner}(c) = \tilde{V} \setminus \texttt{outer}(c)$. From this partition of the set of tiles, we are interested in the frontier between the outer and inner tiles, and how close it is from a perfect circle. This has to do with the coordinates of tiles in the Euclidean space $\mathbb{R}^2$, so let us denote $\texttt{coord}(v)$ the set of coordinates of the bounds of some tile $v \in \tilde{V}$, and $\texttt{body}(v)$ the subset of $\mathbb{R}^2$ covered by the tile. Our convention regarding the sink $s$ is discussed below. Regarding the circle, let us denote $\texttt{B}(r)$ the inside of the circle of radius $r \in \mathbb{R}$ centered at coordinate (0, 0), *i.e.*

$$\texttt{B}(r) = \{(x, y) \in \mathbb{R}^2 \mid \sqrt{x^2 + y^2} \leq r\}.$$

**Fig. 12** Differences between successive identity elements of the sandpile group on P3 cut and project tilings. From left to right, top to bottom, are displayed the identity elements on P3 cut and project tilings of sizes 11 (2900 tiles), 12 (3520 tiles), 13 (4155 tiles), 14 (4790 tiles), 15 (5570 tiles), 16 (6250 tiles), 17 (7140 tiles), 18 (8080 tiles), 19 (8890 tiles) and 20 (9940 tiles), where the part of the configuration which differs from the previous size (sizes $n$ is compared to size $n-1$) are highlighted with redshifted colors

We may therefore denote $\texttt{body}(v) \cap \texttt{B}(r) = \emptyset$ to state that tile $v$ is entirely outside the circle of radius $r$ centered at $(0, 0)$, and $\texttt{body}(v) \subseteq \texttt{B}(r)$ (or equivalently $\texttt{coord}(v) \subseteq \texttt{B}(r)$ since we deal with polygons and the circle is convex) to state that tile $v$ is entirely inside the same circle. We define the *outer radius* as the maximum scalar $r$ such that all outer tiles are outside the circle of radius $r$ around the origin $(0, 0)$ of the tiling,

$$\underline{\texttt{r}}(c) = \max\{r \in \mathbb{R}_+ \mid \forall v \in \texttt{outer}(c) : \texttt{body}(v) \cap \texttt{B}(r) = \emptyset\}$$

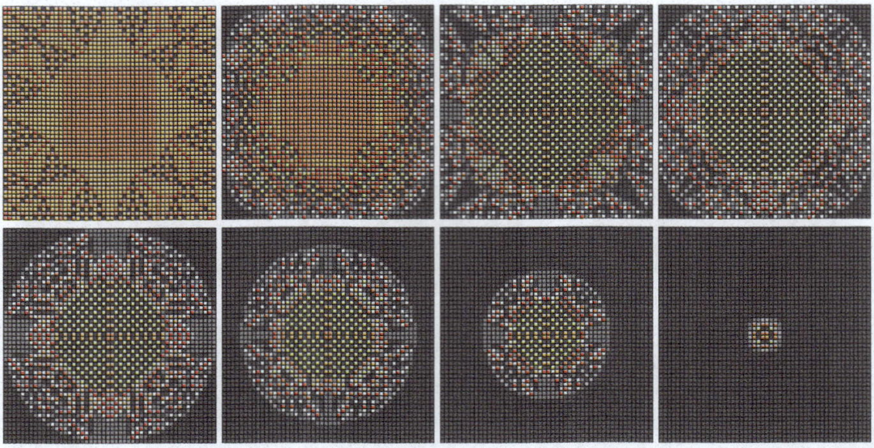

**Fig. 13** Stabilization of $(m+e)^\circ = m$ on a square grid of side length 50. From left to right, top to bottom, are displayed the configurations every 100 time steps (starting with $m+e$ at step 0, ending with time step 700). The process converges to $m$ at step 707

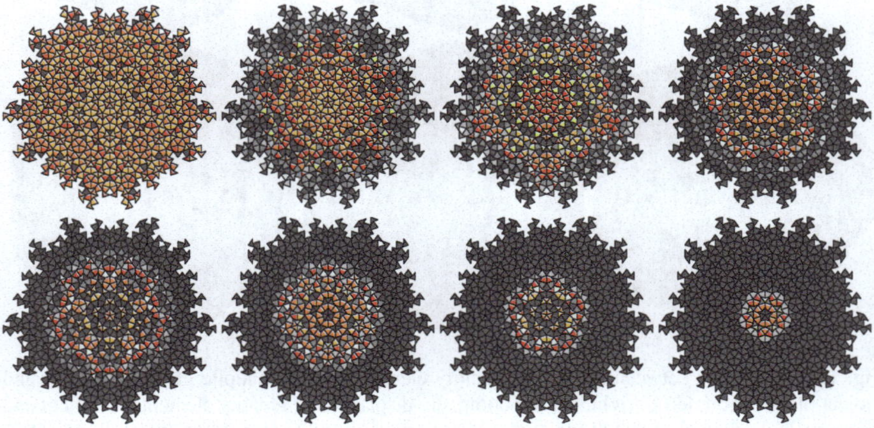

**Fig. 14** Stabilization of $(m+e)^\circ = m$ on 5 iterations of the substitution from a P2 Sun. From left to right, on top are displayed time steps 0, 50, 100, 150 and at the bottom time steps 165, 180, 195, 210. The process converges to $m$ at step 220

and the *inner radius* as the minimum scalar $r$ such that all inner tiles are inside the circle of radius $r$ around the origin (0, 0) of the tiling,

$$\overline{\mathtt{r}}(c) = \min\{r \in \mathbb{R}_+ \mid \forall v \in \mathtt{inner}(c) : \mathtt{body}(v) \subseteq \mathtt{B}(r)\}.$$

In order to deal with the case $\mathtt{outer}(c) = \emptyset$, which may for example be the case on some configurations $(m+e)$, we add the convention that the sink $s$ is an infinite tile covering all the space outside the tiling, whose coordinates are the union

**Fig. 15** Examples of roundness measures, circles of outer radius $\underline{\mathtt{r}}(c)$ in blue, inner radius $\overline{\mathtt{r}}(c)$ in red. Left: square grid of side length 50, after 500 steps from $(m+e)$ (converges to $m$ in 707 steps), $\mathtt{r}(c) \approx 21.633 - 20.224 = 1.409$. Right: 5 iterations of the substition from a P2 Sun, after 150 steps from $(m+e)$ (converges to $m$ in 220 steps), $\mathtt{r}(c) \approx 16.662 - 14.729 = 1.933$

of all bounds from tile edges adjacent to the sink, and that it always belongs to $\mathtt{outer}(c)$. As a consequence the outer radius is upper bounded by the radius of the inscribed circle (inside the finite tiling) with center at (0, 0). The case $\mathtt{inner}(c) = \emptyset$ is not problematic since the minimum defining $\overline{\mathtt{r}}(c)$ is taken on $\mathbb{R}_+$, and would therefore equal 0 as expected. The inner radius is upper bounded by the radius of the circumscribed circle (outside the finite tiling) with center at (0, 0). The *roundness of a configuration c* is then measured as the difference between the inner and outer radii,

$$\mathtt{r}(c) = \overline{\mathtt{r}}(c) - \underline{\mathtt{r}}(c).$$

Note that $0 \leq \mathtt{r}(c)$, and we have $\mathtt{r}(c) = 0$ when the frontier between inner and outer tiles is a perfect circle. Two examples of roundness are given on Fig. 15.

## 4.2 Base Roundnesses

Remark that since all our tiles are polygonal (with three to six sides), we cannot expect to reach roundness 0 (except when the stabilization process has converged to $m$). To get some easy to interpret base values, we consider the diameter of each tile, as the diameter of the circumscribed circle around the tile (smallest radius of a circle having the tile entirely is its interior). See Fig. 16. The greatest diameter of some tiling's tiles can be interpreted as an upper bound on the best achievable roundness, for any radius. Indeed, consider some tiling and a circle $C$ (of radius at most equal to the inscribed radius), then all tiles having the center of their circumscribed circle on

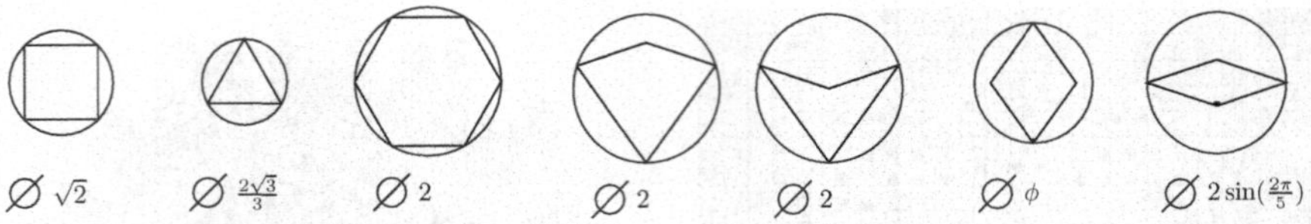


**Fig. 16** Diameters of all tiles, as the diameter of a circumscribed circle (in purple). All tiles at the same scale

**Table 2** Base roundnesses as the greatest diameter of some tiling's tiles

| Tiling | Square grids | Triangular grids | Hexagonal grids | P2 tilings | P3 tilings |
|---|---|---|---|---|---|
| Base roundness | $\sqrt{2}$ $\approx 1.414$ | $\frac{2\sqrt{3}}{3}$ $\approx 1.155$ | 2 | 2 | $2\sin(\frac{2\pi}{5})$ $\approx 1.902$ |

or outside the circle $C$ can be set as outer tiles, and all tiles having the center of their circumscribed circle inside the circle $C$ can be set as inner tiles, which results in a roundness smaller than twice the radius of the greatest tile's radius, i.e. smaller than the greatest tile's diameter. We obtain the *base roundnesses* presented on Table 2.

## 4.3 Plots

We now present plots of roundness measured during the stabilization process $(m + e)^{\circ} = m$ on the different tilings considered in this article. We decompose the stabilization process from $m + e$ to $m$ into two phases:

- **phase 1**: the dynamics is erratic,
- **phase 2**: the set of inner tiles slowly shrinks, until reaching $\texttt{inner}(m) = \emptyset$.

The *beginning of phase 2* is defined as the first step such that the inner radius is smaller or equal to the inscribed radius of the tiling (maximum radius of a circle entirely inside the finite tiling, and centered at the origin). Remark that at the beginning of phase 2, all tiles on the border of the tiling are outer tiles, because the polygonal shape of any inner tile on the border would otherwise lead to the inner radius being greater than the inscribed radius of the tiling. An important observation is that, in all the experiments presented in this article and performed during its preparation, once in phase 2 with all border tiles as outer tiles, then all border tiles *remain* outer tiles,[5] and that the inner radius *remains* smaller or equal to the inscribed radius. Two full examples of roundness plots are given on Figs. 17, 19, and their companion Figs. 18, 20.

[5] They all remain stable with $deg(v) - 1$ grains until reaching $m$. Observe that any outer tile receiving some grain would topple, and that toppling any outer tile would result in toppling the whole maximum stable component it belongs to.

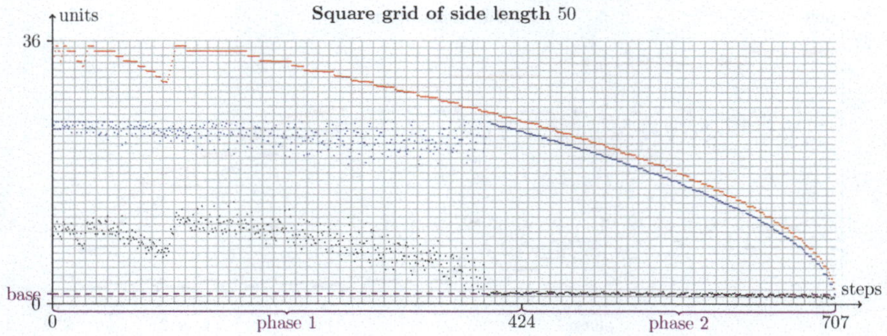


**Fig. 17** Plot of the roundness during the stabilization process $(m+e)^\circ = m$, on the square grid of side length 50. At each step we plot the inner radius $\overline{\mathtt{r}}(c)$ in red, outer radius $\underline{\mathtt{r}}(c)$ in blue, and roundness $\mathtt{r}(c)$ in black. Grid has one row per unit and one column per 10 time steps. For example, at step 3, with $m+e \Rightarrow c^1 \Rightarrow c^2 \Rightarrow c^3$, we observe that $\overline{\mathtt{r}}(c^3) = 25\sqrt{2} \approx 35.355$ (the radius of the circumscribed circle around the whole tiling) and that $\underline{\mathtt{r}}(c^3) = 25$ (the radius of the inscribed circle inside the whole tiling), so that $\mathtt{r}(c^3) = 25(\sqrt{2}-1) \approx 10.355$ (at step 0 some outer tiles near the $x=0$ and $y=0$ axis give different radii). Phase 2 begins at step 424, and configuration $m$ is reached at step 707. The base roundness of this tiling (dashed) is $\sqrt{2}$. See also the companion Fig. 18

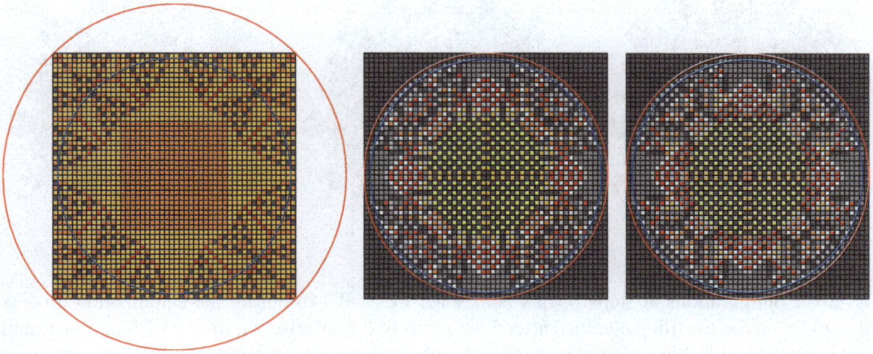

**Fig. 18** Configurations at steps 0 (this is $m+e$), 423 and 424 during the stabilization process $(m+e)^\circ = m$, on the square grid of side length 50. Inner radii in red, outer radii in blue. One can observe the phase transition occurring at step 424: the inner radius becomes smaller or equal to the inscribed radius of the tiling

Experimental results, picturing only phase 2 of the stabilization process from $m+e$ to $m$, are presented on Figs. 21 and 22. Note that during all the experiments we have performed in preparing this article, we have observed similar behaviors for all other sizes.

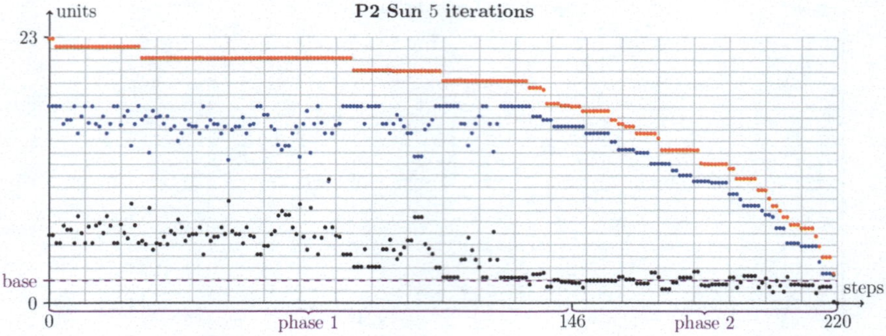


**Fig. 19** Plot of the roundness during the stabilization process $(m+e)^\circ = m$, on the tiling obtained after 5 iterations of the substitution from a P2 Sun. At each step we plot the inner radius $\overline{\mathtt{r}}(c)$ in red, outer radius $\underline{\mathtt{r}}(c)$ in blue, and roundness $\mathtt{r}(c)$ in black. Grid has one row per unit and one column per 10 time steps. For example, at step 0 we observe that $\overline{\mathtt{r}}(m+e) \approx 22.940$ (the radius of the circumscribed circle around the whole tiling) and that $\underline{\mathtt{r}}(m+e) \approx 17.069$ (the radius of the inscribed circle inside the whole tiling), so that $\mathtt{r}(m+e) \approx 5.871$. Phase 2 begins at step 146, and configuration $m$ is reached at step 220. The base roundness of this tiling (dashed) is 2. See also the companion Fig. 20

**Fig. 20** Configurations at steps 0 (this is $m+e$), 145 and 146 during the stabilization process $(m+e)^\circ = m$, on the tiling obtained after 5 iterations of the substitution from a P2 Sun. Inner radii in red, outer radii in blue. One can observe the phase transition occurring at step 146: the inner radius becomes smaller or equal to the inscribed radius of the tiling

## *4.4 Computation Times and Data*

The graphics were computed on our personal machines (standard laptops), with simulation times and data presented on Table 3. Details on the implementation (no parallelization) can be found at https://github.com/huacayacauh/JS-Sandpile/wiki/Roundness.

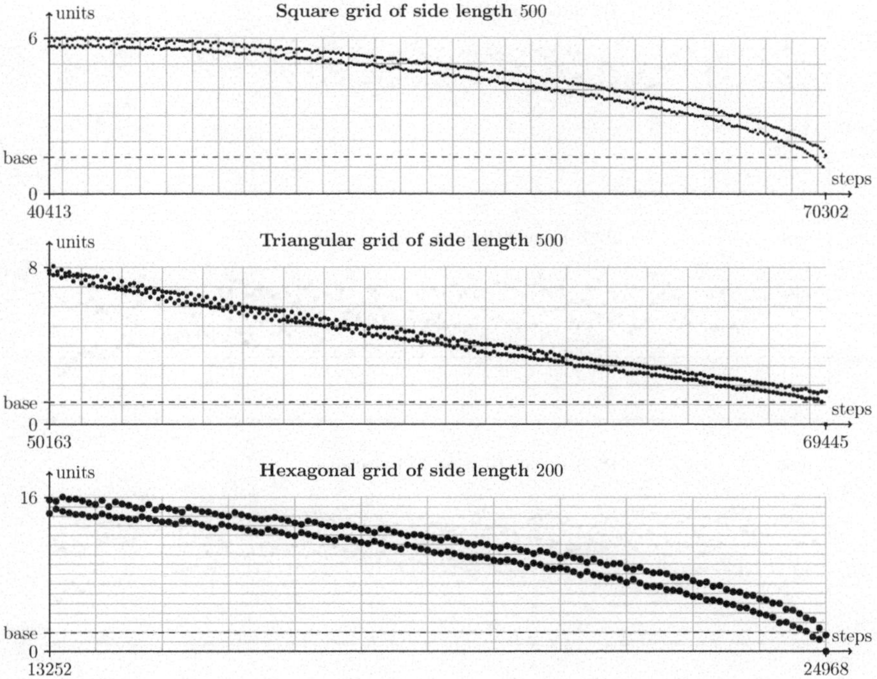


**Fig. 21** Plots of the roundness during phase 2 of the stabilization process $(m+e)^{\circ} = m$, on the square grid of side length 500, triangular grid of side length 500, and hexagonal grid of side length 200. Grids have one row per unit and one column per 1000 time steps. Two points are drawn every 100 time steps: the maximum and minimum roundness values $\mathtt{r}(c)$ observed among these 100 steps

**Table 3** Data regarding the computations generating the graphics presented on Figs. 21 and 22. Running times are given for the computation of identities (from Formula 1) and roundness measures

| Tiling | Size/iter. | Tiles | Identity (seconds) | Roundness (seconds) | Phase 2 begin step | Stabilization step |
|---|---|---|---|---|---|---|
| Squaregrid | 500 | 250000 | 825 | 1294 | 40413 | 70302 |
| Triangulargrid | 500 | 250000 | 1143 | 689 | 50163 | 69445 |
| Hexagonalgrid | 200 | 120601 | 271 | 405 | 13252 | 24968 |
| P2 Sunsubst. | 11 | 327750 | 2931 | 1850 | 42724 | 67326 |
| P2 Starsubst. | 11 | 234410 | 944 | 640 | 27337 | 41918 |
| P3 Sunsubst. | 10 | 266860 | 2227 | 1570 | 42067 | 62333 |
| P3 cut& project | 100 | 249610 | 3145 | 2386 | 39210 | 66495 |

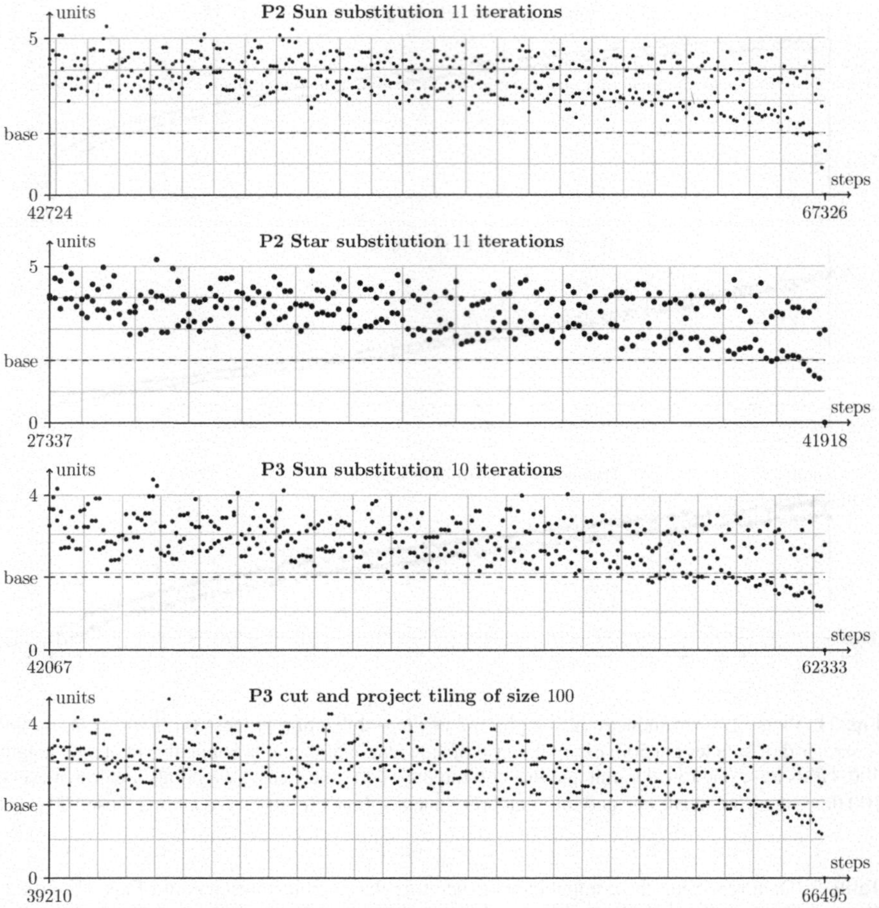


**Fig. 22** Plots of the roundness during phase 2 of the stabilization process $(m + e)^{\circ} = m$, on the P2 Sun after 11 iterations of the substitution, P2 Star after 11 iterations of the substitution, P3 Sun after 10 iterations of the substitution, P3 cut and project tiling of size 100. Grids have one row per unit and one column per 1000 time steps. Two points are drawn every 100 time steps: the maximum and minimum roundness values $\mathtt{r}(c)$ observed among these 100 steps

## *4.5 Analysis of Roundness Measures*

Well, it appears clearly that the circular shapes observed during the stabilization process $(m + e)^{\circ} = m$ are not asymptotically approaching perfect circles. Indeed, in all experiments conducted on large tilings (Figs. 21 and 22), at the beginning of phase 2 the roundness is above the base roundness, whereas the base roundness is an upper bound on the best achievable roundness (see Sect. 4.2).

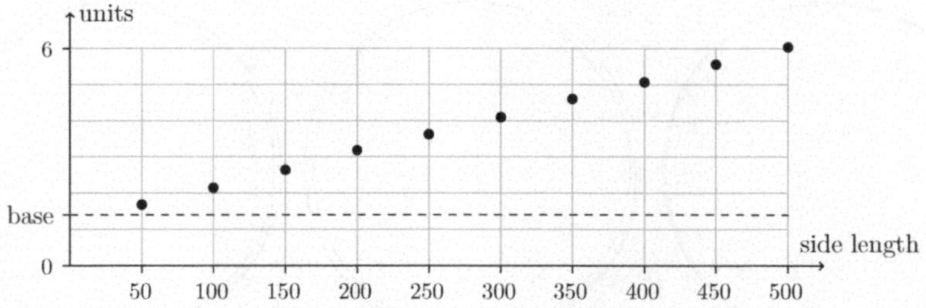


**Fig. 23** Maximum roundness $\mathtt{r}(c)$ measured during phase 2 of the stabilization process $(m+e)^{\circ} = m$ on square grids of various side lengths, illustrating the correlation between roundness and size

As the circular shapes shrink, it is normal to see the roundness decrease until the value 0 on configuration $m$ (all plots reach the minimum roundness 0 at stabilization step). On the other hand, it appears that roundness and tiling size are correlated, meaning that as the size of the tiling increases, configurations at the beginning of phase 2 are less round. This is illustrated on Fig. 23.

Although they do not tend to perfect circles, we may admit that these shapes are close to be, in regard of tiling sizes (Table 3). Increases of roundness are visible to the naked eye on hexagonal and triangular grids which deviate largely from a perfect circle, but are hardly noticeable without measurements on other tilings. Figure 24 illustrates this with the frontier between outer and inner tiles, on the configuration obtained at the beginning of phase 2 during the experiments from Sect. 4.3 (since these configurations are quite large, we picture only the frontier in order to reduce the numerical weight of the present document). If they are not perfect circles, then we may naturally ask: what characterize these frontier shapes for each tiling?

Finally, we see on Fig. 24 (first line) that the frontier on grids (respectively square, triangular and hexagonal) reflect the shape on the border of the tilings, somehow "rounded". Indeed, it appears that the number of corners of the grids are equal to the number of parts where the frontier deviates from a circle (the number of times it goes from the inner radius to the outer radius). Theses symmetries are expected, they come from the symmetry of the grids and therefore the symmetries of the dynamics. A natural attempt would be to experiment the roundness of a square grid cropped to a circle, in order to remove the effect of the border's shape. Despite the fact that the difference between the circumscribed and inscribed radius is smaller than the base roundness, this does not lead to significantly smaller roundness measurements (frontiers are not closer to perfect circles) during phase 2 of the stabilization process $(m+e)^{\circ} = m$, as shown on Fig. 25. It feels that the frontier reflects the anisotropy of the square grid itself rather than that of its border's shape.

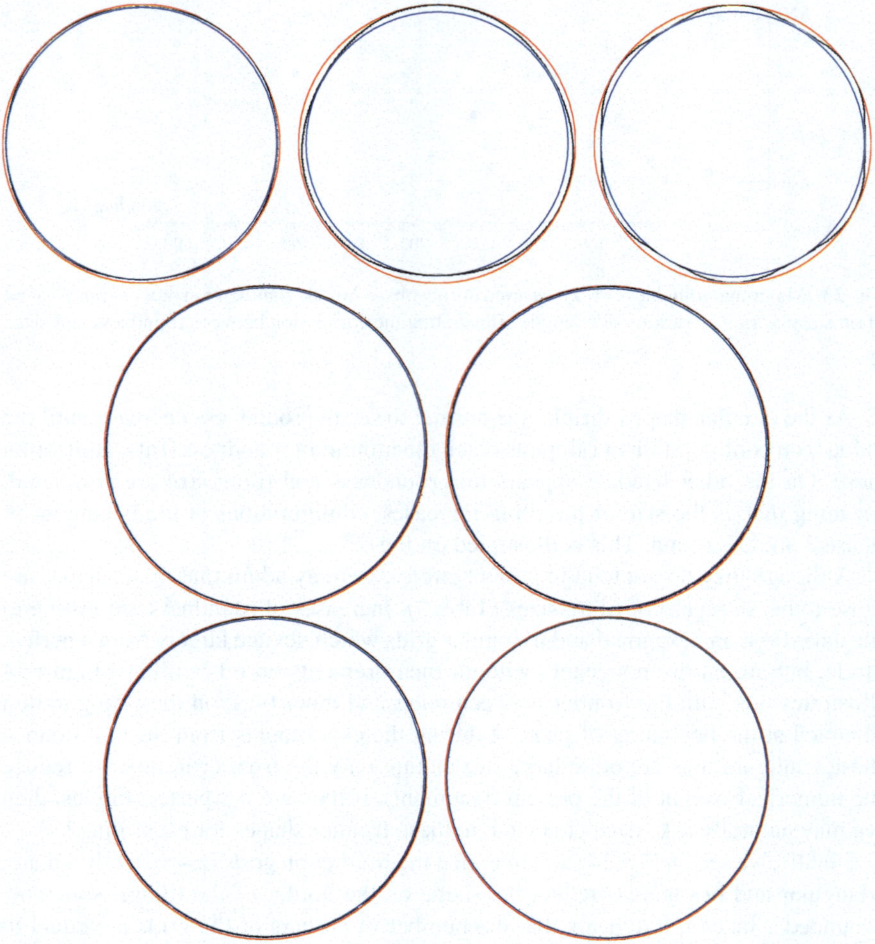

**Fig. 24** Frontier between inner and outer tiles on the configuration obtained at the beginning of phase 2, as the set of edges shared by one inner tile and one outer tile. Outer radius in blue, inner radius in red. Top: square grid of side length 500 at step 40413 ($\mathtt{r}(c) \approx 5.687$), triangular grid of side length 500 at step 50163 ($\mathtt{r}(c) \approx 7.755$), hexagonal grid of side length 200 at step 13252 ($\mathtt{r}(c) \approx 14.470$). Middle: P2 Sun after 11 iterations of the subsitution at step 42724 ($\mathtt{r}(c) \approx 4.246$), P2 Star after 11 iterations of the subsitution at step 27337 ($\mathtt{r}(c) \approx 4.064$). Bottom: P3 Sun after 10 iterations of the subsitution at step 42067 ($\mathtt{r}(c) \approx 3.220$), P3 Sun cut and project tiling of size 100 at step 39210 ($\mathtt{r}(c) \approx 2.452$)

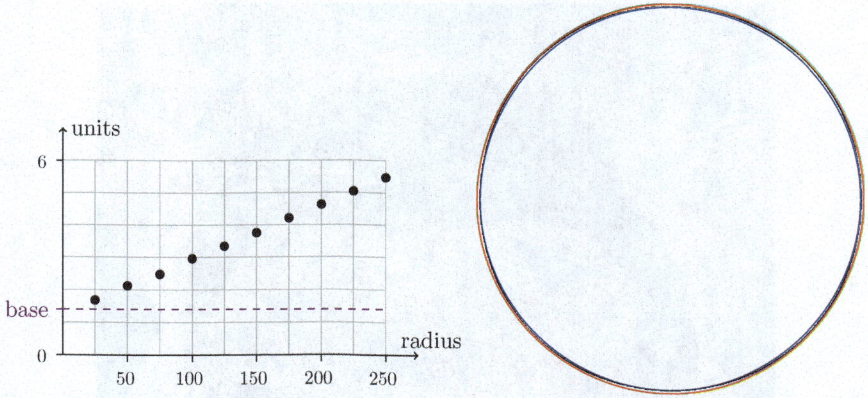


**Fig. 25** Left: maximum roundness $\mathtt{r}(c)$ measured during phase 2 of the stabilization process $(m+e)^{\circ} = m$ on square grids cropped to circles of various radii. Right: frontier between inner and outer tiles on the configuration obtained at the beginning of phase 2 (step 32131, $\mathtt{r}(c) \approx 4.073$) on a square grid cropped to a circle of radius 250 (the border of the tiling is pictured in green)

## 5 Conclusions and Perspectives

The experiments presented in this article are reproducible with *JS-Sandpile* (links in Preamble). The software implements no parallelization mechanism, which would allow to perform larger simulations (e.g. using GPU). Nevertheless we believe that this would not lead to qualitatively different observations.

We have presented some identity elements on Penrose tilings, revealing no obvious structure related to these famous aperiodic tilings. Identities are highly sensitive to the shape of the tiling, and it may be the case that other finite croppings of (infinite) Penrose tilings lead to different observations. We tried to build the most "natural" finite Penrose tilings: Suns and Stars obtained by substitution, along with cut and project from a 5-dimensional hypercube.

The apparent isotropy, observed during the stabilization of the maximum stable configuration plus the identity element of the sandpile group, has been measured through the notion of roundness. Experiments revealed that these frontiers are actually not approaching perfect circles on any tiling under consideration. Two further directions may be investigated. First, if these shapes are not perfect circles, then how to characterize them on each tiling (especially on grids)? Second, which tilings would lead to frontiers approaching perfect circles (if any)? The modest attempt to crop a square grid to a circle fails, suggesting that this may not be easy to achieve on tilings which are intrinsically anisotropic at the tile level.

Penrose tilings exhibit frontiers closer to perfect circles, though they also deviate significantly from their best achievable roundness. It is not very surprising to observe on Penrose tilings a behavior similar to regular lattices, as they are quasi-periodic, and in some sense the "most regular" aperiodic tilings. Let us open the large perspective of

**Fig. 26** Identity element of the sandpile group on a square tiling constructed from a picture of Eric Goles (original photo credit Claudiogonzalo85). Colors try to reflect his welcoming home country: almost white = 0, light purple = 1, chilean red = 2, chilean blue = 3, and background (sink) in black

considering other tilings, for example tilings with higher quasi-periodicity functions (above affine [4, 8, 12][6]), Cayley graphs [53], or hyperbolic planes (e.g. with Poincaré disk models).

Finally, could the stabilization process $(m + e)^{\circ} = m$ shed some light on the enigmatic identity elements on grids? This is really out of reach for our present

[6] The difficulty may be to find constructions from non Wang tiles, because Wang tiles would lead to square grids for the sandpile model to play on (as we remove tile decorations).

knowledge, but we hope that the recent progresses of Levine et al. [35–37, 44, 45] are breaking some scientific locks in the domain of sandpiles.

**Tribute** This work has been prepared as a tribute to the contributions of Eric Goles to the field of discrete dynamical systems, Fig. 26 is for him.

**Acknowledgements** The authors are thankful to Valentin Darrigo for his contributions to *JS-Sandpile*, to Victor Poupet for stressing that the apparent isotropy of the $(m+e)^\circ$ process is surprising at the occasion of a talk given by KP during AUTOMATA'2014 in Himeji, to Thomas Fernique for sharing his expertise (and code!) regarding the cut and project method, and to Christophe Papazian for useful comments on quasi-periodicity functions. The work of JF was conducted while a Master student at Aix-Marseille Université, doing an internship at the LIS laboratory (UMR 7020), both in Marseille, France. The work of KP was funded mainly by his salary as a French State agent and therefore by French taxpayers' taxes, affiliated to Aix-Marseille University, University de Toulon, CNRS, LIS, UMR 7020, Marseille, France and University Côte d'Azur, CNRS, I3S, UMR 7271, Sophia Antipolis, France. Secondary financial support came from ANR-18-CE40-0002 FANs project, ECOS-Sud C16E01 project, and STIC AmSud CoDANet 19-STIC-03 (Campus France 43478PD) project.

## References

1. Baake M, Scholottmann M, Jarvis PD (1991) Quasiperiodic tilings with tenfold symmetry and equivalence with respect to local derivability. J Phys A: Math Gen 24(19):4637–4654. https://doi.org/10.1088/0305-4470/24/19/025
2. Bak P, Tang C, Wiesenfeld K (1987) Self-organized criticality: an explanation of the 1/f noise. Phys Rev Lett 59:381–384. https://doi.org/10.1103/PhysRevLett.59.381
3. Bak P, Tang C, Wiesenfeld K (1988) Self-organized criticality. Phys Rev A 38(1):364–374. https://doi.org/10.1103/PhysRevA.38.364
4. Ballier A, Jeandel E (2010) Computing (or not) Quasi-periodicity functions of tilings. In: Journées automates cellulaires, pp 54–64
5. Berger R (1966) The undecidability of the domino problem. Mem Am Math Soc 66. https://doi.org/10.1090/memo/0066
6. de Bruijn NG (1981) Algebraic theory of penrose's non-periodic tilings of the plane. ii. Indag Math 84(1):53–66. https://doi.org/10.1016/1385-7258(81)90017-2
7. Cairns H (2018) Some halting problems for abelian sandpiles are undecidable in dimension three. SIAM J Discret Math 32(4):2636–2666. https://doi.org/10.1137/16M1091964
8. Cervelle J, Durand B (2004) Tilings: recursivity and regularity. Theor Comput Sci 310(1):469–477. https://doi.org/10.1016/S0304-3975(03)00242-1
9. Delorme M, Mazoyer J, Tougne L (1999) Discrete parabolas and circles on 2D cellular automata. Theor Comput Sci 218(2):347–417. https://doi.org/10.1016/S0304-3975(98)00330-2
10. Dhar D (1990) Self-organized critical state of sandpile automaton models. Phys Rev Lett 64:1613–1616. https://doi.org/10.1103/PhysRevLett.64.1613
11. Dhar D, Ruelle P, Sen S, Verma DN (1995) Algebraic aspects of abelian sandpile models. J Phys A 28(4):805–831. https://doi.org/10.1088/0305-4470/28/4/009
12. Durand B (1999) Tilings and quasiperiodicity. Theor Comput Sci 221(1):61–75. https://doi.org/10.1016/S0304-3975(99)00027-4
13. Durand-Lose JO (1998) Parallel transient time of one-dimensional sand pile. Theor Comput Sci 205(1–2):183–193. https://doi.org/10.1016/S0304-3975(97)00073-X
14. Fast VG, Efimov IR (1991) Stability of vortex rotation in an excitable cellular medium. Phys D: Nonlinear Phenom 49(1):75–81. https://doi.org/10.1016/0167-2789(91)90196-G

15. Fernique T (2007) Pavages, fractions continues et géométrie discrète. PhD thesis, Université Montpellier II
16. Formenti E, Goles E, Martin B (2012) Computational complexity of avalanches in the Kadanoff sandpile model. Fundam Inform 115(1):107–124. https://doi.org/10.3233/FI-2012-643
17. Formenti E, Masson B, Pisokas T (2007) Advances in symmetric sandpiles. Fundam Inform 76(1–2):91–112. https://doi.org/10.5555/2366416.2366423
18. Formenti E, Perrot K (2019) How hard is it to predict sandpiles on lattices? a survey. Fundam Inform 171:189–219. https://doi.org/10.3233/FI-2020-1879
19. Formenti E, Perrot K, Rémila E (2014) Computational complexity of the avalanche problem on one dimensional Kadanoff sandpiles. In: Proceedings of AUTOMATA'2014, LNCS, vol 8996, pp 21–30. https://doi.org/10.1007/978-3-319-18812-6_2
20. Formenti E, Perrot K, Rémila E (2018) Computational complexity of the avalanche problem for one dimensional decreasing sandpiles. J Cell Autom 13:215–228
21. Formenti E, Pham VT, Phan HD, Tran TH (2014) Fixed-point forms of the parallel symmetric sandpile model. Theor Comput Sci 533:1–14. https://doi.org/10.1016/j.tcs.2014.02.051
22. Gajardo A, Goles E (2006) Crossing information in two-dimensional sandpiles. Theor Comput Sci 369(1–3):463–469. https://doi.org/10.1016/j.tcs.2006.09.022
23. Goles E (1992) Sand pile automata. Ann de l'institut Henri Poincaré (A) Physique théorique 56(1):75–90
24. Goles E, Kiwi M (1993) Games on line graphs and sand piles. Theor Comput Sci 115(2):321–349. https://doi.org/10.1016/0304-3975(93)90122-A
25. Goles E, Latapy M, Magnien C, Morvan M, Phan HD (2004) Sandpile models and lattices: a comprehensive survey. Theor Comput Sci 322(2):383–407. https://doi.org/10.1016/j.tcs.2004.03.019
26. Goles E, Maldonado D, Montealegre P, Ollinger N (2017) On the computational complexity of the freezing non-strict majority automata. In: Proceedings of AUTOMATA'2017, pp 109–119. https://doi.org/10.1007/978-3-319-58631-1_9
27. Goles E, Margenstern M (1997) Universality of the chip-firing game. Theor Comput Sci 172(1–2):121–134. https://doi.org/10.1016/S0304-3975(95)00242-1
28. Goles E, Montealegre P (2014) Computational complexity of threshold automata networks under different updating schemes. Theor Comput Sci 559:3–19. https://doi.org/10.1016/j.tcs.2014.09.010
29. Goles E, Montealegre P (2016) A fast parallel algorithm for the robust prediction of the two-dimensional strict majority automaton. In: Proceedings of ACRI'2016, pp 166–175. https://doi.org/10.1007/978-3-319-44365-2_16
30. Goles E, Montealegre P, Perrot K, Theyssier G (2017) On the complexity of two-dimensional signed majority cellular automata. J Comput Syst Sci 91:1–32. https://doi.org/10.1016/j.jcss.2017.07.010
31. Goles E, Montealegre-Barba P, Todinca I (2013) The complexity of the bootstraping percolation and other problems. Theor Comput Sci 504:73–82. https://doi.org/10.1016/j.tcs.2012.08.001
32. Goles E, Morvan M, Phan HD (2002) Sandpiles and order structure of integer partitions. Discret Appl Math 117(1–3):51–64. https://doi.org/10.1016/S0166-218X(01)00178-0
33. Grünbaum B, Shephard GC (1986) Tilings and patterns. WH Freeman & Co
34. Kadanoff LP, Nagel SR, Wu L, Zhou S (1989) Scaling and universality in avalanches. Phys Rev A 39(12):6524–6537. https://doi.org/10.1103/PhysRevA.39.6524
35. Levine L, Pegden W, Smart CK (2016) Apollonian structure in the abelian sandpile. Geom Funct Anal 26:306–336. https://doi.org/10.1007/s00039-016-0358-7
36. Levine L, Pegden W, Smart CK (2017) The apollonian structure of integer superharmonic matrices. Ann Math 186(1):1–67. https://doi.org/10.4007/annals.2017.186.1.1
37. Levine L, Peres Y (2017) Laplacian growth, sandpiles, and scaling limits. Bull Am Math Soc 54(3):355–382. https://doi.org/10.1090/bull/1573
38. Marek M (2013) Grid anisotropy reduction for simulation of growth processes with cellular automaton. Phys D: Nonlinear Phenom 253:73–84. https://doi.org/10.1016/j.physd.2013.03.005

39. Markus M, Hess B (1990) Isotropic cellular automaton for modelling excitable media. Nature 347:56–58. https://doi.org/10.1038/347056a0
40. Miltersen PB (2005) The computational complexity of one-dimensional sandpiles. In: Proceedings of CiE'2005, pp 342–348. https://doi.org/10.1007/11494645_42
41. Moore C (1997) Majority-vote cellular automata, ising dynamics, and P-completeness. J Stat Phys 88(3):795–805. https://doi.org/10.1023/B:JOSS.0000015172.31951.7b
42. Moore C, Nilsson M (1999) The computational complexity of sandpiles. J Stat Phys 96:205–224. https://doi.org/10.1023/A:1004524500416
43. Nguyen VH, Perrot K (2018) Any shape can ultimately cross information on two-dimensional abelian sandpile models. In: Proceedings of AUTOMATA'2018, LNCS, vol 10875, pp 127–142. https://doi.org/10.1007/978-3-319-92675-9_10
44. Pegden W, Smart CK (2013) Convergence of the abelian sandpile. Duke Math J 162(4):627–642. https://doi.org/10.1215/00127094-2079677
45. Pegden W, Smart CK (2020) Stability of patterns in the abelian sandpile. Ann Henri Poincaré 21:1383–1399. https://doi.org/10.1007/s00023-020-00898-1
46. Penrose R (1974) The role of aesthetics in pure and applied mathematical research. Bull Inst Math Its Appl 10(2):266–271
47. Penrose R (1979) Pentaplexity: a class of non-periodic tilings of the plane. Math Intell 2:32–37. https://doi.org/10.1007/BF03024384
48. Perrot K (2013) Les piles de sable Kadanoff. PhD thesis, École normale supérieure de Lyon
49. Perrot K, Phan HD, Pham VT (2011) On the set of fixed points of the parallel symmetric sand pile model. In: Proceedings AUTOMATA'2011, DMTCS. Open Publishing Association, pp 17–28
50. Phan, H.D.: Structures ordonnées et dynamiques de piles de sable. Ph.D. thesis, Université Paris 7 (1999)
51. Phan HD (2008) Two sided sand piles model and unimodal sequences. ITA 42(3):631–646. https://doi.org/10.1051/ita:2008019
52. Robinson RM (1971) Undecidability and nonperiodicity for tilings of the plane. Inven Math 12:177–209. https://doi.org/10.1007/BF01418780
53. Roka Z (1994) Automates cellulaires sur graphes de cayley. PhD thesis, École Normale Supérieure de Lyon
54. Schepers HE, Markus M (1992) Two types of performance of an isotropic cellular automaton: stationary (Turing) patterns and spiral waves. Phys A: Stat Mech Its Appl 188(1):337–343. https://doi.org/10.1016/0378-4371(92)90277-W
55. Schönfisch B (1997) Anisotropy in cellular automata. Biosystems 41(1):29–41. https://doi.org/10.1016/S0303-2647(96)01664-4
56. Sirakoulis GC, Karafyllidis I, Thanailakis A (2005) A cellular automaton for the propagation of circular fronts and its applications. Eng Appl Artif Intell 18(6):731–744. https://doi.org/10.1016/j.engappai.2004.12.008
57. Wang H (1961) Proving theorems by pattern recognition —II. Bell Syst Tech J 40:1–41. https://doi.org/10.1002/j.1538-7305.1961.tb03975.x
58. Weimar JR (1997) Cellular automata for reaction-diffusion systems. Parallel Comput 23(11):1699–1715. https://doi.org/10.1016/S0167-8191(97)00081-1
59. Weimar JR, Tyson JJ, Watson LT (1992) Diffusion and wave propagation in cellular automaton models of excitable media. Phys D: Nonlinear Phenom 55(3):309–327. https://doi.org/10.1016/0167-2789(92)90062-R

# On Boolean Automata Isolated Cycles and Tangential Double-Cycles Dynamics

**Jacques Demongeot, Tarek Melliti, Mathilde Noual, Damien Regnault, and Sylvain Sené**

**Abstract** Our daily social and political life is more and more impacted by social networks. The functioning of our living bodies is deeply dependent on biological regulation networks such as neural, genetic, and protein networks. And the physical world in which we evolve, is also structured by systems of interacting particles. Interaction networks can be seen in all spheres of existence that concern us, and yet, our understanding of interaction networks remains severely limited by our present lack of both theoretical and applied insight into their clockworks. In the past, efforts at understanding interaction networks have mostly been directed towards applications. This has happened at the expense of developing understanding of the generic and fundamental aspects of interaction networks (properties and behaviours due primarily to the fact that a system is an interaction network, as opposed to properties and behaviours rather due to the fact a system is a *genetic* interaction network for instance). Intrinsic properties of interaction networks (*e.g.*, the ways in which they transmit information along entities, their ability to produce this or that kind of global dynamical behaviour depending on local interactions) are thus still not well understood. Lack of fundamental knowledge tends to limit the innovating power of applications. Without more theoretical fundamental knowledge, applications cannot evolve deeply and become more impacting. Hence, it is necessary to better appre-

J. Demongeot (✉)
Université publique, Grenoble, France
e-mail: jacques.demongeot@univ-grenoble-alpes.fr

T. Melliti · D. Regnault
Université publique, Évry, France
e-mail: tarek.melliti@univ-evry.fr

D. Regnault
e-mail: damien.regnault@univ-evry.fr

M. Noual
Freie Universität Berlin, Berlin, Germany
e-mail: m.noual@fu-berlin.de

S. Sené (✉)
Université publique, Marseille, France
e-mail: sylvain.sene@univ-amu.fr

 
A. Adamatzky (ed.), *Automata and Complexity*, Emergence, Complexity and Computation 42, https://doi.org/10.1007/978-3-030-92551-2_11

hend and comprehend the intrinsic properties of interaction networks, notably the relations between their architecture and their dynamics and how they are affected by and set in *time*. In this chapter, we use the elementary mathematical model of Boolean automata networks as a formal archetype of interaction networks. We survey results concerning the role of feedback cycles and the role of intersections between feedback cycles, in shaping the asymptotic dynamical behaviours of interaction networks. We pay special attention to the impact of the automata updating modes.

## 1 Introduction

Interaction networks occupy an important place in our daily life. We see this today in particular with the massive use of social networks, the fundamental implications and mechanisms of which we hardly have any understanding of. And social media is just one example among many other kinds of interaction networks that affect us consequentially. At all levels of our lives there are interaction networks, that can be comprised as sets of entities that interact locally with each other over time.

In an interaction network, local interactions take place. And as a result of these, the network as a whole exhibits global behaviours that generally remain difficult to explain on the sole basis of local processes. Undeniably one of the most telling examples lies at the origin of all living organisms: genes and their regulation, associated with other mechanisms inducing variability (splicing, role of the chromatin, *etc.*). It is currently accepted in biology and medicine, that a better understanding of genetic regulation is a necessary condition for improving our knowledge of life, which in turn could allow us to achieve a more precise comprehension of pathologic mechanisms, and to access more targeted and thereby more efficient therapies.

Whilst the application aspects of interaction networks are obviously of real and tangible importance, and they have progressed quantitatively at a frantic pace over the last twenty years, the more fundamental aspects aiming at understanding and analysing the intrinsic properties of these networks have so far received less attention from the scientific community. Current applications, in particular those emerging in biology (the privileged domain of application of the present chapter), are awaiting significant theoretical advances to continue their qualitative development. Undoubtedly, such theoretical advances may be the fruit of the combination of computer science and discrete mathematics. Together with bioinformatics, viewed here as "the study of computer processes in biotic systems" [29, 30], they seem particularly suited to meet the needs arising from applications. Indeed, real networks, through the entities and interactions that compose them, can naturally be viewed as discrete objects. They can easily be represented by computer models which capture their essence thanks to a high level of abstraction. Where traditional (continuous) mathematical modelling focuses mainly on the quantitative characteristics of networks, the interest of discrete computer science modelling comes from the qualitative nature of the questions that it raises, which makes it possible to realise that the central elements of networks are not the entities themselves but rather the interactions that link them.

This survey adopts the qualitative point of view of fundamental computer science to examine some properties of interaction networks. This choice of approach is consistent with the origins of modern computing. Indeed, investigations of formal neural networks [41] and cellular automata [72] in the 1940s, both strongly inspired by natural processes, helped establish the first links between data processing and biology. Also, the pioneering work of McCulloch and Pitts introduced automata networks as a fundamental model of interaction networks. For reasons discussed later, we choose to rely on the same model here.

We thus here study automata networks, and more specifically Boolean automata networks, from a fundamental point of view interested in developing a qualitative understanding of networks. We focus on understanding how feedback cycles that are parts of network architectures, come to influence the asymptotic dynamical properties of these networks. By way of methods from discrete dynamical systems theory, enumerative combinatorics, algorithms and complexity theory, we explore the behavioural diversity of cycles and their tangential intersections.

Another point of focus in this chapter is how updating modes influence the dynamical behaviours of cycles and cycle intersections (and those of the networks composed partly by them). From both the fundamental and the applied points of view, updating modes are known to be of decisive importance in the shaping of a network behaviour. Updating modes define the way in which the states of the automata of a network are updated as a function of a discrete time. The time being unbounded towards the future, there is an infinite number of different possible updating modes which we can come with and update the automata of a network according to. We focus here on the two modes that are the most customary in the literature: the parallel mode and the asynchronous mode. The parallel mode is a deterministic and periodic mode. At each time step, it updates the states of all the automata in a network. The asynchronous mode is a non-deterministic mode. It allows for any possible series of sequential updates—which update exactly one automaton per time step—to take place. We show how the differences between these two updating modes imply profound differences in the network dynamics.

This chapter is a synthesis of results about interaction cycles and variations of interaction cycles derived since the early 2000s. The list of results presented here is not exhaustive. Details and demonstrations of these results can be found in the literature [16, 42, 47, 49, 63].

We will start with a presentation of the history of automata networks in the context of modern fundamental computer science, including a mention of the links that this science has always entertained with biology. Main definitions and notations will be explained after that. Then, we will briefly present seminal results obtained in the 1980s that highlight the crucial importance of the role of feedback cycles on the dynamical expressiveness and behavioural diversity of interaction networks. This will lead us to focus on the dynamical and combinatorial properties of isolated cycles when they are subjected to the parallel and asynchronous updating modes. And finally, before concluding, we will take a step towards contextualising cycles and look into tangential intersections of cycles, called "tangential double-cycles" for short.

# 2 Automata Networks: Between Fundamental Computer Science and Biology

Automata network research is part of natural computation field in computer science. And the scope of natural computation is twofold. First, it designs and develops models of computations that draw inspiration from natural phenomenology. Second, it manipulates such models so as to build up a firm, albeit necessarily incomplete, grasp of biological reality natural phenomena. As mentioned above, automata networks were initially introduced as a theoretical model of neural networks. And since the late 1960s, the literature evidences that they are also relevant as models of genetic regulation networks. In this section, we provide a comprehensive overview of the interconnections between computer science and biology manifested through automata networks.

## *2.1 Biology as an Inspiration for Modern Computer Science*

Generally, when we are interested in the history of the so-called modern computer science, we go back to the 1930s. This period saw the development of classical computing paradigms such as the recursive functions of Herbrand and Gödel, the main works of which are available in [20], the lambda-calculus developed initially in [7], and also the Turing machines [70]. However, to think of modern computer science only in terms of these paradigms is to forget a whole family of less conventional models that have grounded many developments in computer science of relevance today. This is the family of automata networks that we mainly owe to McCulloch and Pitts, and von Neumann, whose first elements of the theory date from the 1940s. Based on advances at the time, the original works [41, 72] highlight a desire to develop the science of computation while inspiring and advancing the modelling of natural biological phenomena. Thus, McCulloch and Pitts introduced the model of formal neural networks which provides an abstraction of neural interactions. They showed in particular that propositional logic can represent neural events and that these networks can be considered, to a certain extent, as a universal model of computation. A little later, in the late 1940s, von Neumann developed cellular automata in order to "compare natural and artificial automata" and "abstract the logical structure of life". The result of his work was the construction of the first self-reproducing and universal cellular automaton.

Formal neural networks and cellular automata constitute the founding base of the theory of automata networks, an automata network being defined "roughly" as a set of entities (automata) which interact with each other. Their interactions happen in a discrete time, according to transition functions which are local to the entities. The two major differences that distinguish these two models relate to the number of interacting entities and the nature of the network on which the entities are placed. Indeed, cellular automata are defined by default as having an *infinite* number of entities (a.k.a. cells). And the entities are placed on a *regular and homogeneous* network (in general $\mathbb{Z}^d$, with $d > 0$). Formal neural networks (also called threshold Boolean automata networks) have a *finite* number of entities. And the entities are placed on an *irregular and heterogeneous* network. These differences give to each of these models its own characteristics that have enabled strong advances in computer science.

By their infinite nature, cellular automata have been mainly studied for their computability properties. Among them, we find the Turing universality of the self-reproducing automaton [72], of the Game of Life of Conway [5] and of the elementary cellular automaton 110 [8]. Also, in 1971, Smith showed that any Turing machine could be simulated by a cellular automaton defined on $\mathbb{Z}$ [64]. In addition, cellular automata proved to be good mathematical tools for studying the parallel functioning of computers, of which they were at the origin of systolic architectures [39]. Finally, within the framework of dynamical system theory, the desire to understand their behavioural diversity was at the origin of studies of complexity [40, 73]. Formal neural networks have also brought a lot of progress. In their original article, McCulloch and Pitts showed that they can simulate any Boolean function. Kleene resumed this work. Based on their finite nature, he proved that the languages recognised by these objects are regular, which also allowed him to introduce the concept of finite automata [38]. Behavioural characterisation conditions were also given and algebraic methods were then developed within this framework [11, 18, 25, 31]. This last reference highlights in particular, strong links between these networks and the Boolean model of genetic regulation networks introduced in [33, 34]. This latter model is at the origin of numerous works emphasising the interest taken by computer science in the context of research in theoretical biology.

## 2.2 *Computer Science as a Methodological Source for Biology*

Understanding the mechanisms of biological regulation, in all their diversity, is one of the major current problems in molecular biology. This was notably highlighted by Jacob and Monod in the early 1960s, in particular in [32, 45]. However, from the end of the 1960s, an observation was shared by two biologists, Kauffman (biochemist and biophysicist) and Thomas (biochemist and geneticist). The usual methods of treatment stemming from molecular biology are not adapted to treat as a whole such a problem at the genetic level. According to them, the experimental nature of the methods specific to biology at the time can only provide a piecemeal response to this problem and needs to be supplemented by methodological approaches. On the basis of Delbrück's work, according to which there are links between differentiated cell types and the attractors of theoretical network models [12], Kauffman and Thomas proposed to use discrete mathematics to go beyond simple observational knowledge of regulatory systems, advocating in a sense that biology must move towards more general and systematic approaches to living things. This resulted in two articles which organise and federate a whole section of research at the frontier between discrete mathematics and theoretical biology [33, 66].

Kauffman is thus the first to have proposed a model of genetic regulatory networks, based on formal neural networks. This model is known as *Boolean networks*. It is a formalisation of regulation where genes are the vertices of a randomly constructed graph. Genes interact over time (discrete). Their interactions are dictated by local Boolean transition functions. They determine whether the genes can be expressed or not, that is to say, transcribed or not [35]. Originally, this model is based on two strong hypotheses: the interactions are based on an architecture of $k$-regular graphs, namely graphs of which all the vertices have the same number of neighbours; the evolution is perfectly synchronous (or parallel). Relaxations of these hypotheses were subsequently carried out [2, 19]. They gave rise to applications to biological problems such as the analysis of the behaviour of the yeast regulatory network [36], and more generally to the analysis of signalling networks [1, 28]. In 1973, Thomas opposed the parallelism hypothesis and developed another method that sought to be closer to "genetic reality" [66]. This method comes with two new ideas. The first proposes to represent the causal dynamics of genetic regulations by means of an *asynchronous* state transition system. The second idea is to represent the networks themselves by digraphs whose arcs are signed according to the promoting or inhibiting nature of interactions. In [67], Thomas introduces two fundamental conjectures, proven in the discrete framework in [52, 53]. The first one (resp. The second one) states that the presence of a positive cycle (resp. of a negative cycle), composed of an even number (resp. an odd number) of inhibitory edges, in the architecture of the network is necessary for dynamical multi-stationarity (resp. for the existence of an oscillating limit regime). Beyond theoretical research, this method has been widely applied in biology, such as for example to the immune response [37, 43, 62] or to the infection of *Escherichia coli* by λ phage [27, 65, 69].

Although they were initiated by scientists from biology, these two visions emphasise the relevance of automata networks and thus bear the mark of computer science and discrete mathematics. The contribution of mathematician F. Robert played a key role. From the end of the 1960s he pioneered the study of automata networks from a more fundamental and formal point of view. Indeed, Kauffman and Thomas made "arbitrary" choices regarding the ways of updating the automata over time. But Robert was interested in the updating modes as such and their influences on the network behaviours. He formalised the concepts of block-sequential iterations and chaotic iterations [55–59], which make it possible to obtain updating modes that are partly synchronous and asynchronous. This makes perfect sense in theoretical biology since there is no biological argument today to define the temporal organisation of genetic regulations. In addition, the work carried out by Robert and his collaborators made it possible to establish solid theoretical bases (simple and general) for the behavioural study of automata networks [9, 21–23] while keeping in mind their representational capacities for biology [10, 13, 15]. In this theoretical framework, Robert proved the essential role played by cycles in defining the intrinsic behavioural properties of networks: his theorem stipulates that any acyclic network has a trivial behaviour and admits to the temporal asymptote only a single point fixed. Again, Robert's work has found many applications in biology, including modelling the genetic control of the flower development of *Arabidopsis thaliana* [14, 44, 61] and the study of ventral invagination during gastrointestinal morphogenesis in Drosophila [4].

### 2.3 Boolean Automata Networks, A Simple but Complex Model

From a general point of view, automata networks can be used to model any system which satisfies the following three properties:

- It is made up of distinct entities that interact with each other;
- Each entity is characterised by a variable quantity, which precisely calls to be translated in terms of states of the corresponding automaton in the model;
- The events undergone by the system, like the mechanisms that are at their origin, cannot be observed directly or integrally with certainty. Only their consequences are, that is, changes that are fully accomplished.

These three properties impose very few restrictions on the set of systems that can be abstracted and thus modelled by automata networks. These theoretical objects are therefore generic models of a very wide variety of real systems. It is therefore quite easy to understand the reasons that pushed scientists to use them and to keep studying them in the context of "fundamental bioinformatics".

Let us return to entities' characteristic "variable quantity" mentioned above. Translating the quantity in terms of automata states, calls for a first exercise of formalisation. This consists in choosing whether what interests us in the variation of the quantity is of a Boolean, discrete or continuous nature. As an illustration, let us take the example of genetic regulation and choose the action of a gene as a variable quantity. If, in the action of this gene, what interests us is its expression (and its non-expression), then the state of the automaton chosen to model the gene should be Boolean. If it is the different ways that this gene has of acting on the other elements of the system that interests us, then we can choose to match an automaton state with each way. This induces a discrete formalism which can obviously be encoded without loss in a Boolean formalism, since an automaton with $k$ states can be represented by $\log_2(k)$ Boolean automata. Finally, if we measure the action of the gene by means of the concentration of proteins it produces, continuous formalism turns out to be the most natural. On the other hand, it brings a quantitative character. If this aspect is not desired, the tendency will then be to approximate the protein concentration function at intervals in order to fall back into a discrete, even Boolean framework by considering only extreme concentrations for example. We can therefore grant different statuses to the Boolean context depending on whether we see it as a direct modelling of reality or as an approximation or encoding of an intrinsically continuous or discrete modelling. Note that direct Boolean modelling is consistent with the choice to focus on the state changes of the automata rather than on their states themselves. By analogy to mechanics, if we see automata as internal combustion engines, the interest relates to the fact that an engine is capable of going from the "off" state to the "on" state (and vice versa) rather than the amount of electricity supplied by the battery to start or on that released by the candles to cause the explosion and initiate the movement. Under this assumption, Boolean abstraction is necessary and sufficient. Furthermore, in order to place ourselves in the context of modelling in biology, it should be emphasised that the discourse of biologists is generally imbued with syntactic elements of propositional logic. It is not uncommon to hear sentences such as: "in the absence of the repressor $\alpha$, the $\beta$ gene is expressed" or even "if the products of the $\alpha$ and $\beta$ genes form a complex, the latter promotes the expression of the $\gamma$ gene while these genes tend to inhibit its expression when they are in monomeric form". This syntax also fits perfectly with a direct modelling of reality in the Boolean formalism.

In addition, Boolean automata networks derive other interesting benefits from their simplicity of definition. In particular, they provide a framework with clearly defined contours, ideal for tackling fundamental problems around the modelling of interacting entity systems. Given the variety of their nature and the current state of our knowledge, the problems in question could not currently benefit from significantly more elaborate frameworks. This would inevitably lead to delaying the initial questions and to destructuring the problem posed by paying attention to additional problems induced by the set of parameters to be considered and not intrinsically included in the initial problem. For these issues, on the contrary, Boolean automata networks offer only what is essential and facilitate the manipulation of a minimal concept of causality, which is rooted in the notion of state changes. Their merit therefore lies in the reliability of the information they potentially provide, delivered by their very high level of abstraction that also makes it possible to obtain general laws that remain valid in more specific contexts. In other words, it is crucial to understand that this simplicity of definition does not necessarily make them "simplistic", and does not detract from their ability to model complex phenomena that they allow to analyse qualitatively with a surprising subtlety.

## 3 General Definitions and Notations

Informally, a Boolean automata network is comprised of abstract entities that interact with each other. The abstract entities are called automata. Automata have states which can take one of two values: either $0$ (inactive) or $1$ (active). The states of automata can change over the course of a discrete time. They change under the influence of the states of other automata in the network. This section aims to present the formalism of this model, giving the main definitions and useful notations in the rest of the chapter.

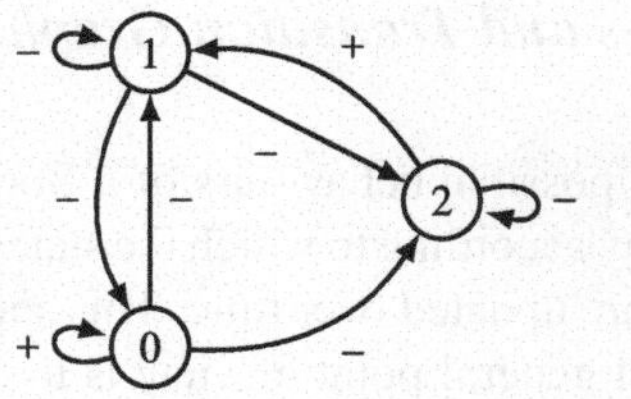


$$\begin{cases} f_0(x) = x_0 \vee \neg x_1 \\ f_1(x) = \neg x_0 \vee \neg x_1 \vee x_2 \\ f_2(x) = \neg x_0 \vee \neg x_1 \vee \neg x_2 \end{cases}$$

**Fig. 1** A Boolean automata network of size 3: its interaction graph (on the left), the ordered set of its local transition functions (on the right)

## *3.1 Boolean Automata Networks*

Let $\mathbb{B} = \{0, 1\}$ and $V = \{0, \dots, n-1\}$ be a set of $n$ Boolean automata such that $\forall i \in V, x_i \in \mathbb{B}$ represents the *state* of automaton $i$. A *configuration* $x$ of a Boolean automata network $f$ of size $n$ assigns a value of $\mathbb{B}$ to each of the automata of $V$ and is classically noted as a vector $x = (x_0, \dots, x_{n-1})$ that is a vertex of the $n$-cube $\mathbb{B}^n$, or as a binary word $x = x_0 \dots x_{n-1}$. Formally, a *Boolean automata network* $f$ of size $n$ whose set of automata is $V$ is an ordered set of $n$ Boolean functions, such that $f = (f_i : \mathbb{B}^n \to \mathbb{B} \mid i \in V)$. Given $i \in V$, $f_i$ is the *local transition function* of automaton $i$. It predetermines its evolution from any configuration $x$: if $i$ is updated in configuration $x$ at time $t$, it goes from state $x_i(t)$ to state $f_i(x(t)) = x_i(t+1)$.

Let $s : \mathbb{B} \to \mathbb{1}$, with $\mathbb{1} = \{-1, 1\}$, defined such that $s(b) = b - (\neg b)$, the function allowing to convert a Boolean number into a signed integer in $\mathbb{1}$. In this chapter, we pay particular attention to the state changes of automata, which leads us to introduce the following notations for all $x$ in $\mathbb{B}^n$:

$$\forall i \in V,\ \bar{x}^{\{i\}} = (x_0, \dots, x_{i-1}, \neg x_i, x_{i+1}, \dots, x_{n-1}).$$

$$\forall W \subseteq V,\ \forall i \notin W,\ \bar{x}^{W \cup \{i\}} = \overline{\bar{x}^{W}}^{\{i\}}.$$

The *sign of an interaction* from $i$ to $j$ in configuration $x$ is defined by $\text{sign}_x(i, j) = s(x_i) \cdot (f_j(x) - f_j(\bar{x}^{\{i\}}))$. The *effective interactions* in $x$ belong to $E(x) = \{(i, j) \in V \times V \mid \text{sign}_x(i, j) \neq 0\}$. From there, we define the *interaction graph* of $f$ as being the oriented graph $G = (V, E)$, where $E = \bigcup_{x \in \mathbb{B}^n} E(x)$ is the set of interactions. In this chapter, the Boolean automata networks discussed (see Sect. 3.3) are special in the sense that their interaction graphs are simple, namely that there can only be one signed interaction $(i, j) \in E$. If it is signed positively (resp. negatively), we say that it is activating (resp. inhibiting) and the state of $j$ tends to mimic (resp. to oppose) that of $i$. In the following, the interaction graphs will be signed for convenience of reading (see Fig. 1).

### 3.2 Updating Modes and Transition Graphs

In order to determine the possible behaviours of a Boolean automata network, it is essential to specify the way according to which the states of the automata (or, abusing language, the automata) are updated over time. This specification is what we call an updating mode. The most general point of view is to consider all the possibilities. This amounts to seeing the evolution of a network as a discrete dynamical system associated with a relation so that, for each configuration, $2^n - 1$ outgoing transitions are taken into account, namely a transition for each subset of automata whose states can be updated. More precisely, for all $W \neq \emptyset \subseteq V$, we define the update function $F_W : \mathbb{B}^n \to \mathbb{B}^n$ such that:

$$\forall x \in \mathbb{B}^n, \forall i \in V, \; F_W(x)_i = \begin{cases} f_i(x) & \text{if } i \in W, \\ x_i & \text{otherwise.} \end{cases}$$

Thus, for the most general updating mode, called the *elementary updating mode*, the global behaviour of the network is given by the elementary transition graph $\mathscr{G}_e = (\mathbb{B}^n, T_e)$, where $T_e = \{(x, F_W(x)) \mid x \in \mathbb{B}^n, W \neq \emptyset \subseteq V\}$, introduced in [47, 63].

The transitions $(x, F_i(x))$ that involve updating a single automaton $i \in V$ are called *asynchronous transitions*. The transitions $(x, F_W(x))$, with $|W| > 1$, that induce the updating of several automata are called *synchronous transitions*. The subgraph $\mathscr{G}_a = (\mathbb{B}^n, T_a)$ of $\mathscr{G}_e$ whose set of arcs $T_a = \{(x, F_{\{i\}}(x)) \mid x \in \mathbb{B}^n, i \in V\}$ equals the set of asynchronous transitions of network is called the *asynchronous transition graph*. This graph defines the asynchronous dynamics of the network that corresponds to its dynamics when it evolves according to the asynchronous updating mode, *i.e.* such that in each configuration, only $n$ transitions are considered, one for each automaton. This updating mode has been widely used in the studies of Thomas and his collaborators [49, 50, 52–54, 67, 68]. As an illustration, the asynchronous transition graph of the network depicted in Fig. 1 is presented in Fig. 2 (left). However, because it is the most "natural" (mathematically speaking) when only the local transition functions are known, and because it allows to give relevant insights through transition graphs of smaller sizes, the *parallel updating mode* occupies a very important place in the literature on discrete dynamical systems in general. When a network evolves in parallel, all of its automata update their states at each time step. In other words, the parallel transition graph of a network is $\mathscr{G}_p = (\mathbb{B}^n, T_p)$, where $T_p = \{(x, F_V(x)) \mid x \in \mathbb{B}^n\}$ is the subset of the perfectly synchronous transitions of $T_e$. The parallel transition graph of the network given in Fig. 1 is presented in Fig. 2 (right).

Let us now specify notations and vocabulary relative to dynamical behaviours of networks. Consider an arbitrary Boolean automata network $f$ of size $n$, an updating mode $\mu$, the associated transition graph $\mathscr{G}_\mu = (\mathbb{B}^n, T_\mu)$, and $x \in \mathbb{B}^n$ one of its possible configurations. A *trajectory* of $x$ is any path in $\mathscr{G}_\mu$ that starts from $x$. A strongly connected component of $\mathscr{G}_\mu$ which does not admit any outgoing transition

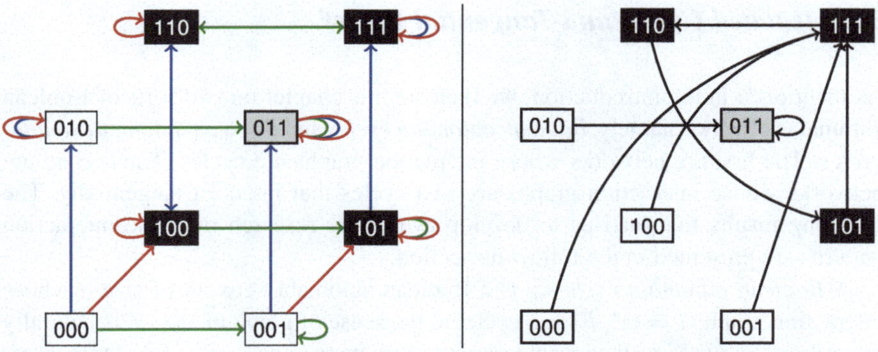


**Fig. 2** Two transition graphs of the Boolean automata network defined in Fig. 1. Left panel: its asynchronous transition graph where every → (resp. → and →) represents an update of automaton 0 (resp. 1 and 2). Right panel: its parallel transition graph. In each graph, stable configurations (a.k.a. fixed points) are depicted in light gray while recurring configurations belonging to stable oscillations (a.k.a. limit cycles) are depicted in dark gray

is an asymptotic behaviour of $(f, \mu)$, that we classically designate as an *attractor* of $(f, \mu)$. A configuration of $\mathbb{B}^n$ that belongs to an attractor is a *recurring configuration*. Given an attractor, its *length* is the number of recurring configurations that compose it. An attractor of length 1 (resp. of length strictly greater than 1) is a *stable configuration*, a.k.a. a *fixed point* (resp. a *stable oscillation*, a.k.a. a *limit cycle*) of $(f, \mu)$. If $\mu$ is a deterministic updating mode, such as the parallel mode, attractors are simple cycles and the term *period* is preferred to refer to their length. Finally, we define the *convergence time of a configuration x* as the length of its shortest trajectory which makes it reach a recurring configuration. The *convergence time of the network* is the greatest convergence time of all its $2^n$ configurations. In the transition graphs presented in this chapter, by convention, we associate the light gray color with stable configurations and the dark gray color with recurrent configurations belonging to stable oscillations. Thus, in Fig. 2, it can be seen that configuration 011 is stable for the asynchronous and parallel updating modes. This is a direct consequence of the fact that a stable configuration that is a fixed point of $f$ (implicitly evolving in parallel) is preserved by every updating mode since it corresponds to the vector of the local fixed points of the local transition functions. We also observe that the asynchronous transition graph admits a stable oscillation of size 4 while the parallel update mode admits a stable oscillation of size 3. This illustrates that stable oscillations are generally not conserved when the updating mode is changed. This is a current important field of study in the context of interaction networks and Boolean automata networks.

### 3.3 Isolated Cycles and Tangential Cycles

As mentioned in the introduction, we focus in this chapter on two sorts of Boolean automata networks, namely, *Boolean automata cycles* and *Boolean automata double-cycles*. The first are networks whose interaction graphs are cycles. The second are networks whose interaction graphs are two cycles that intersect tangentially. The founding results that lead us to develop ever more research on these interaction patterns are presented in the following section.

A *Boolean automata cycle* $\mathscr{C}_n$ is a Boolean automata network of size $n$ whose interaction graph $G = (V, E)$ is a cycle, in the sense of graph theory. $V$ is naturally assimilated to $\mathbb{Z}/n\mathbb{Z}$, so that considering two automata $i$ and $j$ of $V$, $i + j$ represents $i + j \mod n$. Thus, a Boolean automata cycle $\mathscr{C}_n$ is defined as an ordered set of local transition functions of arity 1 that are such that: $\forall i \in V,\ f_i : \mathbb{B}^n \to \mathbb{B}$, and either $f_i(x) = x_{i-1}$ or $f_i(x) = \neg x_{i-1}$. Note that there are two types of Boolean automata cycle, positive and negative. A Boolean automata cycle is a *positive cycle* $\mathscr{C}^+$ (resp. a *negative cycle* $\mathscr{C}^-$) if it is composed of an even number (resp. of an odd number) of inhibiting interactions.

A *Boolean automata double-cycle* $\mathscr{D}_n$, with $n = \ell + r - 1$, is a Boolean automata network of size $n$ composed of two Boolean automata cycles $\mathscr{C}_\ell$ and $\mathscr{C}_r$ which tangentially intersect in one automaton, automaton 0. In the following, for reasons of clarity, we prefer the notation $\mathscr{D}_{\ell,r}$. The set of automata of cycle $\mathscr{C}_\ell$ is $V^{\mathscr{L}} = \mathbb{Z}/\ell\mathbb{Z} = \{0, ..., \ell - 1\}$ and that of cycle $\mathscr{C}_r$ is $V^{\mathscr{R}} = 0 \cup \{\ell - 1 + i \mid i \neq 0 \in \mathbb{Z}/r\mathbb{Z}\}$. In a Boolean automata double-cycle, by definition, all the local transition functions are of arity 1 except that of automaton 0 that is of arity 2. In this work, we only consider locally monotonous Boolean automata double-cycle, which induces that function $f_0$ is defined as $f_0(x) = f_0^{\mathscr{L}}(x) \diamond f_0^{\mathscr{R}}(x) = x_{\ell-1} \diamond x_{n-1}$, where $\diamond \in \{\wedge, \vee\}$. Note that the choice of operator $\diamond$ only changes the position of the configurations on the trajectories. In other words, whatever the chosen updating mode, given two Boolean automata double-cycle $\mathscr{D}_{\ell,r}$ and $\mathscr{D}'_{\ell,r}$ such that $f_0(x) = f_{\ell-1}(x) \wedge f_{n-1}(x)$ and $f'_0(x) = f'_{\ell-1}(x) \vee f'_{n-1}(x)$, their respective transition graphs are identical up to an isomorphism on the configurations. In addition, it is trivial to determine one from the other, replacing the configurations with their opposites. Consequently, in the rest of this chapter, to insist on this property, we will use without loss of generality $f_0(x) = f_{l-1}(x) \vee f_{n-1}(x)$ (resp. $f_0(x) = f_{\ell-1}(x) \wedge f_{n-1}(x)$) for Boolean automata double-cycles evolving according to the parallel (resp. asynchronous) updating mode. Since they are made up of two Boolean automata cycles, it is easy to see that there are three distinct types of Boolean automata double-cycles. A positive Boolean automata double-cycle is composed of two positive Boolean automata cycles and is denoted by $\mathscr{D}^{+,+}_{\ell,r}$; a negative Boolean automata double-cycle is composed of two negative Boolean automata cycles and is denoted by $\mathscr{D}^{-,-}_{\ell,r}$; a mixed Boolean automata double-cycle is composed of a negative Boolean automata cycle $\mathscr{C}^-$ tangentially intersected with a positive Boolean automata cycle $\mathscr{C}^+$, and is denoted by $\mathscr{D}^{-,+}_{\ell,r}$.

Finally, in [47, 63], the authors have shown that the Boolean automata cycles admit canonical representatives, and Boolean automata double-cycles too by induction. The

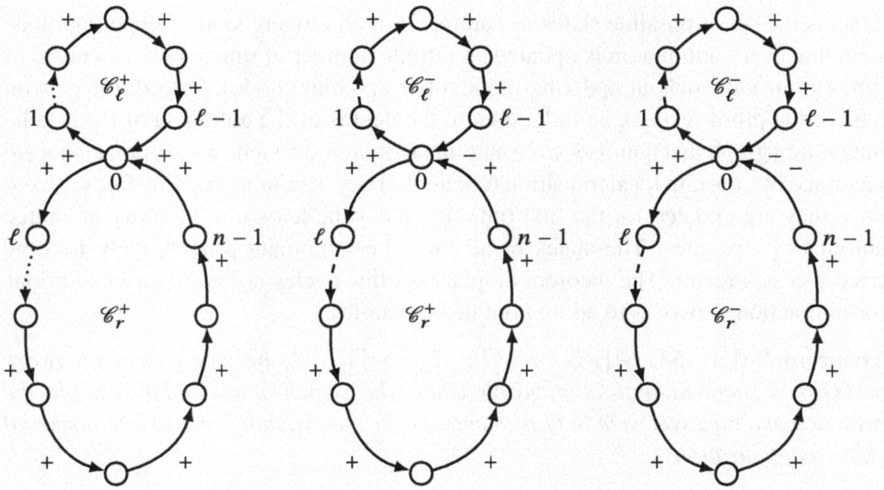


**Fig. 3** The three canonical Boolean automata double-cycles of size $n = \ell + r - 1$: the positive one (on the left), the mixed one (on the center), and the negative one (on the right)

studies presented in the sequel focus on these canonical representatives only. Indeed, canonicity means that two distinct Boolean automata cycles or Boolean automata double-cycles of same sign and same size that evolve following the same updating mode admit the same transition graph up to an isomorphism on the configurations. A positive Boolean automata cycle is said to be canonical when its interaction graph contains only activating interactions. A negative Boolean automata cycle is canonical when its interaction graph admits a single inhibiting interaction, represented by the arc $(n - 1, 0) \in E$. The canonical Boolean automata double-cycles are the canonical Boolean automata cycle compositions, depicted in Fig. 3.

## 4 Seminal Results on Cycles

The works of Robert [57] and Thomas [67] have highlighted three fundamental results that explain the primordial role that cycles play in the behavioural diversity of interaction networks.

**Theorem 1** ([57–59]) *Let $f : \mathbb{B}^n \to \mathbb{B}^n$ be a Boolean automata network of size $n$ and $G$ its associated interaction graph. If $G$ is an acyclic graph, then:*

1. *$f$ has a unique attractor that is a stable configuration, let us say $x$.*
2. *$\mathscr{G}_p$ has a path of length at most $n$ from every configuration $y$ to $x$.*
3. *$\mathscr{G}_a$ is acyclic and has a a geodesic path from every configuration $y$ to $x$.*

This theorem is particularly interesting for two reasons. First, it is easy to extend it to any kind of multi-valued automata networks $g : \prod_{i=1}^{n} X_i \to \prod_{i=1}^{n} X_i$, where $X_i$

denotes the set of possible states of automaton $i$, and to any kind of updating mode such that every automaton is updated an infinite number of times over the course of time (let us call such an updating mode a fair updating mode). Indeed, the general idea of the proof rests on an induction on the depths of the automata of the acyclic interaction graph that admits *source* automata. Source automata are automata that are governed by constant local transition functions. They inevitably become forever fixed once they are updated for the first time. In an acyclic network, the fixity of source automata propagates. The states of all the other automata progressively become fixed too as a result. The theorem emphasizes that cycles are necessary conditions for interaction networks to admit complex dynamics.

**Theorem 2** ([50, 53, 67]) *Let $g : \prod_{i=1}^{n} X_i \to \prod_{i=1}^{n} X_i$ be an automata network and G its associated interaction graph. Under the asynchronous updating mode, the presence of a positive cycle in G is necessary for the dynamics of g to admit several stable configurations.*

This second theorem sheds light on the role of positive cycles on the ability of interaction networks to stabilise in several ways. Although it was originally stated and demonstrated under the asynchronous updating mode hypothesis, this result has been shown to hold for any fair updating mode [47, 63]. The general proof starts from the result of [53] and is made by contradiction under the assumption of the absence of a positive cycle. In this case, either the interaction graph is acyclic and Theorem 1 applies, or it has at least one negative cycle and it is shown that such cycles, whatever the updating mode, cannot remove the local instabilities on all automata.

**Theorem 3** ([52, 67]) *Let $g : \prod_{i=1}^{n} X_i \to \prod_{i=1}^{n} X_i$ be an automata network and G its associated interaction graph. Under the asynchronous updating mode, the presence of a negative cycle is necessary for the dynamics of g to admit a stable oscillation.*

As for this third theorem, it turns out not to be general for all updating modes. To be convinced, it suffices to compute the parallel transition graph of an arbitrary positive Boolean automata cycle of size greater than 2 and to note that it admits at least one stable oscillation. Despite this lack of generality, it should be noted that this theorem underlines the singular role of negative cycles in connection with asymptotic dynamic oscillations, as will be explained below.

Taken together, these three theorems, of which we can say that two of them are laws insofar as their generality makes them sorts of meta-theorems of interaction network theory, emphasise that the feedback cycles are the causes of the dynamical complexity of networks. In other words, they effectively constitute the sources of the behavioural diversity of networks and consequently of the computational expressiveness of networks. This naturally brings us to the following parts of this chapter, about major results concerning the dynamical and combinatorial properties of Boolean automata cycles and Boolean automata double-cycles. The results are illustrated with examples. As discussed above, Robert's result highlighted feedback cycles as kinds of complexity engines. Then, Thomas' results put forward the necessity to distinguish feedback cycles depending on their (positive or negative) nature to understand their influence. The results presented below about Boolean automata double-cycles are a first step further, towards understanding how cycle combinations operate and what their effects are.

# 5 Boolean Automata Cycle Dynamics

In this section, we focus on Boolean automata cycles. In a first part, we present in Theorem 4 the main results related to the dynamics of isolated Boolean automata cycles when the latter evolve according to the parallel updating mode. This theorem requires some preliminary definitions and notations of number theory such as the Dirichlet convolution, the Möbius function and the Euler's totient function. In a second part, we present the results related to isolated Boolean automata cycles when the latter evolve according to the asynchronous updating mode.

## *5.1 Parallel Boolean Automata Cycles*

The first elements on the dynamics of Boolean automata cycles evolving according to the parallel updating mode were introduced in [49]. The full characterisation of their dynamics had to wait until [16]. This characterisation could be obtained thanks to an approach combining discrete dynamical systems theory, enumerative and word combinatorics, particularly appropriate to the nature of the mathematical objects in question.

### 5.1.1 Definitions and Notations of Pertinent Quantities

To describe the results obtained, we give below definitions and notations. First, given an attractor of (minimal) period $p$, we says that all the multiples of $p$ are also periods of this attractor. So if $x \in \mathbb{B}^n$ is a recurring configuration of an attractor of period $p$, then $p$ is the period of $x$ and of any other configuration $y$ such that there exists $t \in \mathbb{N}$ such that $y = F_V^t(x)$. We denote by $\mathscr{X}(p) = \{x \in \mathbb{B}^n \mid x = F_V^p(x)\}$ the set of recurring configurations of period $p$ and by $X(p) = |\mathscr{X}(p)|$ their number. We define the smallest integer $\omega$ that is a period common to all recurring configurations as the *order* of the Boolean automata cycle, which is said *to be reached* when there exists an attractor of minimum period $\omega$.

Let us consider the function one : $n \in \mathbb{N} \mapsto 1$ as well as the *Dirichlet convolution*, denoted by $\star$ [3]. Given two functions $f$ and $g$, $\star$ is the binary operator defined such that $f \star g : n \in \mathbb{N}^* \mapsto \sum_{p|n} f(p) \cdot g(n/p)$. The set of the arithmetic functions with point-to-point addition and Dirichlet convolution is a commutative ring. The identity by the multiplication of this ring is the function $\delta : \mathbb{N}^* \to \mathbb{N}^*$, defined by $\delta(1) = 1$ and for all $n > 1$, $\delta(n) = 0$. The inverse of the function one for the Dirichlet convolution is the Möbius function $\mu$, defined as:

$$\mu : n \in \mathbb{N}^* \mapsto \begin{cases} 0 & \text{if } n \text{ is not square-free,} \\ 1 & \text{if } n > 0 \text{ has an even number of prime factors,} \\ -1 & \text{if } n > 0 \text{ has an odd number of prime factors.} \end{cases}$$

If $n = \prod_{i=0}^{k} p_i$, where the $p_i$s are the distinct prime numbers taken in increasing order, then $\mu(n) = (-1)^k$. In our context, this function is of interest through the Möbius inversion formula that is obtained from $\mathsf{one} \star \mu = \delta$, that is satisfied by all the functions $f$ and $g$, and that is such that: $g = f \star \mathsf{one} \implies f = g \star \mu$. In other words, we have:

$$\forall n \in \mathbb{N}^*,\ g(n) = \sum_{p|n} f(p) \implies f(n) = \sum_{p|n} g(p) \cdot \mu(n/p).$$

Another particularly useful function in the sequel is the *Euler's totient function*, denoted by $\phi$. Given an integer $n \in \mathbb{N}^*$, it associates the number of strictly positive integers less than or equal to $n$ that are prime with $n$, such as: $\phi(n) = |\{m \in \mathbb{N}^* \mid m \leq n \text{ and } m \text{ is prime with } n\}|$. Note that there is a relationship between the Möbius function and the Euler's totient function. Indeed, as $\phi$ satisfies $\forall n \in \mathbb{N}^*,\ n = \phi \star \mathsf{one}(n)$, it respects $\phi = \mu \star \mathsf{id}$, where $\mathsf{id} : n \in \mathbb{N}^* \mapsto n$.

In terms of combinatorics, the asymptotic behaviour of an interaction network can be described by means of four quantities [48] related to each other and given below. Consider that $p$ is a divisor of the order $\omega$ of the Boolean automata cycle studied and the function $\mathsf{inv} : n \in \mathbb{N}^* \to 1/n$. Then we have the following quantities:

- The *number* $\mathtt{X}(p)$ *of configurations of period* $p$ is $\mathtt{X} = \widetilde{\mathtt{X}} \star \mathsf{one}$;
- The *number* $\widetilde{\mathtt{X}}(p)$ *of configurations of minimal period* $p$ is $\widetilde{\mathtt{X}} = \mathtt{X} \star \mu$;
- The *number* $\mathtt{A}(p) = \widetilde{\mathtt{X}}(p)/p$ *of attractors of period* $p$ is $\mathtt{A} = \mathsf{inv}(\mathtt{X} \star \mu)$;
- The *total number* $\mathtt{T}(\omega)$ *of attractors* is $\mathtt{T} = \mathtt{A} \star \mathsf{one} = \mathsf{inv}(\mathtt{X} \star \phi)$.

The last two quantities correspond to well known formulas in the context of Lyndon words and binary necklaces [6, 26, 60]: a Lyndon word $w$ being such that $w < v$ for all nonempty words $v$ such that $w = uv$ and $u$ is nonempty; a binary necklace $w$ of length $n$ being a circular binary word such that for all $i$ in $\mathbb{Z}$, $w_i = w_i \mod n$. In particular, the penultimate defining $\mathtt{A}$ corresponds to the Witt formula counting the number of Lyndon words; the last one defining $\mathtt{T}$ corresponds to the Burnside's orbit-counting lemma. Note that the last formula is satisfied because $\mathsf{inv}$ distributes over $\star$. Finally, note that it suffices to calculate $\mathtt{X}$ to obtain the others.

### 5.1.2 Results

The qualitative characterisation of the dynamics of Boolean automata cycles can be summarised by the following theorem that presents in the form of a table the set of

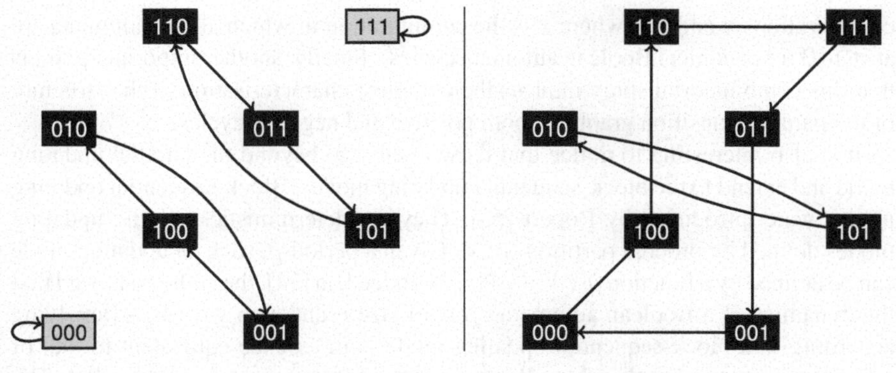


**Fig. 4** Left panel: the parallel transition graph of positive canonical Boolean automata cycle $\mathscr{C}_3^+$. Right panel: the parallel transition graph of negative canonical Boolean automata cycle $\mathscr{C}_3^-$

all the quantities presented above. For reasons of space, the details and the demonstrations related to these results are not presented here. The reader can nevertheless find all the details in [16, 47, 63].

**Theorem 4** ([16]) *The order $\omega$, the numbers $\mathtt{X}(p)$ and $\widetilde{\mathtt{X}}(p)$ of configurations of (minimal) period p, where $\mathtt{X}(\omega)$ is the total number of recurring configurations, as well as the number $\mathtt{A}(p)$ of attractors of period p and the total number $\mathtt{T}(\omega)$ of attractors of positive and negative Boolean automata cycles are:*

| *Positive cycles* $\mathscr{C}_n^+$ | *Negative cycles* $\mathscr{C}_n^-$ |
|---|---|
| $\omega = n$ | $\omega = 2n$ |
| $\mathtt{X}^+(p) = 2^p$ | $\mathtt{X}_n^-(p) = \neg(p\vert n) \cdot 2^{\frac{p}{2}}$ |
| $\widetilde{\mathtt{X}}^+(p) = \sum_{d\vert p} \mu\left(\frac{p}{d}\right) \cdot 2^d$ $= \mathtt{OEIS\ A27375}(p)$ | $\widetilde{\mathtt{X}}_n^-(p) = \sum_{k\vert p\ odd} \mu(k) \cdot 2^{\frac{p}{2k}}$ , |
| $\mathtt{A}^+(p) = \frac{\widetilde{\mathtt{X}}^+(p)}{p}$ $= \mathtt{OEIS\ A1037}(p)$ | $\mathtt{A}_n^-(p) = \frac{\widetilde{\mathtt{X}}_n^-(p)}{p}$ $= \mathtt{OEIS\ A48}(\frac{p}{2})$ |
| $\mathtt{T}^+(\omega) = \frac{1}{n} \sum_{d\vert n} \mathrm{e}\left(\frac{n}{d}\right) \cdot 2^d$ $= \mathtt{OEIS\ A31}(n)$ | $\mathtt{T}^-(\omega) = \frac{1}{2n} \sum_{k\vert 2n\ odd} \mathrm{e}(k) \cdot 2^{\frac{n}{2k}}$ $= \mathtt{OEIS\ A16}(n)$ |

*where $\neg(p|n)$ is 0 if p divides n and 1 otherwise.*

Among the particularly interesting properties of a parallel Boolean automata cycle that emerge from this theorem, it should be noted that all of its configurations are recurring, as illustrated in Fig. 4, and that the total number of attractors, for a given period or not, is exponential according to its size $n$. Moreover, as a corollary, for the two types of cycles, the order is reached, and positive cycles admit two stable configurations $x$ and $\bar{x}^V$ (where $x$ is the configuration in which all the automata are at state 0 for canonical Boolean automata cycles). Finally, another important point is that this combinatorial study induces the complete characterisation of the structure of the parallel transition graphs of both positive and negative cycles.

It is also interesting to notice that these results go beyond the parallel updating mode and extend to the block-sequential updating modes. Block-sequential updating modes were introduced by Robert [55]. They are deterministic periodic updating modes defined by ordered partitions of $V$. Given a period $p$, such an updating mode can be defined by a function $\mu : V \to \mathbb{N}/p\mathbb{N}$. Indeed, in [24], the authors showed that the dynamics of a Boolean automata cycle of size $n$ and sign $s \in \{+, -\}$ evolving according to a block-sequential updating mode is in essence equivalent to that of a Boolean automata cycle of smaller size and same sign evolving in parallel. The proof rests on substitutions of local transition functions according to their execution over time. Hence, to understand the dynamics of a Boolean automata cycle evolving according to a block-sequential updating mode is a matter of understanding the dynamics in parallel of a smaller Boolean automata cycle.

## 5.2 Asynchronous Boolean Automata Cycles

In combinatorial terms, the dynamics of asynchronous Boolean automata cycles are much simpler than that of parallel Boolean automata cycles. It is in [49] that we find the first characterisation of the attractors of asynchronous Boolean automata cycles. The general idea of these results is based on the concept of instability. In a given configuration of the network, an automaton is said to be unstable if the application of its local transition function would then make its state change. A configuration is unstable when it has at least one unstable automaton. In particular, the authors showed that asynchronism makes it possible to reduce the number of instabilities, until there are none (resp. only one) left in the case of positive (resp. negative) Boolean automata cycles. This property implies the existence of at least one stable configuration in the positive case. In the negative case, it implies the absence of stable configurations and thereby the existence of at least one stable oscillation. Theorem 5 derives almost directly from this work. It is illustrated by Fig. 5.

**Theorem 5** ([49]) *A positive Boolean automata cycle $\mathscr{C}_n^+$ has two attractors which are two stable configurations $x$ and $\bar{x}^V$. A negative cycle $\mathscr{C}_n^-$ has a single attractor of length $2n$.*

It is easy to see that the results stated in this theorem are strongly related to those of Theorem 4. Indeed, first, the two stable configurations of asynchronous positive cycles are identical to those of these same cycles in parallel. And second, the unique stable oscillation of length $2n$ of a negative cycle $\mathscr{C}_n^-$ under the asynchronous mode is identical to that of period $\omega$ of parallel negative cycles. The two configurations $x$ and $\bar{x}^V$ which are stable configurations in the case of positive cycles, belong to the stable oscillations in the case of negative cycles.

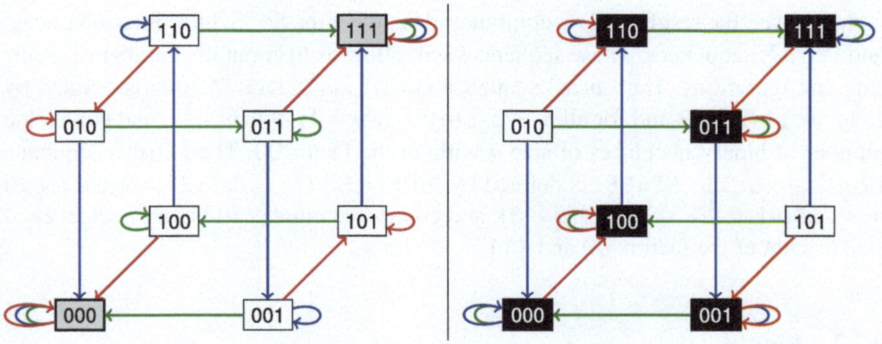


**Fig. 5** Left panel: asynchronous transition graph of positive canonical Boolean automata cycle $\mathscr{C}_3^+$. Right panel: asynchronous transition graph of negative canonical Boolean automata cycle $\mathscr{C}_3^-$. In both graphs, every $\rightarrow$ (resp. $\rightarrow$ and $\rightarrow$) represents an update of automaton 0 (resp. 1 and 2)

Finally, notice that this approach based on instabilities has been generalised in [47, 63] to show the validity of Theorem 2 regardless of the updating mode (see Sect. 4).

## 6 Boolean Automata Double-Cycle Dynamics

Now that the structural and combinatorial properties of the parallel and asynchronous transition graphs of the Boolean automata cycles are established, we present in this section those related to Boolean automata double-cycles.

### *6.1 Parallel Boolean Automata Double-Cycles*

In this section, we focus on Boolean automata double-cycles evolving according to the parallel updating. The method used to obtain the results presented follows the lines of method used for Boolean automata cycles.

#### 6.1.1 Definitions and Notations

Besides the four quantities $\mathtt{X}(p)$, $\widetilde{\mathtt{X}}(p)$, $\mathtt{A}(p)$ and $\mathtt{T}(\omega)$ that we are going to use, we introduce here other definitions and notations to characterise the dynamics of Boolean automata double-cycles that are Boolean automata cycles tangentially interconnected. In this sense, given any Boolean automata double-cycle $\mathscr{D}^s_{\ell,r}$, where $s \in \{(+,+), (-,+), (-,-)\}$, two quantities defined by means of $\ell$ and $r$ will be particularly useful, $\Delta = \gcd(\ell, r)$ and $\Delta_p = \gcd(\Delta, p)$.

Moreover, the results call on combinatorics on words. So we introduce the Lucas' and Perrin's sequences. These sequences will allow us to count the number of recurring configurations. The Lucas' sequence $(\mathtt{L}(n))_{n\in\mathbb{N}^*}$ (OEIS A204) is defined by $\mathtt{L}(1) = 1$, $\mathtt{L}(2) = 3$ and for all $n > 2$, $\mathtt{L}(n) = \mathtt{L}(n-1) + \mathtt{L}(n-2)$, and counts the number of binary necklaces of size $n$ without the factor 00. The Perrin's sequence $(\mathtt{P}(n))_{n\in\mathbb{N}}$ (OEIS A1608) is defined by $\mathtt{P}(0) = 3$, $\mathtt{P}(1) = 0$, $\mathtt{P}(2) = 2$ and for all $n > 2$, $\mathtt{P}(n) = \mathtt{P}(n-2) + \mathtt{P}(n-3)$, and counts the number of binary necklaces of size $n$ without the factors 00 and 111.

### 6.1.2 Results

On the basis of the previous definitions and notations, Theorem 6 below gives the qualitative characterisation of the dynamics of the different types of Boolean automata double-cycles. It will be illustrated on six distinct Boolean automata double-cycles depicted in Fig. 6. The details of the proofs can be found in [46, 47, 63].

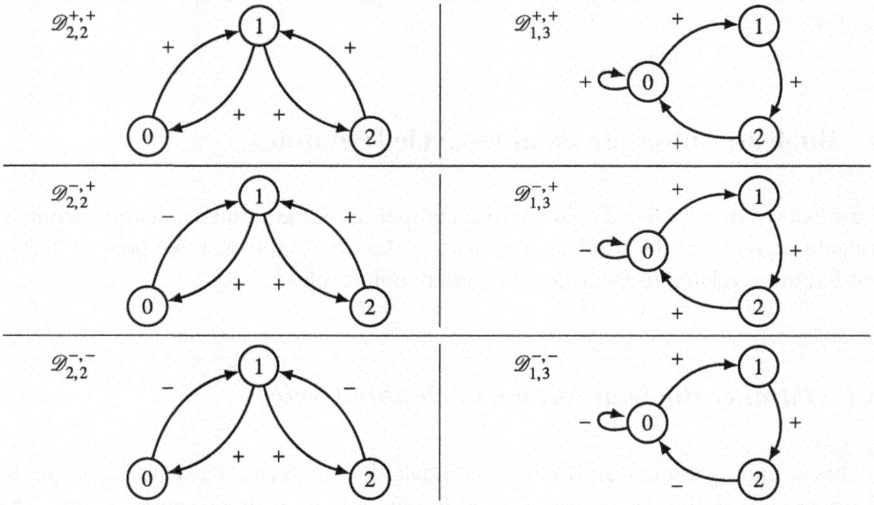


**Fig. 6** Top panel: Interaction graphs of two positive canonical Boolean automata double-cycles. Middle panel: Interaction graphs of two mixed canonical Boolean automata double-cycles. Bottom panel: Interaction graphs of two negative canonical Boolean automata double-cycles. These six networks will serve as examples in the sequel

**Theorem 6** ([46]) *The order $\omega$, the numbers $\mathtt{X}(p)$ and $\widetilde{\mathtt{X}}(p)$ of configurations of (minimal) period $p$, where $\mathtt{X}(\omega)$ represents the total number of recurring configurations, as well as the number $\mathtt{A}(p)$ of asymptotic behaviours of period $p$, and the total number $\mathtt{T}(\omega)$ asymptotic behaviours of the positive, mixed and negative Boolean automata double-cycles are:*

| *Positive double-cycles* $\mathscr{D}^{+,+}_{\ell,r}$ | *Mixed double-cycles* $\mathscr{D}^{-,+}_{\ell,r}$ | *Negative double-cycles* $\mathscr{D}^{-,-}_{\ell,r}$ |
|---|---|---|
| $\omega = \Delta$ | $\omega = r$ | $\begin{cases} \frac{\ell+r}{2} & \textit{if } \frac{\ell+r}{\Delta} = 4 \\ \ell + r & \textit{sinon} \end{cases}$ |
| $X^{+}(p)$ | $X^{-,+}_{\ell}(p) = \neg(p\|\ell) \cdot \mathrm{L}(\frac{p}{\Delta_p})^{\Delta_p}$ | $X^{-,-}_{\Delta}(p) = \neg(p\|\Delta) \cdot \mathrm{P}(\frac{p}{\Delta_p})^{\Delta_p}$ |
| $\widetilde{X}^{+}(p)$ | $\widetilde{X}^{-,+}_{\ell}(p) = \sum_{\substack{d\|p \\ \neg(d\|\ell)}} \mu\left(\frac{p}{d}\right) \cdot \mathrm{L}(\frac{d}{\Delta_d})^{\Delta_d}$ | $\widetilde{X}^{-,-}_{\Delta}(p) = \sum_{\substack{d\|p \\ \neg(d\|\Delta)}} \mu\left(\frac{p}{d}\right) \cdot \mathrm{P}(\frac{d}{\Delta_d})^{\Delta_d}$ |
| $A^{+}(p)$ | $A^{-,+}_{\ell}(p) = \frac{\widetilde{X}^{-,+}_{\ell}(p)}{p}$ | $A^{-,-}_{\Delta}(p) = \frac{\widetilde{X}^{-,-}_{\Delta}(p)}{p}$ |
| $T^{+}(\omega)$ | $T^{-,+}_{\ell}(\omega) = \frac{1}{r} \sum_{\substack{d\|r \\ \neg(d\|\ell)}} \mathrm{e}\left(\frac{r}{d}\right) \cdot \mathrm{L}(\frac{d}{\Delta_d})^{\Delta_d}$ | $T^{-,-}_{\Delta}(\omega) = \frac{1}{n} \sum_{\substack{d\|n \\ \neg(d\|\Delta)}} \mathrm{e}\left(\frac{n}{d}\right) \cdot \mathrm{P}(\frac{d}{\Delta_d})^{\Delta_d}$ |

,

*where* $\neg(p|m)$ *equals* 0 *if* $p$ *divides* $m$ *and* 1 *otherwise.*

Among other things (see Fig. 6), it emerges from these results that *(i)* positive Boolean automata double-cycles have two stable configurations $x$ and $\bar{x}^V$ (with $x = (0, \ldots, 0)$ when they are canonical) and that they have an asymptotic behaviour similar to positive Boolean automata cycles of the same order, *(ii)* mixed Boolean automata double-cycles have a single stable configuration, and *(iii)* negative Boolean automata double-cycles have no stable configurations (Fig. 7).

Also, we remark that, unlike Boolean automata cycles, the order of Boolean automata double-cycles is not necessarily reached. Finally, this theorem highlights again that, like Boolean automata cycles, Boolean automata double-cycles admit an exponential number of attractors according to their size. However, despite its exponential nature, the number of Boolean automata double-cycle attractors is significantly smaller than that of Boolean automata cycles. In other words, the intersections of cycles seem to participate strongly in the reduction of asymptotic degrees of freedom of interaction networks. Based on this idea, studies have been conducted to compare $\mathrm{T}^{+}(n)$ and $\mathrm{T}^{-}(2n)$ of Boolean automata cycles with quantities $\mathrm{T}^{+}(\Delta)$, $\mathrm{T}^{-,+}_{\ell}(r)$ and $\mathrm{T}^{-,-}_{\Delta}(\ell + r)$ of Boolean automata double-cycles.

Consider a network $f$ of order $\omega$, so that $f = \mathscr{C}^{s}_{n}$ or $f = \mathscr{D}^{s,s'}_{\ell,r}$, where $s, s' \in \{+, -\}$. Let $p$ be a divisor of $\omega$. Also note $\mathrm{Q} \in \{\mathrm{X}, \widetilde{\mathrm{X}}, \mathrm{A}, \mathrm{T}\}$ one of the four quantities analysed. It has been demonstrated in [47, 63] that $\mathrm{Q}^{+,+}_{\ell,r}(p) = \mathrm{Q}^{+}(p)$, $\mathrm{X}^{-,+}_{\ell,r}(p) \leq \mathrm{X}^{+}(p)$, $\mathrm{X}^{-,-}_{\ell,r}(p) \leq \mathrm{X}^{+}(p)$ and $\mathrm{Q}^{s,s}_{\ell,\ell}(p) = \mathrm{Q}^{s}_{\ell}(p)$.

However, the number of recurring configurations of a Boolean automata double-cycle, $\mathrm{X}^{s,s}_{\ell,r}(p)$ has been bounded more finely as a function of $\mathrm{X}^{+}_{n}(p)$ and $\mathrm{X}^{-}_{n}(p)$, based on the previous results as well as on the relation of the Lucas sequence with the golden ratio [51], denoted by $\varrho = (1 + \sqrt{5})/2$ (root of $x^2 - x - 1 = 0$) and that of the Perrin sequence with the plastic number [71], denoted by $\xi = \sqrt[3]{\frac{1}{2} + \frac{1}{6}\sqrt{\frac{23}{3}}} + \sqrt[3]{\frac{1}{2} - \frac{1}{6}\sqrt{\frac{23}{3}}}$ (root of $x^3 - x - 1 = 0$). The bounds found led to Theorem 7 that concludes this part by highlighting that a vast majority of recurring configurations have the greatest minimum period possible.

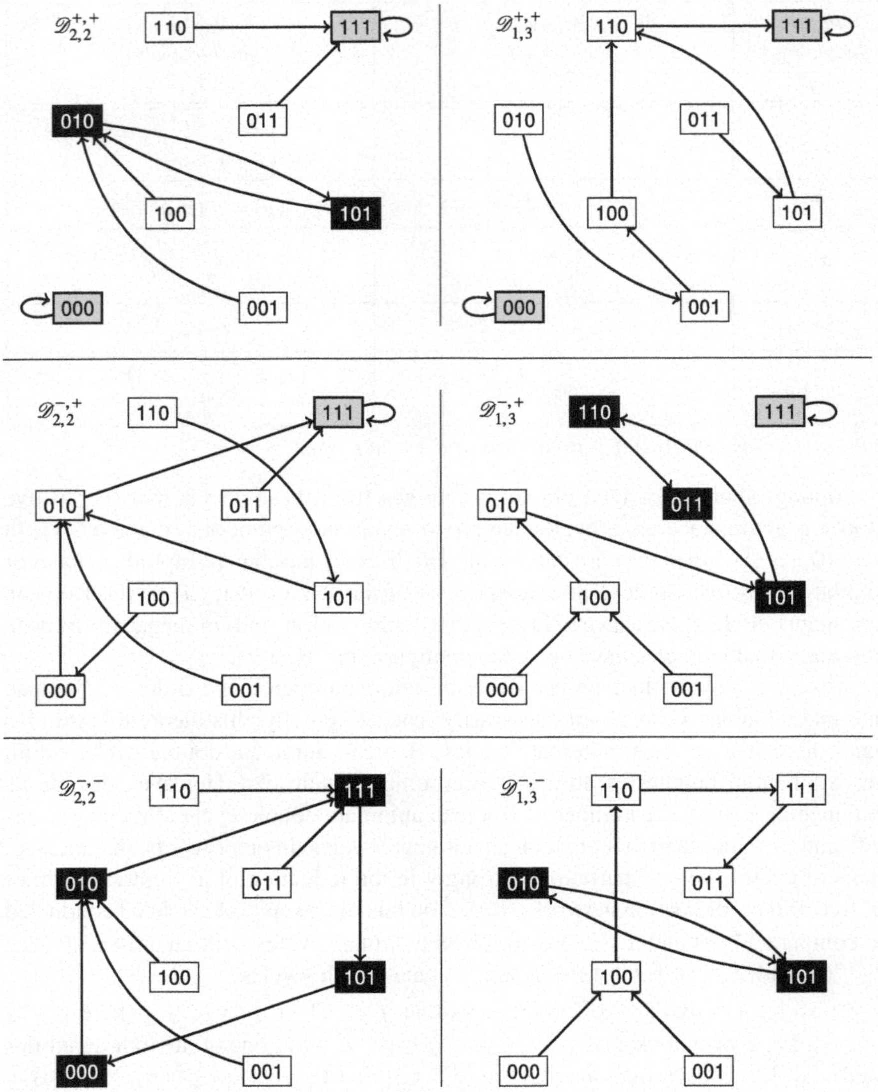


**Fig. 7** Parallel transition graphs of the canonical Boolean automata double-cycles depicted in Fig. 6, where $\diamond = \vee$

**Theorem 7** ([47]) *Let $f$ be a Boolean automata network of order $\omega$. If $f$ is a Boolean automata cycle or a Boolean automata double-cycle such as $f$ is neither $\mathcal{D}^{-,-}_{5,1}$ nor $\mathcal{D}^{-,-}_{1,5}$, then its total number of attractors $T(\omega)$ is bounded by its total number of recurring configurations $X(\omega)$ so that:*

$$\frac{X(\omega)}{\omega} \leq T(\omega) \leq 2 \cdot \frac{X(\omega)}{\omega},$$

*which means that the attractor periods of $f$ are very large:*

$$\sum_{p|\omega} p \cdot \frac{A(p)}{T(\omega)} = \frac{X(\omega)}{T(\omega)} \geq \frac{\omega}{2}.$$

To end this section dedicated to parallel Boolean automata double-cycles, notice that all the results presented here extend naturally to any tangential double-cycles, namely two cycles that admit several automata in common such that these common automata are organised into an isolated path so that each of them have a local transition function of arity 1 except the first one for which the arity is 2. The proof is simple and rests on an induction consisting in transforming each tangential automaton whose local transition function arity is 1 (by following the isolated path in reverse direction) into two copies, one in the left cycle and one in the right cycle. Following this reasoning, it is easy to see that such a tangential double-cycle is equivalent to a Boolean automata double-cycle whose cycles are of bigger sizes and same signs.

Moreover, notice that both Boolean automata cycles and Boolean automata double-cycles admit an exponential number of attractors. This is quite unrealistic if we view these objects as models of genetic regulation networks. Indeed, "real" genetic regulation networks seem to have a number of asymptotic behaviours (cellular types, biological rhythms…) that is polynomial (perhaps linear) according to the number of their genes. Nevertheless, it is important to see that the results obtained show that the number of attractors of Boolean automata double-cycles is drastically smaller than that of Boolean automata cycles. Actually, we think that the polynomial characteristics of the number of attractors of real regulation systems comes from the entanglement of cycles. More precisely, without formalising it, we conjecture that the more there are entangled cycles, the less there are local and global instabilities, the less there are attractors (simply because adding intersections adds dynamical constraints), and thus the less automata networks are sensitive to synchronism.

## *6.2 Asynchronous Boolean Automata Double-Cycles*

The last part of this synthesis is devoted to the dynamics of asynchronous Boolean automata double-cycles. It presents the characterisation established in [42]. It is the counterpart of the study of the dynamics of Boolean automata double-cycles under the parallel updating mode. The study of the asynchronous case takes a new approach. It formalises long sequences of updates by way of algorithmic descriptions. This approach allows an elegant and more detailed description of the dynamics of feedbacks in interaction networks than the previous works. In particular it facilitates the study of convergence times.

### 6.2.1 Definitions and Notations

For the sake of clarity, let us first recall that the study surveyed here focuses on canonical Boolean automata double-cycles.

*States and configurations*
Here, we will use the classical notation $V = \{0, \dots, n-1\}$ for representing the automata of a network of size $n$ and its congruence $V \equiv \{c = c_0, c_1, \dots, c_{n-1}\}$. A configuration $x \in \mathbb{B}^n$ is seen as a vector of two binary words. The first symbol of these two words represents $x_c \equiv x_0$. The null configuration is therefore denoted by $(0^\ell, 0^r)$. Furthermore, we denote by $x^\ell$ (resp. $x^r$) the projection of $x$ on the Boolean automata cycle $\mathscr{C}_\ell$ (resp. $\mathscr{C}_r$). This way, $x = (x^\ell, x^r)$, and in configuration $x$, the state of automaton $c_i^\ell$ is $x_i^\ell$. Notice that $x_0 = x_0^\ell = x_0^r$ since the three notations represent the state of automaton $c$ in $x$.

*Expressiveness measure*
Let $x$ be a configuration of a Boolean automata cycle $\mathscr{C}_n$. Its *expressiveness* is the number of factors 01 that compose it, namely $|\{i \mid 0 \le i \le n-1, x_i = 0 \text{ and } x_{i+1 \bmod n} = 1\}|$. The expressiveness of a configuration $x$ of a Boolean automata double-cycle is the sum of the expressiveness of $x^\ell$ and $x^r$. From this definition follows that, if $\ell$ and $r$ are even, the least expressive configurations are $(0^\ell, 0^r)$ and $(1^\ell, 1^r)$, and that the most expressive ones are $((01)^{\frac{\ell}{2}}, (01)^{\frac{r}{2}})$ and $((10)^{\frac{\ell}{2}}, (10)^{\frac{r}{2}})$.

*Elementary instructions*
Network trajectories can be very long. To study them, we need a way to efficiently describe the sequence of automata updates that they execute. Our human minds must be able to understand from the description, what is the effect of the trajectory on the network, what changes does the network undergo along the trajectory. To do so, we proposed to view these sequences as instructions that make it easier to capture their effect on configurations. Let us therefore consider:

- a Boolean automata double-cycle $\mathscr{D}_n$,
- one of the Boolean automata cycles of $\mathscr{D}_n$, namely $\mathscr{C}$, whose size is noted `size`($\mathscr{C}$),
- the current configuration $x$ of $\mathscr{C}$, and,
- two automata of $\mathscr{C}$ distinct from $c$, namely $c_i$ and $c_j$, such that $i < j$.

With these notations, the following seven basic instructions are defined:

1. `/* update of automaton` $c$ `*/`
   `sync:` $x_c \leftarrow f_c(x)$

- Instruction `sync` is the only instruction that updates automaton $c$ and where both Boolean automata cycles interact with each other. This (key)-instruction will always be called when $c$ can change its state. This instruction can be used either to set $c$ at a desired state or to increase the expressiveness from a configuration. Furthermore, it is the only way to switch a 111 (resp. 000) pattern into a 101 (resp. 010) pattern and, thus, to increase the expressiveness.

2. `/* update of automaton` $c_i$ `*/`
`update`$(c_i)$: $x_{c_i} \leftarrow f_{c_i}(x)$

   Instruction `update` updates an automaton distinct from $c$.

3. `/* incremental updates */`
`incUp`$(\mathscr{C}, i, j)$: `for` $k = i$ `upto` $j$ `do update`$(c_k)$

   Instruction `incUp` updates consecutive automata in increasing order. In fact, `incUp` propagates the state of $c_{i-1}$ along $\mathscr{C}$. Notice that if $j < i$ then no automata are updated. Moreover, since $i \neq 0$ and $j \neq 0$, $c$ cannot be updated with `incUp`. This instruction admits the following property. Let $x'$ be the result of the execution of `incUp`$(\mathscr{C}, i, j)$ on configuration $x$. Then $\forall k \in \{i, \ldots, j\}$, $x'_k = x_{i-1}$ and $\forall k \notin \{i, \ldots, j\}$, $x'_k = x_k$.

4. `/* incremental propagation of` $x_c$ `*/`
`erase`$(\mathscr{C})$: `incUp`$(\mathscr{C}, 1, \texttt{size}(\mathscr{C}) - 1)$

   Instruction `erase` is a particular case of `incUp`. It propagates the state of $c_0$ along $\mathscr{C}$. As a consequence, using `erase` on $\mathscr{C}$ decreases its expressiveness to 0, and thus, is really efficient to reach quickly the least expressive configuration. This instruction admits the following property. Let $x'$ be the result of applying `erase`$(\mathscr{C})$ on configuration $x$. Then we have: $\forall k \in \{0, \ldots, \texttt{size}(\mathscr{C}) - 1\}$, $x'_k = x_0$.

5. `/* incremental propagation of` $x_c$ `with no loss of expressive -ness */`
`expand`$(\mathscr{C})$: `incUp`$(\mathscr{C}, 1, \kappa - 1 \in \mathbb{N})$

   $$\text{where } \kappa = \min_{1 \leq k \leq \texttt{size}(\mathscr{C}-1)} \left\{ k \mid \begin{cases} (x_k = 0) \text{ and } (x_{k+1 \bmod \texttt{size}(\mathscr{C})} = 1) & \text{if } x_c = 1 \\ (x_k = 1) \text{ and } (x_{k+1 \bmod \texttt{size}(\mathscr{C})} = 0) & \text{if } x_c = 0 \end{cases} \right\}.$$

   Instruction `expand` is another particular case of `incUp` that aims at propagating the state of $c_0$ along $\mathscr{C}$ while neither 01 nor 10 patterns are destroyed, which avoids decreasing the expressiveness of $\mathscr{C}$.

6. `/* decremental updates */`
`decUp`$(\mathscr{C}, i, j)$: `for` $k = j$ `downto` $i$ `do update`$(c_k)$

   Instruction `decUp` is the converse of instruction `incUp`, and updates consecutive automata in decreasing order. Once `decUp`$(\mathscr{C}, i, j)$ executed, the information of $c_j$ is lost and that of $c_{i-1}$ is possessed by both $c_{i-1}$ and $c_i$. In fact, `decUp` aims at shifting partially a Boolean automata cycle section. As for `incUp`, if $j < i$ then no automata are updated and $c$ cannot be updated with `decUp`. This instruction admits the following property. Let $x'$ the result of the execution of `decUp`$(\mathscr{C}, i, j)$ on $x$. Then $\forall k \in \{i, \ldots, j\}$, $x'_k = x_{k-1}$ and $\forall k \notin \{i, \ldots, j\}$, $x'_k = x_k$.

7. `/* complete decremental update (except` $c$`) */`
`shift`$(\mathscr{C})$: `decUp`$(\mathscr{C}, 1, \texttt{size}(\mathscr{C}) - 1)$

   Instruction `shift` is a particular case of instruction `decUp`. Once executed, every automaton of $\mathscr{C}$ takes the state of its predecessor, except $c$ whose state does not change. Automaton $c_{\texttt{size}(\mathscr{C})-1}$ excluded, all the information contained along $\mathscr{C}$ is kept safe. This instruction is useful to propagate information along a Boolean automata cycle without loosing too much expressiveness (at most one 01 pattern is destroyed).

**Table 1** The update sequences `copy_c`, `copy` and `copy_p`

**copy_c**$(x, x', \mathscr{C}_m)$

$\eta \leftarrow$ `size`$(\mathscr{C}_m)$;
**if** $(x^m_{\eta-1} = x^m_{\eta-2}$ **and** $x^m_{\eta-1} \neq x'^m_{\eta-1})$ **then**
  $j \leftarrow \max\{k \mid k < \eta - 1 \text{ and } x^m_k \neq x'^m_k\}$;
**else**
  $j \leftarrow \eta$;
**fi**
**for** $(k = \eta - 1)$ **downto** $(j + 1)$ **do**
  `update`$(c^m_{k-1})$;
  `update`$(c^m_k)$;
**od**
**for** $(k = j - 1)$ **downto** $(1)$ **do**
  **if** $(x^m_k \neq x'^m_k)$ **then**
    `update`$(c^m_k)$;
  **fi**
**od**

**copy**$(x, x')$

`copy_c`$(x, x', \mathscr{C}_\ell)$;
`copy_c`$(x, x', \mathscr{C}_r)$;

**copy_p**$(x, x')$

**if** $(x_0 \neq x'_0)$ **then**
  `shift`$(\mathscr{C}_\ell)$;
  `shift`$(\mathscr{C}_r)$;
  `sync`;
**fi**
`copy`$(x, x')$;

### 6.2.2 Results

*More complex instructions*
Let $x$ be a configuration of a Boolean automata double-cycle $\mathscr{D}$. Let us consider an algorithm composed of instructions defining an update sequence `sequence`$(x)$ from $x$. In every algorithm that follows, Boolean automata double-cycle $\mathscr{D}$ is always considered as a global variable and is not mentioned. Abusing language, `sequence`$(x)$ represents the sequence as well as the configuration resulting from its execution.

For the purpose of the study, we introduce three other more complex sequences in Table 1. In addition, Lemma 1 below shows that the `copy` instruction allows to transform $x$ into another configuration $x'$ if $x$ is sufficiently expressive.

**Lemma 1** ([42]) *Let $\mathscr{D}$ be a Boolean automata double-cycle and let $x$ and $x'$ be two of its configurations such that $x_0 = x'_0$. If, for all $m \in \{\ell, r\}$, one of the following properties holds for $x$:*

1. $\forall i \in \{1, \ldots, \mathtt{size}(\mathscr{C}_m) - 1\},\ x^m_i \neq x^m_{i-1}$,
2. $\forall i \in \{1, \ldots, \mathtt{size}(\mathscr{C}_m) - 2\},\ x^m_i \neq x^m_{i-1}$ *and* $x^m_{\mathtt{size}(\mathscr{C}_m)-1} = x'^m_{\mathtt{size}(\mathscr{C}_m)-1}$,
3. $\forall i \in \{1, \ldots, \mathtt{size}(\mathscr{C}_m) - 2\},\ x^m_i \neq x^m_{i-1}$ *and* $\exists p \in \{1, \ldots, \mathtt{size}(\mathscr{C}_m) - 2\}, x^m_p \neq x'^m_p$,

*then* `copy`$(x, x') = x'$ *and this sequence executes at most* $2(\ell + r - 6)$ *updates.*

This lemma gives strong insights about the expressive power of instructions and sequences to reveal possible trajectories between configurations. Now let us focus on

**Table 2** The update sequences `fix0` and `fix1`

`fix0`$(x)$

**if** $(x_0 = 1)$ **then**
  $i \leftarrow \min\{k \mid x_k^\ell = 0\}$;
  `incUp`$(\mathscr{C}_\ell, i+1, \ell-1)$;
  `sync`;
**fi**
`erase`$(\mathscr{C}_\ell)$;
`erase`$(\mathscr{C}_r)$;

`fix1`$(x)$

**if** $(x_0 = 0)$ **then**
  $i \leftarrow \min\{k \mid x_k^\ell = 1\}$;
  `incUp`$(\mathscr{C}_\ell, i+1, \ell-1)$;
  $j \leftarrow \min\{k \mid x_k^r = 1\}$;
  `incUp`$(\mathscr{C}_r, j+1, r-1)$;
  `sync`;
**fi**
`erase`$(\mathscr{C}_\ell)$;
`erase`$(\mathscr{C}_r)$;

the dynamical behaviour of Boolean automata double-cycles, from a general point of view.

*Positive Boolean automata double-cycles*

In Sect. 6.1, we have seen that positive Boolean automata double-cycles behave like positive Boolean automata cycles, namely they have two stable configurations among their attractors. In the asynchronous case, these two stable configurations are the only attractors. The general idea of the demonstration is based on canonical positive Boolean automata double-cycles and establishes that the two sequences `fix0` and `fix1` given in Table 2 allow to transform any configuration with at least one automaton in state 0 into configuration $(0^\ell, 0^r)$, and any configuration with at least one automaton in state 1 in each of its Boolean automata cycles into configuration $(1^\ell, 1^r)$. This result is illustrated at the top of Fig. 8, and is summarised in Theorem 8 below.

**Theorem 8** ([42]) *Let $\mathscr{D}^{+,+}$ be a canonical positive Boolean automata double-cycle where $\diamond = \wedge$ and $x$ one of its unstable configurations. If $x$ admits one automaton in state 0, then `fix0`$(x) = (0^\ell, 0^r)$. Moreover, if in configuration $x$, there is one automaton in state 1 in each Boolean automata cycle, then `fix1`$(x) = (1^\ell, 1^r)$. The convergence time of $\mathscr{D}^{+,+}$ is at most $2(\ell + r) - 5$.*

*Mixed Boolean automata double-cycles*

For *mixed* asynchronous Boolean automata double-cycles, as for positive ones, asynchronism allows to eliminate all local instabilities. Thus, contrary to parallel Boolean automata double-cycles, asynchronous Boolean automata double-cycles have only one attractor that is a stable configuration. In the canonical case, this is evidenced by the `simp` sequence given in Table 3 that provides a way to converge towards this stable configuration from any initial configuration $x$, by reducing progressively its expressiveness. This result is illustrated in Fig. 8 (middle) and is formalised by Theorem 9 below.

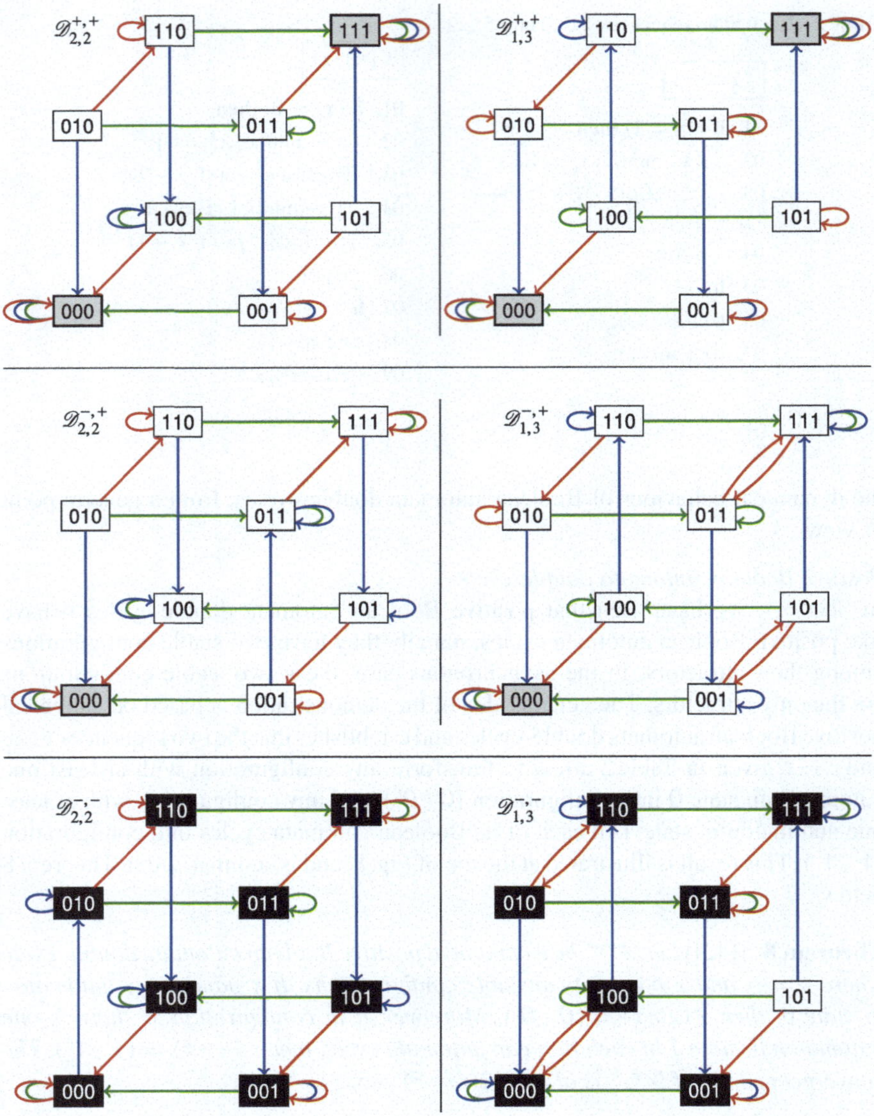


**Fig. 8** Asynchronous transition graphs of the canonical Boolean automata double-cycles depicted in Fig. 6, where $\diamond = \wedge$ and where every $\rightarrow$ (resp. $\rightarrow$ and $\rightarrow$) represents an update of automaton 0 (resp. 1 and 2)

**Table 3** The update sequences simp, comp1 and comp2

**simp($x$)**

```
if (x_0 = 1) then
    erase(C_ℓ);
    sync;
fi
erase(C_ℓ);
erase(C_r);
```

**comp1($x$)**

```
for (i = 1) upto (ℓ − 1) do
    sync;
    expand(C_ℓ);
    erase(C_r);
od
```

**comp2($x$)**

```
if (x^r = 1^r) then
    sync;
    erase(C_r);
fi
sync;
expand(C_r);
for (i = 1) upto (r − 2) do
    shift(C_ℓ);
    sync;
    expand(C_r);
od
```

**Theorem 9** ([42]) *Let $\mathscr{D}^{-,+}$ be a canonical mixed Boolean automata double-cycle, where $\diamond = \wedge$. For any of its configuration $x$, $\texttt{simp}(x) = (0^\ell, 0^r)$ holds. The convergence time of $\mathscr{D}^{-,+}$ is at most $2\ell + r - 2$.*

*Negative Boolean automata double-cycles*

Here, we distinguish between (1) *even* negative Boolean automata double-cycles and (2) *odd* negative Boolean automata double-cycles. The first ones are defined as having two even-sized Boolean automata cycles, the second ones as having at least one odd-sized Boolean automata cycle.

Even negative Boolean automata double-cycles admit a single attractor. This attractor is a stable oscillation of length $2^{\ell+r-1}$. This means that all the configurations are recurring and consequently the convergence time is null. All configurations are reachable. However this result also means that configurations of maximal expressiveness are hard to reach: the number of updates to reach them is quadratic according to the size of the Boolean automata double-cycle. The general idea of the proof follows the following three points:

- Any configuration can reach the less expressive one $(0^\ell, 0^r)$ in linear time. (P1)
- Configuration $(0^\ell, 0^r)$ can reach the highest expressive one $((10)^{\frac{\ell}{2}}, (10)^{\frac{r}{2}})$ in quadratic time. (P2)
- Any configuration can be reached from $((10)^{\frac{\ell}{2}}, (10)^{\frac{r}{2}})$ in linear time. (P3)

Consider P1. It is easy to see that the sequence simp remains effective for reaching $(0^\ell, 0^r)$, which is formalised by Lemma 2 below.

**Lemma 2** ([42]) *For any configuration $x$ of $\mathscr{D}^{-,-}_{\ell,r}$, $\texttt{simp}(x) = (0^\ell, 0^r)$ and executes at most $2\ell + r - 2$ updates.*

Now, consider P2. Implicitly, P2 requires increasing the expressiveness of $(0^\ell, 0^r)$ by successive updates, and finding a trajectory that reaches $((10)^{\frac{\ell}{2}}, (10)^{\frac{r}{2}})$. To do so, we proceed in two stages. First, we increase the expressiveness of $\mathscr{C}_\ell$ using `comp1` (see Lemma 3). Then, we increase the expressiveness of $\mathscr{C}_r$. This second stage is carried out without reducing the expressiveness of $\mathscr{C}_\ell$ by using `comp2` (see Lemma 4). The expected result follows from the composition `comp` = `comp2` ∘ `comp1`. It is formalised in Lemma 5.

**Lemma 3** ([42]) *In an even negative Boolean automata double-cycle $\mathscr{D}^{-,-}_{\ell,r}$, sequence* `comp1`$((0^\ell, 0^r))$ *leads to configuration* $((10)^{\frac{\ell}{2}}, 1^r)$ *and executes at most* $(\ell - 1)(\ell + r - 2)$ *updates.*

**Lemma 4** ([42]) *In an even negative Boolean automata double-cycle $\mathscr{D}^{-,-}_{\ell,r}$, sequence* `comp2`$(((10)^{\frac{\ell}{2}}, 1^r))$ *leads to configuration* $((10)^{\frac{\ell}{2}}, (10)^{\frac{r}{2}})$ *and executes at most* $(r - 2)(\ell + r - 2) + (2r - 1)$ *updates.*

**Lemma 5** ([42]) *In an even negative Boolean automata double-cycle $\mathscr{D}^{-,-}_{\ell,r}$, sequence* `comp`$((0^\ell 0^r))$ *leads to configuration* $((10)^{\frac{\ell}{2}}, (10)^{\frac{r}{2}})$ *and executes at most* $(\ell + r)^2 - 5(\ell - 1) - 3r$ *updates.*

P3 is developed in Lemma 6 that uses the `copy_p` sequence (see Table 1).

**Lemma 6** ([42]) *In an even negative Boolean automata double-cycle $\mathscr{D}^{-,-}_{\ell,r}$, for any configuration $x'$, sequence* `copy_p`$(((10)^{\frac{\ell}{2}}, (10)^{\frac{r}{2}}), x')$ *transforms configuration* $((10)^{\frac{\ell}{2}}, (10)^{\frac{r}{2}})$ *into $x'$ in at most $3(\ell + r - 4) - 1$ updates.*

Starting from Lemmas 2 to 6, whatever the configurations $x$ and $x'$, the composition `copy_p(comp(simp(`$x$`)),` $x'$`)` $= x'$ holds, which proves that there is a unique attractor of length $2^{\ell+r-1}$. From this is derived Theorem 10. It gives some bounds on convergence time.

**Theorem 10** ([42]) *Let $\mathscr{D}^{-,-}_{\ell,r}$ be a canonical negative Boolean automata double-cycle, where $\diamond = \wedge$. $\mathscr{D}^{-,-}_{\ell,r}$ admits a unique attractor of length $2^{\ell+r-1}$. In this stable oscillation, any configuration can be reached from any other in $O(\ell^2 + r^2)$ updates. However, configurations $(0^\ell, 0^r)$ and $(1^\ell, 1^r)$ can be reached from any other one in $O(\ell + r)$ updates, and the configurations $((01)^{\frac{\ell}{2}}, (01)^{\frac{r}{2}})$ and $((10)^{\frac{\ell}{2}}, (10)^{\frac{r}{2}})$ can reach all the others in $O(\ell + r)$ updates.*

Like even negative Boolean automata double-cycles, odd negative Boolean automata double-cycles also admit a single attractor (a stable oscillation). However, all configurations are not necessarily recurring. Indeed, they admit a set $I$ of non-reachable configurations, from which updates are irreversible. The associated result is formalised in Theorem 11 below.

**Theorem 11** ([42]) *Let* $\rho : \mathbb{N} \to \{0, 1\}$, *with* $\rho(k) = \begin{cases} 0 & \text{if } k = 0 \text{ or } k \equiv 1 \mod 2 \\ 1 & \text{otherwise} \end{cases}$.

*Every Boolean automata double-cycle $\mathscr{D}^{-,-}_{\ell,r}$ admits a unique attractor $\mathbb{B}^{\ell+r-1} \setminus I$, where $|I| = \alpha(\ell - 1) \times 2^{r-1} + \alpha(r - 1) \times 2^{\ell-1}$.*

This result, whose proof rests on the characterisation of $I$, generalises Theorem 10.

## 7 Conclusion

In this chapter, we have summarised the major results obtained in recent years on the role of feedbacks in interaction networks. The literature has already established that feedback patterns are "engines of complexity" in the dynamical behaviours of larger networks that contain them. Because of their proven essential character, we have therefore deliberately focused on the dynamics of cycles and double-cycles, by focusing especially on the influence of updating modes. Without going back on the results themselves, this chapter highlights the fundamental differences induced by the "scheduling" of events on the behaviours of the complexity engines of interaction networks: while parallelism tends to render all the asymptotic mathematical diversity of networks, pure asynchronism tends to reduce degrees of freedom of networks, which partly explains why the former is often preferred in theoretical work while the second is often adopted in works that are oriented towards applications in molecular biology. At present, no real knowledge in molecular biology establishes precisely how regulations are implemented over time, even if works state that the chromatin dynamics plays an important role. That is in particular why the type of studies developed in this chapter, focusing on updating modes, remains essential to go further in understanding formal and applied interaction networks. As a consequence, further studies need to be done on the scheduling of updates over time. A first avenue is to focus for instance on more likely updating modes, in agreement with the discussion of what is claimed in [17].

Another natural opening highlighted by these works, deeply rooted in fundamental computer science, consists in developing knowledge on the dynamical properties of interaction networks, including relationships between their architecture and structure. But other equally relevant lines of investigation also call for exploration. One of them was opened up by the study of asynchronous double-cycles and deals with the time complexity of networks. A question that remains currently open is the following: does the time complexity of the networks go hand in hand with their behavioural diversity? If yes, can we find a measure of it? Finally, the work carried out on cycles and double-cycles in parallel, developed in [47, 63] discusses perspectives around computability, modularity/compositionality, and intrinsic universality of networks. These are certainly, like the ever-growing understanding of the influences of updating modes, among the most promising tracks for future works in the field.

**Acknowledgements** This work was funded mainly by our salaries as French or German State agents or pensioner (affiliated to Université Grenoble-Alpes (JD), Université d'Évry (TM and DR), Freie Universität Berlin (MN), and Université d'Aix-Marseille (SS)), and secondarily by the ANR-18-CE40-0002 FANs project (SS).

## References


1. Albert R (2009) Discrete dynamic modeling of cellular signaling networks. Methods Enzym 467:281–306
2. Aldana M (2003) Boolean dynamics of networks with scale-free topology. Phys D 185:45–66
3. Apostol TM (1976) Introduction to analytic number theory. Springer
4. Aracena J, González M, Zuñiga A, Mendez MA, Cambiazo V (2006) Regulatory network for cell shape changes during drosophila ventral furrow formation. J Theor Biol 239:49–62
5. Berlekamp ER, Conway JH, Guy RK (1982) Winning ways for your mathematical plays. Academic Press
6. Berstel J, Perrin D (2007) The origins of combinatorics on words. Eur J Comb 28:996–1022
7. Church A (1932) A set of postulates for the foundation of logic. Ann Math 33:346–366
8. Cook M (2004) Universality in elementary cellular automata. Complex Syst 15:1–40
9. Cosnard M, Demongeot J (1985) On the definitions of attractors. In: Iteration theory and its functional equations, vol 1163 of Lecture notes in mathematics. Springer, pp 23–31
10. Cosnard M, Demongeot J, Le Breton A (eds) (1983) Rhythms in biology and other fields of application, vol 49 of Lecture notes in biomathematics. Springer
11. Cull P (1971) Linear analysis of switching nets. Biol Cybern 8:31–39
12. Delbrück M (1949) Génétique du bactériophage. In: Unités biologiques douées de continuité génétique
13. Demongeot J (1975) Au sujet de quelques modèles stochastiques appliqués à la biologie. PhD thesis, Université scientifique et médicale de Grenoble
14. Demongeot J, Goles E, Morvan M, Noual M, Sené S (2010) Attraction basins as gauges of the robustness against boundary conditions in biological complex systems. PLoS One 5:e11793
15. Demongeot J, Goles E, Tchuente M (eds) Dynamical systems and cellular automata. Academic Press
16. Demongeot J, Noual M, Sené S (2012) Combinatorics of boolean automata circuits dynamics. Discret Appl Math 160:398–415
17. Demongeot J, Sené S (2020) About block-parallel Boolean networks: a position paper. Nat Comput 19:5–13
18. Elspas B (1959) The theory of autonomous linear sequential networks. IRE Trans Circuit Theory 6:45–60
19. Gershenson C (2004) Updating schemes in random Boolean networks: do they really matter? In: Proceedings of artificial life. MIT Press, pp 238–243
20. Gödel KF (1986) Kurt Gödel collected works, volume I—Publications 1929–1936, chapter on undecidable propositions of formal mathematical systems. Oxford University Press, pp 346–372
21. Goles E (1982) Fixed point behavior of threshold functions on a finite set. SIAM J Algebr Discret Methods 3:529–531
22. Goles E, Fogelman-Soulié F, Pellegrin D (1985) Decreasing energy functions as a tool for studying threshold networks. Discret Appl Math 12:261–277
23. Goles E, Martínez S (1990) Neural and automata networks: dynamical behavior and applications, vol 58 of Mathematics and its applications. Kluwer Academic Publishers
24. Goles E, Noual M (2010) Block-sequential update schedules and Boolean automata circuits. In: Proceedings of AUTOMATA. Discrete mathematics and theoretical computer science, pp 41–50
25. Golomb SW (1967) Shift register sequences. Holden-Day Inc
26. Graham RL, Knuth DE, Patashnik O (1989) Concrete mathematics: a foundation for computer science. Addison-Wesley
27. Guet CC, Elowitz MB, Hsing W, Leibler S (2002) Combinatorial synthesis of genetic networks. Science 296:1466–1470
28. Gupta S, Bisht SS, Kukreti R, Jain S, Brahmachari SK (2007) Boolean network analysis of a neurotransmitter signaling pathway. J Theor Biol 244:463–469

29. Hesper B, Hogeweg P (1970) Bioinformatica: EEN werkconcept. Kameleon 1:18–29
30. Hogeweg P, Hesper B (1978) Interactive instruction on population interactions. Comput Biol Med 8:319–327
31. Huffman DA (1959) Canonical forms for information-lossless finite-state logical machines. IRE Trans Inf Theory 5:41–59
32. Jacob F, Monod J (1961) Genetic regulatory mechanisms in the synthesis of proteins. J Mol Biol 3:318–356
33. Kauffman SA (1969) Homeostasis and differentiation in random genetic control networks. Nature 224:177–178
34. Kauffman SA (1969) Metabolic stability and epigenesis in randomly constructed genetic nets. J Theor Biol 22:437–467
35. Kauffman SA (1971) Current topics in developmental biology, vol 6, chapter Gene regulation networks: a theory for their global structures and behaviors. Elsevier, pp 145–181
36. Kauffman SA, Peterson C, Samuelsson B, Troein C (2003) Random boolean network models and the yeast transcriptional network. Proc Natl Acad Sci USA 100:14796–14799
37. Kaufman M, Thomas R (1985) Towards a logical analysis of the immune response. J Theor Biol 114:527–561
38. Kleene SC (1956) Automata studies, vol 34 of Annals of mathematics studies, chapter Representation of events in nerve nets and finite automata. Princeton Universtity Press, pp 3–41
39. Kung HT, Leiserson CE (1980) Introduction to VLSI systems, chapter Algorithms for VLSI processor arrays. Addison-Wesley, pp 271–292
40. Kurka P (1997) Languages, equicontinuity and attractors in cellular automata. Ergod Theory Dyn Syst 17:417–433
41. McCulloch WS, Pitts W (1943) A logical calculus of the ideas immanent in nervous activity. J Math Biophys 5:115–133
42. Melliti T, Noual M, Regnault D, Sené S, Sobieraj J (2015) Asynchronous dynamics of Boolean automata double-cycles. In: Proceedings of UCNC, vol 9252 of Lecture notes in computer science. Springer, pp 250–262
43. Mendoza L (2006) A network model for the control of the differentiation process in Th cells. Biosystems 84:101–114
44. Mendoza L, Alvarez-Buylla ER (1998) Dynamics of the genetic regulatory network for arabidopsis thaliana flower morphogenesis. J Theor Biol 193:307–319
45. Monod J, Changeux J-P, Jacob F (1963) Allosteric proteins and cellular control systems. J Mol Biol 6:306–329
46. Noual M (2012) Dynamics of circuits and intersection circuits. In: Proceedings of LATA, vol 7183 of Lecture notes in computer science. Springer, pp 433–444
47. Noual M (2012) Updating automata networks. PhD thesis, École normale supérieure de Lyon. http://tel.archives-ouvertes.fr/tel-00726560
48. Puri Y, Ward T (2001) Arithmetic and growth of periodic orbits. J Integer Seq 4:01.2.1
49. Remy É, Mossé B, Chaouiya C, Thieffry D (2003) A description of dynamical graphs associated to elementary regulatory circuits. Bioinformatics 19:ii172–ii178
50. Remy É, Ruet P, Thieffry D (2008) Graphic requirements for multistability and attractive cycles in a Boolean dynamical framework. Adv Appl Math 41:335–350
51. Ribenboim P (1996) The new book of prime number records. Springer
52. Richard A (2010) Negative circuits and sustained oscillations in asynchronous automata networks. Adv Appl Math 44:378–392
53. Richard A, Comet J-P (2007) Necessary conditions for multistationarity in discrete dynamical systems. Discret Appl Math 155:2403–2413
54. Richard A, Comet J-P, Bernot G, Thomas R, dynamics (2004) Modeling of biological regulatory networks introduction of singular states in the qualitative. Fundam Inform 65:373–392
55. Robert F (1969) Blocs-H-matrices et convergence des méthodes itératives classiques par blocs. Linear Algebr Its Appl 2:223–265
56. Robert F (1976) Contraction en norme vectorielle: convergence d'itérations chaotiques pour des équations non linéaires de point fixe à plusieurs variables. Linear Algebr Its Appl 13:19–35

57. Robert F (1980) Itérations sur des ensembles finis et automates cellulaires contractants. Linear Algebr Its Appl 29:393–412
58. Robert F (1986) Discrete iterations: a metric study, vol 6 of Springer series in computational mathematics. Springer
59. Robert F (1995) Les systèmes dynamiques discrets, vo 19 of Mathématiques & applications. Springer
60. Ruskey F (2003) Combinatorial generation. Book preliminary working draft
61. Ruz GA, Goles E, Sené S (2018) Reconstruction of Boolean regulatory models of flower development exploiting an evolution strategy. In: Proceedings of CEC. IEEE Press, pp 1–8
62. Saez-Rodriguez J, Simeoni L, Lindquist JA, Hemenway R, Bommhardt U, Arndt B, Haus U-U, Weismantel R, Gilles ED, Klamt S, Schraven B (2007) A logical model provides insights into T cell receptor signaling. PLoS Comput Biol 3:e163
63. Sené S (2012) Sur la bio-informatique des réseaux d'automates. Habilitation thesis, Université d'Évry – Val d'Essonne http://tel.archives-ouvertes.fr/tel-00759287
64. Smith AR (1971) Simple computation-universal cellular spaces. J ACM 18:339–353
65. Thieffry D, Thomas R (1995) Dynamical behaviour of biological regulatory networks—II. Immunity control in bacteriophage lambda. Bull Math Biol 57:277–297
66. Thomas R (1973) Boolean formalization of genetic control circuits. J Theor Biol 42:563–585
67. Thomas R (1981) On the relation between the logical structure of systems and their ability to generate multiple steady states or sustained oscillations. In: Numerical methods in the study of critical phenomena, vol 9 of Springer series in synergetics. Springer, pp 180–193
68. Thomas R (1991) Regulatory networks seen as asynchronous automata: a logical description. J Theor Biol 153:1–23
69. Thomas R, Thieffry D, Kaufman M (1995) Dynamical behaviour of biological regulatory networks—I. Biological role of feedback loops and practical use of the concept of the loop-characteristic state. Bull Math Biol 57:247–276
70. Turing AM (1936) On computable numbers, with an application to the entscheidungsproblem. Proc Lond Math Soc 2:230–265
71. van der Laan H (1960) Le nombre plastique, quinze leÃğons sur l'ordonnance architectonique. E J Brill
72. von Neumann J (1966) Theory of self-reproducing automata. University of Illinois Press
73. Wolfram S (1984) Universality and complexity in cellular automata. Phys D 10:1–35

# Computing the Probability of Getting Infected: On the Counting Complexity of Bootstrap Percolation

**Pedro Montealegre and Martín Ríos-Wilson**

**Abstract** Consider a network where each node has one over two possible states, namely *healthy* or *infected*. Given an initial configuration, the network evolves in discrete time-steps picking uniformly at random a single node and updating its state according to the following rule: if the node is infected, it remains infected. If the node is healthy it switches its state to the one of the strict majority of its neighbors. We address, from the point of view of the computational complexity, the problem of computing the probability that a given healthy node becomes infected in at most a given number of time-steps, given as input network and an initial configuration. We show that this problem is #P-Complete in general, and solvable in polynomial time when the input graph is of degree at most 4.

## 1 Introduction

### The model

Consider a distributed system such as cellular automata, a neural network or other message-passing model network. In these systems, there is a graph where each node is assigned a *state*, which evolves according in discrete time-steps ruled by local interactions. In such systems, it is usually assumed that all interactions take place simultaneously, so the state of each entity in the network is updated in a *synchronous scheme*. Nevertheless, for several applications this assumption may be unrealistic. For instance, from biological viewpoint [5] in which these perfectly simultaneous interactions between cells are fairly rare. In this regard, dynamical properties of asynchronous update scheme, in which at most one node is updated in each time-

P. Montealegre (✉)
Facultad de Ingeniería y Ciencias, Universidad Adolfo Ibáñez, Santiago, Chile
e-mail: p.montealegre@uai.cl

M. Ríos-Wilson
Departamento de Ingeniería Matemática, FCFM, Universidad de Chile, Santiago, Chile
e-mail: mrios@dim.uchile.cl

Aix Marseille Univ, Université de Toulon, CNRS, LIS, Marseille, France

 
A. Adamatzky (ed.), *Automata and Complexity*, Emergence, Complexity and Computation 42, https://doi.org/10.1007/978-3-030-92551-2_12

step has been widely studied in the context of the boolean network formalism from a theoretical viewpoint [15–17] to the most applied front [14, 23]. Also, from the point of view of a computation model, the synchronous update scheme implies the existence of an internal clock that synchronizes each processor which is computing the state of a node. Therefore, different alternative approaches exist to the synchronous update scheme, all of them generically called *asynchronous update schemes*. In literature it is possible to find many families of updating schemes, deterministic or randomized (see for example [6, 12, 18, 19]).

Unfortunately, given initial condition, considering all possible choices for updating schemes creates a degree of freedom in the system that make explode the number of combinations for possible outcomes. This makes unfeasible the exhaustive computational simulations of these systems in networks with a large number of nodes, relegating them to mean field analysis or other empirical approximations [1, 11]. For instance, consider the *fixed random sweep updating scheme* [12, 19], which consists in choosing a random permutation of the nodes at initial time, and from there, at each time-step, one node is updated according to this permutation. The possible number of permutations of $n$ nodes is $n!$, so an exhaustive simulation of a given initial configuration will require the analysis of an exponential number of different updating schemes.

In this chapter, we address the problem of computing the exact probability that a node reaches a given state, in the Bootstrap Percolation model under a fixed random sweep updating scheme. In the Bootstrap Percolation model, nodes in the network have two possible states, namely 0 or 1, that evolve according to the following rule: (1) when a node in state 0 is updated, it evolves taking the state of the strict majority of its neighbors; (2) when a node in state 1 is updated, it remain in state 1. Bootstrap Percolation models are well-studied within the framework of modelling many physical, social and biological phenomena such as magnetic properties of some materials [4] crystal growth [10, 21], alert spreading in distributed networks [20], sand pile formation and disease spreading [3].

**The problem Contagion-Probability**

We define a computational task which, inspired by the relations with the dynamics of Bootstrap Percolation and disease spreading, we call CONTAGION- PROBABILITY. The input of problem CONTAGION- PROBABILITY is a simple undirected $n$-node graph $G$ with vertex set $[n]$, an initial condition $x \in \{0, 1\}^n$ which assigns a state to each node of the graph, a node $v \in [n]$ that we call *objective node* and a *time-span* $t \leq n$. The task consists in computing the probability that node $v$ reaches state 1 in at most $t$ time-steps under a fixed random sweep updating scheme.

Observe that in the definition of CONTAGION- PROBABILITY we restricted the time-span to be bounded by the number of nodes $n$. There are two reasons for considering this restriction. The first reason is that in many percolation processes (such as rumor or disease spreading) is unnatural to consider a time-spans larger than the number of nodes, as the later number could be extremely large. Second, given an initial condition, these processes reach the same fixed point on every updating scheme, as it was shown on [8, Theorem 1]. Moreover, it is possible to reach the fixed point updating at most $n$ nodes.

To study problem CONTAGION- PROBABILITY, we appeal to the theory of computational complexity. Computational complexity theory studies the amount of resources, such as time or space, required to solve a problem in a Turing Machine. This tool is used to classify problems in *complexity classes*, which group problems that can be solved by roughly the same amount of resources. The two classical examples of complexity classes are **P** and **NP** which contain all the decision problems (problems with *yes* or *no* output) that can decided, respectively, in a deterministic or non-deterministic Turing machine in a running time that is polynomial in the size of the input.

**The complexity of Contagion-Probability**

Observe that CONTAGION- PROBABILITY is not a decision problem, as in consists in computing a number in [0, 1]. The natural generalization of classes of decision problems are the *functional problems*. These problems are defined by a function $f : \{0,1\}^* \rightarrow \{0,1\}^*$ (or equivalently $f : \{0,1\}^* \rightarrow \mathbb{N}^*$), and the task is to compute $f(x)$ for a given input $x \in \{0,1\}^*$. The class **FP** is the class of all functions that can be computed by a polynomial-time deterministic Turing machine. In other words, **FP** is the generalization of class **P** to functional problems, hence is known as the class of feasible functional problems.

Then, our goal is to answer in which context CONTAGION- PROBABILITY is in **FP**. Unfortunately, the brute-force algorithm that test all possible permutations of the set of vertices in unfeasible, as this number is exponential in the size of the input. A natural question is whether the brute-force algorithm is optimal, or there exist properties of the dynamic (perhaps restricting the input graph to some class), that can be algorithmically exploited in order to obtain a polynomial-time algorithm.

The class **#P** (pronounced "sharp **P**" or "number **P**") is the set of function problems $f : \{0,1\}^* \rightarrow \mathbb{N}$ such that there is a polynomial deterministic Turing machine $M$ such that, for each $x \in \{0,1\}^*$, $f(x)$ equals the number of accepting branches in the computation graph of $M$ on input $x$. Informally, class **#P** is the set of *counting problems* associated to the number of certificates that have an instance of decision problem in **NP**. It is not hard to see that **FP** is a subset of **#P** (see the preliminaries section for more details), and it is conjectured that the inclusion is proper. The equality of the two classes, i.e. **FP** = **#P**, would have important consequences in the theory of computational complexity. In particular it would imply that **P** = **NP**.

The function problems in **#P** that are the most likely to not belong to **FP** are the **#P**-Complete problems. A function $f \in$ **#P** is **#P**-*Complete* if every other function $g$ in **#P** is computable in polynomial time by a Turing machine with an oracle access to the values of $f$. Roughly, a problem is $f$ is **#P**-Complete if an efficient algorithm computing $f$ can be used to efficiently compute every function in **#P** with at most a polynomial slowdown. In particular, the definition implies that if a **#P**-Complete function belongs to **FP**, then **#P** = **FP**.

Last remark suggests that the counting version of an **NP**-Complete problem is a good candidate for a **#P**-Complete problem. In fact, up to our knowledge, there are no examples of a **NP**-Complete problems whose counting version is not **#P**-Complete [13]. Interestingly, Valiant shows in [22] that there are examples of problems that their decision version is solvable in polynomial-time, while their counting version is **#P**-Complete. For instance, the problem consisting in computing a maximum matching of a given graph is solvable in polynomial time, while the problem of counting all matchings is **#P**-Complete [22].

Let $(G, x, v, t)$ be an instance of CONTAGION- PROBABILITY. We say that an injective function $\sigma : [t] \rightarrow V$ is a *good-sequence* if the objective node $v$ reaches state 1 in the dynamics of Bootstrap Percolation under the updating sequence $\sigma$. Now consider the following decision problem, that we call GOOD- SEQUENCE. This problem receives the same instances that CONTAGION- PROBABILITY, but output *yes* when there exist at least one-good sequence of the given instance. Clearly this problem is in **NP**, as a good sequence is a witness of polynomial size that can be verified in polynomial time by simply simulation of the dynamics of Bootrstap Percolation.

Observe that the output of CONTAGION- PROBABILITY corresponds to the number of good sequences of the input instance, divided by $n!/(n-t)!$, which corresponds to the number of possible injective functions $\sigma : [t] \rightarrow V$. As the latter number can be computed in a running time that is polynomial in $n$ (see the preliminaries section for more details), the difficulty of CONTAGION- PROBABILITY is reduced to the computation of the number of good-sequences. In the following when we refer to CONTAGION- PROBABILITY, we do not distinguish between the computation of the number of good-sequences and the computation of the actual probability. Note that previous observations imply that CONTAGION- PROBABILITY is in **#P**.

**Our results**

We show that CONTAGION- PROBABILITY is **#P**-Complete. Roughly this result implies that in order to compute the output, there is no better algorithm that simply simulate the dynamics on every possible injective function $\sigma : [t] \rightarrow [n]$, and keep the count of the ones that are good-sequences.

Our result is obtained constructing a polynomial-time Turing reduction from a version of #SAT to problem CONTAGION- PROBABILITY. The functional problem #SAT is the counting version of Boolean Satisfability, i.e., given a Boolean formula, the task is to count the number of truth-assignments satisfying it. In fact, our reduction is not from #SAT directly, but instead from a restriction of the problem called #MON- 2- SAT, where the input Boolean formula is in a 2-CNF form and it is monotonic (it does not have any negations of variables).

This result turns to be quite natural, as in [8] it is shown that GOOD- SEQUENCE (i.e. simply decide if the set of good sequences is non empty) is **NP**-Complete. As we said above, up to our knowledge, there are no examples of a **NP**-Complete problems whose counting version is not **#P**-Complete. In the same reference, it is shown that GOOD- SEQUENCE remain **NP**-Complete even when the problem is restricted to graphs of maximum degree 5.

Interestingly, when the input graph has maximum degree 4, GOOD- SEQUENCE is solvable in polynomial time [8]. This leads us to ask for the complexity of CONTAGION- PROBABILITY restricted to that family of graphs. As we mentioned before, there are problems solvable in polynomial time with #**P**-Complete counting versions. We remark that the restriction to graphs of maximum degree 4 contains important instances, as the ones where the dynamics occur on a two-dimensional grid with periodic boundary conditions.

Our second result is that CONTAGION- PROBABILITY restricted to input graphs of maximum degree 4 is solvable in polynomial time, hence belongs to **FP**. Our algorithm is based on a characterization of the initial conditions with at least one good-sequence, given in [8]. This characterization states, roughly, that when a good-sequence exists, the objective node must remain in some tree of the input graph, such that in every good sequence, a *pruning sequence* of its nodes is induced, in which every node of the tree must change its state to 1 before the objective node. In this regard, CONTAGION- PROBABILITY solution uses an algorithm that counts the number of pruning sequences of a given tree as a subroutine.

**Previous work**

This is not the first work that addresses the complexity of the dynamics of bootstrap percolation. In [9] it is studied the STABILITY problem which is the decision problem consisting in deciding, given a graph and an initial condition, whether an objective node reaches state 1 in some time-step, when the states evolve according to the synchronous bootstrap percolation dynamics. This problem is solvable in polynomial time, as the dynamics reach a fixed point in a linear number of synchronous time-steps. Interestingly, in [9] is shown that in graphs of degree at most 4 the problem is in class **NC**, which is the subclass of **P** containing all problems that can be efficiently solved in a parallel machine. Moreover, in graphs of maximum degree at least 5 the STABILITY problem is **P**-Complete, meaning that there is no algorithm solving the problem better than simply simulating the dynamics until a fixed point is reached.

Later, as we mentioned before, in [8] the authors study the complexity of bootstrap percolation on asynchronous updating schemes. To do so, two decision problems are defined. The first one is GOOD- SEQUENCE (which is called ASYNCHRONOUS PREDICTION in [8]). The second one is called ASYNCRONOUS STABILITY, which an asynchronous version of the stability problem. The difference between the two problems is that GOOD- SEQUENCE asks for the existence of a good-sequence for a given time-span, while ASYNCHRONOUS STABILITY asks for the state of the node once the attractor is reached. In that article, it is shown that given an initial condition, the dynamics of bootstrap percolation reaches the same fixed point for every updating scheme. That makes ASYNCRONOUS STABILITY equivalent to STABILITY. Contrarily, problem GOOD- SEQUENCE is solvable in polynomial time when is restricted to graphs maximum degree 4, but **NP**-Complete restricted to graphs of maximum degree 5.

**Structure of the chapter**

The chapter is organized as follows: first, in Sect. 2, we introduce some elementary concepts in graph theory, automata networks dynamics and computational complexity that we will be using in order to show our main results. Then, we present our main results: in Sect. 3 we show that CONTAGION- PROBABILITY problem is #**P**-complete by showing #MON- 2- SAT is Turing reducible to CONTAGION- PROBABILITY, i.e.

#MON- 2- SAT $\in$ **FP**$^{\text{CONTAGION- PROBABILITY}}$ however, in Sect. 4 we show, based on previous results in the characterization of good sequences [8], that problem CONTAGION- PROBABILITY is in **FP** when restricted to graphs of maximum degree 4. Finally, we discuss our results in Sect. 5.

## 2 Preliminaries

In this section we give the main definitions, results and terminology that is used during the next sections. We begin with the key concepts on graph theory. Then we continue with the definitions and results regarding the majority automata. Finally, we give the formal definitions of the complexity classes and algorithmic results. In all this chapter, we denote by $[n]$ the set $\{1, \ldots, n\}$.

### Graph Terminology

A graph $G$ on $n$ vertices and $m$ edges is a pair $(V, E)$ such that $|V| = n$ and $E \subseteq \binom{V}{2}$. All graphs in this article are simple, undirected without self-loops. Moreover, we consider only graphs where the vertex set is $V = [n]$, and use the name vertex and node indistinctly. If $G$ is a graph and the set of vertices and edges is not specified we use the notation $V(G)$ and $E(G)$ for the set of vertices and the set of edges of $G$ respectively. We use the notation $|G|$ or $|V(G)|$ for the number of nodes in $G$. Two nodes are *adjacent*Êif they both belong to the same edge. Given a graph $G = (V, E)$ and a vertex $v$ we call $N_G(v) = \{u \in V : \{u, w\} \in E\}$ to the *neighborhood of* $v$ and $d_G(v) = |N(v)|$ to the *degree of* $v$ in $G$. We write $N(v)$ or $d(v)$ when the context is clear. The maximum and minimum degree of the graph are denoted $\Delta(G)$ and $\delta(G)$, respectively. A graph is ofÊ*maximum degree* $d$ if $\Delta(G) \leq d$.

Let $G = (V, E)$ be a graph and let $S \subset V$. We define the *subgraph induced by* $S$ as the subgraph $G[S] = (S, E[S])$ such that $E[S] = \{\{u, v\} \in E : u, v \in S\}$. A subgraph $P$ of $G$ is said to be a path if $V(P) = \{v_1, \ldots, v_k\}$ where every $v_i$ is different and $E(P) = \{\{v_1, v_2\}, \{v_2, v_3\} \ldots, \{v_{k-1}, v_k\}\}$. We define the *length* of a path $P$ in $G$ as the number of edges in $P$. We say that $P$ is a *cycle* if $k \geq 3$ and $v_k = v_1$. We say that $v$ is *connected* to $v$ if there exist a path $P$ from $u$ to $v$. The maximal sets of pairwise connected vertices are called *connected components*.

A graph without any cycles is called *acyclic*. We say that a graph is a *tree* if it is acyclic and connected. In some parts of this work, we distinguish a certain node $r$ of a tree that we will call its *root*, and consequently we will refer to the graph as a *rooted tree*. A node in a tree of degree 1 is called a *leaf*. The choice of $r$ induces a partial order in the vertices of a tree given by the distance (length of the unique path) between a node $v$ and the root $r$. Given two nodes $v$, $w$ of a tree, we say that $w$ is an *ascendant* of $v$ (and $v$ is a descendant of $w$) if $w$ belongs to the path from $v$ to $r$.

### Asynchronous bootstrap percolation

Given a graph $G = ([n], E)$, we consider that each node has one of two possible *states*, that we denote 0 and 1 and call *healthy* and *infected*, respectively. A *config-*

*uration* Ê of $G$ is a vector $x \in \{0, 1\}^n$ assigning a state to each node. We define the following dynamic over a configuration $x$ (that in the following is called *initial condition*), that we call *asynchronous bootstrap percolation*. First, we fix a positive integer $t \leq n$, that we call *time-span*. Second, we fix an injective function $\sigma : [t] \to [n]$, that we call *updating sequence*. Then, the *trajectory* of $(x, \sigma, t)$ is the sequence of configurations $x^\sigma(0), \ldots, x^\sigma(t)$ such that $x^\sigma(0) = x$ and, for each $0 \leq k < t$ and $i \in [n]$ we define:

$$x_i^\sigma(k+1) = \begin{cases} x_i^\sigma(k) & \text{if } i \neq \sigma(k+1), \\ 1 & \text{if } i = \sigma(k+1) \text{ and } x_i^\sigma(k) = 1, \\ 1 & \text{if } i = \sigma(k+1) \text{ and } x_i^\sigma(k) = 0 \text{ and } \sum_{j \in N(i)} x_j^\sigma(k) > \lfloor d(i)/2 \rfloor, \\ 0 & \text{if } i = \sigma(k+1) \text{ and } x_i^\sigma(k) = 0 \text{ and } \sum_{j \in N(i)} x_j^\sigma(k) \leq \lfloor d(i)/2 \rfloor, \end{cases}$$

In words, in the $k$-th configuration, the state of all nodes except $\sigma(k)$ remain in the same state than in the previous configuration. Node $\sigma(k)$ remains in state 1 if it was in state 1 in the previous configuration, and otherwise it takes the state of the strict majority of its neighbors.

Let us fix a graph $G = ([n], E)$, a vertex $v \in [n]$, a configuration $x$, and a time-span $t$. We say that an updating sequence $\sigma$Ê*activates*Ê$v$ if $x_v^\sigma(t) = 1$. Moreover, in that case we say that $\sigma$ is a *good-sequence* for $G$, $x$, $v$, and $t$. We denote by $\mathsf{Good}(G, x, v, t)$ the set of good-sequences for $G$, $x$, $v$ and $t$. When the context is clear, we omit the specifications of the graph, configuration, node and time-span. We say that $v$ is *stable* if $\mathsf{Good} = \emptyset$, and otherwise we say that the node is *unstable*.

**Contagion-Probability**

We are now ready to formally define problem CONTAGION- PROBABILITY. As we explained in the introduction, this problem asks for the probability of choosing a good-sequence, when an update sequence is picked uniformly at random. Formally, this problem receives as input a graph $G = ([n], E)$, a vertex $v \in [n]$ of $G$, an initial condition $x \in \{0, 1\}^n$ and a time-span $t \leq n$. The task consists in computing $|\mathsf{Good}(G, x, v, t)|$ divided by $n!/(n-t)!$, which is the fraction of good-sequences from the total number of possible updating sequences with time-span $t$.

In the next section, we will give a result stating that number $n!/(n-t)!$ can be computed in time polynomial in $n$, and that the division of two integers can be computed in a running time that is linear in the number of bits required for their binary representation. Therefore, the difficulty of CONTAGION- PROBABILITY rests in the computation of $|\mathsf{Good}(G, x, v, t)|$. For that reason, we abuse notation and refer as CONTAGION- PROBABILITYalso the problem of computing the latter number.

**problem 1** (CONTAGION- PROBABILITY)

Input:
1. an $n$-node graph $G = ([n], E)$.
2. a node $v \in [n]$
3. an initial condition $x \in \{0, 1\}^n$.
4. a natural number $t \leq n$.

Output: $|\mathsf{Good}(G, x, v, t)|$

In order to study how difficult is to solve one of latter problem, we use computational complexity theory. Roughly, we are interested in measuring how difficult is to solve the problem, measuring the amount of resources required to solve in a Turing machine it in the worst case. These resources are measured with respect to the size of the input, in this case this corresponds to the number of bits required to encode $G$, vertex $v$, the configuration $x$ and the time-span $t$. This quantity is $\Theta(n + m + \log n)$, as the graph can be encoded in $\Theta(n + m)$ bits, the name of $v$ and the value of $t$ can be encoded in $\lceil \log n \rceil$ bits, and the configuration $x$ is encoded in $n$ bits.

**Elements of Computational Complexity**

A *functional problem* for a function $f : \{0, 1\}^* \to \mathbb{N}$ is the computational task consisting in computing $f(x)$ for a given input string $x \in \{0, 1\}^*$. Functional problems are a generalization of decision problems, which are functions $f_L : \{0, 1\}^* \to \{0, 1\}$ such that the bit $f_L(x)$ indicates whether $x$ belongs to a set $L \subseteq \{0, 1\}^*$. In this context, such a set $L$ is called a *language*. We abuse the notation and identify a function $f$ with the computational task consisting in computing $f(x)$. Similarly, we identify a decision problem with its associated language.

We say that a Turing machine *solves* a functional problem (or *decides* a decision problem) if, when the machine is initialized with $x$ written in the tape, the machine always halts, and when it does it has left written in the tape $f(x)$. When dealing with a decision problem, we say that the machine *accepts* or *rejects* depending on the bit it writes when it halts. The time-complexity of the problem, is the amount of time-steps that a Turing machine requires to solve the problem in the worst case, measured in terms of the size of the input $n$.

The classical theory of computational complexity defines two main classes of decision problems, namely **P** is the class of problems solvable in polynomial-time (i.e. with time-complexity $n^{O(1)}$) on a deterministic Turing machine; and **NP** are the problems that can solved in polynomial-time on a non-deterministic Turing machine [2]. Equivalently, we can define **NP** as the problems that can *verified in polynomial time*. Formally, a *polynomial-time verifier* $V$ for a language $L$ is a deterministic Turing machine such that, there is a polynomial $p$ satisfying that for every $x \in \{0, 1\}^*$ we have that $x \in L$ if and only if there exists a $u \in \{0, 1\}^{p(|x|)}$ such that $V$ accepts on input $(x, u)$ (i.e. the concatenation of $x$ and $u$). When $V(x, u)$ accepts, we say that $u$ is a *certificate* for $x$.

Clearly $\mathbf{P} \subseteq \mathbf{NP}$ and also it is wide-believed (and probably one of the most famous conjectures in this context) that $\mathbf{P} \neq \mathbf{NP}$, i.e. that the inclusion is proper. In this regard, the problems in **NP** that are most likely to not be contained in **P** are called **NP**-Complete problems. Roughly, a problem in **NP** is **NP**-complete if an efficient solution for it can be transformed into an efficient solution for every other problem in **NP**. Formally, a language $L' \in \mathbf{NP}$ is **NP**-Complete if for every $L \in \mathbf{NP}$ we have that $L'$ is polynomial time reducible to $L$, meaning that there exists a function $f$ computable in polynomial time, such that $x$ belongs to $L$ if and only if $f(x)$ belongs to $L'$.

The canonical **NP**-complete problem is the Boolean satisfiability problem SAT. This problem receives as an input a boolean formula $\mathcal{F}$ on $n$ variables, and the task consists in deciding whether this formula can be satisfied, i.e. it is possible to assign truth-values to its inputs such that the formula evaluates *true*.

### The Complexity of Counting

Some complexity classes for decision problems have natural generalizations to classes of functional problems. For instance, **FP** is the set of functions computable in polynomial time by a deterministic Turing machine. In other words, **FP** is the generalization of **P** to function problems. Unfortunately, it is not easy to generalize into function classes decision problems defined by non-deterministic machines, such as **NP**. A natural approach is to take, for a given instance $x$ of **NP** language $L$, the problem consisting in computing a certificate $y$ for $x$. For example, generalize SAT to a function problem consisting in computing a truth-assignment satisfying a Boolean formula given in the input. The complication is that such a generalization does not define a function but a *binary relation*, as a single instance could have several certificates. In that context, an interesting alliterative is to consider functions that *count* certificates. For instance, consider #SAT as the function problem consisting in compute the number of truth-assignments satisfying a given Boolean formula.

The class **#P** is the set of functions $f$ such that there exists a polynomial-time verifier $V$ such that, for every $x \in \{0, 1\}^*$,

$$f(x) = \#\{y \in \{0, 1\}^{p(|x|)} : V(x, y) \text{ accepts}\},$$

where $p$ is the polynomial defined for $V$. Roughly speaking, class **#P** contains all functions that *count certificates* of **NP** problems. Directly from the definition we have that #SAT $\in$ **#P**. Moreover, CONTAGION- PROBABILITY is also in **#P**.

Observe that **FP** is contained in **#P**. Indeed, for a given function $f \in$ **FP**, we define the polynomial verifier $V_f$ that accepts $(x, y)$ if and only if $y \leq f(x)$, when both strings are interpreted as positive integers. Then, $f(x)$ is exactly the number of certificates for $x$ on verifier $V_f$.

It is conjectured that **#P** $\neq$ **FP**, as clearly **#P** $=$ **FP** implies **NP** $=$ **P**. Moreover, the equality of the two classes would have much stronger (and unlikely) consequences. For instance, Toda's theorem imply that if **FP** $=$ **#P** then the whole polynomial hierarchy collapses in **P** [2, Theorem 9.11].

The functions in **#P** that are the most likely to not belong to **FP** are the **#P**-Complete functions. Roughly a function $f \in$ **#P** is **#P**-complete if an efficient algorithm computing $f$ can be used to efficiently solve all problems in **#P**. For defining this notion of hardness, we need the notion of *Turing reduction*.

We say a Turing machine has *oracle access* to a function $f : \{0, 1\}^* \to \{0, 1\}^*$ if at any given time step, the machine can execute a subroutine that in one time-step chooses a section of the tape containing a string $x \in \{0, 1\}^*$, and writes in the section of the tape the string $f(x)$. For fixed function $f : \{0, 1\}^* \to \{0, 1\}^*$, we call $\mathbf{FP}^f$ the class of functions $g : \{0, 1\}^* \to \mathbb{N}$ that are computable in polynomial-time by a Turing machine with oracle access to $f$. Intuitively, suppose that we have an algorithm computing function $f$. Then $\mathbf{FP}^f$ is the set of functions that can be computed running the algorithm for $f$ a polynomial number of times. In particular, when $f$ is computable in polynomial time $\mathbf{FP}^f = \mathbf{FP}$.

A function $g$ is *Turing reducible in polynomial time* to a function $f$ if $g \in \mathbf{FP}^f$. Then, we say that a function $f \in \#\mathbf{P}$ is #**P**-complete if for all function $g \in \#\mathbf{P}$ we have that $g$ is Turing reducible in polynomial time to $f$, i.e. $g \in \mathbf{FP}^f$. An important example of a #**P**-Complete problem is #SAT [2, Theorem 9.7]. In fact, is natural to expect that an **NP**-Complete decision problem defines a #**P**-Complete counting version. Up to our knowledge, there are no examples of a **NP**-Complete problems whose counting version is not #**P**-Complete [13].

Interestingly, there are #**P**-Complete problems for which the corresponding decision version can be solved in polynomial time. An important example is the *Monotone-2CNF-Satisfability* (MON- 2- SAT) and its counting version #MON- 2- SAT. A $n$-variable Boolean formula $\mathcal{F}$ is a *monotone-2-CNF formula* if $\mathcal{F}$ can be written as

$$\mathcal{F}(z_1, \ldots, z_n) = \bigwedge_{j \in [m]} (c_1^j \vee c_2^j) \text{ with } c_j^1, c_j^2 \in \{z_1, \ldots, z_n\}.$$

In such a case we call $(c_1^j, c_2^j)$ a *clause*. Now consider MON- 2- SAT as the problem SAT restricted to monotone-2-CNF formulas. This problem is trivial, as such a formula can be satisfied assigning all variables *true*. Now consider #MON- 2- SAT as the restriction of #SAT to monotone-2-CNF formulas. Formally,

**problem 2** (#MON- 2- SAT)

Input: a monotone-2 $- CNF$-Boolean formula $\mathcal{F}$ with $n$ variables and $m$ clauses.
Output: $|\{x \in \{\mathbf{True}, \mathbf{False}\}^n : F(x) = \mathbf{True}\}|$

Like #SAT, problem #MON- 2- SAT belongs to #**P**. Moreover, in [22] it is shown that it remains #**P**-Complete.

### The complexity of some arithmetic computations

In our proofs we will require to perform some arithmetic computations of large integers. We finish this section by giving a proposition that compiles results regarding the complexity of arithmetic computations.

**Proposition 1** *Given two integers $x, y \in [n]$:*

- *There is an algorithm computing the product $xy$ in time $O(\log n \log \log n)$*
- *There is an algorithm computing the division $x/y$ in time $O(\log n \log \log^2 n)$*
- *There is an algorithm computing the factorial $x!$ in time $O(n(\log n \log \log n)^2)$.*

*Therefore, when $y \leq x$ the expression $\binom{x}{y} = \frac{x!}{(x-y)!y!}$ can be computed in time $O(n^2)$.*

## 3 CONTAGION-PROBABILITY Is #$P$-Complete

In this section, we show that CONTAGION- PROBABILITY is **#P**-Complete. In order to do that, we reduce #MON- 2- SAT to CONTAGION- PROBABILITY by a polynomial-time Turing reduction. Roughly, we show that given an instance of #MON- 2- SAT, we can produce series of instances of problem CONTAGION- PROBABILITY that represent $\mathcal{F}$, in the sense that the number of truth-assignment satisfying $\mathcal{F}$ can be computed from the number of good sequences of these instances.

Let $\mathcal{F}$ be a monotone 2-CNF formula with $n$ variables and $m$ clauses. We call $Z(\mathcal{F}) = z_1, \dots, z_n$ and $C(\mathcal{F}) = \{C_1, \dots, C_m\}$, respectively the sets of variables and clauses of $\mathcal{F}$. Also, for each $j \in [m]$, we call $c_1^j$ and $c_2^j$ the two variables that participate in clause $C_j$. In other words

$$\mathcal{F}(z_1, \dots, z_n) = \bigwedge_{j \in [m]} (c_1^j \vee c_2^j), \quad \text{with } c_j^1, c_j^2 \in Z(\mathcal{F}).$$

We call $\varphi(\mathcal{F})$ the set of truth-assignments satisfying $\mathcal{F}$. The *weight* of a truth-assignment corresponds to the number of variables that are assigned *true*. For $k \in [n]$ we call $\varphi(\mathcal{F}, k)$ the set of truth-assignments of *weight* $k$ satisfying $\mathcal{F}$. Clearly $|\varphi(\mathcal{F})| = \sum_{k=1}^{n} |\varphi(\mathcal{F}, k)|$.

For $k \in [n]$, we define a graph $G[\mathcal{F}, k]$ on $N = (m + 3)n + 8m + 2k + 2$ nodes, and an initial configuration $x[\mathcal{F}, k] \in \{0, 1\}^N$ as follows.

- First, for each variable $z_i \in Z(\mathcal{F})$ graph $G[\mathcal{F}, k]$ contains a *variable gadget*. This gadget consists in a node $z_i$, called *variable node*, with $m + 1$ pending nodes denoted $z_{i,1}, \dots z_{i,m+1}$ and called *auxiliary variable nodes*. In the initial configuration we assign the variable node inactive and the auxiliary variable nodes to be active.
  Intuitively, the variable nodes are going to simulate the variables of $\mathcal{F}$. They are initially inactive, and are adjacent to a large enough number of auxiliary variable nodes to become active if they are updated, with the aim of associating an updating sequence to a choice of a truth-assignments.
- Second, for each clause $C_j \in C(\mathcal{F})$, graph $G[\mathcal{F}, k]$ contains a *clause gadget*. This gadget consists in a node $v^j$, called *clause node*, and four more nodes that are adjacent to the clause node. Two of these nodes are denoted $u_1^j, u_2^j$ and called *clause variable nodes*; the two remaining nodes are called *clause auxiliary nodes*. Node $u_1^j$ is adjacent to the variable node associated to $c_1^j$, and node $u_2^j$ is adjacent to the variable node associated to $c_2^j$. Finally, the initial configuration fixes $v^j, u_1^j, u_2^j$ as passive, and the auxiliary clause nodes as active.
- Third, graph $G[\mathcal{F}, k]$ contains one *threshold gadget*. This gadget consists in a node $\theta$, called *threshold node*, and $n + 2m + 2k + 1$ other nodes called *auxiliary threshold nodes*, that are adjacent to the threshold node. The threshold node is also adjacent to all variable nodes, and all clause variable nodes. In the initial configuration, the threshold node and $2k$ of auxiliary threshold nodes are inactive, and

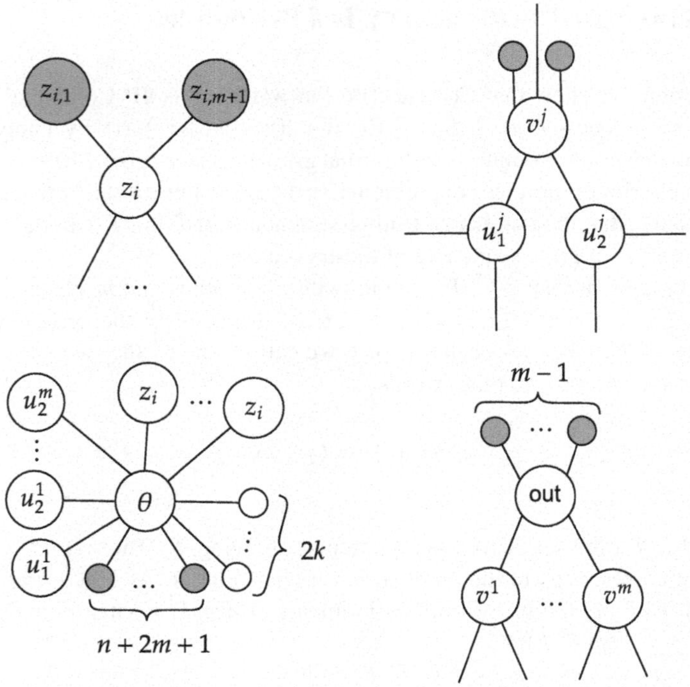


**Fig. 1** Gadgets used to construct graph $G[\mathcal{F}, k]$. From left to right: variable gadget, clause gadget, threshold gadget and output gadget

the remaining $n + 2m + 1$ auxiliary threshold nodes are initially active. Observe that the threshold node has degree $2n + 4m + 2k + 1$.

- Finally, graph $G[\mathcal{F}, k]$ contains one *output gadget*. This gadget consists in a node *out* called *output node* and $m - 1$ other nodes called *auxiliary output nodes*. The output node is also adjacent to every clause node. All nodes in this gadget are initially inactive.

See Fig. 1 for a graphical representation of each gadget. Observe that there are $n$ variable gadgets, each containing $(m + 2)$ nodes; there are $m$ clause gadgets, each containing 5 nodes; one threshold gadget containing $n + 2m + 2k + 2$ nodes; and one output gadget containing $m$ nodes. This sums up a total of $(n + 3)m + 8m + 2k + 2$ nodes in graph $G[\mathcal{F}, k]$.

Let us define the timespan $t(\mathcal{F}, k) = k + 2m + 2$ and call $v[\mathcal{F}, k]$ the output node of $G[\mathcal{F}, k]$. Given a sequence $\sigma$ of $v[\mathcal{F}, k]$ with timespan $t[\mathcal{F}, k]$, we say that $\sigma$ induces a truth-assignment $z$ of $\mathcal{F}$ if $z$ is such that $z_i = true$ if and only if the corresponding variable node is activated in $\sigma$.

Let us call $\rho(\mathcal{F}, k)$ the set of good-sequences for $v[\mathcal{F}, k]$ with timespan $t[\mathcal{F}, k]$, i.e. $|\rho(\mathcal{F}, k)|$ is the output of problem CONTAGION- PROBABILITY on input $(G[\mathcal{F}, k], x[\mathcal{F}, k], v[\mathcal{F}, k], t[\mathcal{F}, k])$.

**Lemma 1** *Suppose that $\sigma \in \rho(\mathcal{F}, k)$. Then, $\sigma$ satisfies that:*

- *For every $i \in \{1, \dots, k\}$, $\sigma(i)$ is a variable node,*
- *$\sigma(k+1)$ is the threshold node,*
- *For every $i \in \{k+2, \dots, 2m+k+1\}$, $\sigma(i)$ is a node in a clause gadget,*
- *$\sigma(2m+k+2)$ is the output node.*

***Proof*** First, observe that the output node of $G[\mathcal{F}, k]$ has degree $2m-1$, with every neighbor initially inactive. Therefore, by the strict majority rule it is necessary to activate $m$ of its neighbors before activating it. As $m-1$ of these neighbors are auxiliary output nodes, it is necessary to activate the $m$ clause nodes before activating the output node. Each clause node has degree 5, where one of the neighbors is the output node, and two neighbors are auxiliary clause nodes, which are initially active. Therefore, the clause node requires that at least one of the corresponding clause variable nodes is activated before it. Remember that the clause variable nodes have degree three, where one neighbor is a clause node, another neighbor is the threshold node, and the remaining one is a variable node. Consider now the time-step on which for the first time a clause-variable node is updated. Then, this node requires that both the threshold node and the adjacent variable node are activated before it. Therefore, threshold must be changed to active during sequence $\sigma$, and must become active before all the clause variable nodes. The threshold node has degree $2n+4m+2k+1$, with $n+2m+1$ adjacent auxiliary threshold nodes initially active and the remaining $n+2m+2k$ neighbors initially inactive. As we explained, $2m$ of its inactive neighbors are still inactive when the threshold node is activated. Moreover, the $2k$ inactive auxiliary threshold nodes are also inactive when the threshold node is activated. Observe that the threshold node requires at least $k$ more active neighbors to become active. More precisely let $p$ the amount of active neighbors that are required to active threshold node. Then, we have that $p$ must be chosen as the minimum number such the amount of active is more than the number of inactive neighbors. Summarizing previous calculations we have $(n+2m+1+p)$ active neighbors and $(n+2m+2k-p)$ inactive neighbors. Then, by asking $p$ to satisfy $(n+2m+1+p)-(n+2m+2k-p)>0$ we deduce $p>k-\frac{1}{2}$ and thus, $p \geq k$. We conclude that the only option is to choose at least $k$ variable nodes.

Wrapping up, we deduce the following order in every sequence activating the output node in $k+2m+2$ steps: first, $k$ variable nodes are activated. Then, the threshold node is activated. Then, all the $m$ clause nodes must be activated, which requires $2m$ steps, one for the clause variable node, and one for the clause node. Finally the output node is activated. We deduce that $\sigma$ satisfies the statements of the lemma (See Figs. 1 and 2). □

**Lemma 2** *Every $\sigma \in \rho(\mathcal{F}, k)$ induces a truth-assignment in $y \in \varphi(k, \mathcal{F})$. For every $y \in \varphi(k, \mathcal{F})$ there is a $\sigma \in \rho(\mathcal{F}, k)$ that induces $y$.*

***Proof*** Let $\sigma$ be a good sequence in $\rho(\mathcal{F}, k)$. Lemma 1 implies that in a good sequence every clause node is activated, implying that at least one variable of each clause is chosen in the $k$ first-steps. We deduce that $\sigma$ induces a truth-assignment of weight

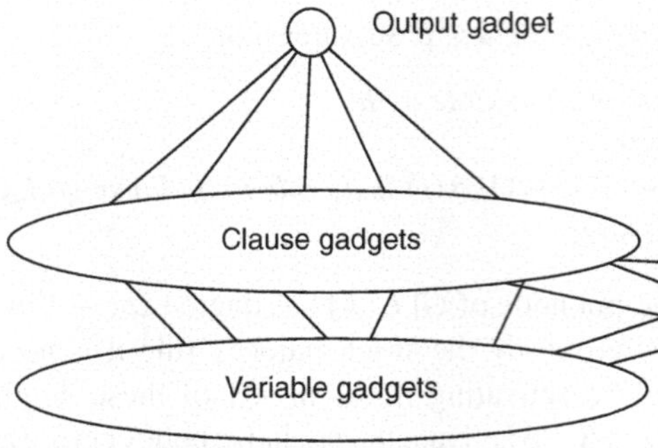


**Fig. 2** Scheme representing graph $G[\mathcal{F}, k]$. Each type of gadget is detailed in Fig. 1

$k$ satisfying $\mathcal{F}$. Conversely, let $y$ be a truth assignment in $\varphi(k, \mathcal{F})$. As this truth-assignment has weight $k$, we can update the corresponding $k$ variable nodes, and update the clause gadgets accordingly. This produces a sequence in $\rho(\mathcal{F}, k)$. □

Observe that Lemma 1 implies that for each truth-assignment $y \in \varphi(\mathcal{F}, k)$, there are several good sequences for $v[\mathcal{F}, k]$ with timespan $t[\mathcal{F}, k]$. Indeed, we know that in the first $k$ time-steps the variable nodes that are $true$ in $y$ are updated. They can be updated in any order, so there are $k!$ ways of updating them. Then, the threshold node is updated. Then, $2m$ nodes are updated to activate all clause nodes. Let us call $\gamma$ the number of ways that the $m$ clause nodes can be activated, once the $k$ variable nodes and the threshold node were activated according to $y$. Unfortunately, the value of $\gamma$ is not easy to compute, as it depends each the truth-assignment $y$. Indeed, the exact value depends on the number of clauses that are fully satisfied by a given truth assignment. A clause that is fully satisfied has two ways to be updated (one for each clause variable node), while a clause that is not fully satisfied has only one.

Therefore, to be able to compute $|\varphi(\mathcal{F}, k)|$ we will require to count with more detail. Let $C = (c_1 \vee c_2)$ be a clause of $\mathcal{F}$. We say that a truth-assignment $y$ *fully satisfies* clause $C$ if $y_{c_1} = y_{c_2} = true$, in other words, when both variables in the clause are true in $y$. For $d \in [m]$ we call $\varphi(\mathcal{F}, k, d)$ the set of truth-assignments of weight $k$ satisfying $\mathcal{F}$, and such that exactly $d$ clauses of $\mathcal{F}$ are fully satisfied. Clearly $|\varphi(\mathcal{F}, k)| = \sum_{d=0}^{m} |\varphi(\mathcal{F}, k, d)|$. Similarly, we define $\rho(\mathcal{F}, k, d)$ as the number of good sequences of $v[\mathcal{F}, k]$ with timespan $t[\mathcal{F}, k]$ that induce truth-assignments in $\varphi(\mathcal{F}, k, d)$. Obviously $|\rho(\mathcal{F}, k)| = \sum_{d=0}^{m} |\rho(\mathcal{F}, k, d)|$.

**Lemma 3** $|\rho(\mathcal{F}, k, d)| = |\varphi(\mathcal{F}, k, d)| \cdot k! \cdot \dfrac{(2m)!}{2^m} \cdot 2^d$

***Proof*** Let us fix $y \in \varphi(\mathcal{F}, k, d)$. From Lemma 2 we know that $y$ induces a good sequence in $\rho(\mathcal{F}, k)$, and then by definition this sequence also belongs to $\rho(\mathcal{F}, k, d)$. From Lemma 1 we know that in the first $k$ steps the variable nodes are updated. As they are independent, there are $k!$ ways of updating the variables that are $true$ in $y$. Once the $k$ variable nodes are activated, we have to continue by activating the threshold node. After that, we know from Lemma 1 that in the next $2m$ steps all the clause nodes must be activated.

For each clause, exactly two nodes are updated in the corresponding clause gadget. Indeed we need 2 time steps to activate each clause node, and if we spend more than two steps in any clause gadget we are not going to be able to activate the output node in the timespan. Let $C$ be the first clause. There are $\binom{2m}{2}$ ways of choosing steps to update one of its clause variable nodes, and the clause node. To update the second clause, we have $\binom{2m-2}{2}$ ways of choosing steps to update the corresponding pair. Repeating this argument we deduce that there are

$$\prod_{i=0}^{m}\binom{2m-2i}{2} = \frac{m!}{2^m}$$

ways of choosing steps for updating a pair of nodes of each clause. Finally, for each fully-satisfied clause there are two possible choices of a clause-variable node to update, giving a total of $2^d$ choices.

Previous calculations imply that there are $k! \cdot \dfrac{(2m)!}{2^m} \cdot 2^d$ possible good sequences that induce $y$. We deduce that there are $|\varphi(\mathcal{F}, k, d)| \cdot k! \cdot \dfrac{(2m)!}{2^m} \cdot 2^d$ good-sequences inducing truth-assignments in $\varphi(\mathcal{F}, k, d)$. □

Now let us call $P(s)$ the degree $m$ polynomial defined by

$$\mathcal{P}(s) = \sum_{d=0}^{m} |\varphi(\mathcal{F}, k, d)| s^d,$$

and observe that $|\rho(\mathcal{F}, k)| = \sum_{d=0}^{m} |\rho(\mathcal{F}, k, d)| = \frac{(2m)!}{2^m} \cdot k! \cdot \mathcal{P}(2)$. We would like to compute the coefficients of $\mathcal{P}(x)$. This is possible when we know an evaluation of the polynomial in a large enough point, as notices by Valiant in [22]. For sake of completeness we give the result and the full proof.

**Lemma 4** *Let $P(x) = \sum_{i=0}^{n} a_i x^i$ be a polynomial with integer coefficients upper bounded by a $A > 2$. Suppose that we know a pair $(x_0, y_0)$ such that $y_0 = P(x_0)$ and $x_0 > A^2$. Then, there exists an algorithm that outputs the coefficients $a_0, \ldots, a_n$ of $P$ in a time that is polynomial in $n(\log(x_0) + \log(y_0) + \log n)$.*

***Proof*** First, observe that $\sum_{i=0}^{j-1} a_i x_0^i < x_0^j$. for every $j \in [n]$. Indeed,

$$\sum_{i=0}^{j-1} a_i x^i < A \sum_{i=0}^{j-1} x^i = A x^{j-1} \sum_{i=0}^{j-1} x^{i-j+1} = A x^{j-1} \sum_{i=0}^{j-1} \frac{1}{x^{j-1-i}}$$

$$< A x^{j-1-1} \sum_{i=0}^{j-1} \frac{1}{A^{2i}} < \frac{3}{2} A x^{j-1} < x_0^j$$

Then, since $a_n = \frac{y_0 - \sum_{i=0}^{n-1} a_i x_0^i}{x_0^n}$ we deduce that, $\frac{y_0}{x_0^n} - \frac{1}{x_0} < a_n \leq \frac{y_0}{x_0^n}$ implying that $a_n = \left\lfloor \frac{y_0}{x_0^n} \right\rfloor$.

We sequentailly obtain the other coefficients taking $(x_0, y_0 - a_n x_0^n)$ as a pair for $P'(x) = \sum_{i=0}^{n-1} a_i x^i$. From from Proposition 1 we deduce that each iteration can be done in time $\log(x_0) + \log(y_0) + \log n$. □

Note that, unfortunately, the expression $|\rho(\mathcal{F}, k)| = \frac{(2m)!}{2^m} \cdot k! \cdot \mathcal{P}(2)$ only allows us to get the value of latter polynomial in $s = 2$ which is not a large-enough value to apply previous lemma. However, we can use a technique that is also inspired in the same paper of Valiant [22], used to show the **#P**-Completeness of a variant of SAT. Let $p$ be a positive integer to be fixed later. Let $\mathcal{F}^p$ be the 2-CNF formula with $n$ variables and $mp$ clauses defined as $\mathcal{F}^p = \mathcal{F} \wedge \mathcal{F} \wedge \cdots \wedge \mathcal{F}$ (repeated $p$ times).

**Lemma 5** $|\rho(\mathcal{F}^p, k)| = k! \cdot \frac{(2mp)!}{2^{mp}} \cdot \mathcal{P}(2^p)$

***Proof*** Observe that $\mathcal{F}$ and $\mathcal{F}^p$ have the same set of variables, and each clause of $\mathcal{F}$ is repeated $p$ times on $\mathcal{F}^p$. Therefore, $\varphi(\mathcal{F}^p, k, pd) = \varphi(\mathcal{F}, k, d)$. Moreover, if $d$ is not a multiple of $p$, then $\varphi(\mathcal{F}^p, k, d) = \emptyset$. Then,

$$\begin{aligned}
|\rho(\mathcal{F}^p, k)| &= \sum_{d=0}^{mp} |\rho(F^p, k, d)| \\
&= \sum_{d=0}^{mp} |\varphi(\mathcal{F}^p, k, d)| \cdot k! \cdot \frac{(2mp)!}{2^{mp}} \cdot 2^d \\
&= \sum_{d=0}^{m} |\varphi(\mathcal{F}^p, k, pd)| \cdot k! \cdot \frac{(2mp)!}{2^{mp}} \cdot 2^{pd} \\
&= \sum_{d=0}^{m} |\varphi(\mathcal{F}, k, d)| \cdot k! \cdot \frac{(2mp)!}{2^{mp}} \cdot 2^{pd} \\
&= k! \cdot \frac{(2mp)!}{2^{mp}} \cdot \mathcal{P}(2^p)
\end{aligned}$$

□

**Theorem 1** CONTAGION- PROBABILITY *is* ***#P****-complete.*

***Proof*** Let $\mathcal{F}$ be an instance of #MON- 2- SAT and consider the following algorithm computing $\varphi(\mathcal{F})$ in polynomial time on a machine with an oracle for CONTAGION- PROBABILITY. For each $k \in [n]$, the algorithm picks $p = 2n + 1$ and constructs the input $(G[\mathcal{F}^p, k], x[\mathcal{F}^p, k], v[\mathcal{F}^p, k], t[F^p, k])$ of

CONTAGION- PROBABILITY and queries the oracle on it, obtaining $|\rho(\mathcal{F}^p, k)|$. Then, the algorithm computes $k! \cdot \frac{(2mp)!}{2^{mp}}$ and divides $|\rho(\mathcal{F}^p, k)|$ by it in order to obtain the value $\mathcal{P}(2^p)$ according to Lemma 5. Then, the algorithm uses as a subroutine the algorithm given by Lemma 4 to obtain all the coefficients of $\mathcal{P}$, which correspond to $\{|\varphi(\mathcal{F}, k, d)|\}_{d \in \{0,\ldots,m\}}$. Finally, the algorithm outputs

$$\varphi(\mathcal{F}) = \sum_{k=1}^{n} \sum_{d=0}^{m} |\varphi(\mathcal{F}, k, d)|$$

From Proposition 1 and Lemma 4, all previous calculations can be done in polynomial time. We deduce that #MON- 2- SAT $\in FP^{\text{CONTAGION- PROBABILITY}}$, implying that CONTAGION- PROBABILITY is #**P**-Complete. □

## 4 Polynomial Time Algorithm for Maximum Degree 4

In this section, we restrict ourselves to the of Bootstrap Percolation in graphs of maximum degree four. We show that in this case, unlike the general case studied in previous section, we can compute the exact probability of infecting some node in polynomial time. Roughly, this means, in this context, that we are able to efficiently count the sequences of nodes, such that, if we update them we change the state of objective node from 0 to 1 in some fixed time $t$. In other words, we show that problem CONTAGION- PROBABILITY is in **FP**. To show this result, we use a characterization given in [8], for the configurations where the objective node is unstable. This characterization involves a topological structure that can be exploited to design an efficient algorithm counting all the good sequences.

In the following, $G = ([n], E)$ is a graph of degree at most 4, $x$ is a configuration of $\{0, 1\}^n$ and $v$ is a node such that $x_v = 0$. We call $G[0]$ the subgraph of $G$ induced by the nodes that are healthy, i.e. $G[0] = G[\{u \in V : x_u = 0\}]$. Observe that $v \in V(G[0])$ and hence we call $G[0, v]$ the connected component of $G[0]$ containing $v$. The following proposition characterizes the stable configurations.

**Proposition 2** ([9]) *Node $v$ is stable in $G$ if and only if $v$ belongs to a path $P$ in $G[0]$ such that an endpoint $w$ of $P$ belongs to a cycle in $G[0]$ or $d_G(w) \leq 2$.*

Suppose that $v$ is a site that is not stable in $G$. For each neighbor $w$ of $v$ in $G$ such that $x_w = 0$, we call $D_w$ the connected component of $G[v, 0] - v$ containing $w$. When $x_w = 1$, we fix $D_w = \emptyset$. The following lemma, characterizes the structure of the configuration around unstable sites.

**Proposition 3** ([8]) *Let $w$ be a neighbor of $v$ such that $x_w = 0$ and $w$ becomes active before $v$ for some good sequence. Then $D_w$ induces a tree of $G[0]$ where all nodes have degree at least 3 in $G$.*

A component $D_w$ which is a tree where all the nodes are of degree at least 3 in $G$ is called a *good tree*. As convention, an empty set is a good tree. Let GoodNeighbors$(v)$ the set of neighbors of $v$ that induce good trees. Observe that $v$ admits a good sequence if the strict majority of its neighbors induce good trees. A tree of $G[0]$ rooted at $v$ is called a *good tree* for $v$, and denoted $T_v$, if $T_v - v$ has $\left\lfloor \frac{d_G(v)}{2} \right\rfloor + 1$ components, each of them being good trees.

**Definition 1** Given a tree $T$ of size $n$ rooted at $r$, a *pruning sequence* of $T$ is a bijective mapping $\rho : [n] \to V(T)$ such that, if $u$ is an ancestor of $v$, then $\rho^{-1}(u) > \rho^{-1}(v)$. In particular $\rho^{-1}(r) = n$. The number of pruning sequences of $T$ is denoted #PRUNE$(T, r)$.

The following lemma links up the good sequences of $v$ with the existence of pruning subsequences of the good trees in the neighborhood of $v$.

**Lemma 6** *A sequence $\sigma$ is a good sequence for $v$ if and only there is a succession $s_1 < \cdots < s_{|T_v|}$ such that $\rho(i) := \sigma(s_i)$ is a pruning sequence a good tree $T_v$ of $v$.*

***Proof*** Suppose first that $\sigma : [t] \to V$ is a good sequence for $v$ and let $\{x(s)\}_{s\in[t]}$ be the succession of configurations such that $x(s) = F(x, \sigma, s)$. Since $\sigma$ is a good sequence, there must exist a step $s^* \in [t]$ in which $v$ becomes infected, i.e. such that $x(s^* - 1)_v = 0$ and $x(s^*)_v = 1$. In step $s^* - 1$ the strict majority of the neighbors of $v$ must be infected.

From Proposition 3, we know that there exist a set $\{w_1, \ldots, w_k\}$ of $k = \left\lfloor \frac{d_G(v)}{2} \right\rfloor + 1$ neighbors of $v$ such that, for all $i \in [k]$, node $w_i$ induces a good tree $D_i$ and $x(s^* - 1)_{w_i} = 1$. Define $T_v = \{v\} \cup \bigcup_{i\in[k]} D_i$, and observe that $T_v$ is a good tree for $v$. Let us call $s_1 < \cdots < s_\ell = s^*$ the sequence on which the nodes of $T_v$ are updated, with $\ell = |T_v|$. More precisely, for each $j \in [\ell]$ we have that $\sigma(s_j) \in D_i$, $x(s_j)_{\sigma(s_j)} = 1$ and $x(s_j - 1)_{\sigma(s_j)} = 0$. Observe that $\sigma(s_j)$ is a leaf of $T_v \setminus \sigma([s_j - 1])$, because for each $s \in \{s_1 < \cdots < s_\ell = s^*\}$ state of node $\sigma(s)$ is updated, meaning that the majority of its children are infected or in state 1 in time $s - 1$ and also $\sigma(s)$ must be an ancestor of $\sigma(s)$ as any node in $T_v$ was initially set to 0. Thus, if $s_j$ is not a leaf of $T_v \setminus \sigma([s_j - 1])$ then, it would have more than one ancestor in $T_v$ which is not possible. We deduce that $\rho(i) := \sigma(s_i)$ is a pruning sequence of $T_v$.

Conversely, if $\rho$ is a pruning sequence of $T_v$, then in step $s_{|T_v|}$ vertex $v$ becomes infected, implying that $\sigma$ is a good. sequence. □

**Lemma 7** *Let $(G, x, v, t)$ be an instance of* CONTAGION- PROBABILITY *such that $G$ is a graph of maximum degree 4 and $v$ is unstable. Then,*

- *If* $|\textsf{GoodNeighbors}(v)| = \left\lfloor \dfrac{d_G(v)}{2} \right\rfloor + 1$, *then*

$$\text{CONTAGION- PROBABILITY}(G, x, v, t) = \alpha(T^*, v, t)$$

- *If* $|\textsf{GoodNeighbors}(v)| > \left\lfloor \dfrac{d_G(v)}{2} \right\rfloor + 1$, *then*

$$\text{CONTAGION- PROBABILITY}(G, x, v, t) = \alpha(T^*, v, t) + \beta(T^*, v, t)$$

*where*

$$\alpha(T, r, t) = \begin{cases} \binom{t}{|T|}\#\text{PRUNE}(T, r)\frac{(n-|T|)!}{(n-|T|-t)!} & \textit{if } |T| \le t \\ 0 & \textit{otherwise} \end{cases},$$

$$\beta(T, r, t) = \sum_{S \subset N(v) s.t |S| = d_G(v)-1} (\alpha(T_S, v, t) - \alpha(T^*, v, t)),$$

$$T^* = \{v\} \cup \bigcup_{w \in \mathsf{GoodNeighbors}(v)} D_w \textit{ and } T_S = \{v\} \cup \bigcup_{w \in S} D_w$$

***Proof*** First, for a given a rooted tree $T$, observe that $\alpha(T, r, t)$ is exactly the number of sequences of length $t$ that contain a pruning sequence of $T$. Indeed, if $|T| > t$ then this number is zero. Otherwise, a sequences containing a pruning of $T$ is constructed picking $|T|$ steps over the $t$ possible choices, and prune $T$ in the chosen steps. In the remaining steps any other node can be updated. The number of ways that $|T|$ steps can be picked over a total of $t$ steps is $\binom{t}{|T|}$. The possible ways of pruning the $T$ on those steps is #PRUNE$(T, v)$. Finally, the number of possible choices for updating other vertices in the remaining steps is $\dfrac{(n-|T|)!}{(n-|T|-t)!}$. We deduce that $\alpha(T, r, v)$ is the product of previous quantities.

Lemma 6 implies that every good sequence for $v$ contain a subsequence that can be mapped into a pruning sequence of a good tree for $v$. When $|\mathsf{GoodNeighbors}(v)| = \left\lfloor \dfrac{d_G(v)}{2} \right\rfloor + 1$, there is only one possible choice of good tree for $v$, which is precisely $T^* = \{v\} \cup \bigcup_{w \in \mathsf{GoodNeighbors}(v)} D_w$. We deduce that the number of good sequences for $v$ is $\alpha(T^*, v, t)$.

When $|\mathsf{GoodNeighbors}(v)| > \left\lfloor \dfrac{d_G(v)}{2} \right\rfloor + 1$, then necessarily the degree of $v$ is 3 or 4. In either case $|Good(v)| = d(v)$ and any good tree for $v$ contains $d(v) - 1$ of its neighbors. Therefore, the choices for good sequences of $v$ is the number of sequences that update contain a prune of $T^*$, plus all the sequences that update some good tree of $v$, but not all $T^*$. The number of sequences that contain a prune of $T^*$ is $\alpha(T^*, v, t)$. For a given set $S$ of three neighbors of $v$, the number of sequences that contain a pruning of the good trees induced by the nodes in $S$, but not a pruning of $T^*$ equals $\alpha(T_S, v, t) - \alpha(T^*, v, t)$. We deduce that CONTAGION- PROBABILITY$(G, x, v, t)$ equals

$$\alpha(T^*, v, t) + \sum_{S \subset N(v) \text{s.t} |S| = d_G(v)-1} (\alpha(T_S, v, t) - \alpha(T^*, v, t)).$$

□

Previous lemma implies that, in order to obtain a polynomial-time algorithm solving problem CONTAGION- PROBABILITY, it is enough to have a polynomial-time algorithm computing the value of #PRUNE$(T, r, t)$, for a given rooted tree $T$. Following lemma states that this is the case.

**Lemma 8** *Given tree $T$ rooted in vertex $r$ with maximum degree* 4, *there is an algorithm computing* #PRUNE$(T, r)$ *time polynomial in* $|T|$.

***Proof*** For a node $u$ in $T$, let us define the *depth* of $u$, denoted $D(u)$, is the distance of $u$ to the root $r$. We also call $M$ the maximum depth of a node in $T$. The function $D(\cdot)$ can be computed in polynomial time simply running a BFS starting at node $r$. The *level* of a node $u$, denoted $L(u)$, equals $M - D(u)$. In other words, vertices at depth $M$ are at level 0. The root is at level $M$.

Observe that all the nodes in level 0 are leafs, but not necessarily all leaves are at that level. For a given $u \in T$, we call $T_u$ the subtree rooted at $u$ containing $u$ and all it descendants. Observe that if $u$ is a leaf of $T$, then #PRUNE$(T_u, u) = 1$. If $u$ is not a leaf, let us call $w_1, \ldots, w_k$ the descendants of $u$ in $T$, with $k = d(u) - 1$. Then,

$$\#\text{PRUNE}(T_u, u) = \frac{(|T_{w_1}| + \cdots + |T_{w_k}|)!}{|T_{w_1}|!, \ldots, |T_{w_k}|!} \#\text{PRUNE}(T_{w_1}, w_1) \cdots \#\text{PRUNE}(T_{w_k}, w_k) \tag{4.1}$$

Indeed, to #PRUNE$(T_u, u)$ we have to prune all the descendants of $u$ before $u$. Observe that a pruning of any two decendants of $u$ are independent. Therefore, a pruning of $T_u$ consist in choosing $|T_w|$ steps and a prune of $T_w$ for each descendant $w$ of $u$. For each $j \in [k]$ let $s_j = |T_{w_j}|$ and $S = s_1 + \cdots + s_k$. Then, we have $\binom{S}{s_1}\binom{S-s_1}{s_2} \cdots \binom{S-\sum_{j=2}^{k-1} s_j}{s_k} = \binom{S}{s_k} \prod_{k=2}^{n} \binom{S-\sum_{j=2}^{k-1} s_j}{s_k} = \frac{(|T_{w_1}|+\cdots+|T_{w_k}|)!}{|T_{w_1}|!,\ldots,|T_{w_k}|!}$ ways of choosing $|T_w|$ steps for every descendant $w$ of $u$. For each such choice, there are

$$\#\text{PRUNE}(T_{w_1}, w_1) \cdots \#\text{PRUNE}(T_{w_k}, w_k)$$

ways of choosing a pruning of each subtree $T_{w_i}$.

Therefore, our algorithm computing #PRUNE$(T, r)$ consists in a dynamic programming scheme over the levels of $T$. In level 0 all nodes $u$ are leafs, and then #PRUNE$(T_u, u) = 1$. If all the nodes al level $i$ are already computed, the value of #PRUNE$(T_u, u)$ for a node in level $i + 1$ can is computed using Eq. 4.1.

Observe that for each $u \in T$, the quantity #PRUNE$(T_u, u)$ is at most $|T|! = O(2^{|T| \log |T|})$. Therefore, from Proposition 1 we deduce that the expression of Eq. 4.1 can be computed in time $O(|T|^2)$. We conclude that #PRUNE$(T, r)$ can be computed in time $O(|T|^3)$. □

**Theorem 2** *There is a polynomial-time algorithm solving* CONTAGION- PROBABILITY *restricted to the graphs of maximum degree* 4.

***Proof*** Let $(G, x, v, t)$ be an input of CONTAGION- PROBABILITY, where $G$ an $n$-node graph of maximum degree 4, $x$ is a configuration of $G$, $v$ is a node such that $x_v = 0$ and $t \leq n$.

The algorithm computes $G[0, v]$, which is the component of $G[0]$ that contains $v$. Let $C_1, \ldots, C_k$, with $k \leq 4$ the connected components of $G[0, v] - v$. Then, it computes $\mathsf{GoodNeighbors}(v)$ verifying which components induce good trees. If $|\mathsf{GoodNeighbors}| < \lfloor \frac{d(v)}{2} \rfloor + 1$ then by Proposition 2 and Proposition 3 algorithm outputs 0 because $v$ is stable. Now suppose $|\mathsf{GoodNeighbors}| \geq \lfloor \frac{d(v)}{2} \rfloor + 1$ then, for each $w \in \mathsf{GoodNeighbors}(v)$, compute #PRUNE$(T_w, w)$, where $T_w$ is the component (good tree) of $G[0, v] - v$ that contains $w$. Finally, the output is computed according to the expressions given in Lemma 7. From Proposition 1 and Lemma 8, we deduce that our algorithm runs in time $O(n^3)$. □

## 5 Conclusion

In this chapter, we have studied the computational complexity of the problem CONTAGION- PROBABILITY. In general, we have shown that CONTAGION- PROBABILITY is #**P**-complete. Roughly, this means there is no better strategy for computing the probability of an inactive node to change to state 1 than simply simulate the system for each possible updating sequence $\sigma$ in order to verify how many of them actually change the state of objective node. However, when we consider networks with maximum degree 4, we have shown, based on previous results that characterize good sequences in terms of underlying graph topology [8], that latter computation can be made in polynomial time. The first natural question that arises in this context is whether this threshold in the maximum degree of the network is tight, or in other words: is CONTAGION- PROBABILITY still #**P**-complete even when restricted to instances of maximum degree 5? Note that gadgets in section 3 strongly use high connectivity in graph $G[\mathcal{F}, k]$ in order to assure that good sequences will respect a fixed order while updating different nodes in the network (first we update variable nodes, then we update threshold gadget, etc.). This is a key aspect of the proof as it helps us to keep control in the amount different sequences associated to one particular assignation satisfying the Boolean formula. This shows that even when one could conjecture a way to implement constant degree gadgets (by for example considering more copies of smaller gadgets) there should also be a way to keep track in all possible good sequences that implement the same assignation satisfying the original Boolean formula.

On the other hand, it could be also interesting to study a version of problem CONTAGION- PROBABILITY but now considering some other model having a different type of local updating rule, e.g. linear functions such as conjunctive or disjuntive networks or totallistic functions in which the next state of every node is given not by the state of the majority of their neighbors but by the sum of their states. In this regard, it is interesting to note that for some of the latter automata networks, there exist efficient algorithms (polynomial or even fast parallel algorithms) in order to compute the dynamics of the system (for example, in the case of conjunctive networks, it suffices to compute powers of some matrix [7]). However, this does not necessarily implies that counting version of Good- Sequence would be efficiently solvable.

In addition, we would like to note that previous results are also interesting in the context of counting decision problems. Within this framework, we are no longer interested in computing the number of all possible polynomial size certificates but to determine whether, for some fixed input, *the majority* of polynomial size strings we give to the verifier along with the input are actually certificates. This notion defines the complexity class **PP**. More precisely a language $L \subseteq \{0, 1\}^*$ is in **PP** if and only if there exist a polynomial function $p : \mathbb{N} \to \mathbb{N}$ and a polynomial-time verifier $V$ such that $x \in L \iff |y \in \{0, 1\}^{p(|x|)} : V(x, y) = 1| \geq \frac{1}{2} \cdot 2^{p(|x|)}$. By considering this formalism, an interesting question arises: is the decision version of Contagion- Probability, which consist on deciding if the probability of transmission for objective node is at least $\frac{1}{2}$, **PP**-complete? (observe that by definition it is in **PP**). In this context, it is known that problem MAJ- SAT, which consist on deciding if the majority of all assignments of some Boolean formula will satisfy it, is **PP**-complete and thus, it could be very interesting to explore as future work if an application of the same techniques we have used in Section 3 might produce a polynomial-time reduction from latter problem to decision version of Contagion- Probability.

Finally, as we have shown that Contagion- Probability is **#P**-complete, it could be interesting to study approximations. In particular, within the framework of applications one could ask whether this problem admits a polynomial randomized approximation scheme.

**Acknowledgements** Eric Goles has been a mentor, colleague, and a dear friend for both of us. This chapter is not only dedicated to Eric's 70th birthday, but also to the passion, love, and enthusiasm that he has taught us for this beautiful research topic. For both of us working with Eric has been a wonderful experience in many aspects. As he likes to say: we have a good time, and in the meantime, we do some mathematics. We also want to acknowledge the financial support given by: ANID via PAI + Convocatoria Nacional Subvención a la Incorporación en la Academia Año 2017 + PAI77170068 (P.M.), FONDECYT 11190482 (P.M.), FONDECYT 1200006 (P.M.), STIC- AmSud CoDANet project 88881.197456/2018-01 (P.M.), ANID via PFCHA/DOCTORADO NACIONAL/2018 – 21180910 + PIA AFB 170001 (M.R.W).

## References

1. Amini H (2010) Bootstrap percolation in living neural networks. J Stat Phys 141(3):459–475
2. Arora S, Barak B (2009) Computational complexity: a modern approach. Cambridge University Press
3. Balogh J, Pete G (1998) Random disease on the square grid. Random Struct Algorithms 13(3–4):409–422
4. Chalupa J, Leath PL, Reich GR (1979) Bootstrap percolation on a bethe lattice. J Phys C: Solid State Phys 12(1):L31
5. Cornforth D, Green DG, Newth D, Kirley M (2003) Do artificial ants march in step? ordered asynchronous processes and modularity in biological systems. In: Proceedings of the eighth international conference on artificial life. MIT Press, pp 28–32

6. Fates NA, Morvan M (1997) An experimental study of robustness to asynchronism for elementary cellular automata. Complex Syst 11:1
7. Goles E, Montealegre P (2014) Computational complexity of threshold automata networks under different updating schemes. Theor Comput Sci 559:3–19
8. Goles E, Montealegre P (2020) The complexity of the asynchronous prediction of the majority automata. Inf Comput 104537
9. Goles E, Montealegre-Barba P, Todinca I (2013) The complexity of the bootstraping percolation and other problems. Theor Comput Sci 504:73–82
10. Gravner J, Griffeath D (1998) Cellular automaton growth on z2: theorems, examples, and problems. Adv Appl Math 21(2):241–304
11. Janson S, Kozma R, Ruszinkó M, Sokolov Y (2016) Bootstrap percolation on a random graph coupled with a lattice. Electron J Comb
12. Kitagawa T (1974) Cell space approaches in biomathematics. Math Biosci 19(1–2):27–71
13. Livne N (2009) A note on -completeness of NP-witnessing relations. Inf Process Lett 109(5):259–261
14. Mendes ND, Henriques R, Remy E, Carneiro J, Monteiro PT, Chaouiya C (2018) Estimating attractor reachability in asynchronous logical models. Front Physiol 9:1161
15. Noual M, Sené S (2018) Synchronism versus asynchronism in monotonic boolean automata networks. Nat Comput 17(2):393–402
16. Remy É, Ruet P, Thieffry D (2008) Graphic requirements for multistability and attractive cycles in a boolean dynamical framework. Adv Appl Math 41(3):335–350
17. Richard A (2010) Negative circuits and sustained oscillations in asynchronous automata networks. Adv Appl Math 44(4):378–392
18. Robert F (2012) Discrete iterations: a metric study, vol 6. Springer Science & Business Media
19. Schönfisch B, de Roos A (1999) Synchronous and asynchronous updating in cellular automata. Biosystems 51(3):123–143
20. Treaster M, Conner W, Gupta I, Nahrstedt K (2006) ContagAlert: using contagion theory for adaptive, distributed alert propagation. In: Fifth IEEE international symposium on network computing and applications (NCA'06). IEEE, pp 126–136
21. Ulam SM (1970) On some mathematical problems connected with patterns of growth of figures. In Bukrs AW (ed) Essays on cellular automata. University of Illinois Press, pp 219–231
22. Valiant LG (1979) The complexity of enumeration and reliability problems. SIAM J Comput 8(3):410–421
23. Desheng Z, Guowu Y, Xiaoyu L, Zhicai W, Feng L, Lei H (2013) An efficient algorithm for computing attractors of synchronous and asynchronous boolean networks. PloS One 8(4):e60593

# Thermodynamics of Small Systems Through a Reversible and Conservative Discrete Automaton

**Marco Montalva-Medel, Sergio Rica, and Felipe Urbina**

**Abstract** The Q2R model is a cellular automaton which is a dynamical variation of the Ising model for ferromagnetism that possesses quite rich and complex dynamics. It has the property of being conservative and reversible but, in practice, it shows irreversible behavior for relatively small system sizes. In this work we review some of its main properties and use it to simulate de behavior of a classical model for irreversible thermodynamical systems: the Ehrenfest's dog-flea model.

**Keywords** Fixed point · Cycle · Cellular automata

## 1 Introduction

The reversible exchange of heat between two distinct reservoirs at different temperatures has remained a central problem of thermodynamics and statistical physics. Despite the reversible character of the equation of motions in mechanics, the nature does not allow to observe a reversible behavior of a macroscopical system. This central question in basic physics has been in the core of debates since the end of the 19th century.

Soon after Boltzmann kinetic theory and his entropy growth H-theorem, Loschmidt and Zermelo's objected the very basis of Boltzmann results. Loschmidt argued that if a dynamical system goes from the set $\Gamma(t)$ in the phase space to $\Gamma(t')$, then by

M. Montalva-Medel · S. Rica
Facultad de Ingeniería y Ciencias, Universidad Adolfo Ibáñez, Avda. Diagonal las Torres 2640, Peñalolén, Santiago, Chile
e-mail: marco.montalva@uai.cl

S. Rica (✉)
Instituto de Física, Pontificia Universidad Católica de Chile, Santiago, Chile
e-mail: sergio.rica@uc.cl

F. Urbina
Facultad de Estudios Interdisciplinarios, Centro de Investigación DAiTA Lab, Universidad Mayor, Santiago, Chile
e-mail: felipe.urbina@umayor.cl

 
A. Adamatzky (ed.), *Automata and Complexity*, Emergence, Complexity and Computation 42, https://doi.org/10.1007/978-3-030-92551-2_13

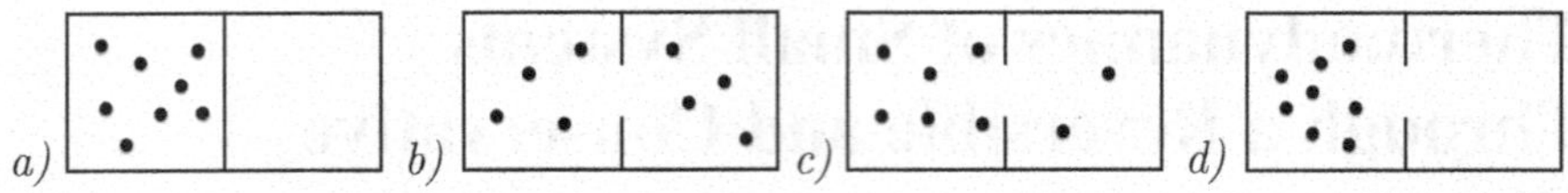


**Fig. 1** Cartoon representation of the Ehrenfest's dog-flea model: **a** Represents an initial state with all particles in the left-hand side container; **b** and **c** show two distinct intermediate configurations that the system displays (**b** is most probable); and, **d** shows a possible recurrence: all particles come back to the left-hand side container

means of Newtonian dynamics it should also exists the inverse trajectory, against the H-theorem [1]. Then Zermelo uses the Poincaré recurrence theorem to argue that if the system is at a time $t$ at a point in the phase space, then it exists a time $t'$ whenever the system will be close enough to previous point in the phase space. After Zermelo [2, 3], the Boltzmann H-theorem is in evident contradiction with the Poincaré recurrence theorem. As Boltzmann itself replies, the recurrence time becomes huge in comparison with all practical times in the usual thermodynamics.

To model the recurrence time paradox, $P$. and $T$. Ehrenfest elaborated particle exchange model [4, 5]. The Ehrenfest model consist of $N$ particles that can be distributed in the left or right side of a container (see Fig. 1 as a scheme), in such a way that we have $N/2+n$ balls are initially at the left container and $N/2-n$ are in the right. As show by Kac [6], the Ehrenfest model maybe mapped into a random walk, moreover average recurrence time is:

$$\langle \tau(N,n) \rangle = \frac{(N/2-n)!(N/2+n)!}{N!}2^N \approx \begin{cases} 2^N & \text{for } n \approx N/2 \\ \sqrt{\frac{\pi N}{2}} & \text{for } n \approx 0. \end{cases}$$

Therefore, if initially the system is filling the left side container $n = N/2$ (Fig. 1a), then the mean recurrence time would be exponentially long, more precisely $2^N$, but if initially the system is equally distributed $n = 0$ (Fig. 1b) then the waiting time scales only as $\sqrt{N}$. Therefore, as one increases the total number of elements $N$, the recurrence time becomes exponentially long. The Ehrenfest model captures the essence of the exponentially long recurrence time. Via an extremely simplified analogy that shows that an improbable initial configuration (all particles in one container) requires an exponentially long time to be back to the same state again (Fig. 1d).

Generically, irreversibility arises from the large number of elements, or more precisely number of degrees of freedom, involved in the dynamical system. If one has ten particle ($N = 10$) and initially all particle are in the left container, it is expected after the Ehrenfest model, that all particles back to the left container after $2^{10} = 1024$ steps. Therefore, irreversibility appears to be a consequence of thermodynamic limit, $N \to \infty$. Nevertheless, in moderate system size the Loschmidt and Zermelo objections may be pertinent.

However, even in moderate system size a thermodynamic description appears to be adequate, in other words, in thermodynamics and statistical physics irreversibility is understood in a statistical sense. In a recent article [7], two of us, developed a master equation approach to a reversible and conservative cellular automaton model (Q2R), which is a dynamical variation of the Ising model for ferromagnetism that possesses quite a rich and complex dynamics. The phase space is composed of a huge number of cycles with exponentially long periods. Following Nicolis and Nicolis [8], a coarse-graining approach is applied to the time series of the total magnetization, leading to a master equation that governs the macroscopic irreversible dynamics of the Q2R automata. The methodology is replicated for various lattice sizes. In the case of small systems, we show that the master equation leads to a tractable probability transfer matrix of moderate size, which provides a master equation for a coarse-grained probability distribution. The method is validated and some explicit examples are discussed.

The present article, examines this problematic in an extremely simplified model, namely the Q2R cellular automaton. The Q2R cellular automaton is a two state Ising-based automaton that rules a reversible evolution of a set of discrete states. By two states, we mean that Q2R is ruled by a duet $(x, y)$ at each time, $x$ and $y$ being elements of $N = L \times L$ regular square lattice with periodic boundary conditions. Remarkably, the evolution preserves an Ising-like energy [9], as well as, a second invariant [10]. Despite the simplicity of the Q2R model its behavior is usually quite rich as it has been reported extensively in the 80s [11, 12].

Furthermore, for a large system size it has been established that the evolution presents an irreversible behavior towards an equilibrium ruled by a micro-canonical ensemble [11, 13]. Moreover, for a set of random initial conditions with different energies one recovers the Ising phase transition ruled by the Onsager canonical partition function [14].

Because the Q2R model is a reversible cellular automaton its phase space is finite. Indeed, for a lattice of size $N$, the phase space is the set of the $2^{2N}$ vertices of a $2N$-dimensional hypercube. However, this phase space is partitioned by different subspaces of constant energies, which, at the same time, are partitioned in a large number of subspaces composed by periodic orbits or fixed points. Notice that because the system is conservative there are neither attractive nor repulsive attractors, all attractor are the fixed points or the cycles.

A given initial condition, with energy, belongs to one of this cycles or a fixed point. As shown numerically, by Herrmann et al. [12] for an energy greater than some value the probability to get a large enough (actually exponentially long) cycle becomes one. However, in the case of small system sizes, the length of the periods are moderate. For instance, for a $2 \times 2$ square lattice, there are $2^8 = 256$ states and the longest orbit is of period 4. In the case, of a $4 \times 4$, the phase space has $2^{32} \approx 4.3 \times 10^9$ elements. In this lattice the longest orbit is $T = 1080$. Moreover, this case can be scrutinized exactly, and we are able to conjecture that the number of states of a given period is exponentially large with the number of sites $N$.

An arbitrary initial condition of energy $E$ falls into one of these cycles and runs until it returns to the initial configuration after a time $T$, which could be exponentially long, and it displays a complex behavior (not chaotic, strictly speaking; see, for instance, Ref. [13, 15]. More importantly, the probability that an initial condition exhibits such a complex behavior is not exponentially small [12]. Moreover, Q2R manifests sensitivity to initial conditions, that is, if one starts with two distinct, but close, initial conditions, then the conditions will evolve into very different cycles as time runs [13]. In some sense, an initial state explores vastly the phase space, justifying the grounds of statistical physics.

In conclusion, the overall picture is that, although for a finite-size system the deterministic automaton Q2R possesses periodic dynamics so it is not ergodic, there is a huge number of initial conditions that explore vastly the configuration space (this is particularly remarkable for initial conditions of random structure). Therefore, one expects that a master equation approach may be successful.

## 2 The Model

The Q2R model, introduced by Vichniac [16], is defined in a regular two dimensional toroidal lattice with even rank $L \times L$, being $N = L^2$ the total number of *nodes*. Special cases of regular and periodic networks in two space dimensions are the square neighbor with a von Neuman neighborhood $|V| = 4$, Fig. 2a, and a hexagonal neighborhood $|V| = 6$, Fig. 2b. However non regular lattices with an even number of neighbors may also be considered.

We associate an index $\boldsymbol{k} \in \{1, \dots, N\}$, as well as a relative position in the lattice specified by two indices $k_1 \in \{1, \dots, L\}$ and $k_2 \in \{1, \dots, L\}$ (the respective row and column indices). Further, a node $\boldsymbol{k}$ is characterized by two possible values $x_{\boldsymbol{k}} = \pm 1$, conforming with the following two-step rule:

$$x_{\boldsymbol{k}}^{t+1} = x_{\boldsymbol{k}}^{t-1} H\left(\sum_{i \in V_{\boldsymbol{k}}} x_i^t\right),$$

where $V_{\boldsymbol{k}}$ denotes an even number neighborhood with periodic boundary conditions. The function $H$ is such a that $H(s = 0) = -1$ and $H(s) = +1$ in all other cases. Thus, the state $x$ belongs to the discrete set $\Omega = \{-1, 1\}^N$ (of size $2^N$).

Using the Hadamard product, which is the multiplication component to component of the state $x \in \Omega$ and $y \in \Omega$, that is $x \odot y \in \Omega$ represents that each component is defined by: $[x \odot y]_{ij} \equiv x_{ij} y_{ij}$. This product is commutative, associative, and it

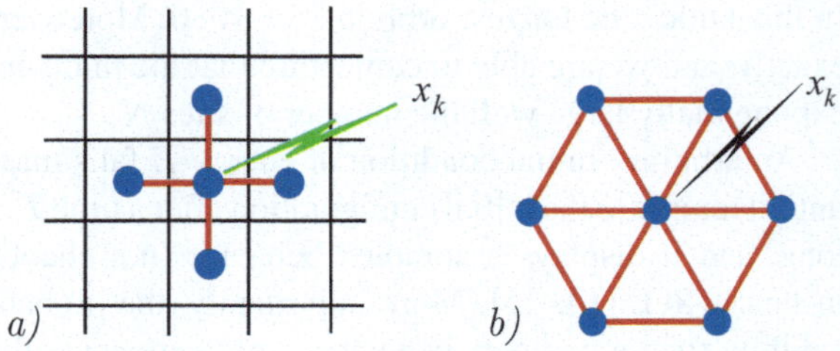


**Fig. 2** **a** Represents a square lattice with the von Neuman neighborhood containing four neighbors; **b** plots a site in a hexagonal lattice with six neighbors

possesses a neutral element, that we denote by $\mathbb{1}$ and corresponds to the state of $\Omega$ composed only by 1s. Moreover, we also define $-\mathbb{1} \in \Omega$ by the states composed only by $-1$s. Given $x \in \Omega$, we will write $-x$ to refer to $-x = [-\mathbb{1}] \odot x$.

Defining the function $\phi : \Omega \to \Omega$ such that, if $x \in \Omega$ then, the $\boldsymbol{k}$-th component of $[\phi(x)]_k = -1$ if the sum of all neighbors of the $\boldsymbol{k}$-th component of $x$ is null, namely $\sum_{i \in V_k} x_i = 0$. Notice that the neighborhood, $V_k$, includes the periodic boundary condition of the lattice. Otherwise, $[\phi(x)]_k = +1$. Therefore, the function $\phi(x)$ is a state in $\Omega$ that has a $-1$ in the sites that $x$ has a null neighborhood.

Then the Q2R rule maybe written in a shorthand notation by

$$x^{t+1} = x^{t-1} \odot \phi\left(x^t\right) \tag{1}$$

However, it is convenient $(x^t, y^t) \in \Omega^2$ at time $t$, and according with the previous definitions we re-write the Q2R model (1) as the following two step deterministic rule:

$$\begin{aligned} y^{t+1} &= x^t \\ x^{t+1} &= y^t \odot \phi\left(x^t\right). \end{aligned} \tag{2}$$

The set of *configurations*, denoted by $\Omega^2$, it is composed by couples of states in $\Omega^2 = \Omega \times \Omega = \{(x, y) \,|\, x \in \Omega \wedge y \in \Omega\}$ (of size $2^{2N}$).

The *evolution* is in parallel, dictated by the rule (2) and is complemented with an initial configuration $(x^0, y^0) \in \Omega^2$. We will write $(x^t, y^t) \to (x^{t+1}, y^{t+1})$ for the one-step evolution from $(x^t, y^t)$ to $(x^{t+1}, y^{t+1}) = (y^t \odot \phi\left(x^t\right), x^t)$.

## 2.1 Fundamental Properties

### 2.1.1 The Phase Space

The *Phase Space of the Q2R model is the set of configurations* $\Omega^2$. Because it is finite, the phase space has two types of invariant sets: *cycles* or *fixed points*. A cycle $\mathcal{C}$ of *period* $T \in \mathbb{N}$ is a sequence dictated by the evolution $(x^0, y^0) \to (x^1, y^1) \to \cdots \to (x^{T-1}, y^{T-1}) \to (x^T, y^T)$ such that all configurations $(x^t, y^t)$ are different, except $(x^0, y^0) = (x^T, y^T)$. We will write $(x, y) \in \mathcal{C}$ if $(x, y)$ is a configuration that is in $\mathcal{C}$ and, more generally, the notation $[(x^t, y^t) \to (x^{t+1}, y^{t+1}) \to \cdots \to (x^{t+\tau}, y^{t+\tau})] \in \mathcal{C}$ will be used to refer to the subsequence of $\mathcal{C}$ that goes from $(x^t, y^t)$ to $(x^{t+\tau}, y^{t+\tau})$, $\tau \leq T$. A fixed point is a cycle of period $T = 1$, i.e., is a configuration $(x, y) \in \Omega^2$ such that $(x, y) \to (x, y)$.

### 2.1.2 Reversibility

Notice that $\phi\left(x^t\right) \odot \phi\left(x^t\right) = \mathbb{1}$, $\forall x^t \in \Omega$. Then Q2R rule may be inverted getting the backward evolution of the system but for the couple $(y^t, x^t)$, that reads:

$$\begin{cases} x^{t-1} = y^t \\ y^{t-1} = x^t \odot \phi\left(y^t\right), \end{cases}$$

which is exactly the same rule (2), displaying the remarkable property of reversibility.

The reversibility property implies the following Reversibility Lemma:

**Lemma 1** (Reversibility) *Let* $x, y, z$ *in* $\Omega$, *then,*

$$[(x, y) \to (z, x)] \Leftrightarrow [(x, z) \to (y, x)].$$

For the proof we refer the reader to [17].

This Reversibility Lemma says that, if there is a one time step evolution between two configurations, then, there is also a one time step evolution between their symmetric configurations, but, in the opposite sense. As a consequence, we have the following generalization:

**Corollary 2** *Let* $x^t$, $y^t$ *in* $\Omega$, $t \in \{0, \ldots, p\}$, $p \in \mathbb{N}$, *then,*

$$\left(x^0, y^0\right) \to \cdots \to \left(x^p, y^p\right) \Leftrightarrow \left(y^p, x^p\right) \to \cdots \to \left(y^0, x^0\right).$$

Figure 3 illustrates the proof of this property.

### 2.1.3 Energy Conservation

Let the energy function be,

$$E[(x^t, y^t)] = -\frac{1}{2} \sum_{\langle k,i \rangle} x_k^t y_i^t,$$

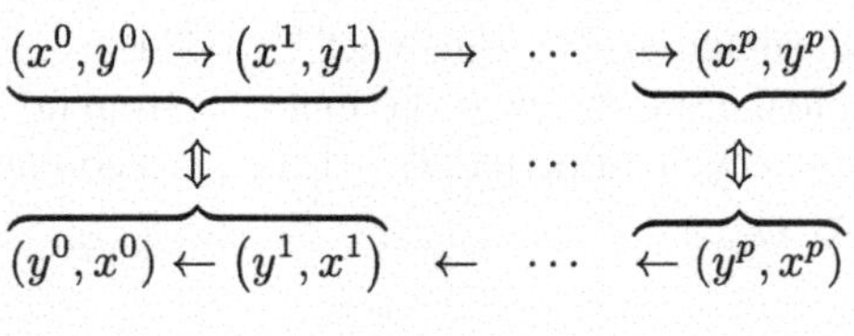


**Fig. 3** Applying successively the Lemma 1 at each step-evolution $(x^t, y^t) \to (x^{t+1}, y^{t+1})$, for $t \in \{0, \ldots, p\}$, $p \in \mathbb{N}$, one constructs the backward evolution between their symmetric configurations

where the symbol $\sum_{\langle k,i\rangle}$ means the sum over all sites $k$ and the sum over all neighbors $i$ of node $k$, i.e., $\sum_{\langle k,i\rangle} \equiv \sum_k \sum_{i\in V_k} \equiv \sum_i \sum_{k\in V_i}$. Such a energy is bounded by $-2N \leq E \leq 2N$ and, as shown by Pomeau [9], it is conserved[1] under the dynamics defined by the Q2R rule (2). That is: $E[(x^t, y^t)] = E[(x^0, y^0)],\ \forall t \in \mathbb{N}$.

### 2.1.4 The Qualitative Dynamics

When the phase space is partitioned in different sub-spaces accordingly to its energy $E$ one observes that the constant energy subspace shares in principle many cycles of different periods, as well as, many different fixed points. An arbitrary initial condition of energy $E$ falls into one of these cycles, and it runs until time $T$, which could be exponentially long, and displaying a complex behavior (not chaotic *stricto-sensu*, see for instance [15]). More importantly, there is numerical evidence that the probability that an initial condition belongs to an exponentially long period cycle, and it exhibits a complex behavior is finite [12].

Moreover, Q2R manifests sensitivity to initial conditions, that is if one starts with two distinct, but close, initial conditions, then, the distance (e.g. the Manhattan distance) between the respective evolutions of the initial states become large as time evolves [13].

## 2.2 *Phase Space Classification of the Q2R Automaton*

In the following we summarizes the main results of Ref. [17].

1. The characterization and existence of fixed points, period-two and period-three cycles. More precisely,

**Proposition 3** *$\{x, x\}$ is a fixed point of Q2R iff $\phi(x) = \mathbb{1}$.*

**Proposition 4** *$\{x, y\}$ is in a period 2 cycle iff $x \neq y \wedge \phi(x) = \mathbb{1} \wedge \phi(y) = \mathbb{1}$.*

**Proposition 5** *Let $\{(x, y), (z, x), (y, z)\} \subseteq \Omega^2$ such that $(x, y) \to (z, x) \to (y, z)$. Then, $\{(x, y), (z, x), (y, z)\} \subseteq P_3 \Leftrightarrow \phi(x) \odot \phi(y) \odot \phi(z) = 1$.*

2. The number of fixed points is the perfect square of an even number $\alpha^2$, with $\alpha = 2k_N$, and $k_N$ is an integer that depends explicitly on the lattice size $N$. In fact, if the lattice is considered as a checkerboard and a configuration corresponds to a set of values over the black (or white) cells, then $\alpha$ is the number of configurations without null neighborhoods.

[1] Other dynamical invariants are known in the literature [10].

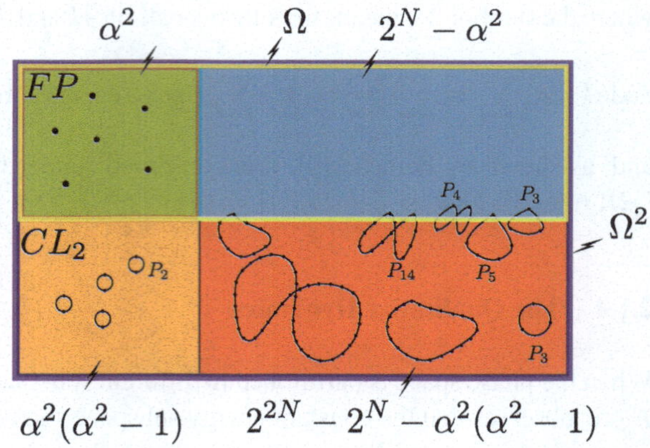


**Fig. 4** Phase space cartoon displaying the dynamical partition of the phase space accordingly to the four possible existing cycles. namely, the symmetric cycles S-I, S-II and S-III or an asymmetric cycle, AS. Some known cardinalities are displayed as well as the fixed points and the period two cycles

3. An algebraic relation between the fixed points and period-two cycles:

**Proposition 6** *If the total number of fixed points is* $|FP| = \alpha^2$*, then, the total number of period 2 cycles is* $|P_2| = \alpha^2(\alpha^2 - 1)$.

4. A complete classification of all invariant sets in four types (Theorem 7). This characterization depends on the specific topological features of each cycle, which may be a symmetric of type S-I, S-II and S-III or an asymmetric cycle, AS (see next Sect. 2.2.1).
5. Lower and upper bounds for the total number of cycles of type S-I and S-II that scale both exponentially with the system size ($\sim 2^N$) (Fig. 4).

### 2.2.1 Theorem on the Classification of Cycles in Q2R

**Definition 1** Let be the following sets:

$$\begin{aligned}
A &= \left\{(x, y) \in \Omega^2 / \phi(x) = \mathbb{1} \wedge x = y\right\} \\
B &= \left\{(x, y) \in \Omega^2 / \phi(x) = \mathbb{1} \wedge x \neq y\right\} \\
C &= \left\{(x, y) \in \Omega^2 / \phi(x) \neq \mathbb{1} \wedge x = y\right\} \\
D &= \left\{(x, y) \in \Omega^2 / \phi(x) \neq \mathbb{1} \wedge x \neq y\right\}.
\end{aligned}$$

We say that $(x, y) \in \Omega^2$ is a *configuration of type* $A$, $B$, $C$ or $D$, if $(x, y)$ belongs to one of the sets $A$, $B$, $C$ or $D$, respectively. We refer to a *evolution* of type $U \to V$ to the one-step evolution of a configuration $(x, y) \in U$ up to $(w, x) \in V$ with $U, V \in \{A, B, C, D\}$.

**Definition 2** We say that, the *symmetric* configuration of $(x, y) \in \Omega^2$ is the configuration $(y, x) \in \Omega^2$. In particular, the symmetric configuration of $(x, x) \in \Omega^2$ is itself, $(x, x)$, and we will call it as a *self-symmetric* configuration. We say that a cycle $\mathcal{C}$ is *symmetric* if satisfy:

$$(x, y) \in \mathcal{C} \Rightarrow (y, x) \in \mathcal{C}.$$

Otherwise, we say that $\mathcal{C}$ is *non-symmetric*.

**Definition 3** A non-symmetric cycle $\mathcal{C}$ is said to be *asymmetric* if satisfy:

$$(x, y) \in \mathcal{C} \Rightarrow (y, x) \notin \mathcal{C}.$$

**Theorem 7** (Classification of cycles in Q2R) *Let $\mathcal{C}$ be a cycle of Q2R with period $T \in \mathbb{N}$. Then $\mathcal{C}$ is of type S-I, S-II, S-III or AS, where:*

- *Type S-I. If $T = 1$ or if there exists $p \in \mathbb{N}_0$ such that $\mathcal{C}$ has the topology $B \to D \to \cdots \to D \to C \to D \to \cdots \to D \to B$, i.e., is symmetric with:*
  - *An odd period $T = 2(p + 1) + 1$.*
  - *Only one configuration of type $C$, only one configuration of type $B$ and $(2p + 1)$ configurations of type $D$.*
- *Type S-II. If there exists $p \in \mathbb{N}_0$ such that $\mathcal{C}$ has the topology $C \to D \to \cdots \to D \to C \to D \to \cdots \to D \to C$, i.e., is symmetric with:*
  - *An even period $T = 2(p + 2)$.*
  - *Only two configurations of type $C$ and $2(p + 1)$ configurations of type $D$.*
- *Type S-III. If $T = 2$ or if there exists $p \in \mathbb{N}_0$ such that $\mathcal{C}$ has the topology $B \to D \to \cdots \to D \to B \to D \to \cdots \to D \to B$, i.e., is symmetric with:*
  - *An even period $T = 2(p + 2)$.*
  - *Only two type $B$ configurations and $2(p + 1)$ type $D$ configurations.*
- *Type AS. If there exists $p \in \mathbb{N} \setminus \{1\}$ such that $\mathcal{C}$ has the topology $D \to \cdots \to D$, i.e., is an asymmetric cycle with:*
  - *Period $T = p + 1$ (it can be even or odd, depending on the value of $p$).*
  - *All its configurations are of type $D$.*

## 3 An Example of the Ehrenfest Model in Q2R

We can mimic the Ehrenfest's dog-flea model in the Q2R by dividing the whole system in two containers that maybe communicated by a gate. To do that we neutralize the boundaries by imposing that the spin state is constant and equal to 0 on the wall. This procedure neutralizes not only the wall but also a boundary layer of width one next to the wall which accordingly with the Q2R rule do not change in time. The states on the bulk are essentially free of the boundaries so rule the fluctuating dynamics of Q2R.

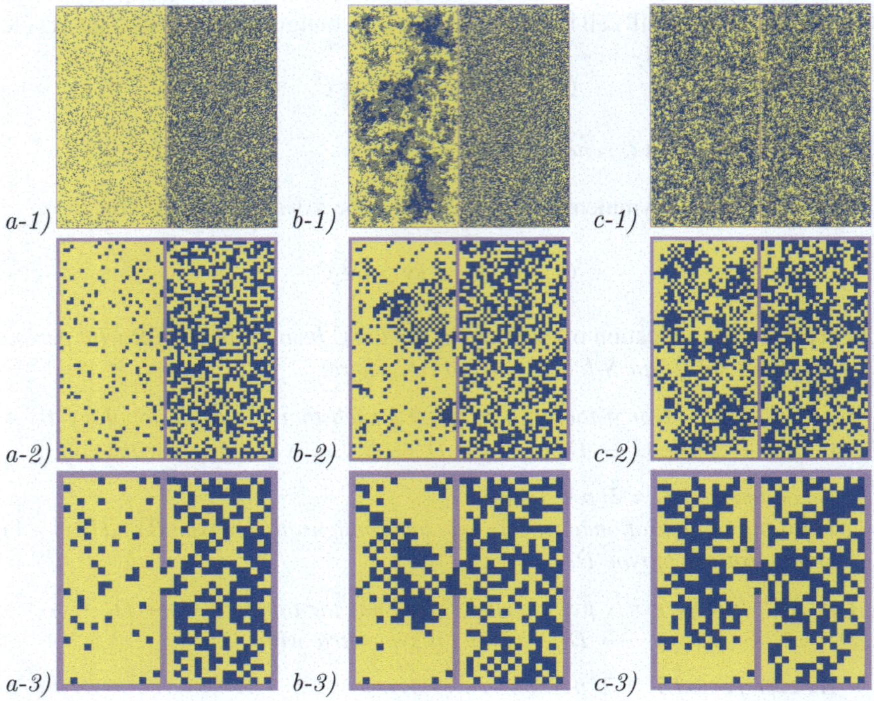


**Fig. 5** Numerical achievement of the Ehrenfest's dog-flea model in the Q2R model. Different columns represents distinct time steps of the evolution. While the rows label the system size. As one moves from the first row (1): $256 \times 256$ (before excluding the walls); the row (2), corresponds to $64 \times 64$; finally, the row (3), is for $32 \times 32$

The Fig. 5 shows the dynamics at three different state: an initial state (column a) which is "cooler" (in the sense that it has less energy) than the right-hand side container. As show the further evolutionary tends to balance irreversible the energies of the two containers. The column (c) shows in general an equilibrium state. As one decreases the number of degrees of freedom from $256 \times 256$ (before excluding the walls) up to $8 \times 8$ (before excluding the walls) one observes that the equilibrium is not longer holds.

More quantitatively Fig. 6 shows the temporal irreversible (despite the fact that the rule as well the global evolution is formally reversible) tendency to an energy equilibrium into the two containers. Although the irreversible behavior appears clear to the eye (one readily cannot imagine how the reverse dynamics may be manifested), the equilibria is not completely fulfilled: in the case of a system of $64 \times 64$ (Fig. 6b) one already notices that the amplitude of fluctuations are large, moreover some times the right-hand side container has less energy than the original "cool" left-hand side container. The case of $32 \times 32$ (Fig. 6c) has a particularity, indeed the initial condition started with the left-bottom sector of the container with only states $x_k = -1$ therefore the existence of such an "alliance" implies that this sector will never change in time.

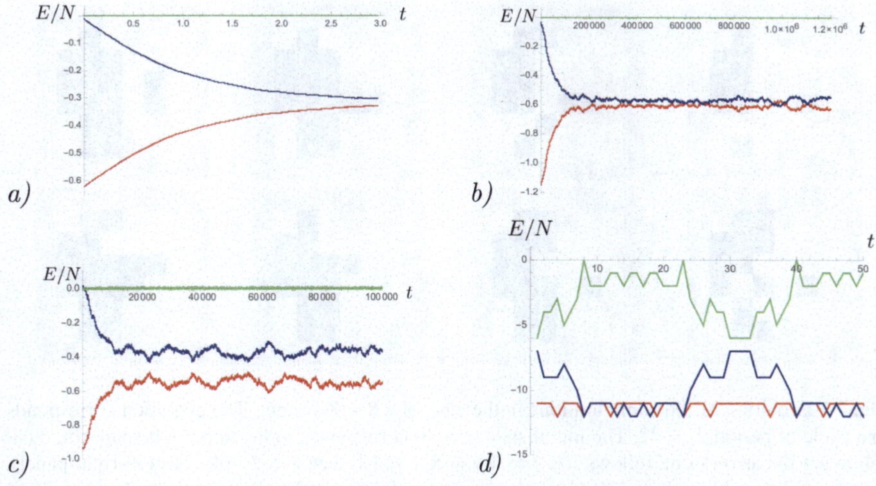


**Fig. 6** Evolution of the energy corresponding to the numerical simulations of Fig. 5. In all plots, the red curves denotes the energy of the left-hand side container; the blue curve denotes the energy of the right-hand side container; and, the green curve plots the energy of the interface and the gate. The sum of all these three energies is constant. All energies are normalized by the total number of sites $N$, which are: **a** $256 \times 256$; **b** $64 \times 64$; **c** $32 \times 32$; and **d** $8 \times 8$. In **c** the energy is not equilibrated in both sides of the container. This happens because of the particularity of the initial state that freeze the states of the left bottom side. The plot **d** corresponds to the periodic sequence displayed in Fig. 7

The first consequence is that the energy of both containers cannot be ultimately equilibrated.

One may wonder if this is against the rules of thermodynamics, and the answer is now the existence of configurations with "alliances" is not generic, so its existence is much less probable.

As the system size diminishes up to $8 \times 8$ (excluding the walls) the recurrence is observed. Indeed Fig. 7 shows the evolution of a period $T = 32$ cycle.

The energy evolution as a function of time is a $T = 32$ periodic function which is shown in Fig. 6d. Although there is an exchange of energy among the two containers there is no obvious flux, moreover because of the periodic character of the evolution one sees the existence of a recurrence after a quite short time.

Qualitatively, in Fig. 6 one observes that there exists a "transition" from the system size (c) $32 \times 32$ in which case the cycle are exponentially long compared to the case (d) $8 \times 8$, in which the period is short. This kind of small degrees of freedom system cannot be represented as a thermodynamical system.

**Discussion**

In conclusion, the scheme of the Ehrenfest's dog-flea model may be applied to the Q2R automaton model which appears to be a good model for a reversible and conservative dynamical system. Moreover, being Q2R a cellular automaton, it posses a great advantage with respect the numerical simulations because there is no approxi-

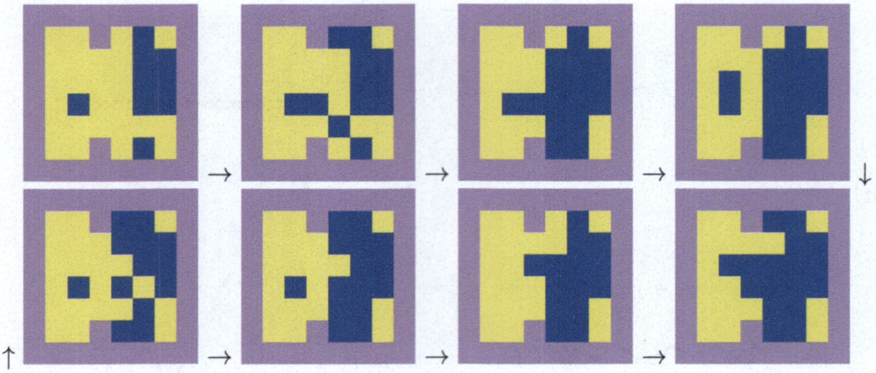

**Fig. 7** Evolution of the configurations in the case of a $8 \times 8$ system. The evolution corresponds to a cycle of period $T = 32$. The initial state ($t = 0$) corresponds to the upper-left snapshot, then, following the arrows one follows for: $t = 4$ $t = 8$ $t = 12$, then $t = 16$ (the bottom-right panel), then $t = 20$, $t = 24$, and, $t = 28$ (the bottom-left panel), finally the system gets back at $t = 32$ to the initial configuration

mation, no round error may be produced multiplying 1 by $-1$. Therefore, the system is both formally and in practice (numerically speaking) reversible and conservative. The old paradoxes pertinent in statistical physics may be revisited in the frame of Q2R which provide a clear example of dynamical system displaying a complex temporal behavior because of the existence of a huge number of periodic orbits with periods that may be exponentially long. In the current paper we review the basic classification of the phase space in for types of cycles. The complex structure of the phase space makes the necessity of a statistical description of the dynamics.

**Tribute to Eric Goles**

*Personal tribute by Marco Montalva-Medel*

I met Eric in August 2011 when he was part of the jury that evaluated my doctoral thesis defense in the University of Concepción. I remember that, when everything was over (very good fortunately), he asked me about my future plans to which I replied that I would like to continue research, so, he tell me something like "if you want, you can do a postdoc with me at Universidad Adolfo Ibáñez" and me, without thinking twice, said yes obviously. Then, in less than a month, I was already installed in Santiago working as a postdoc with him.

The first thing of Eric that surprised me was the speed with which he understood and analyzed a problem; many times throwing thousands of ideas per second, that after thinking them calmly, when I was alone, I realized that they were all true! In this context, we did a couple of interesting works with specific biological networks and also with cellular automata where we continue collaborating today.

During these years where we have interacted not only in the workplace, it is easy to realize that behind this brilliant scientist and multifaceted character, who sometimes loses patience easily, there is a very good, correct and just man, a lover of reading, a man who does not forget his roots or his friends, a man who has a humor that leaves your face hurting from laughing so much, just to name a few of its features. Some non-academic things that I learned by observing Eric and that I confess are now part of me are punctuality and speaking always the truth (although it can bring problems).

I am proud to be a participant in this wonderful initiative in honor of him.

*Personal tribute by Sergio Rica*

I first met Goles on a probably cold and cloudy Tuesday morning in early August 1984. I followed his lectures on Linear Algebra at the School of Engineering and Sciences at the Universidad de Chile, "*La Escuela*". He was in his mid-thirties, his soul was full of energy but with a quite bad character. The smallest noise in the classroom produced a big storm in his mood. It was my first encounter with a profesional mathematician, previously I have gotten lectures only by Teachers, and, say, Engineers amateurs in mathematics. Hence, in my brain, it was a complete mess the logic construction of vectorial-spaces that Goles tried to build every Tuesday and Thursday for us. At some point, to be precise the demonstration of a theorem saying that $n + 1$ vectors in a vectorial-space of dimension $n$ are necessarily linearly dependent, the confusion was the highest: are real tangle-web of symbols ($\forall$, $\exists$, $\in$ , $\cup$, $\cap$, . . . ), theorems, lemmas, etc.

Suddenly, like it happens when an airplane crosses a storm towards sun-shine……, the whole picture was clear (at lest to me), the intentional structure built by the Professor workout in a rational scheme. It was my only class that I followed by Goles. In these times, the lectures were strongly abstract, hard to follow, requiring a lot of thinking. No room for shortcuts. At the University they prepared us as "Submarine Commander", we must solved fast and right quite tricky problems, to prove Theorems, use beautiful tools as the Induction principle, a demonstration by reducing to absurd, etc. Besides his bad character, I think, Goles does not fit today with the current lecturing way.

Probably, today we do not need Engineers as a "Submarine Commanders", certainly most of my classmates forgot the $n + 1$ vectors Theorem, or how to diagonalize a matrix, but to survive this formation gave them a sense of belonging to a school: "La Escuela". I think today, students do not share this sense of belonging. Moreover, they are not proud to belong to some school, neither to be the best. They just pass over.

In the late 80s, occasionally, I met Goles because he was quite close to my former advisor in Chile, Enrique Tirapegui. We met a few times nothing very special neither close. Nevertheless, in 1998 he invited me to a dinner with Ricardo Lagos and scientists. Soon after President Lagos was elected and Goles became Conicyt President and he left the academic life for political-administrative duties for six years. Initially Lagos pushed science, probably because Goles convinced him that Science is important for society, and major programs were created and others improved.

Few times he confessed to me that he was seduced by the "power", seduced by traveling with Lagos, to have lunch with President Chirac at the Élysée, to accompany King Juan Carlos to the Antarctic, to fight for the budget for science. But nothing is perfect, the political time scales are short, and Lagos was only interested in Science for 3 years. I often have seen similar life-histories like Goles but with a sad ending, 6 years of science blackout may be enough for a definite non return back to science. But Goles succeed, he came back to science, with ideas, enthusiasm, and more important with happiness.

After Conicyt, for reasons that are without any interest, he was not welcome to be back School of Engineering, so he got a position at his current institution at the Universidad Adolfo Ibanez. At that time, 2006, the UAI was essentially a College, therefore the idea to do research was not even in the mind of possibilities. Knowing him, he was bored. As I know was Ivan Rapaport, Goles' past student and current Profesor at the University of Chile, started to talk Science with Goles. Ivan saved Goles, who re-started to think on mathematics.

While I lived in Paris in 2008, he knew (by Ivan), that I was interested to be back in Chile. We met at a Bistrot near the *Ecole des hautes études en sciences sociales* (EHESS) at the Boulevard Raspail. He convinced our former Dean Alejandro Jadresic for interviewing me, and finally, I came to UAI in 2009. When I came, the ambiance was very friendly, but none of the young researchers were able to work with Goles. I must say he is quite a time demanding, he needs coffee every 30 minutes, to talk about his thinking, his life, and most probably his favorite subject: himself, "Goles talking about Goles".

We fitted perfectly on a subject that he proposes on the Schelling's segregation model. We were in different buildings so he wrote me letters (see an example below), I did numerical calculations. I was not familiar with this discrete world, neither with rules instead of PDEs, but at the end a PDE is nothing else than a rule. So we amalgamated perfectly. We finish a paper in a few months, and by accident, we start another subject based on a reversible cellular automaton: the Q2R model, which is the main subject of the current paper. He did not pursue the connection of Q2R with the problem of irreversibility, I must say he is interested in his ideas, not other people's ideas, but thanks to this initial collaboration on segregation models I found a nice reversible system that manifests irreversibility much easier to handle than nonlinear reversible PDEs. We still collaborate in segregation-like problems for a while until recently. It is interesting to underline, how *infectious*, has been this problem of segregation, it opens a new way to many different studies and many of them nothing to do with the original problem, in this sense I think this has been a fruitful collaboration.

## References

1. Loschmidt J (1876) Uber das warmegleichgewicht eines systems von korpern mit rucksicht auf die schwere, Sitzungsber. Kais Akad Wiss Wien Math Naturwiss Classe 73:128–142
2. Zermelo E (1896) Über einen Satz der Dynamik und die mechanische wärmetheorie. Annalen der Physik 293(3):485–494
3. Zermelo E (1896) Ueber mechanische Erklärungen irreversibler vorgänge. Eine Antwort auf Hrn Boltzmann's "Entgegnung", Annalen der Physik 295(12):793–801

4. Ehrenfest P, Ehrenfest-Afanassjewa T, Über eine Aufgabe aus der Wahrscheinlichkeitsrechnung, die mit der kinetischen Deutung der Entropievermehrung zusammenhängt, Mathematisch Naturwissenschaftliche Blätter 3
5. Ehrenfest P, Ehrenfest-Afanassjewa T (1907) Über zwei bekannte Einwände gegen das Boltzmannsche H-Theorem, Hirzel
6. Kac M (1947) Random walk and the theory of brownian motion. Amer Math Month 54(7P1):369–391
7. Urbina F, Rica S (2016) Master equation approach to reversible and conservative discrete systems. Phys Rev E 94(6):062140
8. Nicolis G, Nicolis C (1988) Master-equation approach to deterministic chaos. Phys Rev A 38(1):427
9. Pomeau Y (1984) Invariant in cellular automata. J Phys A: Math Gen 17(8):L415
10. Takesue S (1995) Staggered invariants in cellular automata. Complex Syst 9(2):149–168
11. Herrmann H (1986) Fast algorithm for the simulation of ising models. J Stat Phys 45(1–2):145–151
12. Hermann H, Carmesin H, Stauffer D (1987) Periods and clusters in Ising cellular automata. J Phys A: Math Gen 20(14):4939
13. Goles E, Rica S (2011) Irreversibility and spontaneous appearance of coherent behavior in reversible systems. Eur Phys J D 62(1):127–137
14. Onsager L (1944) Crystal statistics. I. A Two-dimensional model with an order-disorder transition. Phys Rev 65:117–149
15. Grassberger P (1986) Long-range effects in an elementary cellular automaton. J Stat Phys 45(1–2):27–39
16. Vichniac GY (1984) Simulating physics with cellular automata. Phys D: Nonlinear Phenom 10(1–2):96–116
17. Montalva-Medel M, Rica S, Urbina F (2020) Phase space classification of an Ising cellular automaton: the Q2R model. Chaos Solitons Fractals 133:109618

# Analyzing Boolean Networks Through Unsupervised Learning

**Gonzalo A. Ruz**

**Abstract** Boolean networks are typically used as simple models of gene regulatory networks. We use a particular class of Boolean networks called threshold Boolean networks defined by a weight matrix, a threshold vector, and an updating mode in this work. We consider the reconstruction of synthetic threshold Boolean networks that contain the same fixed points as the Mendoza and Alvarez-Buylla network of flower development by using an evolution strategy. We propose a characterization by computing topological and dynamical features of the inferred synthetic networks and then applying machine learning, particularly unsupervised learning techniques, to analyze these networks. We discover how these networks are clustered and what features are relevant to discriminate the cluster containing the Mendoza and Alvarez-Buylla network from all the other clusters.

## 1 Introduction

The capacity to reconstruct synthetic gene regulatory network models under the Boolean network [6] formalism enables the possibility to explore the topological and dynamical robustness of the original or wild type network of the biological process being studied. The reconstruction process from data will typically consist of inferring Boolean networks that satisfy some Boolean trajectory associated with a biological process, for example, modeling cell-cycle in yeast [5]. There are some well-known inference algorithms for Boolean networks, such as REVEAL [8] and the Best-Fit extension algorithm [7]. But when considering a particular class of Boolean networks, namely, threshold Boolean networks, where a weight matrix and a threshold vector characterizes the network, the use of computational intelligence and related techniques have become popular. In [11] simulated annealing was used to infer synthetic networks that contained the six fixed points or steady states of the original threshold Boolean network model of the dynamical behavior of the flower development

G. A. Ruz (✉)
Facultad de Ingeniería y Ciencias, Universidad Adolfo Ibáñez, Santiago, Chile

Center of Applied Ecology and Sustainability (CAPES), Santiago, Chile
e-mail: gonzalo.ruz@uai.cl

 
A. Adamatzky (ed.), *Automata and Complexity*, Emergence, Complexity
and Computation 42, https://doi.org/10.1007/978-3-030-92551-2_14

process in *Arabidopsis thaliana* plants. Out of five hundred runs, the network inference framework found 114 networks containing the desired fixed points. Synthetic Boolean network models of the mammalian cell-cycle network [4] were obtained by using a swarm intelligence optimization technique called the bees algorithm [10, 13] in [12, 15]. Genetic algorithms (GAs) were used to reconstruct the gene regulatory network of *Arabidopsis thaliana* saline stress response. The GA framework was able to successfully infer 1000 threshold Boolean networks that contained the desired Boolean trajectory. A consensus network using the inferred networks was useful to identify the regulations or interactions among the genes that were more plausible [17]. Reference [19] presents the problem of inferring bistable *lac* operon Boolean regulatory networks using three different evolutionary computation approaches: differential evolution, genetic algorithms, and particle swarm optimization. The results showed that the three algorithms could find solutions, being differential evolution the most effective, whereas genetic algorithms were the least effective and efficient in runtime. Particle swarm optimization obtained a good trade-off between effectiveness versus efficiency. Reference [20] used a neural network approach for inferring threshold Boolean network for generating a model of bacterial quorum-sensing systems, and [21] used differential evolution for the reconstruction of a gene regulatory network of the induced systemic resistance defense response in *Arabidopsis thaliana* plants.

An interesting topic within threshold Boolean network reconstruction is the possibility to study the neutral space [16], i.e., to infer synthetic networks that emulate the functionality of a wild type network or a base model. Here functionality is typically related to a dynamical property of a given network. For example, sampling the neutral space of a wild type network could consist of obtaining networks with different wirings but sharing the same asymptotic behavior as the wild type network. A typical visualization of the neutral space (consisting of the wild type network plus the sample of functionally equivalent networks) is through the neutral graph. A neutral graph (also known as a neutral network or a metagraph) [1–3] is an undirected graph where each node represents a regulatory network. Suppose two nodes are connected in the neutral graph. In that case, this means that the Hamming distance (number of entries in which two adjacency matrices differ) between the interaction (adjacency) matrix of one network and the other is one. The connectivity of a neutral graph can give an insight into the topological robustness of the regulatory networks. A neutral graph with large connected components can be considered to have high robustness. In contrast, a neutral graph with many small connected components (or disconnected) can be considered as having low robustness [2]. For sampling the neutral space, a common approach is to conduct a neighborhood search around the wild type network via an evolution strategy [14].

Although the neutral graph visualization allows a qualitative analysis of the neutral space, other complementary tools for analyzing the neutral space are needed. One approach could be using machine learning, particularly unsupervised learning typically used for descriptive modeling. Two typical applications of unsupervised learning include high-dimensional data visualization and clustering.

In this work, we present a neutral space analysis of the threshold Boolean network model of the dynamical behavior of the flower development process in *Arabidopsis thaliana* plants. We will infer one thousand networks containing the same fixed points as the original model using the evolution strategy of [14]. Then, we cluster the resulting networks by using the k-means algorithm and visualize the resulting clusters through an embedding technique called t-SNE [22]. Each cluster is analyzed topologically and dynamically. Overall, we obtained a more quantitative analysis through the use of unsupervised learning techniques. The rest of this work is organized as follows. Section 2 describes the threshold Boolean network used as the wild type, the unsupervised learning algorithms used in this work, and the evolution strategy to infer networks. The methods appear in Sect. 3 and the results and discussions in Sect. 4. The conclusions and future directions are presented in Sect. 5.

## 2 Background

### 2.1 Threshold Boolean Networks

In [9], Mendoza and Alvarez-Buylla proposed a threshold Boolean network that captured the dynamics of the floral development in *Arabidopsis thaliana*. The model consists of 12 interacting chemical species, designated by EMF1, TFL1, LFY, AP1, CAL, LUG, UFO, BFU, AG, AP3, PI, and SUP. The BFU species, is a dimer of the AP3 and PI proteins, all the rest are proteins as well. So in this model we have $n = 12$, and each node $x_i$ from $i = 1, \ldots, n$ will update its value using the following rule:

$$x_i(t+1) = H\left(\sum_{j=1}^{n} w_{ij} x_j(t) - \theta_i\right) \tag{1}$$

$$H(z) = \begin{cases} 1, & \text{if } z > 0 \\ 0, & \text{if } z \leq 0 \end{cases}$$

with $w_{ij}$ the weight of the edge coming from node $j$ into the node $i$, and $\theta_i$ the activation threshold of node $i$. The weights and thresholds are the network's parameters (see Fig. 1).

If we start in any of the $2^{12} = 4096$ possible configurations, and use the parallel updating scheme, then the network will converge to one of the possible thirteen attractors. Seven of these, are limit cycles of length two each, which have no biological meaning. The remaining six are the following fixed points: 1) {000100000000}, 2) {000100010110}, 3) {000000001000}, 4) {000000011110}, 5) {110000000000}, 6) {110000010110}. Each one has associated a cell type: 1) sepal, 2) petal, 3) carpel, 4) stamen, 5) inflorescence, 6) mutant (unobserved cell).

For simplicity we will refer to this network from now on as the original network.

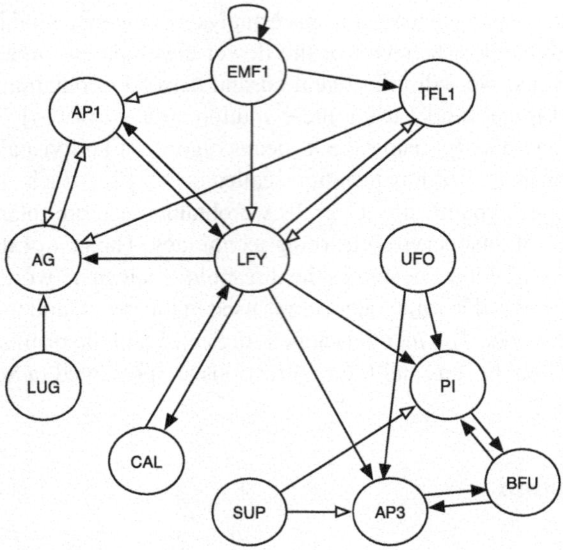


$$W = \begin{pmatrix} & EMF1 & TFL1 & LFY & AP1 & CAL & LUG & UFO & BFU & AG & AP3 & PI & SUP \\ EMF1 & 1 & 0 & 0 & 0 & 0 & 0 & 0 & 0 & 0 & 0 & 0 & 0 \\ TFL1 & 1 & 0 & -2 & 0 & 0 & 0 & 0 & 0 & 0 & 0 & 0 & 0 \\ LFY & -2 & -1 & 0 & 2 & 1 & 0 & 0 & 0 & 0 & 0 & 0 & 0 \\ AP1 & -1 & 0 & 5 & 0 & 0 & 0 & 0 & 0 & -1 & 0 & 0 & 0 \\ CAL & 0 & 0 & 2 & 0 & 0 & 0 & 0 & 0 & 0 & 0 & 0 & 0 \\ LUG & 0 & 0 & 0 & 0 & 0 & 0 & 0 & 0 & 0 & 0 & 0 & 0 \\ UFO & 0 & 0 & 0 & 0 & 0 & 0 & 0 & 0 & 0 & 0 & 0 & 0 \\ BFU & 0 & 0 & 0 & 0 & 0 & 0 & 0 & 0 & 0 & 1 & 1 & 0 \\ AG & 0 & -2 & 1 & -2 & 0 & -1 & 0 & 0 & 0 & 0 & 0 & 0 \\ AP3 & 0 & 0 & 3 & 0 & 0 & 0 & 2 & 1 & 0 & 0 & 0 & -2 \\ PI & 0 & 0 & 4 & 0 & 0 & 0 & 1 & 1 & 0 & 0 & 0 & -1 \\ SUP & 0 & 0 & 0 & 0 & 0 & 0 & 0 & 0 & 0 & 0 & 0 & 0 \end{pmatrix} \quad \Theta = \begin{pmatrix} 0 \\ 0 \\ 3 \\ -1 \\ 1 \\ 0 \\ 0 \\ 1 \\ -1 \\ 0 \\ 0 \\ 0 \end{pmatrix}$$

**Fig. 1** Original Mendoza and Alvarez-Buylla network. Activations (resp. repressions) are represented by full arrows (resp. empty arrows). Below, the matrix $W$ of size $12 \times 12$ contains the interaction weights between genes and $\Theta$ is the thresholds vector

## 2.2 *Unsupervised Learning*

Unsupervised learning deals with analyzing unlabeled data, i.e., data that has not been previously classified into a particular class. Two common tasks in unsupervised learning are clustering and high-dimensional data visualization. For clustering, one of the most popular algorithms corresponds to k-means. The algorithm works in the following way,

1. Select k points randomly as the initial centroids.
2. **repeat**
3. Form k clusters by assigning all points to the closest centroid.
4. Recompute the centroid of each cluster.
5. **until** The centroids do not change

The above algorithm minimizes the objective function known as the sum of the squared error (SSE), defined by

$$\mathrm{SSE} = \sum_{i=1}^{k} \sum_{\mathbf{x} \in C_i} dist(\mathbf{c}_i, \mathbf{x})^2 \tag{2}$$

where $\mathbf{x}$ is an n-dimensional data point that belongs to cluster $C_i$ and $\mathbf{c}_i$ is the centroid (mean) of cluster $C_i$. Also, $dist$ is the standard Euclidean ($L_2$) distance between two objects in Euclidean space.

For high-dimensional data visualization there are linear projection techniques such as Principal Component Analysis (PCA), and non-linear techniques such as t-Distributed Stochastic Neighbor Embedding (t-SNE). We will consider the latter. Basically, the t-SNE algorithm calculates a similarity measure between pairs of examples in the high dimensional space (step 1) and in the low dimensional space (step 2). It then tries to optimize these two similarity measures using a cost function, in particular, the Kullback-Liebler divergence (step 3). Details of the three steps of t-SNE can be found in [22].

## 2.3 Evolution Strategy

To reconstruct synthetic networks starting from the *A. thaliana* network described previously we will use an evolution strategy (ES) developed in [14] and used recently in [18]. A flow chart of the ES is shown in Fig. 2, where it can be seen that the main variation operator is mutation. The initial candidate solutions (networks) are generated using as a seed the *A. thaliana* network. Edges are removed or added from the original network $ngh$ times (a user defined parameter) to generate a candidate solution, as well as the respective threshold vector is changed. The fitness function is the mean squared error between the dynamics (output) of the candidate network and the dynamics of the original network, given the same input. Therefore, it is a minimization problem, where we want to find networks with the least error. After the fitness value is computed for each candidate solution, these are ranked in a descending order. Then, the top $m\%$ are selected (another user defined parameter) to perform mutation, in a similar way as the candidate networks were generated, but now using the top ranked networks as seeds to generate new solutions. These new solutions plus the top $m\%$ are completed with random candidate networks, using as seed the original network, to generate the new population. More details of the algorithm can be found in [16].

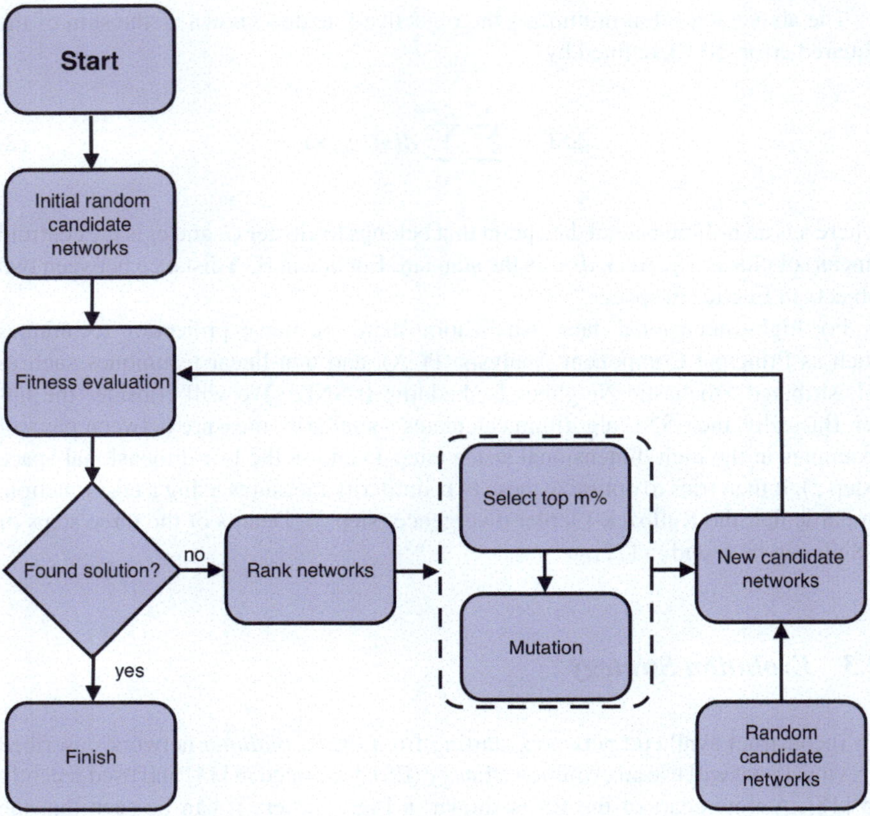


**Fig. 2** Evolution strategy (ES) flow chart to search for synthetic networks

# 3 Methods

We first begin by searching for one thousand synthetic networks with the same fixed points of the original *A. thaliana* network, for this we use the ES described with the following parameter values. The $ngh$ parameter is selected randomly between 1 and 30 for each candidate network. The elements of the weight matrices and the threshold vectors were constrained to the following integer range $[-5, -4, \ldots, 4, 5]$. Also, $popSize = 20$, $m\% = 30\%$ and max iterations =100.

It is important to point out that when the fitness function reaches 0, we can assure that the candidate solution will in fact have the six fixed points, but we cannot assure that these will be the only ones existing, since there might be additional fixed points present. In order to avoid this situation, whenever a candidate solution obtains a fitness value of 0, we check how many fixed points the networks has. If the network has more than six, than we penalize the solution by adding 0.1 to the fitness value.

By this way, we can assure that, when the fitness value is 0, the network will only have the desired six fixed points and no others.

Then each network is characterized by a 162-dimensional feature vector. Features 1 to 144 corresponds to the flattened weight matrix. Features 145 to 156 corresponds to the threshold vector, and finally, features 157 to 162 corresponds to the basin of attraction of each of the six fixed points. It is important to point out that this feature vector contains simultaneously, topological and dynamical information of each network. We will also characterize the original network with this feature vector. At the end of this feature construction process, we end up with a $1001 \times 162$ input data matrix to analyze using unsupervised learning techniques.

The next step is to discover how many clusters naturally form when the one thousand synthetic networks plus the original network are characterized by the 162-dimensional feature vector. For this, first the input matrix needs to be normalized. We used the min-max normalization method. Then to discover a plausible number of clusters, we employed the elbow method which consists in plotting the sum of the squared error (SSE) as the number of clusters (k in k-means) increases. The idea is that in this plot, at the beginning, SSE drops quickly but then at some point (the elbow) it starts decreasing slower, therefore, the increase of k does not contribute significantly in reducing the SSE.

Once an appropriate number for k is found, we then visualize the 162-dimensional points and how they cluster by performing a non-linear projection of these points into a 2-dimensional space using t-SNE, where each point is labeled with the corresponding cluster number. This will allow to see how the networks are distributed, as well as which clusters are near the cluster where the original network if found, as well as networks which are far away from the original network.

To analyze the discriminatory power of the features we will consider box plots per cluster. In particular, for topology related features we will consider the number of positive and negative edges for each network. For the dynamics, we will consider the basin of attraction size for each fixed point.

The synthetic network inference using SE and the feature construction were carried out using the open-source R software environment for statistical computing. Whereas the unsupervised learning techniques and box plot analysis were carried out using Python with the libraries: Scikit-learn, NumPy, and pandas. In both cases, running on a 2.6 GHz Intel Core i7 and 16 GB-RAM computer.

## 4 Results and Discussions

Figure 3 shows the SSE versus number of clusters plot, we notice that the elbow occurs around $k = 20$. With $k = 20$ there are a couple of clusters with significantly smaller number of examples in comparison with the rest. When we consider $k = 18$, we obtained a number of examples per cluster more reliable.

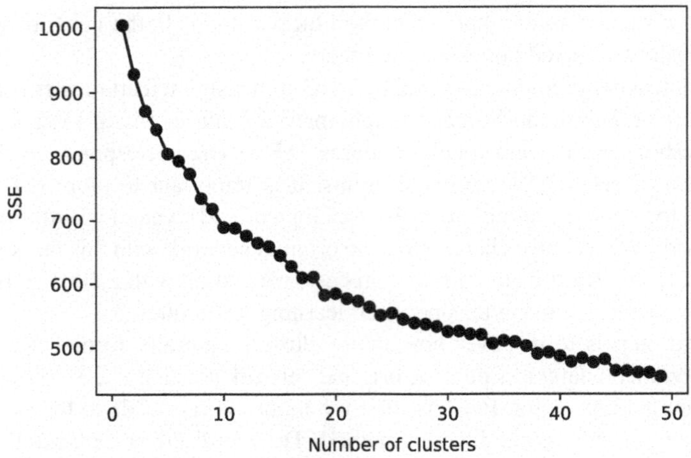


**Fig. 3** The sum of the squared error versus the number of clusters

We will consider 18 clusters. The t-SNE projection is shown in Fig. 4. We notice that the largest cluster corresponds to cluster number 1, it also presents the highest dispersion. The original network is in cluster 4, which contains only 17 networks. We notice that neighbouring clusters to cluster 4 correspond to clusters 8, 16, and 9. All the rest of the clusters are located at lager distances, in particular, cluster 13 is the furthest from cluster 4.

Figure 5 shows the box plots per cluster of the positive (Pos) number of edges that the networks have. We notice that cluster 5 contains slightly larger values, with the highest median value. Nevertheless, this characteristic does not discriminate cluster 4, which contains the original model, from the rest.

On the other hand, Fig. 6 shows the distribution of the negative edges per clusters, where it is clear that the median value of cluster 4 is the highest.

We notice from Fig. 7, that the size of the basin of attraction of fixed point 1 (BAF1) of the networks in cluster 4 is in general larger than the basin of attraction for that fixed points of the networks in the other clusters. The distribution is completely different from the sizes of the basin of attraction in the other clusters.

This behaviour is also present for the size of the basin of attraction of fixed point 2 (BAF2) (Fig. 8), the size of the basin of attraction of fixed point 3 (BAF3) (Fig. 9), the size of the basin of attraction of fixed point 5 (BAF5) (Fig. 11), and the size of the basin of attraction of fixed point 6 (BAF6) (Fig. 12). It is less evident for the size of the basin of attraction of fixed point 4 (BAF4) shown in Fig. 10.

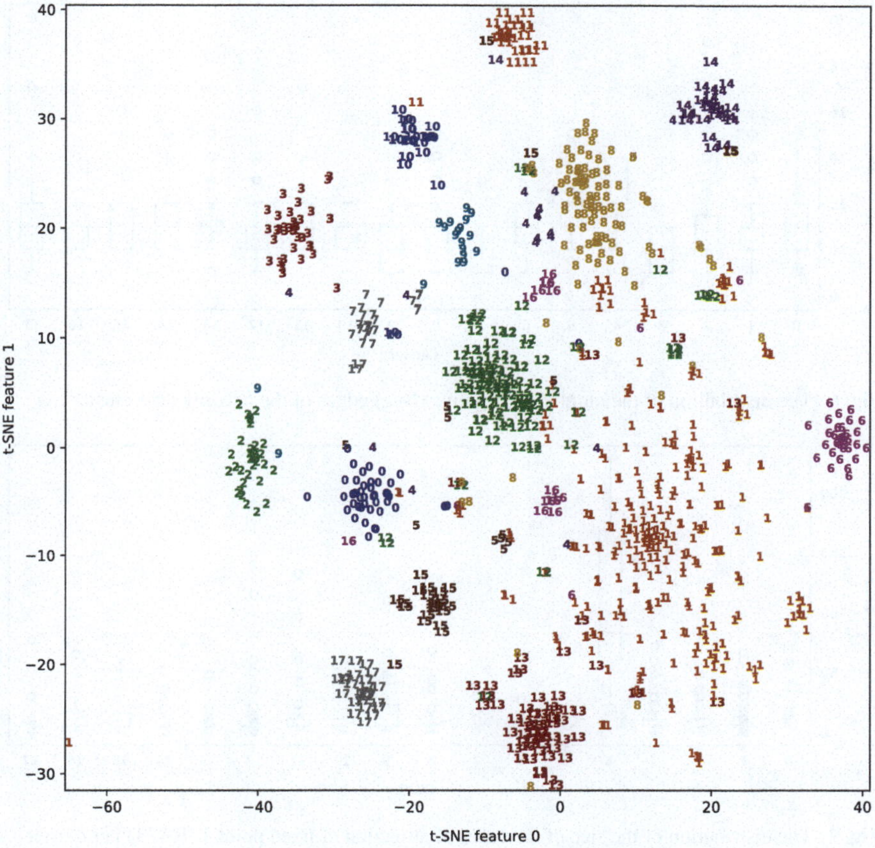


**Fig. 4** t-SNE projection with points labeled with their corresponding cluster number

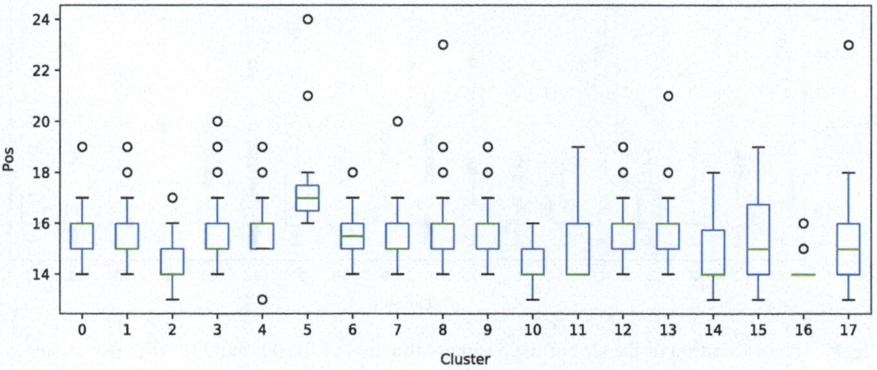


**Fig. 5** The distribution of the number of positive (PoS) edges of the networks per cluster

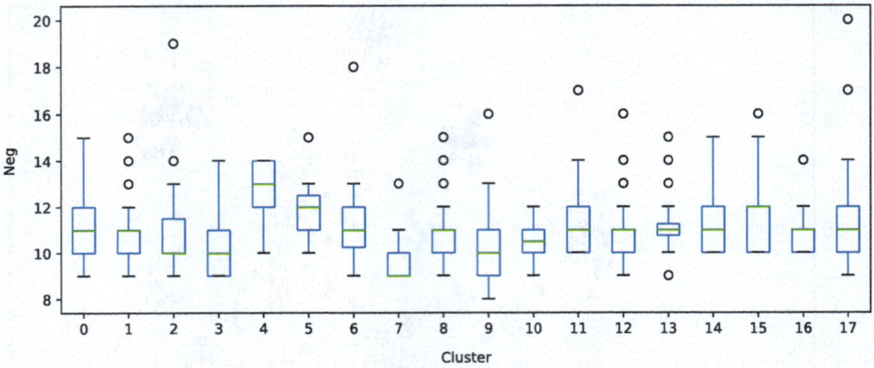


**Fig. 6** The distribution of the number of negative (Neg) edges of the networks per cluster

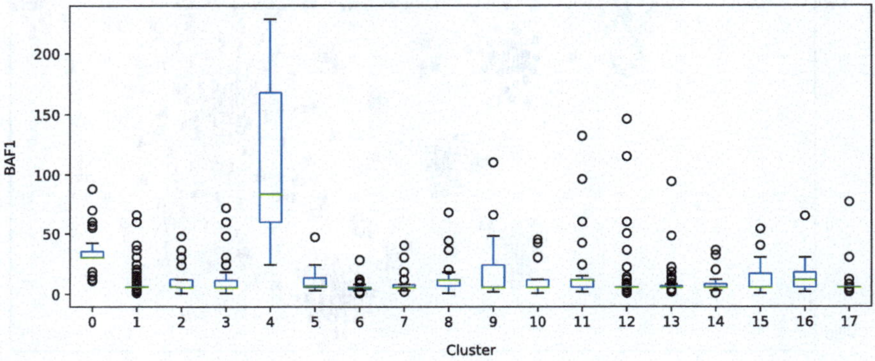


**Fig. 7** The distribution of the size of the basin of attraction of fixed point 1 (BAF1) per cluster

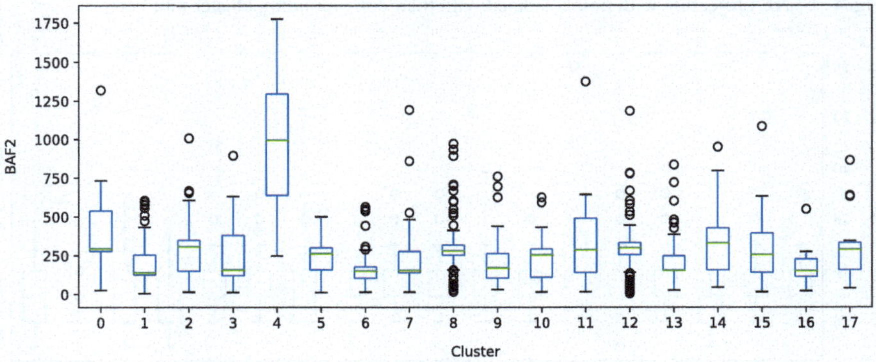


**Fig. 8** The distribution of the size of the basin of attraction of fixed point 2 (BAF2) per cluster

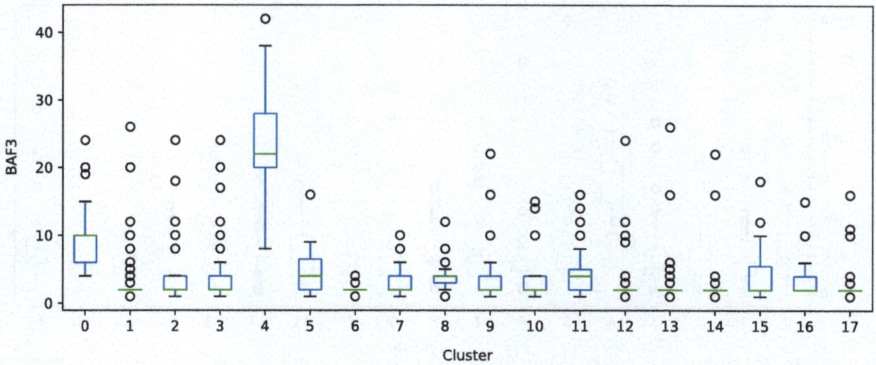


**Fig. 9** The distribution of the size of the basin of attraction of fixed point 3 (BAF3) per cluster

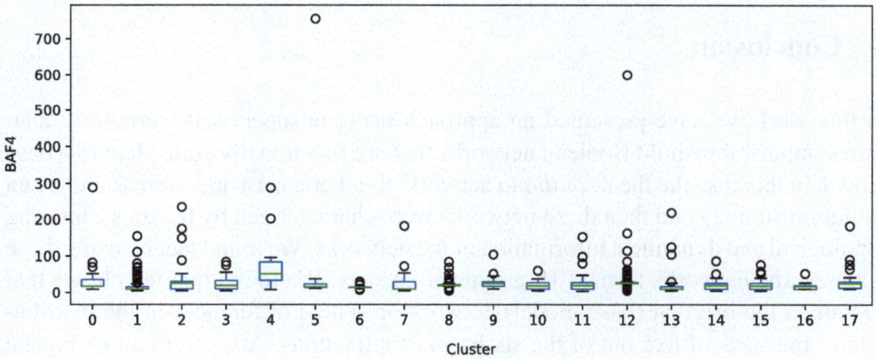


**Fig. 10** The distribution of the size of the basin of attraction of fixed point 4 (BAF4) per cluster

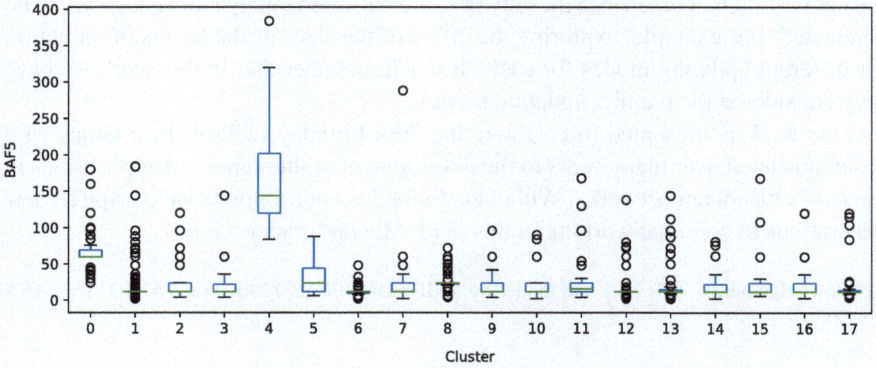


**Fig. 11** The distribution of the size of the basin of attraction of fixed point 5 (BAF5) per cluster

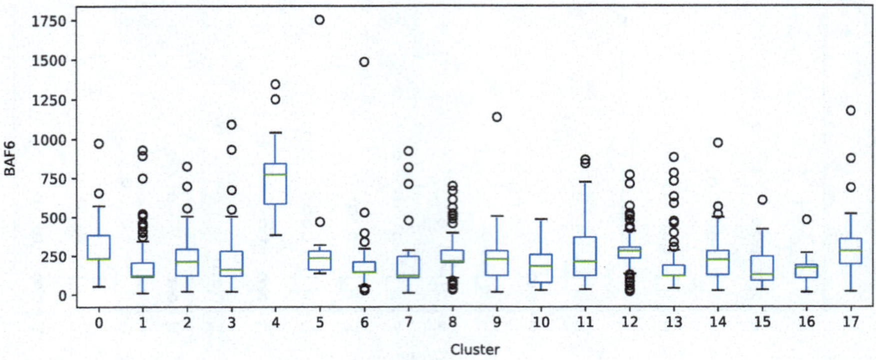


**Fig. 12** The distribution of the size of the basin of attraction of fixed point 6 (BAF6) per cluster

# 5 Conclusion

In this work we have presented an approach using unsupervised learning to analyze synthetic threshold Boolean networks that are functionally equivalent to a base model, in this case the the *A. thaliana* network. Synthetic networks were found by an evolution strategy and then these networks were characterized by features capturing topological and dynamical information of the networks. We found that by using these features, the networks formed 18 groups or clusters. We found that the cluster that contained the original (base model) network presented differences in the distribution of the sizes of five out of the six basin of attractions. Also, from a topological aspect, the cluster that contained the original network showed differences with the other clusters, when analyzing the distribution of the number of negative edges of the networks. Of course, once the networks are clustered, many more analysis can be conducted. For example, exploring the effect of the sizes of the basins of attraction for different updating modes for each cluster (remember that in this work we have only considered the parallel updating mode).

This work is presented to celebrate the 70th birthday of Prof. Eric Goles who has contributed over many years to the development of theorems and applications in the field of Boolean networks. Without a doubt, his contributions have inspired new generations to continue working in this field. Muchas gracias Goles!

**Acknowledgements** This work was funded by ANID FONDECYT 1180706 and ANID PIA/BASAL FB0002.

## References


1. Boldhaus G, Klemm K (2010) Regulatory networks and connected components of the neutral space. Eur Phys J B 77:233–237
2. Ciliberti S, Martin OC, Wagner A (2007) Innovation and robustness in complex regulatory gene networks. PNAS 104:13591–13596
3. Ciliberti S, Martin OC, Wagner A (2007) Robustness can evolve gradually in complex regulatory gene networks with varying topology. PLoS Comput Biol 3:e15
4. Fauré A, Naldi A, Chaouiya C, Thieffry D (2006) Dynamical analysis of a generic Boolean model for the control of the mammalian cell cycle. Bioinformatics 22:e124–e131
5. Goles E, Montalva M, Ruz GA (2013) Deconstruction and dynamical robustness of regulatory networks: application to the yeast cell cycle networks. Bull Math Biol 75:939–966
6. Kauffman SA (1969) Metabolic stability and epigenesis in randomly constructed genetic nets. J Theor Biol 22:437–467
7. Lahdesmaki H, Shmulevich I, Yli-Harja O (2003) On learning gene regulatory networks under the boolean network model. Mach Learn 25:147–167
8. Liang S, Fuhrman S, Somogyi R (1998) Reveal, a general reverse engineering algorithm for inference of genetic network architectures. In: Pac Symp Biocomput, pp 18–29
9. Mendoza L, Alvarez-Buylla ER (1998) Dynamics of the genetic regulatory network for arabidopsis thaliana flower morphogenesis. J Theor Biol 193:307–319
10. Pham DT, Castellani M (2009) The bees algorithm: modelling foraging behaviour to solve continuous optimization problems. Proc IMechE Part C: J Mech Eng Sci 223:2919–2938
11. Ruz GA, Goles E (2010) Learning gene regulatory networks with predefined attractors for sequential updating schemes using simulated annealing. In: Proceedings of IEEE the ninth international conference on machine learning and applications (ICMLA 2010), pp 889–894
12. Ruz GA, Goles E (2012) Reconstruction and update robustness of the mammalian cell cycle network. In: 2012 ieee symposium on computational intelligence and computational biology, CIBCB 2012, pp 397–403
13. Ruz GA, Goles E (2013) Learning gene regulatory networks using the bees algorithm. Neural Comput Appl 22:63–70
14. Ruz GA, Goles E (2014) Neutral graph of regulatory Boolean networks using evolutionary computation. In: The 2014 ieee conference on computational intelligence in bioinformatics and computational biology (CIBCB 2014), pp 1–8
15. Ruz GA, Goles E, Montalva M, Fogel GB (2014) Dynamical and topological robustness of the mammalian cell cycle network: a reverse engineering approach. Biosystems 115:23–32
16. Ruz GA, Timmermann T, Barrera J, Goles E (2014) Neutral space analysis for a boolean network model of the fission yeast cell cycle network. Biol Res 47:64
17. Ruz GA, Timmermann T, Goles E (2015) Reconstruction of a GRN model of salt stress response in Arabidopsis using genetic algorithms. In: The 2015 ieee conference on computational intelligence in bioinformatics and computational biology (CIBCB 2015), pp 1–8
18. Ruz GA, Timmermann T, Goles E (2016) Neutral space analysis of gene regulatory network models of salt stress response in arabidopsis using evolutionary computation. In: The 2016 ieee congress on evolutionary computation (IEEE CEC 2016), pp 4281–4288
19. Ruz GA, Ashlock D, Ledger T, Goles E (2017) Inferring bistable lac operon Boolean regulatory networks using evolutionary computation. In: The 2017 ieee conference on computational intelligence in bioinformatics and computational biology (CIBCB 2017), pp 1–8
20. Ruz GA, Zúñiga A, Goles E (2018) A Boolean network model of bacterial quorum-sensing systems. Int J Data Min Bioinform 21:123–144
21. Timmermann T, González B, Ruz GA (2020) Reconstruction of a gene regulatory network of the induced systemic resistance defense response in Arabidopsis using boolean networks. BMC Bioinform 21:142
22. van der Maaten L, Hinton G (2008) Visualizing data using t-SNE. J Mach Learn Res 9(86):2579–2605

# Existence and Non Existence of Limit Cycles in Boolean Networks

**Lilian Salinas, Luis Gómez, and Julio Aracena**

**Abstract** Boolean networks have been used as models of gene regulation networks and other biological systems. One key element in these models is the update schedule, which indicates the order in which states are to be updated. The presence of any limit cycle in the dynamics of a network depends on the update scheme used. Here, we study the complexity of the problems of determining the existence of a block-sequential update schedule for a given Boolean network such that it yields any limit cycle (LCE) and does not yield any limit cycle (LCNE). Besides, we prove that in AND-OR networks LCE is NP-hard and LCNE is coNP-hard. Finally, we show that both problems are polynomial in symmetric AND-OR networks. For these networks, we find a polynomial characterization for the existence of limit cycles in terms of the interaction digraph.

## 1 Introduction

A Boolean network is a system of $n$ interacting Boolean variables, which evolve, in a discrete time, according to a predefined rule. They have applications in many areas, including circuit theory, computer science and social systems [14, 26]. In particular, from the seminal works of Kauffman [16, 17] and Thomas [24, 25], they are extensively used as models of gene networks. In this context, the limit cycles of a network are often associated with mitotic cycles in cells [4, 6].

L. Salinas
Departamento de Ingeniería Informática y Ciencias de la Computación, Facultad de Ingeniería, Universidad de Concepción, Concepción, Chile
e-mail: lilisalinas@udec.cl

L. Gómez
Departamento de Estadística, Facultad de Ciencias, Universidad del Bío-Bío, Concepción, Chile
e-mail: lgomez@ubiobio.cl

J. Aracena (✉)
Departamento de Ingeniería Matemática, Facultad de Ciencias Físicas y Matemáticas, Universidad de Concepción, Concepción, Chile
e-mail: jaracena@ing-mat.udec.cl

 
A. Adamatzky (ed.), *Automata and Complexity*, Emergence, Complexity and Computation 42, https://doi.org/10.1007/978-3-030-92551-2_15

The update schedule in a Boolean network, that is the order in which each node is updated, is of great importance in its dynamical behavior. In general, Boolean networks are usually studied with synchronous (parallel) or sequential schemes. A generalization of these schemes, known as block-sequential update schedules, was introduced by Robert [21, 22], and they are currently used in the modeling of regulatory networks [10, 23].

Many analytic studies have been done about the limit cycles of a Boolean network with different block-sequential update schedules [1–3, 5, 8, 11, 13, 18, 19]. Most of them show that the limit cycles are very sensitive to changes in the updating scheme of a network. In particular, some of these articles exhibit examples of Boolean networks where the existence of limit cycles depends on the used update schedule [1–3, 5, 19].

An important issue that arises in the modeling of genetic regulatory networks with Boolean networks, especially in the case of the construction of networks with some prescribed dynamical property, is to define what update schedule to use. Thus, a natural previous question is: Given a Boolean network, does there exist a block-sequential update schedule such that the network updated under it yields any limit cycle? The solution to this problem, named Limit Cycle Existence problem (LCE), could enable us for instance to know if a given Boolean network can be used to model a cellular cycle. Other directly related problem is deciding for a given Boolean network the existence of a block-sequential update schedule such that the network updated under it has only fixed points as attractors, named Limit Cycle Nonexistence problem (LCNE). An affirmative instance of LCNE updated under a scheme that yields only fixed points guarantees that any trajectory of the dynamical behavior ends in a fixed point of the network.

To our knowledge both problems LCE and LCNE have not been sufficiently studied so far. However, thanks to works about the dynamical behavior of Boolean networks with different block-sequential update schedules (as those mentioned above), we know the answer to LCE and LCNE for certain families of networks. In particular, L. Gómez in his Ph.D. thesis proved that LCE is NP-hard in AND-OR networks and in Boolean networks with symmetric interaction digraph [13]. These results are included in this chapter with an improved demonstration. More recently, in [3] are exhibited some very close results like the NP-completeness of the problem of existence of a limit cycle of length $k$ in a Boolean network with block-sequential update.

On the other hand, by result in [9] we know that some symmetric threshold networks, including the symmetric AND-OR networks, with any sequential update schedule have only fixed points as attractors, i.e. they are affirmative instances of LCNE. Besides, for the OR (AND) symmetric networks was proved that they can cycle if and only if the interaction digraph is bipartite [7, 8]. Thus, LCE is polynomial in OR (AND) symmetric networks.

In this chapter we prove that LCE and LCNE are both NP-hard problems in Boolean networks with symmetric interaction digraph. Besides, we prove that in AND-OR networks the LCE problem is NP-hard and the LCNE problem is coNP-hard. Nevertheless, we show that LCE and LCNE are both polynomial problems in networks verifying the two conditions, that is in symmetric AND-OR networks. For this family of networks, we prove that there exists a limit cycle in a network iterated with a block-sequential update schedule if and only if there exists a limit cycle under parallel scheme. This last condition is equivalent to a topological property on the interaction digraph which can be verified in polynomial time. This result is somewhat surprising, because these networks can have limit cycles of length super-polynomial [11] with block-sequential update schedules different from the synchronous and sequential schemes, and only of length two with parallel schedule [12].

## 2 Definition and Notation

A *Boolean network* on a finite set $V$ of $n$ elements is a dynamical system defined by an activation function $f : \{0, 1\}^n \to \{0, 1\}^n$, where its component functions $f_v : \{0, 1\}^n \to \{0, 1\}$ are Boolean functions, called *local activation functions*, verifying that $\forall x \in \{0, 1\}^n, \forall v \in V, f(x)_v = f_v(x)$.

An *update schedule* is a function $s : V \to \{1, \dots, n\}$ such that $s(V) = \{1, \dots, m\}$ for some $m \leqslant n$. A *block* of an update schedule $s$ is a set $B_i = \{v \in V : s(v) = i\}$, $1 \leqslant i \leqslant m$. In this way, an update schedule $s$ is usually denoted by $s = B_1 B_2 \cdots B_m$ and also known as a *block-sequential update schedule*. Particular cases of block-sequential update schedules are the *synchronous* or *parallel*, which is given by an update schedule $s^p$ such that $\forall v \in V, s^p(v) = 1$ (just one block with cardinality $n$); and the *sequential* which corresponds to a bijective function ($n$ blocks of cardinality 1).

The dynamics of a Boolean network $f$ on the set $V$ of elements with an update schedule $s = B_1 B_2 \cdots B_m$ is given by the function:

$$f^s = f^{B_m} \circ f^{B_{m-1}} \circ \cdots \circ f^{B_1},$$

where $\forall i \in \{1, \dots, m\}$, $\forall x \in \{0, 1\}^n$, $\forall v \in V$:

$$f_v^{B_i}(x) = \begin{cases} f_v(x) & \text{if } v \in B_i, \\ x_v & \text{otherwise.} \end{cases}$$

Since $\{0, 1\}^n$ is a finite set, we have two limit behaviors for the iteration of a network $f$ with scheme $s$:

- *Fixed Point*. We define a fixed point as a stable state of the dynamical system, i.e. $x \in \{0, 1\}^n$ is a fixed point if $f(x) = x$.
- *Limit Cycle*. We define a cycle of length $p > 1$ as the vector sequence $[x^k]_{k=0}^p = [x^0, \dots, x^{p-1}, x^0]$ such that $x^k \in \{0, 1\}^n$, $x^k$ are pairwise distinct and $f^s(x^k) = x^{k+1}$, for all $k \in \{0, \dots, p-1\}$ and $x^p \equiv x^0$. We note that any cyclic permutation of a sequence represents the same limit cycle.
  The set of limit cycles of $f$ updated under $s$ is denoted by $\text{LC}(f, s)$. Besides, we say that $f$ cycles under $s$ if $\text{LC}(f, s) \neq \emptyset$.

Fixed points and limit cycles are called *attractors* of the network. A node is said to be *frozen* for a limit cycle if its state value does not change on it.

Given a digraph $G$, the node set of $G$ is referred to as $V(G)$, and its arc set as $A(G)$. An arc $(v, v) \in A(G)$ is called *loop* of $G$. $G$ is said to be *symmetric* if $\forall (u, v) \in A(G), (v, u) \in A(G)$. Given a node $v \in V(G)$, the set of incoming nodes to $v$ is denoted by $N_G^-(v) = \{u \in V(G) : (u, v) \in A(G)\}$.

Given $U \subseteq V(G)$, $G[U]$ is the digraph obtained from $G$ by removing all nodes in $V(G) \setminus U$ and all arcs incoming to or outgoing from these nodes.

Given $x = (x_v)_{v \in V} \in \{0, 1\}^n$ and $u \in V$, we define $\bar{x}^u \in \{0, 1\}^n$ as:

$$\forall v \in V, \quad \bar{x}^u_v = \begin{cases} x_v & \text{if } v \neq u \\ \neg x_u & \text{if } v = u \end{cases}$$

where $\forall x_u \in \{0, 1\}$ , $\neg x_u = 1 \Leftrightarrow x_u = 0$. We also define $\bar{x} \in \{0, 1\}^n$ as: $\forall v \in V, \ \bar{x}_v = \neg x_v$.

The digraph associated to a Boolean network $f$, called *interaction digraph*, is the directed graph $G^f = (V, A)$, where $(u, v) \in A$ if and only if $f_v$ depends on $x_u$, i.e., if there exists $x \in \{0, 1\}^n$ such that $f_v(x) \neq f_v(\bar{x}^u)$. Note that if $f_v$ is constant, then $N_f^-(v) = \emptyset$. See an example of interaction digraph in Fig. 1. We say that a Boolean network $f$ is symmetric if its interaction digraph $G^f$ is symmetric.

Given a finite set $U$ of $k$ elements, a Boolean function $f : \{0, 1\}^k \to \{0, 1\}$ is said to be an *AND function*, denoted $f(x) = \bigwedge_{v \in U} x_v$, where $f(x) = 1$ if and only if $\forall v \in U, x_v = 1$. Analogously, $f$ is an *OR function*, denoted $f(x) = \bigvee_{v \in U} x_v$, where $f(x) = 1$ if and only if $\exists v \in U, x_v = 1$.

In this way, we say that $f$ is an AND-OR *network* if each local activation function is either an AND or an OR function. In this case, we define $V_{\text{AND}}(f) \subseteq V(G^f)$ ($V_{\text{OR}}(f) \subseteq V(G^f)$) as the nodes that have an AND (OR) local activation function. In particular, we say that $f$ is an *OR network* if each local activation function is an OR function.

An AND-OR network $f$ can be completely described by its interaction digraph, labeling AND and OR nodes differently (in the figures of this chapter, white nodes represent OR nodes, dark gray nodes represent AND nodes, and light gray nodes represent nodes that are neither AND nor OR nodes). That is, given $G = (V, A)$ a digraph and $\{V_{\text{AND}}, V_{\text{OR}}\}$ a partition of $V$, with $|V| = n$, we define $f : \{0, 1\}^n \to \{0, 1\}^n$ as follows:

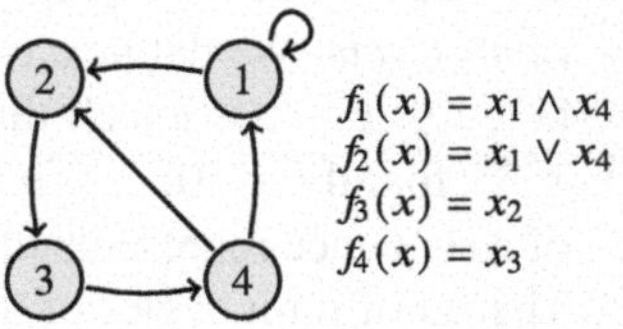


**Fig. 1** Interaction digraph associated to a Boolean network

$$\forall v \in V,\ f_v(x) = \begin{cases} \bigwedge\limits_{u \in N_f^-(v)} x_u & \text{if } v \in V_{\text{AND}} \\ \bigvee\limits_{u \in N_f^-(v)} x_u & \text{if } v \in V_{\text{OR}} \end{cases}$$

Note that if $N_f^-(v) = \emptyset$, by definition we have that: $f_v(x) = 1$ if $v \in V_{\text{AND}}$ (because for all $v \in N_f^-$, $x_v = 1$) and $f_v(x) = 0$ if $v \in V_{\text{OR}}$.

### 2.1 *LCE and LCNE Problems*

As explained above, we are interested in the algorithmic complexity of the following two decision problems:

> LIMIT CYCLE EXISTENCE (LCE): Given a Boolean network $f$. Does there exist an update schedule $s$ such that $\text{LC}(f, s) \neq \emptyset$?
>
> LIMIT CYCLE NONEXISTENCE (LCNE): Given a Boolean network $f$. Does there exist an update schedule $s$ such that $\text{LC}(f, s) = \emptyset$?

Note that LCNE is not the complement of LCE. Indeed, for instance, $f : \{0, 1\}^2 \to \{0, 1\}^2$ defined by $f(x_1, x_2) = (x_2, x_1)$ verifies that $\text{LC}(f, s^p) = [(1, 0), (0, 1), (1, 0)]$ and $\text{LC}(f, s) = \emptyset$, with $s = \{1\}\{2\}$, i.e. $f$ is an affirmative instance of both LCE and LCNE.

## 3 Complexity

### 3.1 *Limit Cycle Existence Problem*

In this section we study the complexity of the LCE problem. A particular and directly related problem is to determine the existence of a limit cycle for a given Boolean network with a fixed update schedule. This latter problem was proved to be NP-hard for threshold networks with asymmetric weight matrix [20] and for AND-OR networks [15], in both cases with synchronous scheme. In this way, the symmetry of the interaction digraph and the AND-OR functions in Boolean networks seem to play an important role in the complexity of these problems.

Next, in Theorem 1 we prove that LCE is NP-hard in networks with symmetric interaction digraph and in Theorem 2 for AND-OR networks. Besides, we show that LCE is polynomial in symmetric AND-OR networks, which corresponds to a generalization of the result proved by [8] in symmetric OR networks.

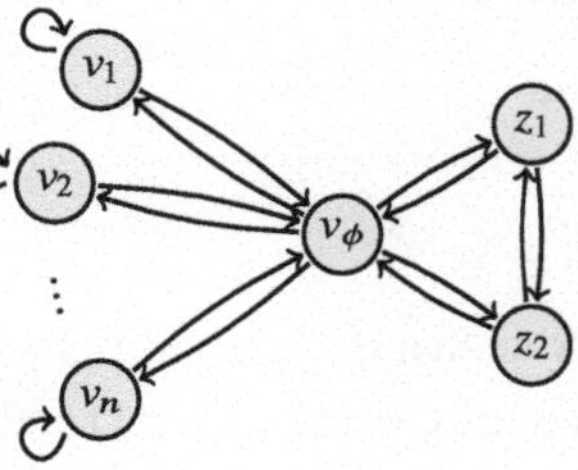


**Fig. 2** Interaction digraph of the transformation defined in Theorem 1

**Theorem 1** LCE *is NP-hard in symmetric Boolean networks*

***Proof*** We show that SAT $\leqslant_p$ LCE.

Given a Boolean formula $\phi$ in variables $w_1, \ldots, w_n$, we consider $V = \left\{v_1, \ldots, v_n, v_\phi, z_1, z_2\right\}$ and we define $f : \{0, 1\}^{n+3} \to \{0, 1\}^{n+3}$ as follows (see Fig. 2):

$$
\begin{aligned}
f_{v_i}(x) &= x_{v_i} \wedge x_{v_\phi} && \forall i \in \{1, \ldots, n\} \\
f_{v_\phi}(x) &= \phi\left(x_{v_1}, \ldots, x_{v_n}\right) \wedge (x_{z_1} \vee x_{z_2}) \\
f_{z_1}(x) &= x_{v_\phi} \wedge x_{z_2} \\
f_{z_2}(x) &= x_{v_\phi} \wedge x_{z_1}
\end{aligned}
$$

We now prove that $\phi$ is satisfiable if and only if there exists $s$ such that LC$(f, s)$ $\neq \emptyset$.

Let us suppose that $\phi$ is satisfiable, and let $w$ be such that $\phi(w) = 1$. If we consider the update schedule $s = \left\{v_1, \ldots, v_n, v_\phi\right\}\{z_1, z_2\}$, then it is clear that $C = [(w, 1, 0, 1), (w, 1, 1, 0), (w, 1, 0, 1)] \in \mathrm{LC}(f, s)$, where $(w, 1, 0, 1)$ and $(w, 1, 1, 0)$ are state vectors of the vector $(x_{v_1}, \ldots, x_{v_n}, x_{v_\phi}, x_{z_1}, x_{z_2}) \in \{0, 1\}^{n+3}$.

Now, let us suppose $\phi$ is not satisfiable, then $\forall w \in \{0, 1\}^n$, $\phi(w) = 0$. Hence, for every update schedule $s$, we have that

$$
\forall x \in \{0, 1\}^{n+3}, \; f^s_{v_\phi}(x) = 0.
$$

Therefore, $\forall x \in \{0, 1\}^{n+3}$ :

$$
\begin{aligned}
\forall i \in \{1, \ldots, n\}, \quad f^s_{v_i}(f^s(x)) &= f^s(x)_{v_i} \wedge 0 = 0, \\
f^s_{z_1}(f^s(x)) &= 0 \wedge x_{z_2} = 0, \\
f^s_{z_2}(f^s(x)) &= 0 \wedge x_{z_1} = 0.
\end{aligned}
$$

Thus, LC$(f, s) = \emptyset$, for every update schedule $s$. $\square$

**Table 1** Definition of $f$ of Theorem 2

| $v \in V$ | Type | $N_f^-(v)$ |
|---|---|---|
| $\forall i \in \{1, \ldots, n\}, v_i$ | AND | $\{v_i\}$ |
| $\forall i \in \{1, \ldots, n\}, \bar{v}_i$ | AND | $\{\bar{v}_i\}$ |
| $\forall i \in \{1, \ldots, n\}, o_i$ | OR | $\{v_i, \bar{v}_i\}$ |
| $\forall i \in \{1, \ldots, n\}, a_i$ | AND | $\{v_i, \bar{v}_i\}$ |
| $A$ | AND | $\{o_1, \ldots, o_n\}$ |
| $O$ | OR | $\{a_1, \ldots, a_n\}$ |
| $\forall j \in \{1, \ldots, m\}, v_{C_j}$ | OR | $\{v_i : w_i \in C_j\} \cup \{\bar{v}_i : \neg w_i \in C_j\}$ |
| $v_\phi$ | AND | $\{v_{C_1}, \ldots, v_{C_m}\}$ |
| $z_1$ | AND | $\{z_2, v_\phi, A\}$ |
| $z_2$ | OR | $\{z_3, O\}$ |
| $z_3$ | OR | $\{z_1\}$ |

**Theorem 2** LCE *is NP-hard in AND-OR networks*

***Proof*** We show that 3-SAT $\leqslant_p$ LCE in AND-OR networks.

Given a formula $\phi$ in conjunctive normal form where each clause is limited to at most three literals, in variables $w_1, \ldots, w_n$ with clauses $C_1, \ldots, C_m$, we define $f : \{0, 1\}^{4n+m+6} \to \{0, 1\}^{4n+m+6}$ according to Table 1. See $G^f$ in Fig. 3.

Here, $\forall i \in \{1, \ldots, n\}$, nodes $v_i$ represent literals $w_i$ and nodes $\bar{v}_i$ represent literals $\neg w_i$.

Next, we prove that $\phi$ is satisfiable if and only if there exists $s$ such that $\text{LC}(f, s) \neq \emptyset$.

Let us suppose there exists $s$ an update schedule such that $C = \left[x^k\right]_{k=0}^p \in \text{LC}(f, s)$. We note that only nodes $z_1$, $z_2$ and $z_3$ can be non frozen in a limit cycle. For that, it is necessary to verify:

$$f_A(x^0) = 1, \tag{1}$$

$$f_O(x^0) = 0. \tag{2}$$

$$f_{v_\phi}(x^0) = 1, \tag{3}$$

Equation (1) implies for all $i \in \{1, \ldots, n\}$, $o_i = 1$ and Eq. (2) implies that for all $i \in \{1, \ldots, n\}$, $a_i = 0$. Therefore, if for all $i \in \{1, \ldots, n\}$, $o_i = 1$ and $a_i = 0$, then $\forall i \in \{1, \ldots, n\}, x_{\bar{v}_i} = \neg x_{v_i}$. Of this way, from (3), we have that $\phi(x^0_{v_0}, \ldots, x^0_{v_n}) = 1$.

For the converse, let us suppose that $\phi$ is satisfiable, then $\exists \hat{w} \in \{0, 1\}^n, \phi(\hat{w}) = 1$, and if we consider the update schedule $s$ and the limit cycle $C = \left[x^0, x^1, x^0\right]$ as described in Table 2. Then, $C \in \text{LC}(f, s)$. □

Now we study the case of symmetric AND-OR networks. In Theorem 3 we characterize the symmetric AND-OR networks $f$ for which there exists an update schedule $s$ such that $f$ updated under $s$ has a limit cycle. Since this characterization is testable

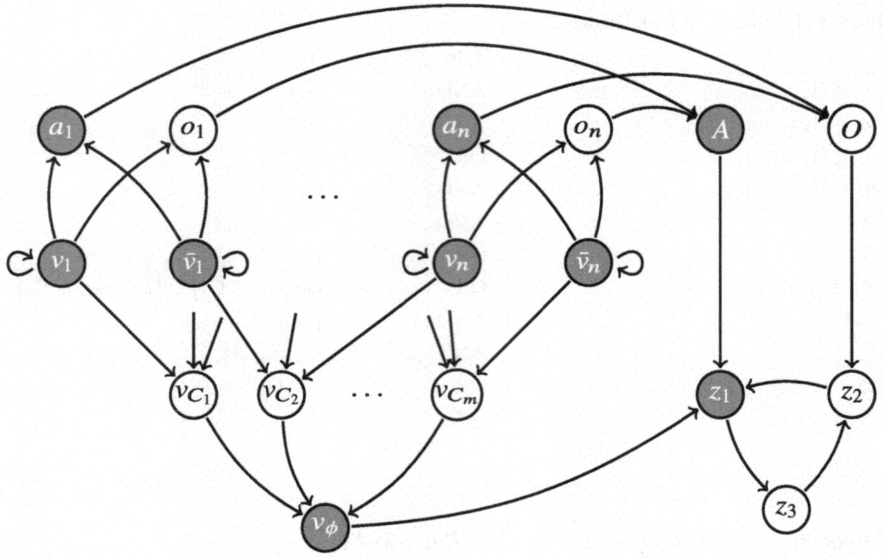


**Fig. 3** Interaction digraph of the transformation $f$ in Theorem 2

**Table 2** Definition of $C$ in Theorem 2

| $v \in V$ | $v_i$ | $\bar{v}_i$ | $o_i$ | $a_i$ | $C_j$ | $A$ | $O$ | $v_\phi$ | $z_1$ | $z_2$ | $z_3$ |
|---|---|---|---|---|---|---|---|---|---|---|---|
| $s(v)$ | 1 | 1 | 1 | 1 | 1 | 1 | 1 | 1 | 2 | 3 | 2 |
| $x_v^0$ | $\hat{w}_{v_i}$ | $\neg\hat{w}_{v_i}$ | 1 | 0 | 1 | 1 | 0 | 1 | 1 | 0 | 0 |
| $x_v^1$ | $\hat{w}_{v_i}$ | $\neg\hat{w}_{v_i}$ | 1 | 0 | 1 | 1 | 0 | 1 | 0 | 1 | 1 |

in polynomial time, we conclude that LCE is polynomial in symmetric AND-OR networks.

Other interesting result is Proposition 2, which states that a symmetric AND-OR network cycles with some update schedule if and only if it cycles under the parallel scheme. However, the lengths of limit cycles can be very different in each case, because in the parallel case they are of length two [12] and in the case of others block-sequential update schedules they can be of length super-polynomial [11].

Next, we give some required definitions.

**Definition 1** Let $f$ be a symmetric AND-OR network.

- We denote each non trivial connected component of $G[V_{\mathrm{OR}}(f)]$ by $G_1^{\mathrm{OR}}, \ldots, G_{k_{\mathrm{OR}}}^{\mathrm{OR}}$. We call them OR components of $G^f$.
- We denote each non trivial connected component of $G[V_{\mathrm{AND}}(f)]$ by $G_1^{\mathrm{AND}}, \ldots, G_{k_{\mathrm{AND}}}^{\mathrm{AND}}$. We call them AND components of $G^f$.
- We define the alternated nodes as

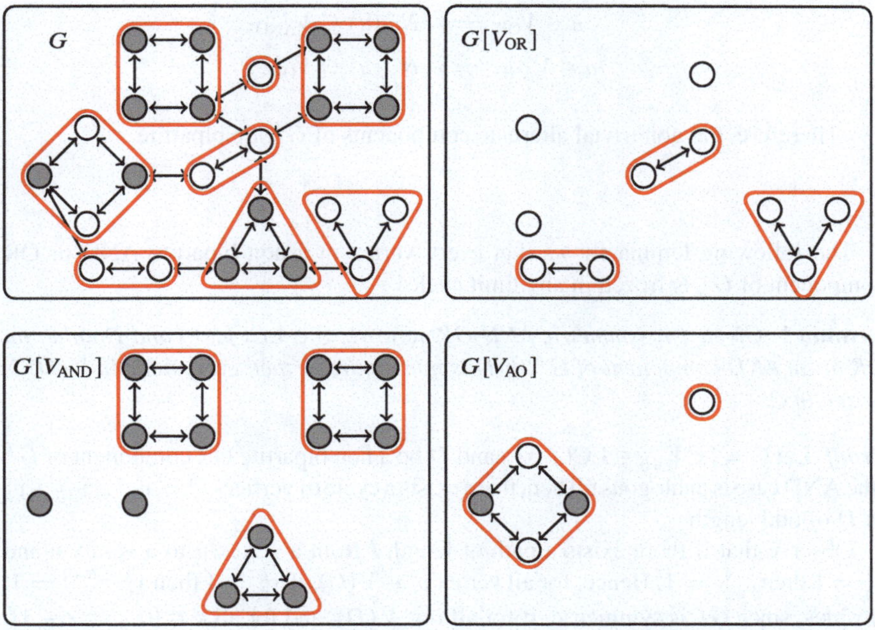


**Fig. 4** Example of Definition 1. The subdigraphs surrounded by a red line are the components of an AOA decomposition

$$V_{\text{AO}} = V \setminus \left( \bigcup_{i=1}^{k_{\text{OR}}} V(G_i^{\text{OR}}) \cup \bigcup_{i=1}^{k_{\text{AND}}} V(G_i^{\text{AND}}) \right)$$

and we denote by $G_1^{\text{AO}}, \ldots, G_{k_{\text{AO}}}^{\text{AO}}$, to the connected components of $G[V_{\text{AO}}]$. We call them alternated components of $G^f$.

- We call to the set $\{G_1^{\text{OR}}, \ldots, G_{k_{\text{OR}}}^{\text{OR}}, G_1^{\text{AND}}, \ldots, G_{k_{\text{AND}}}^{\text{AND}}, G_1^{\text{AO}}, \ldots, G_{k_{\text{AO}}}^{\text{AO}}\}$, an AOA (AND-OR ALTERNATED) decomposition of $G^f$ (see example Fig. 4).
  Here, a trivial connected component of $G^f$ is a subdigraph $G'$ of $G^f$ such that $|V(G')| = 1$ and $A(G') = \emptyset$.

> **Important**

1. The set $\left\{V(G_1^{\text{OR}}), \ldots, V(G_{k_{\text{OR}}}^{\text{OR}}), V(G_1^{\text{AND}}), \ldots, V(G_{k_{\text{AND}}}^{\text{AND}}), V(G_1^{\text{AO}}), \ldots, V(G_{k_{\text{AO}}}^{\text{AO}})\right\}$ is a partition of $V(G^f)$.
2. Given $i \in \{1, \ldots, k_{\text{AO}}\}$, we note that $\forall u \in V(G_i^{\text{AO}})$:

$$u \in V_{\text{OR}} \implies N_f^-(u) \subseteq V_{\text{AND}},$$
$$u \in V_{\text{AND}} \implies N_f^-(u) \subseteq V_{\text{OR}}.$$

Therefore, the non trivial alternate components of $G^f$ are bipartite.

---

The following lemma shows that every vertex in a non bipartite AND or OR component of $G^f$ is frozen in any limit cycle.

**Lemma 1** *Given $f$ a symmetric AND-OR network, $C \in \text{LC}(f, s^p)$ and $D$ either an OR or an AND component of $G^f$. If $D$ is non bipartite, then every node in $V(D)$ is frozen in $C$.*

***Proof*** Let $C = [x^k]_{k=0}^r \in \text{LC}(f, s^p)$ and $D$ be a non bipartite OR component of $G^f$ (the AND case is analogous). Then, there exists a cycle of vertices $C = v_1 \dots v_{2N+1}v_1$ in $D$ of odd length.

Observe that if there exists a path of length $l$ from a vertex $u$ to a vertex $v$ and $x_u^t = 1$ then $x_v^{t+l} = 1$. Hence, for all vertex $v_i \in V(C)$, if $x_{v_i}^k = 1$ then $x_{v_i}^{k+2N+1} = 1$. Besides, since $G^f$ is symmetric, if for all $v \in V(D)$, and for all $k \in \{0, \dots, r-1\}$, if $x_v^k = 1$ then $x_v^{k+2} = 1$. Therefore, if there exists $v \in V(C)$ and $k \in \{0, \dots, r-1\}$ such that $x_v^k = 1$, then for all $v \in V(C)$ and for all $k \in \{0, \dots, r-1\}$, $x_v^k = 1$. Thus, every vertex in the cycle $C$ is frozen. Finally, since $G^f$ is strongly connected, the result holds. □

Observe that, the neighbor vertices of $V(D)$ are not involved in the property of every node in $V(D)$ is frozen in $C$, but in the value of the vertices of $V(D)$ in $C$.

Next proposition gives a polynomial testable characterization of when a symmetric AND-OR network can have limit cycles with the parallel scheme.

**Proposition 1** *Let $f$ be a symmetric AND-OR network. Then, $\text{LC}(f, s^p) \neq \emptyset$ if and only if there exists a bipartite component in the AOA decomposition of $G^f$.*

***Proof*** Let us suppose that $\text{LC}(f, s^p) \neq \emptyset$.

If there does not exist a bipartite component in the AOA decomposition of $G^f$, then by Lemma 1, every vertex in each OR and AND component of $G^f$ is frozen in any cycle. Also, each alternate component is trivial, and hence frozen in any cycle. Therefore, $f$ updated under $s^p$ has no limit cycle, which is a contradiction.

For the converse, Let $V' \subseteq V$ the union of bipartite components of AOA decomposition of $G^f$. Since $G[V']$ is bipartite, let $V_1, V_2$ be the bipartition of $V'$. We define the limit cycle of length two $[x^0, x^1, x^0]$ of $f$ updated under the scheme $s_p$ in Table 3.

In first place, note from cases 1 and 2 that vertices in the bipartition cycle with period 2. From cases 3 and 4, we see that vertices in non bipartite OR components (AND components respectively) of the decomposition are frozen at value 1 (0 respectively). Observe that this type of vertices do not affect the dynamical behavior of other vertices in the network. Cases 5 to 10 correspond to the trivial components

**Table 3** Limit cycle in Proposition 1

| | | | $x_v^0$ | $x_v^1$ |
|---|---|---|---|---|
| 1 | $v \in V_1$ | | 1 | 0 |
| 2 | $v \in V_2$ | | 0 | 1 |
| 3 | $v \in \left( \bigcup_{i=1}^{k_{\text{OR}}} V(G_i^{\text{OR}}) \right) \setminus V'$ | | 1 | 1 |
| 4 | $v \in \left( \bigcup_{i=1}^{k_{\text{AND}}} V(G_i^{\text{AND}}) \right) \setminus V'$ | | 0 | 0 |
| 5 | $v \in V_{\text{AO}} \setminus V'$, | $N^-(v) \cap V_1 \neq \emptyset, N^-(v) \cap V_2 = \emptyset$ | 0 | 1 |
| 6 | $v \in V_{\text{AO}} \setminus V'$, | $N^-(v) \cap V_1 = \emptyset, N^-(v) \cap V_2 \neq \emptyset$ | 1 | 0 |
| 7 | $v \in (V_{\text{AO}} \setminus V') \cap V_{\text{AND}}$, | $N^-(v) \cap V_1 \neq \emptyset, N^-(v) \cap V_2 \neq \emptyset$ | 0 | 0 |
| 8 | $v \in (V_{\text{AO}} \setminus V') \cap V_{\text{OR}}$, | $N^-(v) \cap V_1 \neq \emptyset, N^-(v) \cap V_2 \neq \emptyset$ | 1 | 1 |
| 9 | $v \in (V_{\text{AO}} \setminus V') \cap V_{\text{AND}}$, | $N^-(v) \cap V_1 = \emptyset, N^-(v) \cap V_2 = \emptyset$ | 1 | 1 |
| 10 | $v \in (V_{\text{AO}} \setminus V') \cap V_{\text{OR}}$, | $N^-(v) \cap V_1 = \emptyset, N^-(v) \cap V_2 = \emptyset$ | 0 | 0 |

of the AOA decomposition; the vertices in cases 5 and 6 cycle, because if they are neighbors of vertices in only one partition their behavior is similar to those in the other partition. Vertices in cases 7 and 8 are connected to vertices in both sides of the bipartition, of this way they frozen at value 0 if the local activation function is AND and in 1 for an OR local activation function. Finally, cases 9 and 10 are vertices not connected to nodes in cases 3 and 4, since vertices in $V_{\text{AND}}$ are connected to vertices in $V_{\text{OR}}$ they will be frozen at value 1, and analogously vertices in $V_{\text{OR}}$ will be frozen at value 0. □

**Proposition 2** *Let $f$ be a symmetric AND-OR network. If* $\text{LC}(f, s^p) = \emptyset$, *then for every update schedule* $s \neq s^p$, $\text{LC}(f, s) = \emptyset$.

***Proof*** We prove that if $\text{LC}(f, s^p) = \emptyset$, then for each update schedule $s \neq s^p$, $\text{LC}(f, s) = \emptyset$. For that, we will assume that $G^f$ is connected. The general case is direct from it.

Let $s \neq s^p$ be an update schedule. We note that, since $G^f$ is symmetric and connected, then it is strongly connected. Thus, $s \neq s^p$ if an only if there exists $(u, v) \in A(G^f)$ such that $s(u) < s(v)$.

Let $C = [x^k]_{k=0}^r \in \text{LC}(f, s)$. Since $\text{LC}(f, s^p) = \emptyset$, then all elements in the AOA decomposition of $G^f$ are not bipartite, by Proposition 1. Therefore, by Lemma 1, every OR and AND component updated in parallel is frozen in $C$. Besides, by the equivalence proved above, there are only trivial alternated components of $G^f$.

Now, let $u, v \in V_{\text{OR}}$ (the AND case is analogous) such that $(u, v) \in A(G^f)$ and $s(u) < s(v)$. If $u$ is not frozen in $C$ at value 0, then there exists $k \in \{0, \ldots, r-1\}$ such that $x_u^k = 1$. Since $v \in V_{\text{OR}}$, $u \in N_f^-(v)$ and $s(u) < s(v)$, $x_v^k = 1$. Because of the symmetry of the network $v \in N_f^-(v)$, and therefore $x_u^{k+1} = 1$. In this way, vertices $u$ and $v$ become frozen nodes in $C$ at value 1 as well as every node in the same connected component. In either case, all vertices in the OR component are frozen in $C$.

Finally, all nodes in alternated components have only frozen neighbors, so they are also frozen. Therefore, every node is frozen in $C$, which is a contradiction. □

**Theorem 3** *Let $f$ be a symmetric AND-OR network. Then, there exists an update schedule $s$ such that* $\mathrm{LC}(f, s) \neq \emptyset$ *if and only if there exists a bipartite component in the AOA decomposition of $G^f$.*

***Proof*** Straightforward from Propositions 1 and 2. □

**Corollary 1** LCE *is polynomial in symmetric AND-OR networks.*

***Proof*** In Theorem 3 we characterized the existence of solution of this problem and such characterization is testable in polynomial time. □

## 3.2 Limit Cycle Nonexistence Problem

In this section, we study the complexity of deciding when there exists an update schedule that yields only fixed points as attractors for a given Boolean network (LCNE problem). As mentioned in Sect. 1, the LCNE problem has a known polynomial solution in some classes of networks. In particular, in [9] was proved that threshold networks with symmetric weight matrix and without negative loops do not have any limit cycle when it is updated with any sequential update schedule. However, this is not true in general, even in networks with symmetric interaction digraph. Lemma 2 shows a Boolean network with symmetric interaction digraph such that every block-sequential update yields a limit cycle in its dynamics.

Previously, we introduce a Boolean network of $n$ elements with a limit cycle of length $2^n$ which be will useful in some results below.

**Definition 2** Let $H : \{0, 1\}^n \to \{0, 1\}^n$ be the Boolean network defined on the set $V = \{1, \ldots, n\}$ by:

$$\forall i \in \{1, \ldots, n-1\}, \quad H_i(x) = x_i \veebar \bigwedge_{j=i+1}^{n} x_j$$

$$H_n(x) = \neg x_n$$

It is easy to check that network $H$ has one limit cycle of length $2^n$. In fact, if we consider the bijection $g : \{0, 1\}^n \to \{0, \ldots, 2^n - 1\}$ where every $x \in \{0, 1\}^n$ is the binary representation of $g(x)$, $H(x) = g^{-1}(g(x_1, \ldots, x_n) + 1 \pmod{2^n})$ (see Fig. 5).

Ir order to obtain a Boolean network with a limit cycle of length $2^n$ for any given $n$ and symmetric interaction digraph we slightly modify the network $H$ as shown in the following lemma.

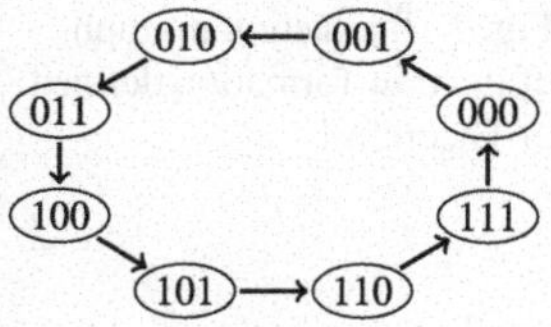


**Fig. 5** Dynamical behavior of $H$ with $n = 3$

**Lemma 2** *Let* $\tilde{f} : \{0,1\}^{n+1} \to \{0,1\}^{n+1}$ *be such that:*

$$\tilde{f}_{v_0}(x) = \bigwedge_{j=0}^{n} x_{v_j},$$

$$\forall i \in \{1, \dots, n-1\}, \quad \tilde{f}_{v_i}(x) = x_{v_i} \veebar \left( \bigwedge_{j=0}^{i-1} x_{v_j} \vee \bigwedge_{j=i+1}^{n} x_{v_j} \right),$$

$$\tilde{f}_{v_n}(x) = \neg x_{v_n} \vee \bigwedge_{j=0}^{n} x_{v_j}.$$

*Then,* $\tilde{f}$ *is a Boolean network with symmetric interaction digraph and such that updated under the schedule* $\tilde{s} = \{v_0\}\{v_1, \dots, v_n\}$ *has no fixed points and only a limit cycle of length* $2^n$*, where* $v_0$ *is frozen at value 0.*

***Proof*** An example of the dynamical behavior of $\tilde{f}$, with $n = 3$, updated under $\tilde{s}$ can be observed in Fig. 6.

It is easy to see that if we consider the bijection $g : \{0,1\}^n \to \{0, \dots, 2^n - 1\}$ such that every $x \in \{0,1\}^n$ is the binary representation of $g(x)$, then for every $(x_1, \dots, x_n) \in \{0,1\}^n$, $\tilde{f}^{\tilde{s}}(0, x_1, \dots, x_n) = \big(0, g^{-1}(g(x_1, \dots, x_n) + 1 \pmod{2^n})\big)$. This induces a cycle of length $2^n$ with the node $v_0$ frozen at value 0.

On the other hand, if $x_0 = 1$ and $(x_1, \dots, x_n) \neq (1, \dots, 1) = \mathbf{1}$, then $\tilde{f}^{\tilde{s}}(1, x_1 \dots, x_n) = (0, g^{-1}(g(x_1, \dots, x_n) + 1 \pmod{2^n}))$. Besides, $\tilde{f}^{\tilde{s}}(1, \mathbf{1}) = (1, 0, \dots, 0, 1)$ and $\tilde{f}^{\tilde{s}}(1, 0, \dots, 0, 1) = (0, g^{-1}(g(0, \dots, 1) + 1 \pmod{2^n}))$. Hence, any initial state reaches the limit cycle mentioned above.

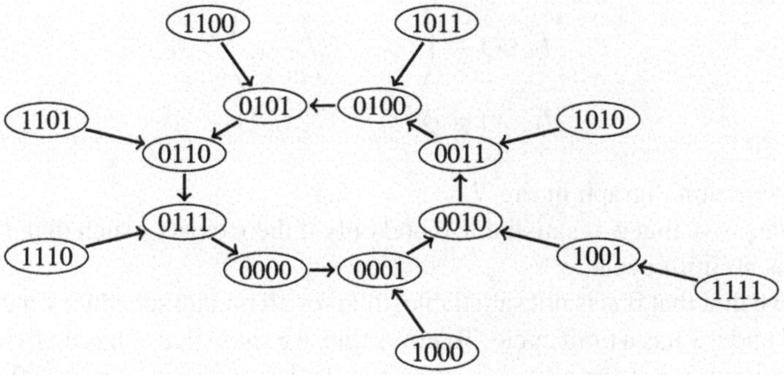


**Fig. 6** Dynamics of $\tilde{f}^{\tilde{s}}$ with $n = 3$

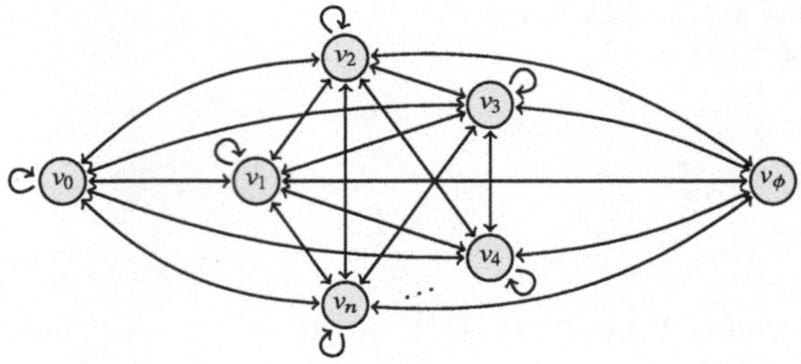


**Fig. 7** Interaction digraph of the transformation defined in Theorem 4

Finally, it is easy to check that the interaction digraph of $\tilde{f}$ is complete and therefore symmetrical. □

Note that since the network $\tilde{f}$ defined in Lemma 2 has no fixed points, it cycles under any update schedule and with the scheme $\tilde{s}$ has a limit cycle containing all configurations in the set $\{0, 1\}^n$. These are important properties used to prove the following theorem.

**Theorem 4** LCNE *is NP-hard in symmetric Boolean networks*

***Proof*** We show that SAT $\leqslant_p$ LCNE.

First, by using the network $\tilde{f}$ and the update schedule $\tilde{s}$ as defined in Lemma 2, we construct a symmetric Boolean network $f$ such that if some state satisfies $\phi$ then it evolves to a fixed point, and if $\phi$ is not satisfiable then it has only limit cycles under any update schedule.

Now, given $\phi$ a Boolean formula in variables $w_1, \ldots, w_n$ we define the set $V = \{v_0, v_1, \ldots, v_n, v_\phi\}$ and the function $f : \{0, 1\}^{n+2} \to \{0, 1\}^{n+2}$ as follows:

$$f_{v_0}(x) = \bigwedge_{j=0}^{n} x_{v_j},$$

$$\forall i \in \{1, \ldots, n-1\}, \quad f_{v_i}(x) = x_{v_i} \veebar \left( \left( \bigwedge_{j=0}^{i-1} x_{v_j} \vee \bigwedge_{j=i+1}^{n} x_{v_j} \right) \wedge \neg x_{v_\phi} \right),$$

$$f_{v_n}(x) = \left( \neg x_{v_n} \vee \bigwedge_{j=0}^{n} x_{v_j} \right) \veebar x_{v_\phi},$$

$$f_{v_\phi}(x) = \phi\left(x_{v_1}, \ldots, x_{v_n}\right).$$

See the interaction digraph in Fig. 7.

Next, we prove that $\phi$ is satisfiable if and only if there exists $s$ such that $f$ updated under $s$ has no limit cycle.

We prove first that if $\phi$ is not satisfiable, then for all update schedule $s$ the network $f$ updated under $s$ has a limit cycle. To prove this, we show that $f$ has no fixed points.

Let us suppose that for all $(w_1, \ldots, w_n) \in \{0, 1\}^n$, $\phi(w_1, \ldots, w_n) = 0$, then the dynamical behavior of $f$ with update schedule $s^* = \{v_\phi, v_0\}\{v_1, \ldots, v_n\}$ is given by

**Table 4** Transitions of function $f^{s^*}$ in Theorem 4

| $(x_{v0}, y, x_{v_\phi})$ | $y \in \{0,1\}^n$ | $f^{s^*}(x_{v0}, y, x_{v_\phi})$ |
|---|---|---|
| $(x_{v0}, y, x_{v_\phi})$ | $y \in S_\phi \wedge (x_{v0}, y) \neq (1, 1 \ldots, 1))$ | $(0, y, 1)$ |
| $(x_{v0}, y, x_{v_\phi})$ | $y \in S_\phi \wedge (x_{v0}, y) = (1, 1 \ldots, 1))$ | $(1, 1, \ldots, 1, 0, 1)$ |
| $(x_{v0}, y, x_{v_\phi})$ | $y \notin S_\phi$ | $(f_c^{s_c}(x_{v0}, y), 0)$ |

$\tilde{f}^{\tilde{s}}$ and $v_\phi$ frozen at 0, i.e. $f^{s^*}(x_{v_0}, \ldots, x_{v_n}, x_{v_\phi}) = (\tilde{f}^{\tilde{s}}(x_{v_0}, \ldots, x_{v_n}), 0)$. Therefore, by Lemma 2 $f^{s^*}$ has no fixed point. Thus, for any update schedule $s$ we have that $\mathrm{LC}(f, s) \neq \emptyset$.

Now we prove that if $\phi$ is satisfiable then there exists $s$ such that $\mathrm{LC}(f, s) = \emptyset$. Let us suppose that $S_\phi = \{w \in \{0,1\}^n : \phi(w) = 1\} \neq \emptyset$ and let be $s^* = \{v_0, v_\phi\}$ $\{v_1, \ldots, v_n\}$, we show that $\mathrm{LC}(f, s^*) = \emptyset$.

First, observe that the dynamical behavior of $f$ updated under $s^*$ is described by the transitions of $f^{s^*}$ shown in Table 4. From it, we can note that all $y \in S_\phi$, $(0, y, 1)$ is a fixed point of $f$ and $f^{s^*}(0, y, 0) = (0, y, 1)$. In the case $(x_{v_0}, y, x_{v_\phi})$ when $y \notin S_\phi$ the network $f$ with $s^*$ behaves as $\tilde{f}^{\tilde{s}}$, that is $f^{s^*}(x_{v_0}, y, x_{v_\phi}) = (\tilde{f}^{\tilde{s}}(x_{v_0}, y), 0)$. Since $\tilde{f}^{\tilde{s}}$ reaches every state $(0, z) \in \{0,1\}^{n+1}$ at some time, $f^{s^*}$ will reach some state $(0, z, 0)$, where $z \in S_\phi$, and therefore it will reach the fixed point $(0, z, 1)$. Of this way, we observe that from every state in $\{0,1\}^{n+2}$ the network $f$ updated under $s^*$ will reach a fixed point $(0, y, 1)$, $y \in S_\phi$. Hence, $\mathrm{LC}(f, s^*) = \emptyset$. □

Next, we prove that LCNE is coNP-hard in AND-OR networks. Previously, we introduce an AND-OR network that always has at least one limit cycle with any update schedule.

**Lemma 3** *There exists an AND-OR function $f$ such that for every update schedule $s$, $\mathrm{LC}(f, s) \neq \emptyset$.*

***Proof*** Let us consider the AND-OR function $\bar{f} : \{0,1\}^{11} \to \{0,1\}^{11}$ as defined in Fig. 8. We note that vertices 1 and 2 will remain constant trough any trajectory. Therefore, we can decompose its dynamics into the dynamical behaviors of the four AND-OR functions associated of $\bar{f}$: $\bar{f}^{00}$, $\bar{f}^{10}$, $\bar{f}^{01}$, $\bar{f}^{11}$, where $\bar{f}^{ij}$ is the value of $\bar{f}$ with fixed values: $x_{r_1} = i$ and $x_{r_2} = j$.

In Table 5 a and b are shown the four AND-OR functions associated to $\bar{f}$. The corresponding interaction digraphs of $\bar{f}^{10}$ and $\bar{f}^{01}$ are shown in Fig. 9a and b, respectively.

Now, let us suppose that there exists an update schedule that generates only fixed points as attractors for $\bar{f}^{00}$, $\bar{f}^{11}$, $\bar{f}^{01}$ and $\bar{f}^{10}$ simultaneously. Since $\bar{f}^{00}$ and $\bar{f}^{11}$ has no limit cycles under any update schedule, we just need to focus only on $\bar{f}^{01}$ and $\bar{f}^{10}$.

If we analyze $\bar{f}^{10}$ (case $\bar{f}^{01}$ is analogous), we note that the only update schedules $s$ that have no limit cycles are those in which there exists only one arc $(r_i, r_j)$ with $r_i, r_j \in \{r_3, r_4, r_5, r_6, r_7, r_8\}$ such that $s(r_i) \geqslant s(r_j)$. Without lost of generality, we can suppose that such arc is $(r_3, r_6)$, i.e. $s(r_3) \geqslant s(r_6)$. On the other hand,

$$\begin{aligned}
\bar{f}_{r_1}(x) &= x_{r_1}\\
\bar{f}_{r_2}(x) &= x_{r_2}\\
\bar{f}_{r_3}(x) &= x_{r_8} \vee x_{r_{10}}\\
\bar{f}_{r_4}(x) &= x_{r_6} \vee x_{r_{11}}\\
\bar{f}_{r_5}(x) &= x_{r_7} \vee x_{r_9}\\
\bar{f}_{r_6}(x) &= x_{r_3} \wedge x_{r_1}\\
\bar{f}_{r_7}(x) &= x_{r_4} \wedge x_{r_1}\\
\bar{f}_{r_8}(x) &= x_{r_5} \wedge x_{r_1}\\
\bar{f}_{r_9}(x) &= x_{r_3} \wedge x_{r_2}\\
\bar{f}_{r_{10}}(x) &= x_{r_4} \wedge x_{r_2}\\
\bar{f}_{r_{11}}(x) &= x_{r_5} \wedge x_{r_2}
\end{aligned}$$

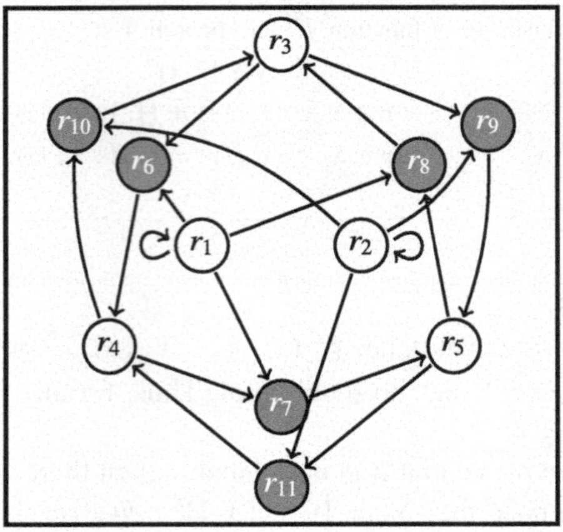


**Fig. 8** AND-OR function $f$ and its interaction digraph

**Table 5** AND-OR functions described in Lemma 3

| | $\bar{f}^{00}$ | $\bar{f}^{11}$ | $\bar{f}^{01}$ | $\bar{f}^{10}$ |
|---|---|---|---|---|
| (a) | $\bar{f}_{r_1}(x)=0$<br>$\bar{f}_{r_2}(x)=0$<br>$\bar{f}_{r_3}(x)=x_{r_8}\vee x_{r_{10}}$<br>$\bar{f}_{r_4}(x)=x_{r_6}\vee x_{r_{11}}$<br>$\bar{f}_{r_5}(x)=x_{r_7}\vee x_{r_9}$<br>$\bar{f}_{r_6}(x)=0$<br>$\bar{f}_{r_7}(x)=0$<br>$\bar{f}_{r_8}(x)=0$<br>$\bar{f}_{r_9}(x)=0$<br>$\bar{f}_{r_{10}}(x)=0$<br>$\bar{f}_{r_{11}}(x)=0$ | $\bar{f}_{r_1}(x)=1$<br>$\bar{f}_{r_2}(x)=1$<br>$\bar{f}_{r_3}(x)=x_{r_8}\vee x_{r_{10}}$<br>$\bar{f}_{r_4}(x)=x_{r_6}\vee x_{r_{11}}$<br>$\bar{f}_{r_5}(x)=x_{r_7}\vee x_{r_9}$<br>$\bar{f}_{r_6}(x)=x_{r_3}$<br>$\bar{f}_{r_7}(x)=x_{r_4}$<br>$\bar{f}_{r_8}(x)=x_{r_5}$<br>$\bar{f}_{r_9}(x)=x_{r_3}$<br>$\bar{f}_{r_{10}}(x)=x_{r_4}$<br>$\bar{f}_{r_{11}}(x)=x_{r_5}$ | $\bar{f}_{r_1}(x)=0$<br>$\bar{f}_{r_2}(x)=1$<br>$\bar{f}_{r_3}(x)=x_{r_8}\vee x_{r_{10}}$<br>$\bar{f}_{r_4}(x)=x_{r_6}\vee x_{r_{11}}$<br>$\bar{f}_{r_5}(x)=x_{r_7}\vee x_{r_9}$<br>$\bar{f}_{r_6}(x)=0$<br>$\bar{f}_{r_7}(x)=0$<br>$\bar{f}_{r_8}(x)=0$<br>$\bar{f}_{r_9}(x)=x_{r_3}$<br>$\bar{f}_{r_{10}}(x)=x_{r_4}$<br>$\bar{f}_{r_{11}}(x)=x_{r_5}$ | $\bar{f}_{r_1}(x)=1$<br>$\bar{f}_{r_2}(x)=0$<br>$\bar{f}_{r_3}(x)=x_{r_8}\vee x_{r_{10}}$<br>$\bar{f}_{r_4}(x)=x_{r_6}\vee x_{r_{11}}$<br>$\bar{f}_{r_5}(x)=x_{r_7}\vee x_{r_9}$<br>$\bar{f}_{r_6}(x)=x_{r_3}$<br>$\bar{f}_{r_7}(x)=x_{r_4}$<br>$\bar{f}_{r_8}(x)=x_{r_5}$<br>$\bar{f}_{r_9}(x)=0$<br>$\bar{f}_{r_{10}}(x)=0$<br>$\bar{f}_{r_{11}}(x)=0$ |
| (b) | $\bar{f}_{r_1}(x)=0$<br>$\bar{f}_{r_2}(x)=0$<br>$\bar{f}_{r_3}(x)=0$<br>$\bar{f}_{r_4}(x)=0$<br>$\bar{f}_{r_5}(x)=0$<br>$\bar{f}_{r_6}(x)=0$<br>$\bar{f}_{r_7}(x)=0$<br>$\bar{f}_{r_8}(x)=0$<br>$\bar{f}_{r_9}(x)=0$<br>$\bar{f}_{r_{10}}(x)=0$<br>$\bar{f}_{r_{11}}(x)=0$ | $\bar{f}_{r_1}(x)=1$<br>$\bar{f}_{r_2}(x)=1$<br>$\bar{f}_{r_3}(x)=x_{r_8}\vee x_{r_{10}}$<br>$\bar{f}_{r_4}(x)=x_{r_6}\vee x_{r_{11}}$<br>$\bar{f}_{r_5}(x)=x_{r_7}\vee x_{r_9}$<br>$\bar{f}_{r_6}(x)=x_{r_3}$<br>$\bar{f}_{r_7}(x)=x_{r_4}$<br>$\bar{f}_{r_8}(x)=x_{r_5}$<br>$\bar{f}_{r_9}(x)=x_{r_3}$<br>$\bar{f}_{r_{10}}(x)=x_{r_4}$<br>$\bar{f}_{r_{11}}(x)=x_{r_5}$ | $\bar{f}_{r_1}(x)=0$<br>$\bar{f}_{r_2}(x)=1$<br>$\bar{f}_{r_3}(x)=x_{r_{10}}$<br>$\bar{f}_{r_4}(x)=x_{r_{11}}$<br>$\bar{f}_{r_5}(x)=x_{r_9}$<br>$\bar{f}_{r_6}(x)=0$<br>$\bar{f}_{r_7}(x)=0$<br>$\bar{f}_{r_8}(x)=0$<br>$\bar{f}_{r_9}(x)=x_{r_3}$<br>$\bar{f}_{r_{10}}(x)=x_{r_4}$<br>$\bar{f}_{r_{11}}(x)=x_{r_5}$ | $\bar{f}_{r_1}(x)=1$<br>$\bar{f}_{r_2}(x)=0$<br>$\bar{f}_{r_3}(x)=x_{r_8}$<br>$\bar{f}_{r_4}(x)=x_{r_6}$<br>$\bar{f}_{r_5}(x)=x_{r_7}$<br>$\bar{f}_{r_6}(x)=x_{r_3}$<br>$\bar{f}_{r_7}(x)=x_{r_4}$<br>$\bar{f}_{r_8}(x)=x_{r_5}$<br>$\bar{f}_{r_9}(x)=0$<br>$\bar{f}_{r_{10}}(x)=0$<br>$\bar{f}_{r_{11}}(x)=0$ |

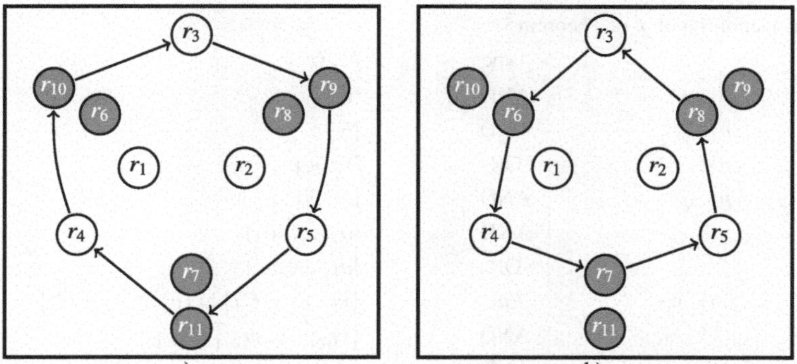


**Fig. 9** Interaction digraphs of **a** $\bar{f}^{01}$ and **b** $\bar{f}^{10}$ described in Lemma 3

because $s(r_4) < s(r_7) \wedge s(r_7) < s(r_5)$, then $s(r_5) \geqslant s(r_{11}) \vee s(r_{11}) \geqslant s(r_4)$. Otherwise, $s(r_4) < s(r_7) < s(r_5) < s(r_{11}) < s(r_4)$, which is a contradiction. Analogously, we deduce that $s(r_3) \geqslant s(r_9) \vee s(r_9) \geqslant s(r_5)$. Hence, $G^{\bar{f}^{01}}$ will have necessarily two different arcs $(a, b)$ and $(c, d)$ such that $s(a) \geqslant s(b)$ and $s(c) \geqslant s(d)$, and thus $G^{\bar{f}^{01}}$ induce a limit cycle for $(\bar{f}, s)$. Therefore, there is no update schedule $s$ such that $\mathrm{LC}(\bar{f}, s) = \emptyset$. □

**Theorem 5** LCNE *is coNP-hard in AND-OR networks.*

***Proof*** We prove that SAT $\leqslant_P$ LCNE in AND-OR networks. Let $\phi$ be a Boolean formula in the variables $\{w_1, \ldots, w_n\}$. We define $f$ as exhibited in Table 6, where the network induced by the vertex set $\{r_1, \ldots, r_{11}\}$ is like $\bar{f}$ defined in Lemma 3.

Notice that for any update schedule and for any limit cycle of the network the states of the vertices $v_i, \bar{v}_i, a_i, o_i, A, O, C_j, v_\phi, r'_1, r'_2, r_1$ and $r_2$ are frozen nodes. Notice that, if the component $A$ is fixed in $x_A = 0$, then $x_{r'_1} = x_{r'_2} = 0$, and $x_{r_1} = x_{r_2} = x_O$. In this case, the vertices $r_i$ converge to a fixed point. The case where $x_{v_\phi} = 0$ is analogous, and if $x_O = 1$ then $x_{r_1} = x_{r_2} = 1$, and in this case the vertices $r_i$ also converge to a fixed point. Therefore, the only way to get a limit cycle in this network is that the frozen vertices satisfy:

1. $\forall i \in \{1, \ldots, n\},\ x_{v_i} = \neg x_{\bar{v}_i}$,
2. $\phi(x_{v_1}, \ldots, x_{v_n}) = 1$,
3. $\forall i \in \{1, \ldots, n\},\ x_{a_i} = 0$,
4. $\forall i \in \{1, \ldots, n\},\ x_{o_i} = 1$,
5. $\forall j \in \{1, \ldots, m\},\ x_{C_j} = 1$,
6. $x_A = 1$, $x_O = 0$ and $x_{v_\phi} = 1$.

In this way the components $r_i$ have the dynamical behavior of the network $\bar{f}$ defined in Lemma 3.

Let us suppose $\phi$ is a satisfiable conjunctive normal form, then there exists $w \in \{0, 1\}^n$ such that if $\phi(w_1, \ldots, w_n) = 1$, in this way we hold the conditions above. Since $\mathrm{LC}(f, s) \neq \emptyset$ for all $s$, therefore $f \notin$ LCNE.

**Table 6** Definition of $f$ of Theorem 5

| $v \in V$ | Type | $N_f^-(v)$ |
|---|---|---|
| $\forall i \in \{1, \dots, n\}, v_i$ | AND | $\{v_i\}$ |
| $\forall i \in \{1, \dots, n\}, \bar{v}_i$ | AND | $\{\bar{v}_i\}$ |
| $\forall i \in \{1, \dots, n\}, o_i$ | OR | $\{v_i, \bar{v}_i\}$ |
| $\forall i \in \{1, \dots, n\}, a_i$ | AND | $\{v_i, \bar{v}_i\}$ |
| $A$ | AND | $\{o_1, \dots, o_n\}$ |
| $O$ | OR | $\{a_1, \dots, a_n\}$ |
| $\forall j \in \{1, \dots, m\}, v_{C_j}$ | OR | $\{v_i : w_i \in C_j\} \cup \{\bar{v}_i : \neg w_i \in C_j\}$ |
| $v_\phi$ | AND | $\{v_{C_1}, \dots, v_{C_m}\}$ |
| $r_1'$ | AND | $\{A, v_\phi, r_1'\}$ |
| $r_2'$ | AND | $\{A, v_\phi, r_2'\}$ |
| $r_1$ | OR | $\{O, r_1'\}$ |
| $r_2$ | OR | $\{O, r_2'\}$ |
| $\forall i \in \{3, 4, 5\}, r_i$ | OR | $N_f^-(r_i)$ |
| $\forall i \in \{6, \dots, 11\}, r_i$ | AND | $N_f^-(r_i)$ |

If $\phi$ is not satisfiable then the conditions will never hold, thus the network $f$ will have only fixed points with every update schedule, therefore there exists an update schedule such that $\mathrm{LC}(f, s) = \emptyset$, then $f \in \mathrm{LCNE}$. □

## 4 Conclusions

We have studied the algorithmic complexity of two problems about the existence of some block-sequential update schedule for a given Boolean network which yields a limit cycle (LCE) or does not yield any limit cycle (LCNE). We proved that both problems are NP-hard even in networks having a symmetric interaction digraph. Besides, we prove that LCE is also NP-hard in AND-OR networks, and that it is polynomial in symmetric AND-OR networks, because in such networks the existence of a limit cycle depends on a structural property of the interaction digraph, which can be verified in polynomial time. On the other hand, we proved that LCNE is coNP-hard in AND-OR networks. However, it is known that LCNE is polynomial in symmetric AND-OR networks [9].

In conclusion, although the family of Boolean networks to be studied is restricted, considering only those either with a symmetric interaction digraph or with local activation functions of type OR or AND, both problems studied (LCE and LCNE) do not seem to be possible to solve them efficiently unless NP = P. However, both conditions are sufficient to efficiently determine the solution to both problems.

**Acknowledgements** This work was supported by Centro de Modelamiento Matemático (CMM), ACE2100010 and FB210005, BASAL funds for center of excellence from ANID-Chile.

## References

1. Aracena J, Gómez L, Salinas L (2013) Limit cycles and update digraphs in Boolean networks. Discret Appl Math 161:1–2
2. Aracena J, Goles E, Moreira A, Salinas L (2009) On the robustness of update schedules in Boolean networks. Biosystems 97:1–8
3. Bridoux F, Gaze-Maillot C, Perrot K, Sené S (2020) Complexity of limit-cycle problems in boolean networks (2020). arXiv:2001.07391
4. Davidich MI, Bornholdt S (2008) Boolean network model predicts cell cycle sequence of fission yeast. PloS one 3(2):e1672
5. Demongeot J, Elena A, Sené S (2008) Robustness in regulatory networks: a multi-disciplinary approach. Acta Biotheor 56(1–2):27–49
6. Fauré A, Naldi A, Chaouiya C, Thieffry D (2006) Dynamical analysis of a generic boolean model for the control of the mammalian cell cycle. Bioinformatics 22(14):e124–e131
7. Goles E, Hernández G (2000) Dynamical behavior of Kauffman networks with and-or gates. J Biol Syst 8:151–175
8. Goles E, Noual M (2012) Disjunctive networks and update schedules. Adv Appl Math 48:646–662
9. Goles E (1982) Fixed point behavior of threshold functions on a finite set. SIAM J Algebr Discrete Methods 3(4):529–531
10. Goles E, Montalva M, Ruz GA (2013) Deconstruction and dynamical robustness of regulatory networks: application to the yeast cell cycle networks. Bull Math Biol 75(6):939–966
11. Goles E, Montealegre P (2014) Computational complexity of threshold automata networks under different updating schemes. Theoret Comput Sci 559:3–19
12. Goles E, Olivos J (1980) Periodic behaviour of generalized threshold functions. Discret Math 30(2):187–189
13. Gómez L (2015) Dynamics of discrete networks with deterministic updates schedules. Application to genetic regulatory networks. PhD thesis in mathematical engineering, Universidad de Concepción, Concepción, Chile
14. Green DG, Leishman TG, Sadedin S (2007) The emergence of social consensus in Boolean networks. In: IEEE symposium on artificial life, 2007. ALIFE'07. IEEE, pp 402–408
15. Just W (2006) The steady state system problem is np-hard even for monotone quadratic boolean dynamical systems. Submitted to Annals of Combinatorics
16. Kauffman SA (1969) Metabolic stability and epigenesis in randomly connected nets. J Theor Biol 22:437–467
17. Kauffman SA (1993) The origins of order: self-organization and selection in evolution. Oxford University Press, New York
18. Macauley M, Mortveit HS (2009) Cycle equivalence of graph dynamical systems. Nonlinearity 22(2):421
19. Mortveit HS (2012) Limit cycle structure for block-sequential threshold systems. In: Cellular automata. Springer, pp 672–678
20. Porat S (1989) Stability and looping in connectionist models with asymmetric weights. Biol Cybern 60(5):335–344
21. Robert F (1986) Discrete iterations: a metric study. Springer, Berlin
22. Robert F (1995) Les systemes dynamiques discrets, vol 19. Springer Science & Business Media, Berlin
23. Ruz GA, Timmermann T, Barrera J, Goles E (2014) Neutral space analysis for a boolean network model of the fission yeast cell cycle network. Biol Res 47(1):64

24. Thomas R (1973) Boolean formalization of genetic control circuits. J Theor Biol 42:563–585
25. Thomas R (1980) On the relation between the logical structure of systems and their ability to generate multiple steady states or sustained oscillations. Springer Ser Synergetics 9:180–193
26. Tocci R, Widmer N (2001) Digital systems: principles and applications, 7th edn. Prentice-Hall, Hoboken

# A Survey on the Stability of (Extended) Linear Sand Pile Model

**Thi Ha Duong Phan**

*dedicated to the 70th Anniversary of Eric Goles.*

**Abstract** We give a survey of our works on the natural extensions of the well-known Sand Pile Model. These extensions consist of adding outside grains on random columns, allowing sand grains to move from left to right and from right to left, considering cycle graphs and the extension to infinity. We study the reachable configurations and fixed points of each model and show how to compute the set of fixed points, the time of convergence and the distribution of fixed points.

## 1 Introduction

The Sand Piles Model (SPM) was introduced in 1987 by Bak, Tang and Wiesenfeld as a sample model of the Self organized criticality (SOC) phenomena [1]. The authors simulated the behavior of a sand pile which builds up when sand is dropped on a line. A configuration is modeled as a sequence of columns consisting of cubic sand grains such that the height of columns is decreasing from left to right. In this model, a sand grain can fall down from a column to its right neighbors if the difference of height of the two columns is at least two. This model is investigated in many works in physics, combinatorics and computer science [8, 15, 17, 23, 31].

Independtly, a similar model—the Chip Firing Game—was defined by Björner, Lovász and Shor in 1991 [2, 3]. Formally, a $CFG$ is a model consisting of a directed (or undirected) multi-graph $G$ (also called *support graph*), the set of *configurations* on $G$ and an *evolution rule* on this set.

T. H. D. Phan (✉)
Institute of Mathematics - Vietnam Academy of Science and Technology, 18 Hoang Quoc Viet, Hanoi, Vietnam
e-mail: phanhaduong@math.ac.vn

 
A. Adamatzky (ed.), *Automata and Complexity*, Emergence, Complexity and Computation 42, https://doi.org/10.1007/978-3-030-92551-2_16

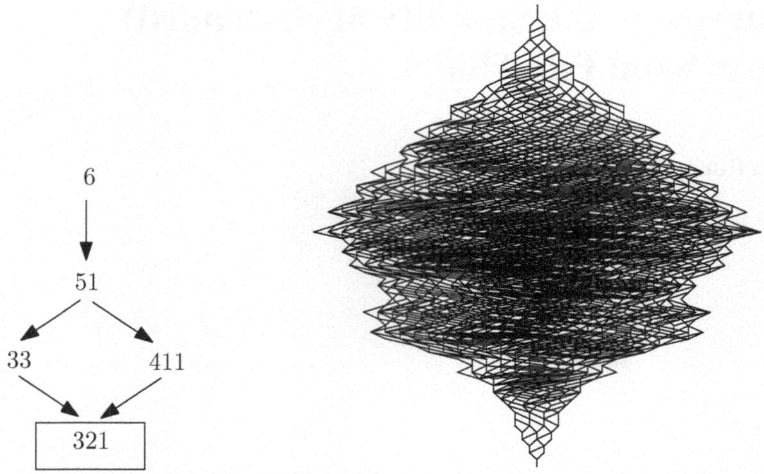


**Fig. 1** Example of configuration spaces of Sand piles model: $SPM(6)$ and $SPM(30)$

In this paper, as the support graph has a linear structure, for convenient, we call the model (Linear) Sand Pile Model.

This model can be defined mathematically as follows (Fig. 1).

**Definition 1** Sand pile model, with respect to a positive integer $n$, denoted by $SPM(n)$, is a model where configurations are partitions of $n$ such that:

- Initial configuration: $(\underline{n})$ (that means the position 0 has value $n$, or equivalently $a_0 = n$ and $a_i = 0$ for all $i \neq 0$).
- Local right vertical rule $\mathcal{R}$: for all $i \geq 0$, $(\ldots, a_i, a_{i+1}, \ldots) \to (\ldots, a_i - 1, a_{i+1} + 1, \ldots)$ if $a_i \geq a_{i+1} + 2$.
- Global rule: at each step, we apply once the $\mathcal{R}$ rule.

On the other hand, in 1973, Brylawski described a model to generate all partitions of a given arbitrary integer $n$ [4], this model can be considered as an extension of $SPM$ because it is nothing but the $SPM$ with an adding horizontal rule which allow grain to slide along a plateau.

**Definition 2** Brylawski's model, with respect to a positive integer $n$, denoted by $L_B(n)$, is a model where configurations are partitions of $n$ such that:

- Initial configuration: $(\underline{n})$.
- Local right vertical rule $\mathcal{R}$: $(\ldots, a_i, a_{i+1}, \ldots) \to (\ldots, a_i - 1, a_{i+1} + 1, \ldots)$ if $a_i \geq a_{i+1} + 2$.
- Local right horizontal rule $\mathcal{H}$: $(\ldots, p + 1, p, \ldots, p, p - 1, \ldots) \to (\ldots, p, p, \ldots, p, p, \ldots)$.
- Global rule: at each step, we apply once the $\mathcal{R}$ rule, or once the $\mathcal{H}$ rule.

From these first steps, $SPM$ and Brylawski's model have been extended in many different directions [6, 9–12, 15–17]. In particular, the problem is derived from real-life questions and focusing on the following issues.

- Reachability problem: study the necessary and sufficient conditions for a configuration to be reachable from another one by applying a sequence of transition rules. In most cases, based on the reachable relation, one can define an order relation.
- Configuration space: the set of all reachable configurations is called space configuration. This set equipped with the order relation can have many interesting structures.
- Stability problem: Determine if the model proceeds to stable configurations (called also "fixed points") or runs non-stop. This property is related to the structure of the configuration space.
- Convergence problem: Almost all models are non-deterministic then if we consider all the cases, it is possible that different sequences of transitions proceed to various fixed points. The uniqueness of the fixed point is usually proven when the configuration space has a lattice structure. Otherwise, the model can have many fixed points.
- Characterization of fixed points: If there is several fixed points (and even if it is unique), then it is important to find out their characterisation.
- Stabilization time of the model. If the model diverges, it is clear that the time to reach different fixed points are variety. But even if the model converges, there are different convergent times which depend on the local behavious of the model. Hence evaluate the upper and lower bounds for convergence time is also an object of study.

We define the order relation (if it exists) on the space configuration of a model by: a configuration $a$ is greater than a configuration $b$ if $b$ is reachable from $a$ by applying a sequence of transition rules (which is the inverse of the definition in Chap. 2).

Both configuration spaces of $SPM(n)$ and $L_B(n)$ have an order relation, moreover they have a lattice structure.

The following results were established for $SPM$ [15, 17, 18].

- Reachability of $SPM$ [15]. A partition of $n$ is a reachable configuration of $SPM(n)$ if and only if it does not contains subsequences of the two following forms: $(p, p, p)$ or $(p+1, p+1, p, p-1, \dots, p-q+1, p-q, p-q)$ (we call this condition $SPM$ *condition*).
- Fixed point of SPM. The model $SPM(n)$ has an unique fixed point, which is $(p, p-1, \dots, q, +1, q, q, q-1, \dots, 2, 1)$ where $p$ and $q$ are uniquely defined by $p(p+1)/2 \leq n < (p+1)(p+2)/2$ and $q = n - p(p+1)/2$ ($q$ can be equal to 0).
- Lattice structure of the configuration space of SPM. The configuration space of $SPM$ equipped with the reachability order is a lattice.
- Length of chains in $SPM$. Every chain between two configurations in $SPM(n)$ has the same length.

And results achieved for $L_B$ ([4, 19]):

- The *exhaustive property* of the model: all partitions are reachable. This property is almost always true for all natural extensions of the model.
- Therefore, the unique fixed point of the model is easy to determine, it is $(1, 1, \ldots, 1)$.
- Lattice structure of the configuration space of $L_B(n)$. The reachability order is the dominance order, that mean $b$ is reachable from $a$ by $L_B$ transition if and only if $b$ is smaller than $a$ by dominance ordering: for all $1 \leq i$, $\sum_{j=1}^{i} b_j \leq \sum_{j=1}^{i} a_j$.
- Length of chains in $L_B$. The property of the $SPM$ that all chains between two given configurations have the same length is no longer true for $L_B$ model. Therefore the problem of finding longest and shortest chains is particularly interesting. While it is pretty simple to find shortest chains, finding longest chains is quite complicated and requires a technical proof which is based on two energies, these energies are defined for the vertical and the horizontal rules respectively.
  Shortest chains. A shortest chain in $L_B$ can be constructed as follows: applying the V-transition at the first position to obtain the partition $(n-1, 1)$. Then apply alternatively V-transition at the first position and H-transition at the second position $n-3$ times to obtain the partition $(2, 1, 1, \ldots, 1)$. At this state, apply the H-transition at the first position, and obtain the fixed point $(1, 1, \ldots, 1)$. This chain has length $2n-4$.
  Longest chains: in [19], the authors proved that longest chains are chains of a sequence of V-transitions followed by a sequence of H-transitions. And their length is $\theta(n^{3/2})$.

These two models were extended by different approaches. First, by parameterizing the horizontal rule, we defined the Ice pile model [17]. Then by considering that $m$ sand grains can fall down in the same time (for a given $m$), we introduced the model $CFG(n, m)$ [18]. Independently, a similar model of $CFG(n, m)$ was studied by the physicist Kadanoff [21]. The enumeration of the number of reachable configurations of these models was widely studied [13, 28]. On the other hand, a parallel version of $SPM$ was studied in [10, 11]. This model is deterministic and converges to the fixed point of the classical SPM. For this parallel model, the reachability problem was studied by mean of language theory.

In this paper, we give a survey of our works, in collaborations with Enrico Formenti, Kevin Perrot, Pham Van Trung and Tran Thi Thu Huong, on the following natural extensions of two models $SPM$ and $L_B$.

- We investigate the study of all stable configurations when outside grains are added on random columns [30].
- A very natural extension of $SPM$ is the symmetric $SPM$ on which sand grains can move from left to right and from right to left. Here, the model lost its properties of existence and uniqueness of the fixed point, therefore this raises the question of how to compute the set of fixed points, the time of convergence and the distribution of fixed points [14, 27, 29].
- We explore the Sand Pile Model and Chip Firing Game on cycle graphs. We study the reachable configurations and fixed points of each model and the similarities between these models [5].

- Finally, we investigate the extension to infinity of Brylawski model, this model gives a method to generate partitions of all integers by rule of a dynamical model. Moreover, this model has a recursive structure from which one can deduce some interesting enumerative formula on partitions [24–26].

## 2 The Stability of SPM

In this section, we consider a more extended Sand Piles Model where outside grains are added on random columns. More precisely, each time the model reach a stable configuration, one grain is added to a random column, and the model evolves to reach another stable configuration, and so on. We investigate the study of all such stable configurations.

First, we give a formal definition of this model and prove that the set of all stable configurations has a lattice structure which is a sub-lattice of the well-known Young lattice. Then we compute explicitly the smallest and greatest times to reach a stable configuration from the initial configuration, and the smallest and greatest times to reach a stable configuration from another stable configuration. These times illustrate the behaviour of the model under outside actions. The key idea of this computation is the introduction of the notion "energy". Indeed, for each configuration, we treat each of its grain by defining the energy of a grain being the greatest number of its possible moves.

### *2.1 Extended Sand Piles Model and Its Stable Configurations*

As well as in almost works of SPM, we represent configurations of this model by integer partitions. So let us first give some preliminary notions:

**Definition 3** (i) A *partition* is an integer sequence $a = (a_1, a_2, \ldots, a_k)$ such that $a_1 \geq a_2 \geq \ldots \geq a_k > 0$ (by convention, $a_j = 0$ for all $j > k$ and $a_0 = \infty$). We call $a_i$ *part* of partition $a$; and $k$ *length* of $a$, and write $l(a) = k$. We say that $a$ is a partition of $n$, or $n$ is the *weight* of $a$, and write $w(a) = n$, if $\sum_{i=1}^{i=k} a_i = n$.

(ii) A *smooth partition* is a partition such that all differences between two consecutive parts are at most 1.

(iv) *Young's lattice* is the lattice of all partitions ordered by containment [32] (i.e. $a \leq b$ if and only if $a_i \geq b_i$ for all $i = 1, 2, \ldots, \min\{l(a), l(b)\}$).

From this definition, one can see that a stable configuration is represented by a smooth partition. So, in the following, we say partition (resp. smooth partition) for configuration (resp. stable configuration).

The Extended Sand Piles Model (ESPM) is a discrete dynamical model where all configurations are partitions and the initial configuration being the partition (0). This model consists of two evolution (or transition) rules:

Fig. 2 First elements of the poset $ESPM$

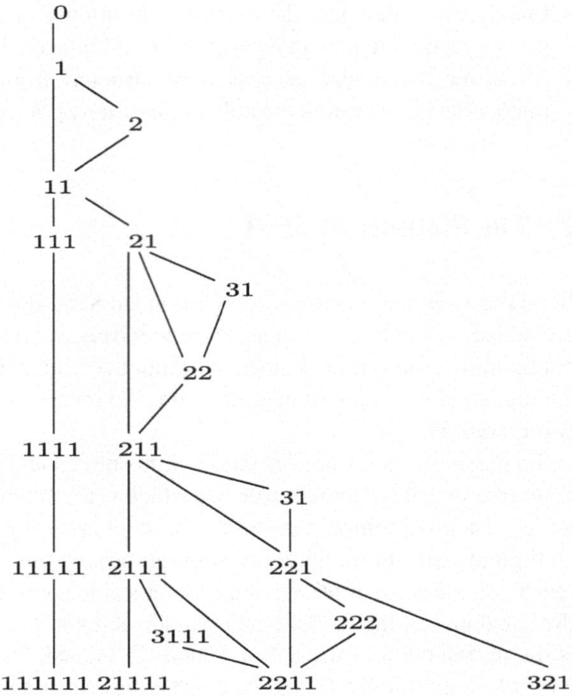


- *Falling rule (inside action)*: one grain on the column $i$ can fall down to the column $i+1$ if the height difference between the column $i$ and the column $i+1$ is greater than or equal to 2.
- *Adding rule (outside action)*: one grain can be added to one column of a smooth partition such that the obtained one is still a partition.

We denote also $ESPM$ the set of all reachable partitions from the initial (0). We call a *chain* in this model a sequence of transitions. By convention, a chain of one element (with no transitions) is of length 0. More particularly, a chain between two smooth partitions is called an *avalanche chain*. Finally, we denote by $a^{\downarrow i}$ the integer sequence obtained from $a$ by increasing part $i$ of $a$ by 1.

Figure 2 shows first elements of $ESPM$. One can see that $ESPM$ does not contain all partitions. However, we prove that this model contains all smooth partitions.

**Proposition 1** *All smooth partitions are reachable from the initial partition.*

In order to study the behaviour of the model under outside actions, we investigate the set of all stable configurations and the relations between them. We denote the induced subposet of all smooth partitions of the poset $ESPM$ by $(SESPM, \leq_S)$. We will describe the nature of order relation in $SESPM$.

First, we analyze the movement of a grain when it is added from outside to a stable configuration. So, let $a = (a_1, \ldots, a_k)$ be a smooth partition. One grain is added on

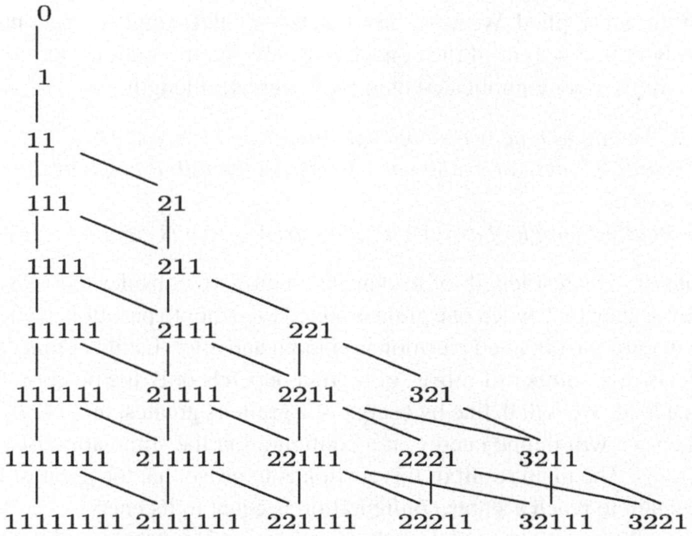


**Fig. 3** First elements of the poset $SESPM$

column $i$ of $a$ with the condition that $a_i < a_{i-1}$. After that, if $a_i = a_{i+1}$, this grain stays at column $i$ and does not move anymore. Otherwise, this grain move to a new position $j > i$ such that $a_i, a_{i+1}, \ldots, a_j$ is a consecutive decreasing integers and that $a_j = a_{j+1}$. Finally, this grain stays at column $j$ and does not move anymore. The obtained configuration $b$ of this sequence of moves of this grain is the same as the configuration obtained by only one move: adding a grain directly on position $j$. Hence, this analyze proves the following result: In the $SESPM$, an element $b$ is an immediate successor of an element $a$ if and only if $b$ can be obtained from $a$ by adding one grain at some column. Figure 3 shows some first elements of the poset $SESPM$. To finish, we discuss about the relation between $SESPM$ and the Young lattice. Due to the characterization of the containment order, we know that the poset $SESPM$ is a suborder of the Young lattice. Futhermore we prove that this relation is in fact a sublattice relation.

**Theorem 1** *The poset $SESPM$ is ordered by containment, moreover it is a sublattice of the Young lattice.*

## 2.2 Avalanche Chains

The purpose of this subsection is to describe the needed time to reach a stable configuration in the Extended Sand Piles Model. We know that there are probably different sequences of evolutions to reach a stable configuration from another stable configuration. Their sizes may be quite different and depend on the columns in which evolution rules are applied. We next show that the smallest length of avalanche chain depends only on the weight of the considered stable configurations. Otherwise, the problem is much more complicated than for the greatest length.

**Theorem 2** *Let a and b be two smooth partitions and $b <_S a$. Then*

(i) *The smallest length of avalanche chain from the initial configuration* (0) *to a is equal to* $w(a)$.

(ii) *The smallest length of avalanche chain from a to b is equal to* $w(b) - w(a)$.

To compute the greatest length of avalanche chains, we consider the movement of grains. We constate that, when one grain is added to a smooth partition, it slides down to a position until the obtained partition is smooth and after that this grain can not be moved. Hence the number of moves of a grain depends only the position (column) where it is added. We will define by energy of a grain its greatest number of possible moves. Then we will define energy of a configuration the summation of energy of all of its grains. The main result of this section is to prove that the greatest length of avalanche chain to reach a stable configuration is equal to its energy.

Let us recall that, in our model, each configuration is represented by a partition, or more precisely, by its Ferrers diagram, where each grain is represented by a case $(i, j)$ where $i$ (resp. $j$) is the column (resp. row) index. So, let us denoted by $F(a)$ the diagram of a partition $a = (a_1, a_2, \ldots, a_k)$, and we write $(i, j) \in F(a)$ for all case $(i, j)$ such that $1 \le i \le k$ and $1 \le j \le a_i$ (see Fig. 4 as an example). We say that $i$ is a *smooth column* of $a$ if $i = 1$ or $a_i = a_{i-1}$ for $i > 1$. Moreover, for a case $(i, j)$, we define *diagonal* $D(i, j)$ the set of all case $(i', j')$ such that $i' + j' = i + j$ and $1 \le j' \le j$ (see Fig. 5). We give now the formal definition and some properties of energy.

**Definition 4** Let $a = (a_1, a_2, \ldots, a_k)$ be a smooth partition.

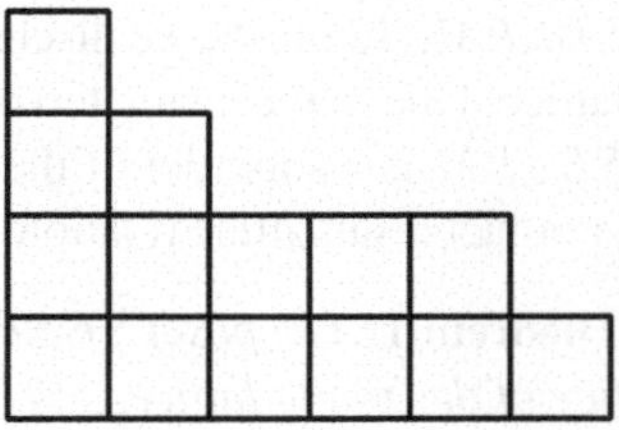

**Fig. 4** The representation of the Ferrer diagram of the partition $a = (4, 3, 2, 2, 2, 1)$

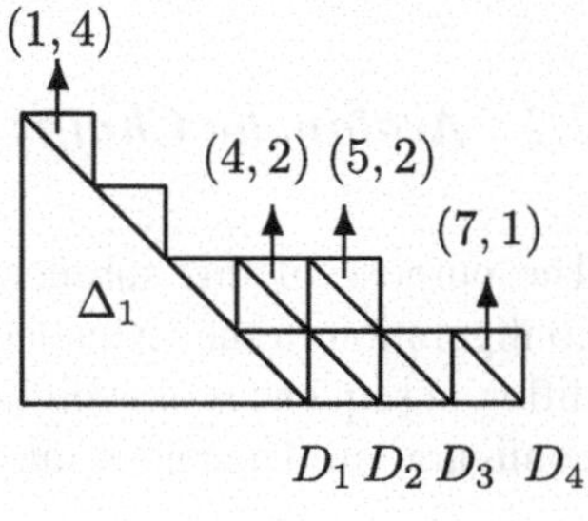


**Fig. 5** Smooth columns (1, 4, 5, 7) and corresponding diagonals $D_1, D_2, D_3, D_4$ of $b = (4, 3, 2, 2, 2, 1, 1)$

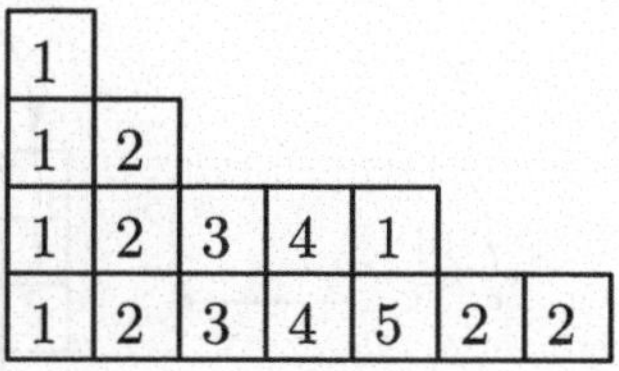


**Fig. 6** Representation of the partition $b = (4, 3, 2, 2, 2, 1, 1)$, the number in each case is the energy of the corresponding grain. The greatest length from $(0)$ to $(b)$ is 34

(i) The energy $e_a(i, j)$ is the greatest possible moves that a grain can do to reach the position $(i, j)$.

(ii) The energy $E(a)$ of $a$ is $E(a) = \sum_{(i,j)\in F(a)} e_a(i, j)$.

**Lemma 1** *Let* $a = (a_1, a_2, \ldots, a_k)$ *be a smooth partition.*

(i) *We have:* $e_a(i, j) = i + 1 - \min\{r : a_r < a_{r-1}$ *and* $a_r + r \geq j + i - 1\}$.

(ii) *Moreover, if* $(i, j) \in F(a)$ *and* $(i - 1, j + 1) \in F(a)$ *then*

$$e_a(i - 1, j + 1) = e_a(i, j) - 1.$$

Now, we want to compute explicitly the energy of a smooth partition $a = (a_1, \ldots, a_k)$. Let $1 = i_1 < i_2 < \cdots < i_\ell$ be all smooth columns of $a$. And let $D_i$ the diagonal $(i, a_i)$. It is evident that we can decompose $F(a)$ as the following disjoint union:

$$F(a) = \Delta_1 \bigsqcup D_2 \bigsqcup \ldots \bigsqcup D_\ell$$

where $\Delta_1$ is the set of all case $(i, j)$ such that $1 \leq i, j$ and $i + j \leq a_1 + 1$. We then compute the energy of $a$ in each of such subset.

**Proposition 2** *Let* $a$ *be a smooth partition, and let* $1 = i_1 < i_2 < \cdots < i_\ell$ *be all smooth columns of* $a$. *We have (Fig. 6):*

$$E(a) = \frac{a_1(a_1 + 1)(a_1 + 2)}{6} + \sum_{r=2}^{\ell} i_r a_{i_r} - \sum_{r=3}^{\ell} i_{r-1} a_{i_r} + \sum_{r=2}^{\ell} \frac{a_{i_r}(a_{i_r} - 1)}{2}.$$

We state now the main result of this subsection.

**Theorem 3** *Let* $a$ *be a smooth partition. Then the greatest length of avalanche chains from* $(0)$ *to* $a$ *is equal to* $E(a)$ *(Figs. 7 and 8).*

From this theorem we can study avalanche chain between two stable configurations.

**Corollary 1** *Let* $b \leq_S a$ *be two smooth partitions. Then the greatest length of avalanche chains from* $a$ *to* $b$ *is* $\sum_{(i,j)\in F(b)-F(a)} e_b(i, j)$.

Nevertheless, it is important to note that the greatest length from $a$ to $b$ is not equal to $E(b) - E(a)$ because for $(i, j) \in F(b)$, we have not $e_a(i, j) = e_b(i, j)$ (see Fig. 9 for an encounter example). Moreover, it is easy to see that $e_b(i, j) \geq e_a(i, j)$. This

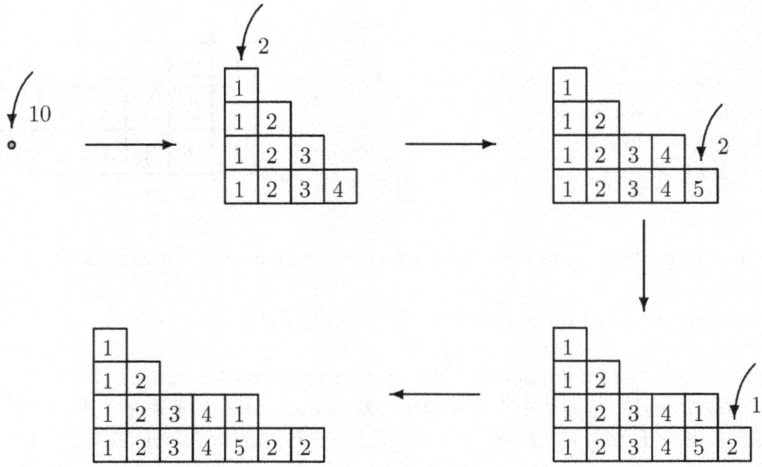

**Fig. 7** A greatest chain from 0 to $b$. Each arrow $\rightarrow^k$ to a column $i$ means that $k$ grains are added to column $i$

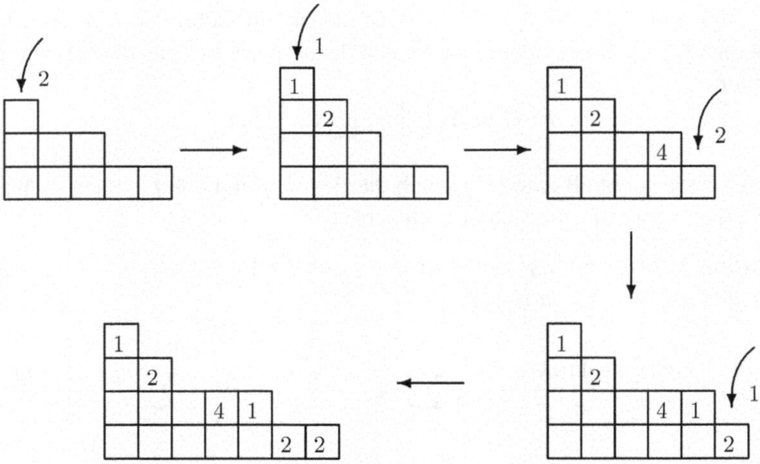

**Fig. 8** A greatest chain from $a$ to $b$. Each arrow $\rightarrow^k$ to a column $i$ means that $k$ grains are added to column $i$. The greatest length from $a$ to $b$ is 12

implies that the greatest length of avalanche chains from $a$ to $b$ is smaller than or equal to the difference of the one from (0) to $b$ and the one from (0) to $a$. This result is opposite to the result in the case of smallest length where the egality is hold. Furthermore, we remark that the avalanche chain of greatest length from (0) to $a$ is unique. Indeed, from the proof of Theorem 3, we constate that the grain $G$ at position $(i, j)$ on the diagonal $D_r$ for $r \geq 2$ (resp. $\Delta_1$) has exactly $e_a(i, j)$ transitions if and only if $G$ is added at the column $i_{r-1} + 1$ (resp. 1) and then it slides diagonally and

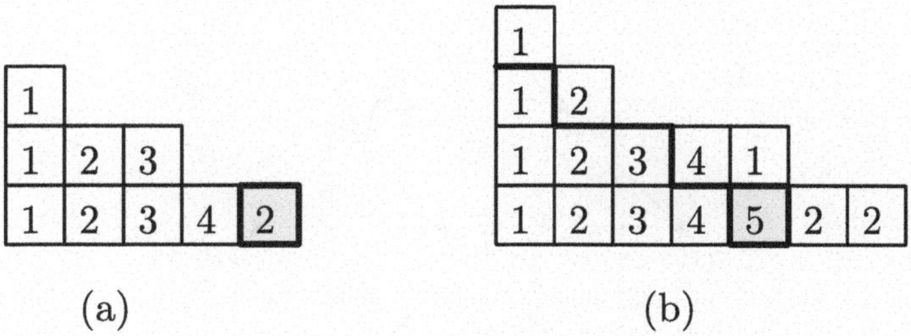


**Fig. 9** **a** Energy tableau of $a$; **b** energy tableau of $b$ and $2 = e_a(5, 1) \neq e_b(5, 1) = 5$

stops at position $(i, j)$. So the diagonal $D_{r-1}$ must be fulfilled,moreover the grain at position $(i + 1, j - 1)$ must be presented before the adding of the grain $G$. So by recurrence we claim that the avalanche chain of greatest length from (0) to $a$ must be defined explicitly as in the proof of Theorem 3, hence it is unique.

However, there are many avalanche chains of greatest length from $a$ to $b$. For instance, if we take $a = (2, 2, 1, 1, 1)$ and $b = (2, 2, 2, 1, 1, 1)$. Then by Corollary 1 we have $l(a, b) = 2$. Moreover, we have two following avalanche chains of length 2:

$$a = (2, 2, 1, 1, 1) \rightarrow (2, 2, 2, 1, 1) \rightarrow (2, 2, 2, 1, 1, 1) = b$$

$$\text{and } a = (2, 2, 1, 1, 1) \rightarrow (2, 2, 1, 1, 1) \rightarrow (2, 2, 2, 1, 1, 1) = b.$$

## 3 Symmetric Sand Pile Model and Unimodal Sequences

Sand Pile Model was first introduced in the context of SOC phenomena. In order to bring it closer to real physical models, we consider the model such that grains can fall to the both sides (left and right). This generalized model is called two sided sand piles model [29] or symmetric sand piles model [12], and denoted by SSPM (Figs. 10, 11, 12 and 13).

**Definition 5** SSPM is a model defined by:

- Initial configuration: $(\underline{n})$.
- Local left vertical rule $\mathcal{L}$: $(\ldots, a_{i-1}, a_i, \ldots) \rightarrow (\ldots, a_{i-1} + 1, a_i - 1, \ldots)$ if $a_{i-1} + 2 \leq a_i$.
- Local right vertical rule $\mathcal{R}$: $(\ldots, a_i, a_{i+1}, \ldots) \rightarrow (\ldots, a_i - 1, a_{i+1} + 1, \ldots)$ if $a_i \geq a_{i+1} + 2$.
- Global rule: we apply the $\mathcal{L}$ rule once, or the $\mathcal{R}$ rule once.

When studying this model, we formulate the characterization of reachable configurations and of fixed points. Note that a configuration is defined by its form and its position. If at the beginning, we have a pile of $n$ sand grains at position 0, then a reachable configuration $a = (a_p, a_{p+1}, \ldots a_{p+k-1})$ can have grains in negative and positive positions. We call *position* of $a$ the smallest index $p$ such that $a_p > 0$, and

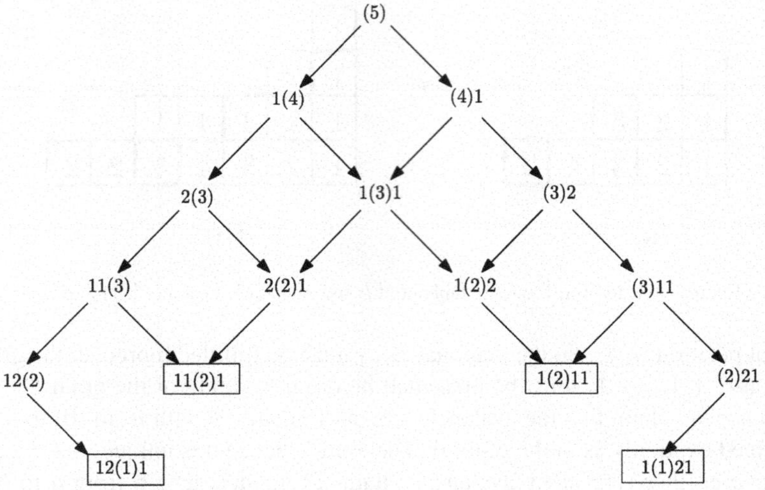


**Fig. 10** Configuration space $SSPM(5)$

**Fig. 11** Configuration space $PSSPM(5)$

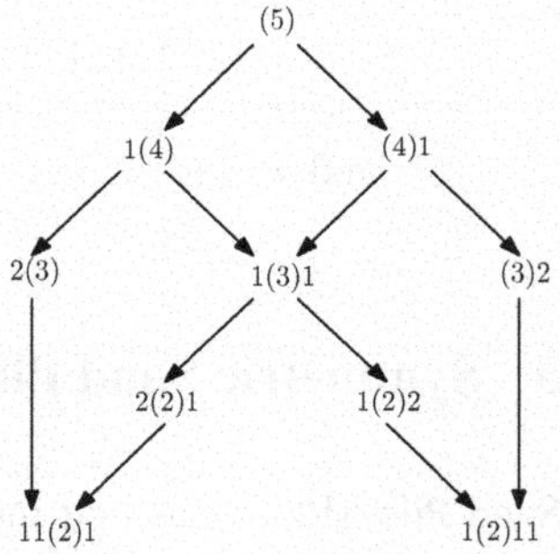


we call the *form* of $a$ the sequence $b = (b_1, \ldots, b_k)$ such that $b_i = a_{p+1-i}$ for all $1 \leq i \leq k$.

It is easy to see that a configuration can be represented by a unimodal sequence which is defined as follows.

**Definition 6** A *unimodal sequence* is a sequence of positive integers $(a_1, a_2, \ldots, a_k)$ such that there exists an index $1 \leq i \leq k$ satisfying the condition $a_1 \leq a_2 \leq \cdots a_{i_1} \leq a_i \geq a_{i+1} \geq \cdots \geq a_{k-1} \geq a_k$. The quantities defined by

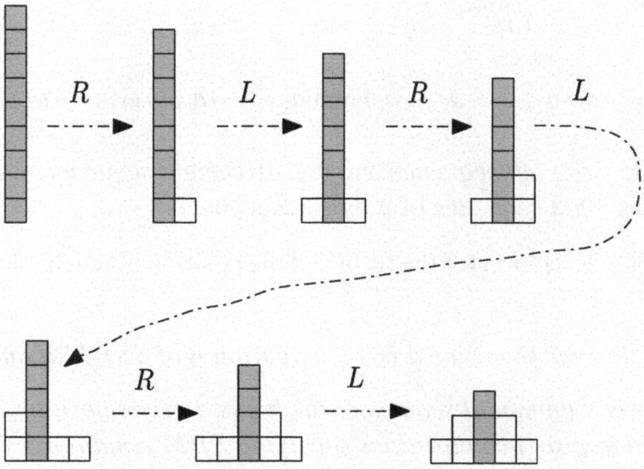


**Fig. 12** 6 first steps of Alternating Procedure from (9). The arrow together with the direction $R$ or $L$ (Right or Left) corresponding to the direction along which the transition at column 0 (dark column) is applied

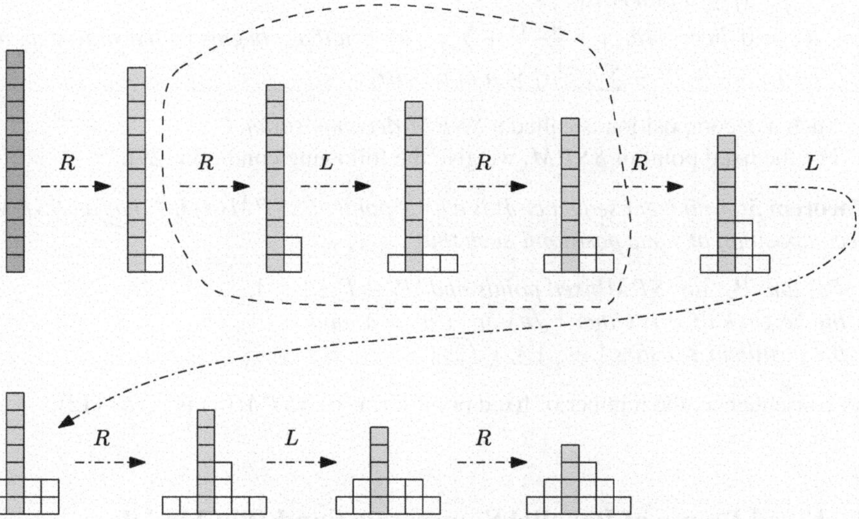


**Fig. 13** 9 first steps of Pseudo-Alternating Procedure on (13)

$$h(a) = \max\{a_i\}_{i=1}^{k} \quad \text{and} \quad w(a) = \sum_{i=1}^{k} a_i$$

are respectively called *the height* and *the weight* of $a$. We say also that $a$ a unimodal sequence of $w(a)$.

Given an index $1 \leq i \leq k$, we denote

$$a_{<i} := (a_1, \ldots, a_{i-1}) \text{ and } a_{>i} := (a_{i+1}, \ldots, a_k),$$

$$a_{\leq i} := (a_1, \ldots, a_{i-1}, a_i) \text{ and } a_{\geq i} := (a_i, a_{i+1}, \ldots, a_k),$$

and call them the *strict left sequence* and the *strict right sequence* of $a$ by $i$, the *left sequence* and the *right sequence* of $a$ by $i$, respectively.

We give a characterization for the form and the position of reachable configuration [29].

**Theorem 4** *An integer sequence $a$ is a configuration of $SSPM$ if and only if*

- *The form of $a$ is a unimodal sequence which has a decomposition $a = (a_{<i}, a_{\geq i})$ where $a_{<i}$ and $a_{\geq i}$ are two partitions satisfying $SPM$ condition.*
- *the position $i$ satisfies:*
  - *if $i \geq 0$ then: $ia_i + \frac{i(i+1)}{2} + \sum_{j\geq i} a_j \leq n$ if $a_{\geq i}$ begins with a slide step (subsequence of the form $(p, p_1, \ldots, q+1, q, q)$ with $p \geq q > 0$), or $ia_i + \frac{i(i-1)}{2} + \sum_{j\geq i} a_j \leq n$ otherwise.*
  - *if $i < 0$ then: $-ia_{i-1} + \frac{i(i-1)}{2} + \sum_{j<i} a_j \leq n$ if $a_{<i}$ begins with a slide step, or $-ia_{i-1} + \frac{i(i+1)}{2} + \sum_{j<i} a_j \leq n$ otherwise.*

Such a decomposition is called a *$SSPM$ decomposition.*

For the fixed point of $SSPM$, we give the following condition [29].

**Theorem 5** *An integer sequence $P$ is a fixed point of $SSPM(n)$ if $P$ has an $SSPM$ decomposition at some position $i$ such that:*

- *$P_{<i}$ and $P_{\geq i}$ are $SPM$ fixed points and $|P_i - P_{i-1}| \leq 1$,*
- *the height $k$ of $P$ is either $\lfloor\sqrt{n}\rfloor$ or $\lfloor\sqrt{n}\rfloor - 1$, and*
- *the position $i$ satisfies $k + |i| \leq \lfloor\sqrt{2n}\rfloor$.*

As consequence, the number of fixed point forms of $SSPM(n)$ is $\lceil\sqrt{n}\rceil$ [12].

## 4 Fixed Points of Parallel Symmetric Sand Pile Model

The Parallel Symmetric Sand Pile Model is a variant of the Symmetric Sand Pile Model where we allow to apply at the same time all possible transitions [14, 27].

**Definition 7** PSSPM is a model defined with the same initial configuration and local rule as SSPM, and with the following global rule.

- Global rule: we apply $\mathcal{L}$ and $\mathcal{R}$ in parallel on all possible columns. We apply at most once of the two rules on each column.

## 4.1 Forms of Fixed Points

Remark that while the parallel $SPM$ is deterministic, the parallel $SSPM$ is not because there may have columns from which grains can fall down on both sides. Recall that $SPM$ has a unique fixed point which implies that PSPM have the same fixed point as $SPM$. But $SSPM$ can have more than one fixed points therefore $PSSPM$ may not have the same set of fixed points as SSPM. Actually, there exists fixed points of $SSPM$ which is not reachable in $PSSPM$ because its position is far from the position 0. Nevertheless, it is surprising that the set of forms of fixed points of $PSSPM$ is the same as that of SSPM. And this fact is the main result of this subsection [14].

**Theorem 6** *The set of fixed point forms of $PSSPM(n)$ is equal to that of $SSPM(n)$. Consequently, there is $\lceil\sqrt{n}\rceil$ fixed point forms of $PSSPM(n)$.*

This theorem need a very long and technical proof but the idea is very constructive and can be presented as follows.

For a fixed point $P$ of $SSPM(n)$, we construct a sequence of $PSSPM$ transitions to obtain a fixed point having the same form as $P$. Because we are interested in the form of $P$ but not in its position, we can suppose that the center column of $P$ is at position 0 (the center of a configuration $a$ is the position $i$ satisfying the condition that $|w(a_{<i}) - w(a_{\geq i})|$ get the minimum value). In the constructed evolution, column 0 is always a highest one, so the choice of $PSSPM$ rules in each step is in fact the choice of the transition's direction at column 0.

- For a symmetric fixed point $P$, i.e. $(P_{<0})^{-1} = P_{>0}$, the evolution is an Alternating Procedure, described as follows: at odd steps, the rule $R$ is applied at position 0, and at even steps, the rule $L$ is applied at position 0. From $(n)$ this procedure will converge to the symmetric fixed point $P$.
- For $P$ not symmetric, we can suppose that the column 0 is the center of $P$, i.e.

$$d = |w(P_{>0}) - w(P_{<0})| = \min_i |w(P_{>i}) - w(P_{<i})|.$$

Without loss of generality we may assume that $w(P_{>0}) - w(P_{<0}) > 0$. The evolution by $PSSPM$ rule is composed of three procedures:

(i) Pseudo-Alternating Procedure: a procedure from $(n)$ to the configuration $Q = \big(1, 2, \ldots, d-1, (n-d^2), d, d-1, \ldots, 2, 1\big)$. Note that $w(Q_{>0}) - w(Q_{<0})$ is exactly $d$.

(ii) Alternating Procedure: a procedure from $Q$ to the configuration $R$ on which we could not apply any more the Alternating Procedure.

(iii) Deterministic procedure: a deterministic procedure from $R$ to $P$, where at each of its step, on each position, only one rule can be applied.

## 4.2 Positions of Fixed Points

We have already a characterization of the form of fixed points of $PSSPM$, but what about their positions? Remark that for $SSPM$ one can obtained a fixed point at a position very far on the left (or on the right) when one applies as much as possible left transition (or right transition respectively). But for $PSSPM$ all transitions are applied at the same time at each step, so one can not apply as much left transition as he wants to. This explains why many fixed points of $SSPM$ very far on the left can not be reached by $PSSPM$. Nevertheless, we can prove that all fixed points of $SSPM$ whose the position is between the leftmost and the rightmost fixed points of $PSSPM$ can be reached also by PSSPM. This is related to the continuity of fixed points of PSSPM. The notion of leftmost, rightmost and continuity will be explained clearly by using relation $\lhd$, a notion of closeness between configurations [27].

**Definition 8** Let $\Delta(a,b)$ be the sequence of differences between configurations $a$ and $b$, $\Delta_i(a,b) = a_i - b_i$.

We define a notion of similarity or closeness between configurations, denoted by the following relations:

$$a \lhd b \iff \Delta(a,b) \in 0^*\bar{1}0^*10^*$$
$$a \overset{*}{\lhd} b \iff \Delta(a,b) \in (0^*\bar{1}0^*10^*)^*$$

where $\bar{1}$ is a minus one value. As a convention $\epsilon = 0^\omega$, so that $a = b$ implies $a \overset{*}{\lhd} b$.

**Notation** We use the symbols $\leq_{lex}$ to denote the lexicographic order over configurations. Note that $a \overset{*}{\lhd} b \Rightarrow a \leq_{lex} b$ and $a \lhd b \Rightarrow a <_{lex} b$.

**Theorem 7** ([27]) *Let*

$$\pi_0 <_{lex} \pi_1 <_{lex} \cdots <_{lex} \pi_{k-1} <_{lex} \pi_k$$

*be the sequence of all fixed points of PSSPM(n) ordered lexicographically. Then this sequence has the following strong relation:*

$$\pi_0 \lhd \pi_1 \lhd \cdots \lhd \pi_{k-1} \lhd \pi_k.$$

*Moreover, for any fixed point $\pi$ of SSPM(n) such that $\pi_0 \leq_{lex} \pi \leq_{lex} \pi_k$, there exists an index $i$, $0 \leq i \leq k$, such that $\pi_i = \pi$.*

# 5 Signed Chip Firing Game and Symmetric Sandpile Model on the Cycles

We explore the Sandpile Model and Chip Firing Game and an extension of these models on cycle graphs. These problems also have a strong relationship to the class of problems on cycles such as games of cards [7, 16, 20]. Furthermore, we are also interested in the signed versions of these models, i.e., we allow the vertices to contain negative numbers of chips for $CFG$ and the sandpiles to have negative heights for SPM. This also reflects deeply some natural phenomena: between sandpiles there may be holes (of negative heights), and besides the delivering chips from vertices containing many chips, it is dually possible receiving chips from vertices lacking (negative enough) chips [22]. We give the characterization of reachable configurations and of fixed points of each model. At the end, we give an explicit formula for the number of their fixed points [5].

## *5.1 SPM, CFG, $SSPM$ and SCFG on Cycles: Definitions and Notations*

Let $C_n$ be a cycle graph of n vertices $\{1, 2, ..., n\}$ ($n \geq 3$). Each integer sequence $(a_1, a_2, \ldots, a_n)$ on vertices of $C_n$ is called *circular distribution* and we say that vertex $i$ contains $a_i$ chips (note that $a_i$ may be negative). We identify two circular distributions if they differ by a rotation of the cycle.

**Definition 9** Let $k$ be a non-negative integer. The *Sandpile model on* $C_n$ of weight $k$ (and its configuration space), denoted by $SPM(C_n, k)$, is described as follows:

(i) The initial configuration is $(k, 0, 0, \ldots, 0)$,
(ii) The evolution rule is the *right rule* as follows: a vertex gives one chip to its right neighbor vertex if it has at least 2 higher than this neighbor.

**Definition 10** Let $k$ be a non-negative integer. The *Symmetric Sandpile model on* $C_n$ of weight $k$ (and its configuration space), denoted by $SSPM(C_n, k)$, is described as follows:

(i) The initial configuration is $(k, 0, 0, \ldots, 0)$,
(ii) The evolution rule: addition to the right rule in $SPM(C_n, k)$, there is also the *left rule*, that means a vertex gives one chip to its left neighbor vertex if it has at least 2 higher than this left neighbor.

**Definition 11** Let $k$ be a non-negative integer. The *Chip Firing Game on* $C_n$ of weight $k$ (and its configuration space), denoted by $CFG(C_n, k)$, is described as follows:

(i) The initial configuration is $(k, 0, 0, \ldots, 0, -k)$,

(ii) The evolution rule is the *positive rule* as follows: a vertex containing at least 2 chips gives one chip to each of its two neighbors.

**Definition 12** Let $k$ be a non-negative integer. The *Signed Chip Firing Game on* $C_n$ of weight $k$ (and its configuration space), denoted by $SCFG(C_n, k)$, is described as follows:

(i) The initial configuration is $(k, 0, 0, \ldots, 0, -k)$.
(ii) The evolution rule: addition to the positive rule in $CFG(C_n, k)$, there is also the *negative rule*, that means a vertex containing at most -2 chips receives one chip from each of its two neighbors.

**Notations**

- We define $SPM(C_n)$ the disjoint union of $SPM(C_n, k)$ for $k \geq 0$, and similarly for $SSPM(C_n)$, $CFG(C_n)$, $SCFG(C_n)$.
- Let $a$ and $b$ be two distributions of non-negative integers on $C_n$, we write $a \stackrel{(i,r)}{\longrightarrow} b$ (resp. $a \stackrel{(i,l)}{\longrightarrow} b$) if $b$ is obtained from $a$ by applying the rule at the vertex $i$ on the right (resp. left); and $a \stackrel{(i,+)}{\longrightarrow} b$ (resp. $a \stackrel{(i,-)}{\longrightarrow} b$) if $b$ is obtained from $a$ by applying the positive rule (resp. negative rule) at the vertex $i$.

**Remark** It is straightforward from the definitions that

- The configurations of $SPM(C_n)$ and $SSPM(C_n)$ are circular distributions of non-negative integers whereas the ones of $CFG(C_n)$ and $SCFG(C_n)$ are circular distributions of integers (may be negative),
- We have the two following inclusions

$$SPM(C_n, k) \subset SSPM(C_n, k) \text{ and } CFG(C_n, k) \subset SCFG(C_n, k).$$

Recall that two models are called isomorphic if there exists a bijection between their configuration spaces and this bijection preserves their evolution rule.

Now, let $a = (a_1, \ldots, a_n)$ be a circular distribution on $C_n$. We define

$$d(a) = (a_1 - a_2, \ldots, a_{n-1} - a_n, a_n - a_1).$$

It is straightforward that $d$ is a well-defined map from the set of circular distributions on $C_n$ to itself. Furthermore, we have the following result.

**Proposition 3** *Under the map $d$ two models $SPM(C_n, k)$ and $CFG(C_n, k)$ are isomorphic; and two models $SSPM(C_n, k)$ and $SCFG(C_n, k)$ are isomorphic.*

It is remarkable that although $d$ is bijective from $SSPM(C_n, k)$ (resp. $SPM$ $(C_n, k)$) to $SCFG(C_n, k)$ (resp. $CFG(C_n, k)$), it is not bijective from $SSPM(C_n)$ (resp. $SPM(C_n)$) to $SCFG(C_n)$ (resp. $CFG(C_n)$). Moreover, while $SSPM(C_n, k)$ and $SPM(C_n, k)$) are absolutely disjoint for different values $k$, $SCFG(C_n, k)$ and $CFG(C_n, k)$ may overlap each other, especially for values $k$ differing by a multiple of $n$. Then a configuration of $SCFG(C_n)$ may correspond to many configurations of SSPM(Cn) whose weights differ by a multiple of $n$.

Next, we study a characterization for the configurations of the four models. Let $a = (a_1, a_2, \ldots)$ be a sequence of positive integers. A pair $(a_i, a_{i+1})$ is called a *cliff* (resp. *plateau*) of a at position $i$ if $a_i - a_{i+1} \geq 2$ (resp. $a_i - a_{i+1} = 0$).

**Theorem 8** *Let a be a circular distribution on $C_n$. Then a is a configuration of $SPM(C_n, k)$ if and only if there is a rotation vertices of $C_n$ such that a (in the sequence form) is a configuration of $SPM(k)$ with the length at most n.*

**Corollary 2** *Let $a = (a_1, a_2, \ldots)$ be a circular distribution. Then a is a configuration of $CFG(C_n, k)$ if and only if $d^{-1}(a)$ is a configuration of $SPM(C_n, k)$.*

**Corollary 3** *The unique fixed point of $SPM(C_n, k)$ is of the form*

- *$(p, p-1, \ldots, q, q, q-1, \ldots, 1, 0, \ldots)$ if $k \leq \frac{n(n-1)}{2}$, where*

$$p = \left\lfloor \frac{3 + \sqrt{9 + 8k}}{2} \right\rfloor \text{ and } q = k - \frac{p(p+1)}{2}.$$

- *$(p, p-1, \ldots, q, q, q-1, \ldots, p-n+3, p-n+2)$ if $k \geq \frac{n(n-1)}{2} + 1$, where*

$$p = \left\lfloor \frac{2k + (n-2)(n+1)}{2n} \right\rfloor \text{ and } q = k - \frac{(2p - n + 2)(n-1)}{2}.$$

*Here $[x]$ is the largest integer not greater than x.*

Next, we give a characterization for configurations of the $SSPM$s as well as $SCFG$s on $C_n$. To do this we first present the concept of 2-decomposable configurations on the cycle which is different a bit from the definition of LR-decomposition (left-right decomposition) on the line defined in [29].

**Definition 13** Let $a = (a_1, a_2, \ldots, a_n)$ be a circular distribution on $C_n$, then $a$ is called 2-decomposable at $(i, j)$ with $1 \leq i \leq j \leq n$ if $(a_i, a_{i+1}, \ldots, a_j)$ and $(a_{i-1}, a_{i-2}, \ldots, a_1, a_n, \ldots, a_j{+}1)$ are $SPM$ configurations. Furthermore, $a$ is called 2-decomposable if there exist two indices $1 \leq i \leq j \leq n$ such that $a$ is 2-decomposable at $(i, j)$.

**Theorem 9** *Let a be a circular distribution on $C_n$. Then a is a configuration of $SSPM(C_n)$ if and only if a is 2-decomposable.*

**Corollary 4** *Let $a = (a_1, a_2, \ldots)$ be a circular distribution. Then a is a configuration of $SCFG(C_n, k)$ if and only if $d^{-1}(a)$ is 2-decomposable.*

### 5.2 *Fixed Points of $CFG(C_n)$ and $SCFG(C_n)$*

Although we have a criterion for the configurations of $SCFG(C_n)$, it requires us to calculate their inverse images by $d$ and then to check their 2-decomposability in $SSPM(C_n)$. In this section, we present a simple and direct characterization for the fixed points (not all their configurations) of $SCFG(C_n)$. Based on this characterization, we give an enumeration for these fixed points. We first classify the configurations of $CFG(C_n)$ and those of $SCFG(C_n)$ [5].

**Proposition 4** *Let $k, l$ be positive integers.*

*(i) If $k \neq l \mod n$ then*

$$CFG(C_n, k) \cap CFG(C_n, l) = \emptyset$$

*and*

$$SCFG(C_n, k) \cap SCFG(C_n, l) = \emptyset.$$

*Consequently, the intersection of the set of fixed points of $SCFG(C_n, k)$ and that of $SCFG(C_n, l)$ is empty.*

*(ii) If $k = l \mod n$ and $k, l \geq (\frac{n+1}{2})^2$ then the set of fixed points of $CFG(C_n, k)$ (resp. $SCFG(C_n, k)$) is equal to that of $CFG(C_n, l)$ (resp. $SCFG(C_n, l)$).*

As we remarked in the previous section that for large enough values of $k$ in a residue class modulo $n$, although the set of fixed points of $SSPM(C_n, k)$ (resp. $SPM(C_n, k)$) are disjoint, the heights of their columns differ up-to a constant. In other words, if $(a_1, \ldots, a_n)$ is a fixed point of $SSPM(C_n, k)$ (resp. $SPM(C_n, k)$), then $(a_1 + 1, \ldots, an + 1)$ is a fixed point of $SSPM(C_n, k + n)$ (resp. $SPM(C_n, k + n)$). Hence their images by $d$ in $SCFG(C_n, k)$ (resp. $CFG(Cn, k)$) and in $SCFG(C_n, k + n)$ (resp. $CFG(C_n, k + n)$) coincide. By Corollary 2.12, $CFG(C_n, k)$ has a unique fixed point whereas $SCFG(C_n, k)$ may have many fixed points. The set of fixed points of $SCFG(C_n)$ includes the fixed points of $SCFG(C_n, k)$ for small values of $k$ and the $n$ distinct residue classes of fixed points of $SCFG(C_n, k)$ for large values of $k$. For a small $k$, their fixed points can be found directly by taking the inverse images of d of 2-decomposable fixed points. We next characterize and enumerate the fixed points of $SCFG(C_n)$ for all $k \geq (\frac{n+1}{2})^2$.

For convenience, we denote by $FP(SCFG(C_n, k)$ the set of fixed points of $SCFG(C_n, k)$ and

$$FP(SCFG(C_n)) = \cap_{k \geq (\frac{n+1}{2})^2} FP(SCFG(C_n, k)).$$

Recall that each fixed point of $SCFG(C_n)$ is a circular distribution on $C_n$ and its chips at vertices are $0, 1, -1$. By a rotation, first we can consider $FP(SCFG(C_n))$ as words on the alphabet $\{0, 1, \bar{1}\}$ where the letter $\bar{1}$ is understood as $-1$.

**Theorem 10** *The set $FP(SCFG(C_n))$ is determined as follows*

- $FP(SCFG(C_3)) = \{(000); (10\bar{1}); (1\bar{1}0)\}$.
- $FP(SCFG(C_4)) = \{(0000); (1\bar{1}00); (10\bar{1}0); (100\bar{1}); (11\bar{1}\bar{1})\}$.
- $FP(SCFG(C_n))$, *with* $n \geq 5$, *consists of the words* $w$ *on the alphabet* $\{0, 1, \bar{1}\}$ *satisfying the following properties:*
  - $w$ *starts from* 1*;*
  - *in* $w$*, the number of occurrences of* 1 *is equal to that of* $\bar{1}$*;*
  - $w$ *avoids the subsequences:* $\bar{1}1$, 1001, $\bar{1}00\bar{1}$ *and* 00000*;*
  - *If* $w$ *has 4 occurrences of* 0 *then it must end by* 0 *and does not contain the sub-word* $1\bar{1}$.

**Theorem 11** *The cardinality of $FP(SCFG(C_n))$ is*

- *3 if* $n = 3$*;*
- *5 if* $n = 4$*;*
- $\frac{(n-1)^2}{2}$ *if* $n$ *is odd and* $n \geq 5$*;*
- $\frac{n(n-2)}{2}$ *if* $n$ *is even and* $n \geq 6$.

## 6 Extension of Brylwaski's Model

In the previous section, we presented many kinds of extensions of $SPM$ model. For each kind of extension, it is natural to think about a similar extension of Brylawski's model. Recall that for the classical models, while the configuration space of $SPM$ are clearly characterized with explicit criteria, we also have the exhaustive property of Brylawski's model: all integer partitions are reachable. Is we adapt the similar notion of slide transition in each extension of the $SPM$ model, we have the corresponding extension of Brylawski's model. And this is very interesting that the exhaustive property remains for these extensions. It is also very surprising that the shortest and longest chains in extended models have the same length as in the classical model of Brylawski.

Let us recall here the definition of Brylawski's model.

**Definition 14** Brylawski's model, with respect to a positive integer $n$, denoted by $L_B(n)$, is a model where configurations are partitions of $n$ such that:

- Initial configuration: $(\underline{n})$.
- Local right vertical rule $\mathcal{R}$: $(\ldots, a_i, a_{i+1}, \ldots) \rightarrow (\ldots, a_i - 1, a_{i+1} + 1, \ldots)$ if $a_i \geq a_{i+1} + 2$.
- Local right horizontal rule $\mathcal{H}$: $(\ldots, p+1, p, \ldots, p, p-1, \ldots) \rightarrow (\ldots, p, p, \ldots, p, p, \ldots)$.
- Global rule: apply the $\mathcal{R}$ rule once, or the $\mathcal{H}$ rule once.

### 6.1 *The Symmetric Brylawski'model*

Let us first investigate to the extended symmetric Brylawski's model ($SB_L$) [9].

**Definition 15** The symmetric Brylawski's model, with respect to $n$, denoted by $SB_L(n)$, is a model defined by:

- Initial configuration: $(\underline{n})$.
- Local left vertical rule $\mathcal{L}$: $(\ldots, a_{i-1}, a_i, \ldots) \to (\ldots, a_{i-1}+1, a_i-1, \ldots)$ if $a_{i-1}+2 \leq a_i$.
- Local right vertical rule $\mathcal{R}$: $(\ldots, a_i, a_{i+1}, \ldots) \to (\ldots, a_i-1, a_{i+1}+1, \ldots)$ if $a_i \geq a_{i+1}+2$.
- Local left horizontal rule $\mathcal{L}_H$: $(\ldots, p+1, p, \ldots, p, p-1, \ldots) \to (\ldots, p, p, \ldots, p, p, \ldots)$.
- Local right horizontal rule $\mathcal{R}_H$: $(\ldots, p-1, p, \ldots, p, p+1, \ldots) \to (\ldots, p, p, \ldots, p, p, \ldots)$.
- Global rule: we apply the $\mathcal{L}$ rule once, or the $\mathcal{R}$ rule once.

The exhaustive property of $SL_B$ is presented as follows.

**Lemma 2** *The set of configuration forms of $SL_B(n)$ are the set of all unimodal sequences of weight $n$.*

We have the following straightforward result on fixed points.

**Corollary 5** *$SL_B(n)$ has $n$ fixed points of form $(1, \ldots, 1)$ where the first position can take value from $-n+1$ to* 0.

And results on shortest and longest chains.

**Proposition 5** *For $n \geq 4$, the shortest chains in $SL_B(n)$ have length $2n-5$ and the longest chains in $SL_B(n)$ have the same length as that in $L_B(n)$ which is $\theta(n^{3/2})$.*

## 7 Infinite Extension

All the previous extensions have a common property: the total number of sand grains is unchanged. Now, we investigate an extended models where the total number of grains can be changed. Furthermore, we consider model where the first column has not a fix number but an infinite number of grains. It is natural to ask how one can construct the lattice $L_B(n+1)$ from the lattice $L_B(n)$. We construct a linear time algorithm which gives a translation from $L_B(n)$ to $L_B(n+1)$ and which reserves the lattice structure. From this algorithm, one can construct the lattice $L_B$ with an arbitrary number of grains at the first position. And by the way, we can see that $L_B(\infty)$ is the limit of $L_B(n)$ where $n$ goes to infinity. To do that, we consider the model with three transition rules: the vertical rule, the horizontal rule and the adding rule (adding one grain at the first position). It is obvious that all reachable configurations are still integer partitions, but what is interesting is that all integer partitions are reachable in this models [24].

## 7.1 Constructing $L_B(n+1)$ from $L_B(n)$

Before entering the core of algorithms, we need one more notation. If the $k$-tuple $a = (a_1, a_2, \ldots, a_k)$ is a partition, then the $k$-tuple $(a_1, a_2, \ldots, a_{i-1}, a_i + 1, a_{i+1}, \ldots, a_k)$ is denoted by $a^{\downarrow i}$. In other words, $a^{\downarrow i}$ is obtained from $a$ by adding one grain on its $i$th column. Notice that the $k$-tuple obtained this way is not necessarily a partition. If $S$ is a set of partitions, then $S^{\downarrow i}$ denotes the set $\{a^{\downarrow i} | a \in S\}$. Finally, we denote by $Succ(a)$ the set of configurations directly reachable from $a$, *i.e.* the set $\{b \mid a \xrightarrow{i} b \text{ for some } i\}$.

Write $d_i(a) = a_i - a_{i+1}$ with the convention that $a_{k+1} = 0$. We say that $a$ has a *cliff* at position $i$ if $d_i(a) \geq 2$. If there exists an $\ell \geq i$ such that $d_j(a) = 0$ for all $i \leq j < \ell$ and $d_\ell(a) = 1$, then we say that $a$ has a *slippery plateau* at $i$. Likewise, $a$ has a *non-slippery plateau* at $i$ if $d_j(a) = 0$ for all $i \leq j < \ell$ and it has a cliff at $\ell$. The integer $\ell - i$ is called the *length* of the plateau at $i$. Note that in the special case $\ell = i$, the plateau is of length 0.

The set of elements of $L_B(n)$ that begin with a cliff, a slippery plateau of length $\ell$ and a non-slippery plateau of length $\ell$ are denoted by $C, SP_\ell, nSP_\ell$ respectively.

Let $a = (a_1, a_2, \ldots, a_k)$ be a partition. It is clear that $a^{\downarrow 1}$ is again a partition. This define an embedding $\pi : L_B(n) \to L_B(n)^{\downarrow 1} \subset L_B(n+1)$ which can be proved, by using infimum formula of $L_B(n)$ and $L_B(n+1)$, as a lattice map.

**Lemma 3** *$L_B(n)^{\downarrow 1}$ is a sub-lattice of $L_B(n+1)$.*

Our next result characterizes the remaining elements of $L_B(n+1)$ that are not in $L_B(n)^{\downarrow 1}$.

**Theorem 12** *For all $n \geq 1$, we have $L_B(n+1) = L_B(n)^{\downarrow 1} \sqcup_\ell SP_\ell^{\downarrow \ell+1}$.*

***Proof*** It is easy to check that each element in one of the sets $L_B(n)^{\downarrow 1}$ and $SP_\ell^{\downarrow \ell+1}$ is an element of $L_B(n+1)$, and that these sets are disjoint. Now let us consider an element $b$ of $L_B(n+1)$. If $b$ begins with a cliff or a step then $b$ is in $L_B(n)^{\downarrow 1}$. If $b$ begins with a plateau of length $\ell + 1$, $\ell \geq 2$, then $b$ is in $SP_\ell^{\downarrow \ell+1}$. □

Finally, we describe an algorithm to compute the successors of any given element of $L_B(n+1)$, thus giving a complete construction of $L_B(n+1)$ from $L_B(n)$.

**Proposition 6** *Let $x$ be an element of $L_B(n+1)$.*

1. *Suppose $x = a^{\downarrow 1} \in L_B(n)^{\downarrow 1}$.*
   **) If $a$ is in $C$ or $nSP$ then $Succ(a^{\downarrow 1}) = Succ(a)^{\downarrow 1}$,*
   **) If $a$ is in $SP_l$ then $Succ(a^{\downarrow 1}) = Succ(a)^{\downarrow 1} \cup \{a^{\downarrow \ell+1}\}$,*
2. *If $x = a^{\downarrow \ell+1} \in SP_\ell^{\downarrow \ell+1}$ for some $a \in SP_\ell$, then*
   **) If $a$ has a cliff at $\ell + 1$ or a non-slippery plateau at $\ell + 1$, then $Succ(a^{\downarrow \ell+1}) = Succ(a)^{\downarrow \ell+1}$,*
   **) If $a$ has a slippery plateau at $\ell + 1$, let $b$ such that $a \xrightarrow{\ell} b$ in $L_B(n)$, then $Succ(a^{\downarrow \ell+1}) = (Succ(a) \setminus \{b\})^{\downarrow \ell+1} \cup \{b^{\downarrow \ell}\}$.* □

Proposition 6 makes it possible to write an algorithm to construct the lattice $L_B(n)$ in linear time (with respect to its size).

## 7.2 The Infinite Lattice $L_B(\infty)$

Imagine that $(\infty)$ is the initial configuration where the first column contains infinitely many grains and all the other columns contain no grains. Then the transitions $V$ and $H$ can be performed on $(\infty)$ just as if it is finite, and we call $L_B(\infty)$ the set of all the configurations reachable from $(\infty)$. A typical element $a$ of $L_B(\infty)$ has the form $(\infty, a_2, a_3, \ldots, a_k)$. As in the previous section, we find that the dominance ordering on $L_B(\infty)$ (when the first component is ignored) is equivalent to the order induced by the dynamical model.

For any two elements $a = (\infty, a_2, \ldots, a_k)$ and $b = (\infty, b_2, \ldots, b_\ell)$ of $L_B(\infty)$, we define $c$ by: $c_i = max(\sum_{j \geq i} a_j, \sum_{j \geq i} b_j) - \sum_{j>i} c_j$ for all $i$ such that $2 \leq i \leq max(k, \ell)$. One can check that $c$ is an element of $L_B(\infty)$, *i.e.* $c_1 = \infty$ and $c_i > c_{i+1}$ for all $i > 1$, and then $c = a \wedge b$. This implies that:

**Theorem 13** *The set $L_B(\infty)$ is a lattice.* □

Now for any $n > 1$, there are two canonical embeddings of $L_B(n)$ in $L_B(\infty)$, defined by

$$\begin{array}{llll} \pi: & L_B(n) & \longrightarrow & L_B(\infty) \\ & a = (a_1, a_2, \ldots, a_k) & \mapsto & \pi(a) = (\infty, a_2, \ldots, a_k), \end{array}$$

$$\begin{array}{llll} \chi: & L_B(n) & \longrightarrow & L_B(\infty) \\ & a = (a_1, a_2, \ldots, a_k) & \mapsto & \chi(a) = (\infty, a_1, a_2, \ldots, a_k). \end{array}$$

The following result is straightforward:

**Proposition 7** *Both $\pi$ and $\chi$ are embedding of lattices.*

By using the embedding $\Pi$, one can consider $L_B(\infty)$ as the limit of $L_B(n)$ when $n$ goes to infinity.

## 7.3 The Infinite Binary Tree $T_B(\infty)$

As shown in our procedure to construct $L_B(n+1)$ from $L_B(n)$, each element $a$ of $L_B(n+1)$ is obtained from an element $a'$ of $L_B(n)$ by adding of one grain: $a = a'^{\downarrow i}$ for some integer $i$. We will now represent this relation by a tree where $a \in L_B(n+1)$ is the son of $a' \in L_B(n)$ if and only if $a = a'^{\downarrow i}$ and we label with $i$ the edge $a' \longrightarrow a$ in this tree. We denote this tree by $ST(\infty)$. The root of this tree is the empty partition (). We will show two ways to find the partitions of a given integer $n$ in $T(\infty)$, which will make it possible to give an efficient and simple algorithm to compute them. Moreover, the recursive structure of this tree will allow us to obtain a recursive formula for the cardinal of $L_B(n)$ and some special classes of partitions.

From the construction of $L_B(n+1)$ from $L_B(n)$, it follows that the nodes of this tree are the elements of $\bigsqcup_{n \geq 0} L_B(n)$, and that each node $a$ has at least one son,

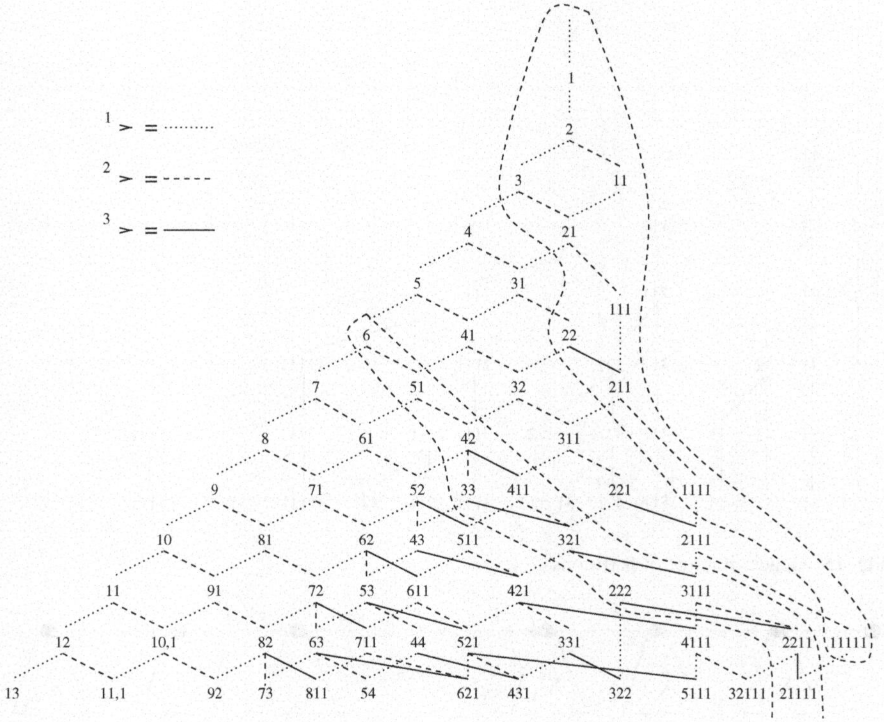


**Fig. 14** The first elements and transitions of $L_B(\infty)$. As shown on this figure for $n = 6$, we have two ways to find parts of $L_B(\infty)$ isomorphic to $L_B(n)$ for any $n$

$a^{\downarrow_1}$, and one more if $a$ begins with a slippery plateau of length $l$: the element $a^{\downarrow_{\ell+1}}$. Therefore, $T(\infty)$ is a binary tree. We will call *left son* the first of two sons, and *right son* the other (if it exists). We call *the level n of the tree* the set of elements of depth $n$. The first levels of $T(\infty)$ are shown in Fig. 14 (Fig. 15).

Like in the case of $L_B(\infty)$, there are two ways to find the elements of $L_B(n)$ in $ST(\infty)$. Based on the construction of $L_B(n+1)$ from $L_B(n)$ as given above, it is straightforward that:

**Proposition 8** *The level n of $ST(\infty)$ is exactly the set of the elements of $L_B(n)$.* □

We will now give a recursive description of $T(\infty)$. We first define a certain kind of subtrees of $T(\infty)$. Then, we show how the whole structure of $T(\infty)$ can be described in terms of such subtrees.

**Definition 16** We will call $X_k$ *subtree* any subtree $T$ of $T(\infty)$ which is rooted at an element $a = (\underbrace{i, \ldots, i}_{k}, a_{k+1}, \ldots)$ with $a_{k+1} \leq i - 1$ and which is either the whole subtree of $T(\infty)$ rooted at $a$ in the case $a$ has only one son, or $a$ and its left subtree otherwise. Moreover, we define $X_0$ as a simple node. □

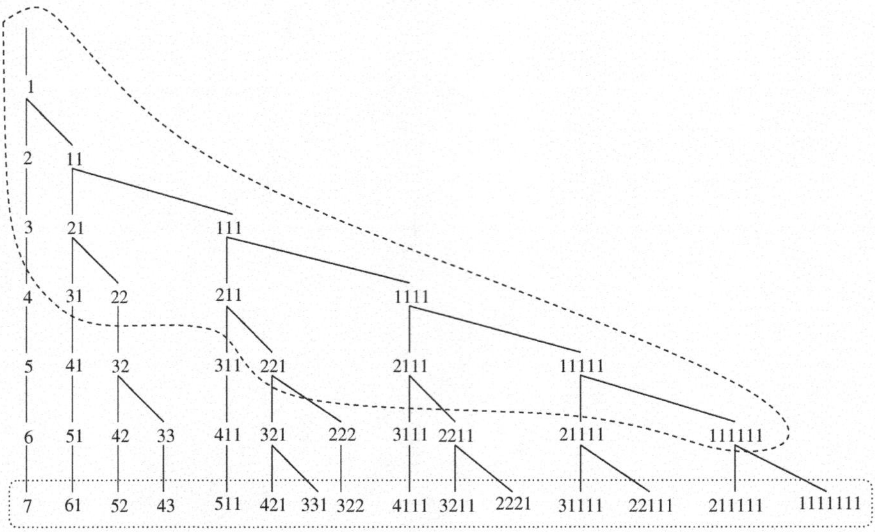


**Fig. 15** Generating tree of partitions

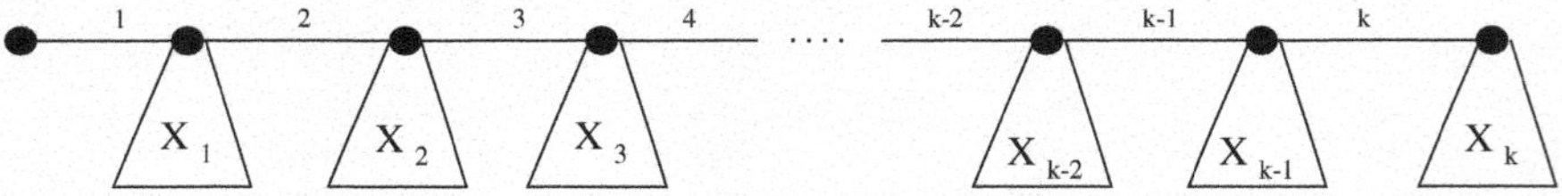


**Fig. 16** Self-referencing structure of $X_k$ subtrees

The next proposition shows that all the $X_k$ subtrees are isomorphic (see Fig. 16).

**Proposition 9** *An $X_k$ subtree, with $k \geq 1$, is composed by a chain of $k + 1$ nodes (the rightmost chain) whose edges are labeled $1, 2, \ldots, k$ and whose $i$th node is the root of an $X_{i-1}$ subtree for all $i$ between $1$ and $k + 1$.*

This recursive structure and the above propositions allow us to give a compact representation of the tree by a chain (see Fig. 17).

**Theorem 14** *The tree $T(\infty)$ can be represented by the infinite chain defined as follows: the $i$th node of this chain, $(\underbrace{1, \ldots, 1}_{i-1})$, is linked to the following node in the chain by an edge labeled with $i$ and is the root of an $X_{i-1}$ subtree.*

Moreover, we can prove a stronger property for each subtree in this chain:

**Corollary 6** *The $X_k$ subtree of $T(\infty)$ with root $(1, \ldots, 1)$ contains exactly the partitions of length $k$.*

We can now state our last result:

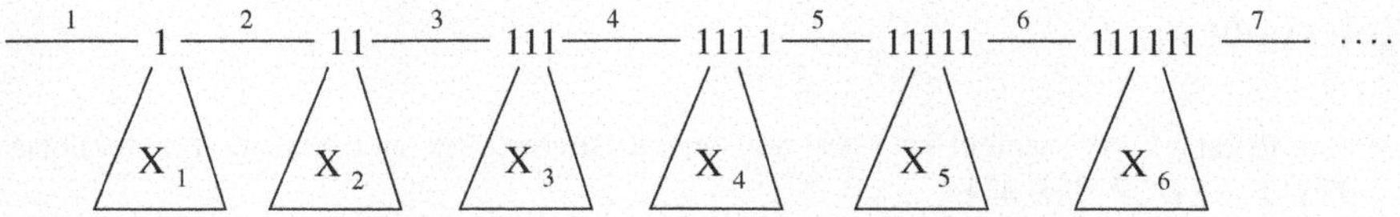


**Fig. 17** Representation of $T_P$ as a chain

**Corollary 7** *Let $c(\ell, k)$ denote the number of paths in an $X_k$ tree originating from the root and having length $\ell$. We have:*

$$c(\ell, k) = \begin{cases} 1 & \text{if } \ell = 0 \text{ or } k = 1; \\ \sum_{i=1}^{inf(\ell,k)} c(\ell - i, i) & \text{otherwise.} \end{cases}$$

*Moreover, $|L_B(n)| = c(n, n)$ and the number of partitions of $n$ with length exactly $k$ is $c(n - k, k)$.* □

# 8 Conclusion

In this work, we seek characterization of reachable configurations and of stable configurations of many extensions of Sand Pile Model. In SPM and their extensions, a configuration is represented by an integer partition or by a unimodal sequence. In almost all cases, the initial configuration is just a 1-part partition. There are also models where the initial configuration is an arbitrary integer partition, some enumerations for this more general models were studied in [28], but questions about convergence time, the structure of configuration space, etc., remained open. It will be interesting to investigate similar problems for other classes of graphs such as trees or planar graphs.

The SPMs taken together when the number of grains is arbitrarily large form another lattice, called the infinite SPM. This yields an infinite lattice as well as an infinite tree on the set of all integer partitions. There are different ways to label the edges of this tree, each labeling gives rise to a generating function on the set of corresponding partitions. This approach is potentially useful in constructing partition identities. We provide some examples and discuss some questions around this point of view.

**Acknowledgements** This work was supported by the Vietnam National Foundation for Science ans Technology Development under the grant number NAFOSTED 101.99-2016.16 and by the Vietnam Institute for Advanced Study in Mathematics.

## References

1. Bak P, Tang C, Wiesenfeld K (1987) Self-organized criticality: an explanation of 1/f noise. Phys Rev Lett 59:381–384
2. Björner A, Lovász L (1992) Chip firing games on directed graphs. J Algebraic Combin 1:305–328
3. Björner A, Lovász L, Shor W (1991) Chip-firing games on graphs. Eur J Combin 12:283–291
4. Brylawski T (1973) The lattice of interger partitions. Discret Math 6:201–219
5. Cori R, Phan THD, Tran TTH (2013) Signed chip firing games and symmetric sandpile models on the cycles. RAIRO Inf Théor Appl 47(2):133–146
6. Cori R, Rossin D (2000) On the sandpile group of dual graphs. Eur J Combin 21:447–459
7. Desel J, Kindler E, Vesper T, Walter R (1995) A simplified proof for the self-stabilizing protocol: a game of cards. Inf Proc Lett 54:327–328
8. Dhar D (1990) Self-organized critical state of sandpile automaton models. Phys Rev Lett 64:1613–1616
9. Duchi E, Mantaci R, Phan THD, Rossin D (2006) Bidimensional sand pile and ice pile models. Pure Math Appl (PU.M.A.) 17(1-2):71–96
10. Durand-Lose J (1996) Grain sorting in the one dimensional sand pile model. Complex Syst 10(3):195–206
11. Durand-Lose J (1998) Parallel transient time of one-dimensional sand pile. Theor Comput Sci 205(1–2):183–193
12. Formenti E, Masson B, Pisokas T (2007) Advances in symmetric sandpiles. Fundam Inf 76(1–2):91–112
13. Formenti E, Perrot K, Rémila E (2014) Computational complexity of the avalanche problem on one dimensional kadanoff sandpiles. In: Proceedings of automata '2014, (LNCS), vol 8996, pp 21–30
14. Formenti E, Pham TV, Duong TH, Phan THD, Tran TTH (2014) Fixed-point forms of the parallel symmetric sandpile model. Theor Comput Sci 533:1–14
15. Goles E, Kiwi MA (1993) Games on line graphs and sand piles. Theor Comput Sci 115:321–349
16. Goles E, Morvan M, Phan HD (2002) Lattice structure and convergence of a game of cards. Ann. Combin 6:327–335
17. Goles E, Morvan M, Phan HD (2002) Sandpiles and order structure of integer partitions. Discret Appl Math 117:51–64
18. Goles E, Morvan M, Phan HD (2002) The structure of linear chip firing game and related models. Theor Comput Sci 270:827–841
19. Greene C, Kleiman DJ (1986) Longest chains in the lattice of integer partitions ordered by majorization. Eur J Combin 7:1–10
20. Huang S-T (1993) Leader election in uniform rings. ACM Trans Program Lang Syst 15(3):563–573
21. Kadanoff LP, Nagel SR, Wu L, Zhou SM (1989) Scaling and universality in avalanches. Phys Rev A 39(12):6524–6537
22. Karmakar R, Manna SS (2005) Particle hole symmetry in a sandpile model. J Stat Mech: Theory and Exp 2005(01):L01002
23. Latapy M, Mataci R, Morvan M, Phan HD (2001) Structure of some sand piles model. Theor Comput Sci 262:525–556
24. Latapy M, Phan THD (2009) The lattice of integer partitions and its infinite extension. Discret Math 309(6):1357–1367
25. Le MH, Phan THD (2009) Integer partitions in discrete dynamical models and ECO method. Vietnam J Math 37(2–3):273–293
26. Le MH, Phan THD Strict partitions and discrete dynamical systems. Theor Comput Sci
27. Perrot K, Pham TV, Phan THD (2012) On the set of fixed points of the parallel symmetric sand pile model. In: Automata 2011 - 17th International Workshop on Cellular Automata and Discrete Complex Systems, Discrete Mathematics & Theoretical Computer Science, pp 17–28

28. Perrot K, Rémila E (2013) Kadanoff sand pile model. avalanche structure and wave shape. Theor Comput Sci 504:52–72
29. Phan THD (2008) Two sided sand piles model and unimodal sequences. RAIRO Inf Théor Appl 42(3):631–646
30. Phan THD, Tran TTH (2010) On the stability of sand piles model. Theor Comput Sci 411(3):594–601
31. Spencer J (1986) Balancing vectors in the max norm. Combinatorica 6:55–65
32. Stanley RP (1999) Enumerative combinatorics, vol 2. Cambridge University Press, Cambridge

# Multiple Fibonacci Trees

**Maurice Margenstern**

**Abstract** In this paper, we synthetise the studies of three papers we deposited on *arXiv*: see [7–9]. The paper considers the extension of the representation used in many of the author's papers about cellular automata in the hyperbolic plane. In those papers, a particular representation of the natural numbers allows us together with the construction of a corresponding tree to define a coordinate system which can be used to navigate in the corresponding tiling of the hyperbolic plane. We extend those considerations to many other kinds of trees for the same purpose and we also extend the technique used for the tilings $\{5, 4\}$ and $\{7, 3\}$ to the tilings $\{p, 4\}$ and $\{p + 2, 3\}$ of the same hyperbolic plane.

## 1 Introduction

Paper [7] investigated the question what happens if the rules generating the standard Fibonacci tree are applied to a tree whose root is a black node. The question was investigated with what is called in this paper the leftmost assignment: in the generating rules, the black son is always the first one. In that paper too, the question was raised of what happens if instead the standard Fibonacci sequence we consider what was called the golden sequence which is associated with the square of the golden number while the standard Fibonacci sequence is associated with the golden number itself. The question was considered for both the white and the black Fibonacci trees.

Paper [8] generalizes the context of the same questions. Instead of considering trees more or less connected with the tilings $\{5, 4\}$ and $\{7, 3\}$ of the hyperbolic plane, that paper considers the trees which span the tilings $\{p, 4\}$ and $\{p + 2, 3\}$ of the same plane. Those trees are finitely generated by rules which generalize the rules of the case $p = 5$ to which paper [7] limited its study. Also the new trees entail the consideration of new families of numbers, we called metallic numbers in [8], which allow representations of the natural numbers which look like the golden family of the case $p = 5$. Paper [9] performs two new steps in the generalization. On one hand, it considers the definition of the generating rules themselves. Instead of considering

M. Margenstern (✉)
LGIMP, Department of Computer Science, Université de Lorraine, Nancy, France



A. Adamatzky (ed.), *Automata and Complexity*, Emergence, Complexity and Computation 42, https://doi.org/10.1007/978-3-030-92551-2_17

that the black son is the first one in each rule, it considers various possibilities if the position of that son is changed, whether the change is always the same or if the change is also submitted to variations. On another hand, it considers the representation of the numbers. In paper [8], the metallic sequences are defined with digits in $\{0 \ldots p-3\}$. What if the digits are constrained to be in $\{1 \ldots p-2\}$, considering only positive integers which is the case for the numbers we attach to the tiles except the central one which is the single tile attached to 0?

Section 2 recalls definitions about trees and about the trees considered in the present paper. Section 3 recalls the results about the metallic numbers and the standard representation of positive numbers it can be inferred from them. The section also considers the representation where the set of digits is $\{1 \ldots p-2\}$. Section 4 defines an assignment, a way of applying rules in the construction process of a tree and the section studies the properties of the representations studies in Sect. 3 with respect to various assignments. Section 7 investigates the contribution of the paper and the problems which remain open. As the detailed proofs can be found in [7–9], the proofs are omitted or shortened to their main ideas when it is possible to do so.

## 2 Metallic Trees

In this section, Sect. 2.1 recalls the vocabulary we shall use in the considerations of the trees which appear in the paper and in the properties connected with them, considering, in particular, the numbering we may attach to the nodes of a finitely generated tree. In Sect. 2.2, we consider the metallic trees which we shall study in this paper.

### 2.1 Preliminary Definitions and Properties of Trees

In the paper, we only consider infinite trees with a bounded branching at each node. Let $\mathcal{T}$ be such a tree. Number its nodes from the root which receives 1, then, level by level and, on each level, from left to right with the conditions that for each node, the numbers of its sons are consecutive numbers. We then say that $\mathcal{T}$ is **numbered** or that it is endowed with its **natural numbering**. In what follows, we shall consider numbered trees only. Clearly, a sub-tree $\mathcal{S}$ of $\mathcal{T}$ can also be numbered in the just above described way but it can also be numbered by the numbers of its nodes in $\mathcal{T}$. In that case, a node $\nu$ may receive two numbers: $n_{\mathcal{S}}$, the number defined in $\mathcal{S}$ as a numbered tree and $n_{\mathcal{T}}$, its number as a node of $\mathcal{T}$. A node may have no son, it is then called a **leaf**. A **path** from $\mu$ to $\nu$ is a finite sequence of nodes $\{\lambda_i\}_{i \in [0 \ldots k]}$, if it exists, such that $\lambda_0 = \mu$, $\lambda_k = \nu$ and, for all $i$ with $i \in [0 \ldots k-1]$, $\lambda_{i+1}$ is a son of $\lambda_i$. A **branch** of $\mathcal{T}$ is a maximal finite or infinite sequence of paths $\{\pi_i\}$ from the root of $\mathcal{T}$ to nodes of that tree such that for all $i, j, \pi_i \subseteq \pi_j$ or $\pi_j \subseteq \pi_i$. Accordingly, a branch connects the root to a leaf or it is infinite. It is clear that for any node, they are connected to the root by a unique path. The **length** of the path from a node to one of its sons is always 1. Let $k$ be the length of a path from $\mu$ to $\nu$. Then, if $\nu$ is not a leaf, the length of the path from $\mu$ to any son of $\nu$ is $k+1$. The length of the path leading from the root to a node $\nu$ of $\mathcal{T}$ is called the **distance** of $\nu$ to the root $\rho$ and it is denoted by dist$\{\rho, \nu\}$. We also define dist$\{\rho, \rho\} = 0$. The **level** $k$ of $\mathcal{T}$ is the set of its nodes which are at the distance $k$ from its root. Denote it by $\mathcal{L}_{k,\mathcal{T}}$. Define $\mathcal{T}_n$ as the set of levels $k$ of $\mathcal{T}$ with $k \leq n$. Say that the **height** of $\mathcal{T}_n$ is $n$. By definition, $\mathcal{T}_n$ is a sub-tree of $\mathcal{T}$. For each node $\nu$ of $\mathcal{T}$, $\lambda_{\mathcal{T}}(\nu)$ is its level in $\mathcal{T}$, i.e. its distance from the root, and $\sigma_{\mathcal{T}}(\nu)$ is the number of its sons. Clearly, if $\nu \in \mathcal{T}_n$ and if $\lambda_{\mathcal{T}}(\nu) = n$, then $\sigma(\nu) = 0$. If $\mathcal{S}$ is a sub-tree of $\mathcal{T}$, denote it by $\mathcal{S} \lhd \mathcal{T}$, and if $\nu \in \mathcal{S}$, then $\lambda_{\mathcal{S}}(\nu) \leq \lambda_{\mathcal{T}}(\nu)$ and the numbers may be not equal.

Consider two infinite numbered trees $T_1$ and $T_2$. Say that $T_1$ and $T_2$ are **isomorphic** if there is a bijection $\beta$ from $T_1$ onto $T_2$ such that:

$$\begin{aligned} &f(n_{T_1}) = n_{T_2} \quad \text{for any } n \in \mathbb{N}. \\ &\lambda_{T_2}(f(n_{T_1})) = \lambda_{T_1}(n). \\ &\sigma_{T_2}(f(n_{T_1})) = \sigma_{T_1}(n). \end{aligned} \tag{1}$$

Clearly, if $T_1$ and $T_2$ are two infinite numbered trees, they are isomorphic if and only if there is a bijection from the nodes of $T_1$ into those of $T_2$ such that a node of $T_1$ and its image in $T_2$ have the same number, they are on the same level of their respective trees and they have the same number of sons.

## 2.2 White and Black Metallic Trees

We call **metallic tree** an infinite tree constituted by two kind of nodes, **b**- and **w**-ones called **black** and **white** respectively, finitely generated by the following rules:

$$\mathbf{b} \rightarrow \mathbf{b}\mathbf{w}^{p-4} \qquad \text{and} \qquad \mathbf{w} \rightarrow \mathbf{b}\mathbf{w}^{p-3}. \tag{2}$$

with $p \geq 5$.

The property for a node to be white or black is called its **literal status**. We also associate to the node its **numerical status**: 0 or 1 depending on whether the node is white or black respectively. If it is not specified, **status** will refer to the literal one.

We shall mainly investigate two kinds of infinite metallic trees. When the root of the tree is a white, black node, we call such a metallic tree a **white**, **black metallic tree** respectively. We denote the infinite white metallic tree by $\mathcal{W}$ and we endow it with its natural numbering. We do the same with the infinite black metallic tree $\mathcal{B}$. Note that we can construct a bijective morphism between $\mathcal{B}$ and a part $\mathbb{B}$ of $\mathcal{W}$ as follows. The morphism is the identity on $\mathbb{B}$ and we fix the following conditions:

$$\sigma_{\mathbb{B}}(1) = \sigma_{\mathcal{W}}(1) - 1,$$
$$\sigma_{\mathbb{B}}(n) = \sigma_{\mathcal{W}}(n), \text{ for all positive integer } n.$$

Moreover, the nodes numbered by $n \in [1 \ldots p-2]$ in $\mathcal{W}$ also belong to $\mathbb{B}$ and receive the same numbers in the natural numbering of $\mathbb{B}$. This morphism allows us to identify $\mathbb{B}$ with $\mathcal{B}$, so that in our sequel, we shall speak of $\mathcal{B}$ only. From what we just said, it is plain that for a node $\nu \in \mathcal{B}$, if $\nu_{\mathcal{B}} > p-2$, then $\nu_{\mathcal{B}} < \nu_{\mathcal{W}}$. We get a hint to the connection between $\nu_{\mathcal{B}}$ and $\nu_{\mathcal{W}}$ in Sect. 3.1. Later on, we shall use $\rightleftharpoons$ to introduce a notation for an expression.

We may wonder whether the simplicity of the rules (2) allow us to give a precise connection between the number of a node and those of its sons, whether in $\mathcal{W}$ or in $\mathcal{B}$. These questions were partially studied in [8]. We shall turn to them in Sect. 4. But before, we need to recall the introduction of an appropriate representation of the numbers used to number the nodes of a metallic tree. It is the goal of Sect. 3 to which we now turn.

## 3 Metallic Numbers

These numbers are introduced by the computation of the number of nodes which lie on a given level of a metallic tree. We consider that point in Sect. 3.1. The sequence allows us to represent the numbers. We consider some basic properties of the standard representation in Sect. 3.2. We study the same properties in the representation where the digits are restricted to $\{1 \ldots p-2\}$ in Sect. 3.3.

### 3.1 The Metallic Sequences

Let $m_n$, $b_n$ be the number of nodes on $\mathcal{L}_{n,\mathcal{W}}$ and $\mathcal{L}_{n,\mathcal{B}}$ respectively. We also define $M_n$, and $B_n$ as the number of nodes of $\mathcal{W}_n$ and $\mathcal{B}_n$ respectively. It appears that these numbers are defined by a simple induction equation as stated in the following statement:

**Theorem 1** ([2, 8]) *Consider the numbers $m_n$ defined as the number of nodes on $\mathcal{L}_{n,\mathcal{W}}$, where $\mathcal{W}$ is the white metallic tree. The numbers $m_n$ satisfy the following induction equation:*

$$m_{n+2} = (p-2)m_{n+1} - m_n \textit{ with } m_0 = 1 \textit{ and } m_{-1} = 0. \tag{3}$$

*We call* ***white metallic sequence*** *the sequence $\{m_n\}_{n\in\mathbb{N}}$.*

See the proof in [8] for instance.

As the black metallic tree is defined by the same rules, we may conclude that the same equation rules the sequence $\{b_n\}_{n\in\mathbb{N}}$:

**Theorem 2** *The sequence $\{b_n\}_{n\in\mathbb{N}}$ of the number of nodes on $\mathcal{L}_{n,\mathcal{B}}$ satisfies the equation:*

$$b_{n+2} = (p-2)b_{n+1} - b_n \text{ with } b_1 = p-3 \text{ and } b_0 = 1. \tag{4}$$

*We call* ***black metallic sequence*** *the sequence $\{b_n\}_{n\in\mathbb{N}}$.*

Note that we could define the white metallic sequence by the initial conditions $m_1 = p-2$ and $m_0 = 1$. In our sequel we shall say **metallic sequence** instead of **white metallic sequence** for a reason which will be made more clear in a while.

Before turning to the properties of the integers with respect to the metallic numbers, we have to consider the numbers $M_n$ and $B_n$ already introduced with respect to the finite trees $\mathcal{W}_n$ and $\mathcal{B}_n$.

**Theorem 3** (see [3]) *On the level $k$ of $\mathcal{W}$, with non-negative $k$, the rightmost node has the number $M_k$, so that the leftmost node on the level $k+1$ has the number $M_k + 1$.*

*On the level $k$ of $\mathcal{B}$ with non-negative $k$, the rightmost node has the number $m_k$, so that the leftmost node on the level $k+1$ has the number $m_k + 1$.*

*The sequence $\{M_n\}_{n\in\mathbb{N}}$ satisfies the following induction equation:*

$$M_{n+2} = (p-2)M_{n+1} - M_n + 1, \tag{5}$$

*with the initial conditions $M_0 = 1$ and $M_{-1} = 0$, while the sequence $\{B_n\}_{n\in\mathbb{N}}$ satisfy the Eq. (3) with the same initial conditions, which means that $B_n = m_n$ for any non-negative $n$. We also have:*

$$M_{n+1} = B_{n+1} + M_n \quad m_{n+1} = b_{n+1} + m_n \tag{6}$$

See the proof in [8].

## 3.2 Metallic Codes for the Nodes of the Metallic Trees

Let us go back to the sequence $\{m_n\}_{n\in\mathbb{N}}$ of metallic numbers. It is clear that the sequence defined by (3) is increasing starting from $m_1$: from (3), we get that $m_{n+2} > (p-3)m_{n+1}$ if we assume that $m_n < m_{n+1}$. As $p \geq 5$, we get that the sequence is increasing starting from $m_1$. Now, as the sequence is increasing, it is known that any positive integer $n$ can be written as a sum of distinct metallic numbers whose terms are defined by Theorem 1:

$$n = \sum_{i=0}^{k} a_i m_i \text{ with } a_i \in \{0\ldots p-3\}. \tag{7}$$

The sum of $a_i m_i$'s in (7) is called the **metallic representation** of $n$ and the $m_i$'s in (7) are the **metallic components** of $n$.

From now on, we use bold characters for the digits of a metallic representation of a number. In particular, we define **d** to represent $p-3$, **c** to represent $p-4$ and **e** to represent $p-5$ when $p>5$. Of course, **0**, **1**, **2** and **3** represent 0, 1, 2 and 3 respectively.

First, note that the representation (7) is not unique.

**Lemma 1** ([3, 8, 10]) *For any integers n and h with* $0 \leq h \leq n$, *we have:*

$$(p-3)m_{n+1} + \sum_{k=h+1}^{n} (p-4)m_k + (p-3)m_h$$

$$= (p-3)m_{n+1} + \sum_{k=h+2}^{n} (p-4)m_k + (p-3)m_{h+1} - m_h + m_{h-1} \qquad (8)$$

**Corollary 1** ([3, 10]) *For any positive integer n, we have:*

$$(p-3)m_{n+1} + \sum_{k=1}^{n}(p-4)m_k + (p-3)m_0 = m_{n+2}$$

$$\textit{so that} \quad \mathbf{d}\mathbf{c}^n\mathbf{d} = 10^{n+2} \qquad (9)$$

See the proofs in [8] for instance.

Let us write the $a_i$'s of (6) as a word $\mathbf{a}_k \ldots \mathbf{a}_1 \mathbf{a}_0$ which we call a **metallic word** for $n$ as the digits $a_i$ which occur in (6) are not necessarily unique for a given $n$. They can be made unique by adding the following condition on the corresponding metallic word for $n$: the pattern $\mathbf{dc}^*\mathbf{d}$ is ruled out from that word. It is called the **forbidden pattern**. Lemma 1 proves that property which is also proved in [3, 10].

When a metallic representation for $n$ does not contain the forbidden pattern it is called the **metallic code** of $n$ which we denote by $[n]$. We shall write $\nu = ([\nu])$ when we wish to restore the number from its metallic code. Let us call **signature** of $\nu$ the rightmost digit of $[\nu] = \mathbf{a}_k \ldots \mathbf{a}_1\mathbf{a}_0$ and denote it by $\mathbf{sg}(\nu)$. Let $\sigma_1$, $\sigma_2$, ..., $\sigma_k$ with $k = p-2$ or $k = p-3$ be the sons of $\nu$. We call **sons signature** of $\nu$ the word $\mathbf{s}_1 \ldots \mathbf{s}_k$, where $\mathbf{s}_i = \mathbf{sg}(\sigma_i)$. We shall denote the literal status of $\nu$ by $\ell s(\nu)$ and its numerical one by $sn(\nu)$.

### 3.3 The Non-zero Metallic Codes

It is known that given a basis $b$ with $b \geq 3$ any positive number $n$ can be written

$$n = \sum_{i=0}^{k} a_i b^i \ \text{ with } \ a_i \in \{1 \ldots b\} \qquad (10)$$

Let $a^+ = a + 1$ and $a^- = a - 1$ for any positive integer $a$. The representation (10) was used by Quine in order to encode any finite sequence of natural numbers: writing $n$ as in (10), $b$ is used as a separator and the other digits which lie in $[1 \ldots b^-]$ can be interpreted as the representations of positive numbers in the base $b^-$ which requires $b \geq 2$. Smullyan, see [11], makes use of such a representation in order to prove Gödel's theorem on the incompleteness of Peano arithmetics.

The question is: taking the metallic numbers $m_n$ as a basis, is it possible to have a representation which rules out all 0's in the representation of a positive number? The answer is given by the following proposition:

**Theorem 4** *Let n be a positive natural number. Then it is possible to write n as:*

$$n = \sum_{i=0}^{k} a_i\, m_i \quad \text{with } a_i \in \{1 \ldots b\}, \tag{11}$$

*where* $b = p - 2$.

The easy proof on induction can be found in [9].

When we use the representation (11) to write a positive integer, we use **x** to denote the digit whose value is $b = p{-}2$. We know that the representation with the metallic numbers of a positive integer is not necessarily unique. This is why we needed to select a criterion in order to ensure the uniqueness of the metallic code. It is also the case that the representation (11) is not unique, despite the fact that it satisfies the constraint of no **0** among the digits. To see that point, we need the following result which enlarges a lemma from [8]:

**Lemma 2** *For any integers n and h with* $0 \leq h \leq n$*, we have:*

$$(p-2)m_{n+1} + \sum_{k=h+1}^{n} (p-3)m_k + (p-2)m_h$$

$$= (p-2)m_{n+1} + \sum_{k=h+2}^{n} (p-3)m_k + (p-2)m_{h+1} + m_{h-1} \tag{12}$$

**Corollary 2** *For any positive integer n, it has a unique representation (11) provided that the pattern* **xd*****x** *is ruled out and is called a* ***forbidden pattern****. A metallic code where the* **0***-digit is ruled out and where the forbidden pattern does not occur is called a* ***non-zero metallic code****,* **nzm***-code for short. The* **nzm***-code of n is denoted by* $[\nu]_{nz}$.

*In order to get a* $[\nu]_{nz}$ *from (11), we apply the conversion rules:*

$$\mathbf{nxd}^k\mathbf{xm} = \mathbf{n}^+\mathbf{1}^{k+2}\mathbf{m}^+ \qquad \text{and} \qquad \mathbf{n}^+\mathbf{0}^{k+2}\mathbf{m}^+ = \mathbf{ndc}^k\mathbf{dm} \tag{13}$$

*where* **n**, **m** *are non-zero digits in* $\{\mathbf{1} \ldots \mathbf{x}\}$, $\mathbf{n}^+ = \mathbf{n}\oplus\mathbf{1}$ *and* $\mathbf{m}^+ = \mathbf{m}\oplus\mathbf{1}$, *with* $\mathbf{a}\oplus\mathbf{1} = \mathbf{a} + 1$ *if* $a + 1 < p - 2$ *and* $\mathbf{a}\oplus\mathbf{1} = \mathbf{0}$ *if* $\mathbf{a} = p - 3$.

We call the forbidden pattern defined in Corollary 2 the **nzm**-forbidden pattern in order to distinguish it from that of Lemma 1. Note that the forbidden pattern of Lemma 1 is no more forbidden in an **nzm**-code. Of course, the application of (13) may be repeated in (11) as long as all occurrences of the **nzm**-forbidden pattern are replaced by their permitted equivalent expression given in (13).

In [8], we gave algorithms for incrementing or decrementing a number given by its metallic code, the algorithm returning a metallic code too. In [9], the corresponding algorithms are given with respect to the **nzm**-codes.

We just mention the key points of these algorithms.

Firstly, for metallic codes. Then, let us look at incrementation. If the signature is not **d**, there is no difficulty. Otherwise, as there is no forbidden pattern in the code, we have a suffix $\mathbf{hc}^k$ with $\mathbf{h} \neq \mathbf{d}$. In that case, $\mathbf{c}^k$ is replaced by $\mathbf{c}^{k-2}\mathbf{d0}$. Otherwise, we have a suffix $\mathbf{hdc}^k$ with $\mathbf{h} \neq \mathbf{d}$, which is replaced by $\mathbf{j0}^{k+1}$, where $\mathbf{j} = \mathbf{h{+}1}$.

Now, let us look at decrementation. If the signature is not zero, there is no difficulty. Otherwise, we have a suffix $\mathbf{a0}^k$, where $\mathbf{a} \neq \mathbf{0}$. Then, $\mathbf{a0}^k$ is replaced by $\mathbf{hdc}^{k-1}$, where $\mathbf{h} = \mathbf{a}-\mathbf{1}$.

Both algorithms are direct consequences of Corollary 1.

Secondly, consider the **nzm**-codes. We begin with incrementation: no problem if the signature is less than **d**. Otherwise, we have a suffix $\mathbf{hxd}^k$ where $\mathbf{h} < \mathbf{x}$. It is replaced by $\mathbf{j1}^{k+1}$ where $\mathbf{j} = \mathbf{h{+}1}$.

The decrementation algorithm works exactly in the opposite way. No problem if the signature is not **1**. Otherwise, we have a suffix $\mathbf{a1}^k$, where $\mathbf{a} > \mathbf{1}$. The suffix is replaced by $\mathbf{hxd}^{k-1}$, where $\mathbf{h} = \mathbf{a}{-}\mathbf{1}$.

This time, both just mentioned algorithms are direct consequences of Corollary 2.

Note that the just mentioned corollaries are tightly connected: it can be remarked that if we subtract the forbidden pattern for the metallic code $\mathbf{dc}^k\mathbf{d}$ from the forbidden **nzm**-code $\mathbf{xd}^k\ \mathbf{x}$ we get $\mathbf{1}^{k+2}$.

Also note that the algorithms we have given in natural language can easily be translated into formalised ones.

## 4 Metallic Trees and Their Assignments

In this section, we consider the notion of assignment which we define in Sect. 4.1. There, we also consider a tool to compare assignments. In Sect. 4.2, we focus our attention on a particular assignment we call penultimate. In Sect. 4.3, we characterise a property shared by the assignments, a property which we shall discover with the penultimate one. In Sect. 4.4, we look at another assignment, the mid-assignment which allows us to have a new look on the particular assignments investigated in the previous sections. In Sect. 4.5, we shall investigate the properties of the assignments in the frame of the **nzm**-codes.

### 4.1 Assignments in Metallic Trees

In Sect. 2.1, we recalled the definition of the **natural numbering** of $\mathcal{W}$ and $\mathcal{B}$. Consider those trees. We can see each of them as an infinite sequence $\mathcal{L}_n$ of finite sequences of numbers defined by $\mathcal{L}_{n+1} = \{U_n + 1 \ldots U_{n+1}\}$, where $U_n = M_n$ for all $n$ or $U_n = B_n$ for all $n$, depending on whether we consider $\mathcal{W}$ or $\mathcal{B}$. In both cases, we can see the application of the rules (2) as an application $\alpha$ which, to each node $\nu$ of the level $n$ associates three numbers $\ell_\nu$, $s_\nu$ and $b_\nu$ such that $s_\nu$ is the numeral status of $\nu$ under $\alpha$, $\ell_\nu$ is the leftmost node of an interval $I_\nu$ of $\mathcal{L}_{n+1}$ with the conditions:

$$\text{for all } \nu, \qquad I_\nu \cap I_{n+1} = \emptyset \qquad \text{and} \qquad \sum_{\nu \in \mathcal{L}_n} |I_\nu| = m_{n+1}, \tag{14}$$

and $b_\nu$ is the position of the black node associated to $\nu$ among the nodes of the interval $I_\nu$, the leftmost position being 1. The nodes belonging to $I_\nu$ are called the **sons of** $\nu$ **under** $\alpha$, for short: $\alpha$-**sons**. They are simply called sons only when it is clear which assignment is considered. The conditions (14) can equivalently be stated as:

$$\begin{gathered}
\text{for all } \nu \text{ with } \nu \in \mathcal{L}_n, \alpha(\nu) = (\alpha_\ell(\nu), \alpha_b(\nu), \alpha_s(\nu)),\\
\text{with } \alpha_\ell(\nu) \in \mathcal{L}_{n+1}, \alpha_s(\nu) \in \{0, 1\}, \alpha_b(\nu) \in \{1 \ldots p-2-\alpha_{s(\nu)}\},\\
\text{for all } \nu, \sum_{k=\alpha_\ell(\nu)}^{\alpha_\ell(\nu+1)-1} \alpha_s(\nu) = 1 \text{ and, for any } \nu \in [M_{n-1}+1 \ldots M_n],\\
\text{for any positive } \nu, \alpha_s(\alpha_b(\nu)) = 1,\\
\alpha_\ell(\nu+1) = \alpha_\ell(\nu) + p - 2 - \alpha_s(\nu), \text{ and } \alpha_\ell(M_n) = M_{n+1} - p + 3 + \alpha_s(M_n).
\end{gathered} \tag{15}$$

We call **assignment** an application $\alpha$ which satisfies (15). We denote by $\mathcal{W}_\alpha$ the white metallic tree $\mathcal{W}$ dotted with the assignment $\alpha$: it means that, starting from the root, the status of each node $\nu$ under $\alpha$ is defined by $\alpha_s(\nu)$ and that the position of the black son of $\nu$ among its $\alpha$-sons is defined by $\alpha_b(\nu)$. When $\alpha$ is associated with the rules (2), we additionally have that $\alpha_b(\nu) = 1$ for all node $\nu$. That assignation is called the **leftmost assignment**. It was called the **standard assignment** in [2] which considers white metallic trees only in the case when $p = 5$. In [10], another assignment was considered, defined by:

$$\alpha(\nu) = (\ell_\nu, p - 3 - sn(\nu), sn(\nu)) \text{ for all } \nu. \tag{16}$$

It is not difficult to see that whether $\ell s(\nu)$ is **b** or **w**, the black son is the penultimate son of $\nu$. For this reason, we call (16) the **penultimate assignment**. Similarly, we define the **rightmost assignment** by

$$\alpha(\nu) = (\ell_\nu, p - 2 - sn(\nu), sn(\nu)) \text{ for all } \nu. \tag{17}$$

Say that an assignment $\alpha$ is an **a-assignment** if and only if for any node $\nu$, one of its sons exactly has **a** as its signature. It means that for one son of $\nu$ and for one of them only, its code has **a** among its suffixes. We say that an assignment $\alpha$ has the **preferred son property** if, for any node $\nu$, exactly one of its sons has the code $[\nu]\mathbf{0}$. Note that an assignment which possesses the preferred son property is also a **0**-assignment. Accordingly, the preferred son property assumes that we consider the representation of the numbers by their metallic codes.

Say that an assignment $\alpha$ is a **b-a-assignment** if and only if all black nodes of the tree and only them have **a** as their signature. We can note that the notion of **b-a**-assignment is meaningful also in the case of the representation of the numbers by their **nzm**-code when it exists. We shall see a bit later that there are many values of **a** for which the **b-a**-assignment exists.

In order to establish the property characterised in Sect. 4.3, we consider the following tool which measures the distance between two assignments as follows. Let $\alpha$ and $\beta$ be two assignments of the white metallic tree. Call **apartness between** $\alpha$ **and** $\beta$ denoted by $\delta_{\alpha\beta}$ the function defined by $\delta_{\alpha\beta}(\nu) = \beta_\ell(\nu) - \alpha_\ell(\nu)$ for any node $\nu$ of $\mathcal{W}$. We have the easy property:

**Lemma 3** *Let $\alpha$, $\beta$ and $\gamma$ be three assignments on the white metallic tree. For any node $\nu$ of $\mathcal{W}$ we have:*

$$\delta_{\alpha\beta}(\nu) = \delta_{\gamma\beta}(\nu) - \delta_{\gamma\alpha}(\nu). \tag{18}$$

Consider the metallic codes which are associated to the numbers by (7), see Sect. 3.2. Consider the metallic codes of the nodes which lie on $\mathcal{L}_1$. One of them only has the signature **0**: it is the node numbered by $p-2$ whose metallic code is **10**. Consider the nodes on $\mathcal{L}_2$. Their numbers grow from $M_1 + 1$ up to $M_2$ and the metallic codes go from **12** to **111**. The nodes whose signature is **0** are: **20**, **30**, …, **c0**, **d0**, **100**, **110**. We can see that the distance between two consecutive such nodes is $p-2$ except for **d0** with **100** whose distance is $p-3$. More generally, call **0-node** any node of $\mathcal{W}$ whose signature is **0**. On $\mathcal{L}_{n+1}$, the **0**-nodes run from $\mathbf{1}^{n-1}\mathbf{20}$ up to $\mathbf{1}^{n+1}\mathbf{0}$. Note that if we erase the last digit of those codes, we get the codes from $\mathbf{1}^{n-1}\mathbf{2}$ up to $\mathbf{1}^{n+1}$, i.e. the metallic codes of the nodes on $\mathcal{L}_n$. Accordingly, the number of **0**-nodes on $\mathcal{L}_{n+1}$ is the number of nodes on $\mathcal{L}_n$. Moreover, we can observe that the distance between two consecutive **0**-nodes on a level is $p-2$ of $p-3$. When it is $p-3$? On the level $\mathcal{L}_2$, the distance $p-3$ occurs between **d0** and **100**. Indeed, $\mathbf{100}\ominus\mathbf{1}=\mathbf{dc}$ and the distance between **d0** and **dc** is $p-4$. More generally, we can state:

**Lemma 4** *On the level $\mathcal{L}_n$ let $\mu$ and $\nu$ be two consecutive* **0**-*nodes, with $\mu < \nu$. Then $\nu - \mu = p - 3$ if and only if $[\nu] = [\omega]\mathbf{0}^k$ with $k \geq 2$. When it is not the case, $\nu - \mu = p - 2$.*

The reader is referred to [9] for the proof which is based on a carefull examination of the decrementation algorithm for the metallic codes.

Is there a connection between those values between two consecutive **0**-nodes and the smaller occurrence of the smaller distance with the distinction between white and black nodes which have $p-2$ and $p-3$ nodes respectively?

That issue is addressed by the next sub-section.

## 4.2 The Penultimate Assignment

Say that an assignment $\alpha$ possesses the **preferred son** property if and only if for any node $\nu$ of $\mathcal{W}$, the signature of one of its sons under $\alpha$ and one of them only is **0** and if the metallic code of that son is $[\nu]\mathbf{0}$. When an assignment $\alpha$ possesses the preferred son property, for any node $\nu$, the node whose signature is $[\nu]\mathbf{0}$ is called its **preferred son** under $\alpha$. Note that if $\alpha$ and $\beta$ are two assignments which possess the preferred son property, for each node, the preferred son under $\alpha$ and that under $\beta$ coincide.

We can state:

**Theorem 5** *The penultimate assignment possesses the preferred son property and it is the* **b-0***-assignment.*

***Proof*** It is based on the following property on the signatures and on the metallic codes of the sons with respect to those of the node.

**Lemma 5** *Let $\nu$ be a node of $\mathcal{W}$ equipped with the penultimate assignment $\pi$. The signatures of its sons under $\pi$ is defined by the following rules:*

$$\mathbf{b0} \rightarrow \mathbf{w2}(\mathbf{wa})^{p-7}\mathbf{wc}.\mathbf{b0}.\mathbf{w1},\ \mathbf{wa} \rightarrow \mathbf{w2}(\mathbf{wa})^{p-6}\mathbf{wd}.\mathbf{b0}.\mathbf{w1}, \tag{19}$$

*The metallic codes of the sons of $\nu$ under $\pi$ are given by the following table, where $\mathbf{a}_k \ldots \mathbf{a}_1\mathbf{a}_0 \rightleftharpoons [\nu]$ and $\mathbf{b}_k \ldots \mathbf{b}_0 \rightleftharpoons [[\mathbf{a}_k \ldots \mathbf{a}_0]\ominus\mathbf{1}]$ and* **a** *is in* $\{\mathbf{1} \ldots \mathbf{d}\}$ *in lines* **4–6**.

| $\nu$ | *range* | *son* | *metalliccode* | **ref.** |
|---|---|---|---|---|
| **b0** | $1 \ldots p-5$ | $h$ | $\mathbf{b}_k \ldots \mathbf{b}_0\mathbf{h}^+$ | 1 |
| | $p-4$ | | $\mathbf{a}_k \ldots \mathbf{a}_0\mathbf{0}$ | 2 |
| | $p-3$ | | $\mathbf{a}_k \ldots \mathbf{a}_0\mathbf{1}$ | 3 |
| **wa** | $1 \ldots p-4$ | $h$ | $\mathbf{b}_k \ldots \mathbf{b}_0\mathbf{h}^+$ | **4** |
| | $p-3$ | | $\mathbf{a}_k \ldots \mathbf{a}_0\mathbf{0}$ | **5** |
| | $p-2$ | | $\mathbf{a}_k \ldots \mathbf{a}_0\mathbf{1}$ | **6** |

(20)

Clearly, Theorem 5 follows from Lemma 5. The reader is referred to [9] for the proof which is based on complete induction on $\nu$, distinguishing the case when the node is white and when it is black.

As proved in [9], we can transport an assignment $\alpha$ on a particular sub-tree of $\mathcal{W}$:

**Theorem 6** *Consider the application $\varphi$ such that $\varphi([\nu]\mathbf{0}) = ([\nu])$, defining a bijection of $\mathcal{W}$ on the set $\mathcal{T}$ of the* **0***-nodes of $\mathcal{W}$. Define the sons of a node $\nu$ of $\mathcal{T}$ as the* **0***-nodes of $\mathcal{W}$ which are in the subtree $\mathcal{S}$ of $\mathcal{W}$ rooted at $([\nu]\mathbf{0})$ and which are on the level 2 of $\mathcal{S}$. Define the status of $\nu$ in $\mathcal{T}$ as the status of $([\nu]\mathbf{0})$. Then, the bijection $\varphi$ defines an isomorphism of $\mathcal{W}$ onto $\mathcal{T}$ which transports the assignment $\alpha$ onto $\mathcal{T}$ whatever $\alpha$.*

### 4.3 Assignments and the Preferred Son Property

In this subsection, we shall see that all assignments on the white metallic tree possess the preferred son property.

To that purpose, we compare the assignments to the same one: the leftmost assignment which we denote by $\lambda$. The reason of this choice lies in the following property:

**Lemma 6** *For any node $\nu$ of $\mathcal{W}$, we have:*

$$0 \leq \delta_{\lambda\alpha}(\nu) \leq 1 \tag{21}$$

We recall that $st_\alpha(\nu)$ is the status of the node $\nu$ under $\alpha$. We refer the reader to [9] for the proof of the lemma which is based on the following easy property:

**Lemma 7** *Let $\lambda$ be the leftmost assignment on $\mathcal{W}$ and let $\alpha$ be an assignment on $\mathcal{W}$. If $st_\lambda(\nu) = st_\alpha(\nu)$ then, $\delta_{\lambda\alpha}(\nu+1) = \delta_{\lambda\alpha}(\nu)$. Otherwise, if $st_\lambda(\nu) = \mathbf{w}$ then $\delta_{\lambda\alpha}(\nu+1) = \delta_{\lambda\alpha}(\nu) - 1$ and if $st_\lambda(\nu) = \mathbf{b}$, then $\delta_{\lambda\alpha}(\nu+1) = \delta_{\lambda\alpha}(\nu) + 1$.*

by proving the stronger property:

$$\delta_{\lambda\alpha}(\alpha_\ell)(\nu) = \delta_{\lambda\alpha}(\alpha_\ell)(\nu+1) = 0 \tag{22}$$

which, in some sense, does not depend on the status of $\nu$.

**Theorem 7** *Any assignment $\alpha$ on $\mathcal{W}$ does possess the preferred son property. In $\mathcal{W}_\alpha$, whatever $\alpha$, we have that $m_{n+1}$ is the preferred son of $m_n$. Call the sequence of nodes $\{m_n\}_{n\in\mathbb{N}}$ the **0-branch**.*

The detailed proof is in [9]. It starts by proving that $\lambda$, the leftmost assignment, possesses the preferred son property. It is obtained by proving an analog of the rules (19) and Eq. (20) for $\lambda$, namely:

**Lemma 8** *Let $\nu$ be a node of $\mathcal{W}$ equipped with the leftmost assignment $\lambda$. The signatures of its sons under $\lambda$ is defined by the following rules:*

$$\begin{array}{c}\mathbf{b1}, \mathbf{b2} \rightarrow \mathbf{b2}(\mathbf{wa})^{p-6}\mathbf{wd}.\mathbf{w0},\ \mathbf{wa} \rightarrow \mathbf{b1}(\mathbf{wa})^{p-5}\mathbf{wd}.\mathbf{w0}, \\ \mathbf{w0} \rightarrow \mathbf{b1}(\mathbf{wa})^{p-6}\mathbf{wc}.\mathbf{w0}.\mathbf{w1},\ \mathbf{w1} \rightarrow \mathbf{b2}(\mathbf{wa})^{p-6}\mathbf{wd}.\mathbf{w0}.\mathbf{w1}.\end{array} \tag{23}$$

*The metallic codes of the $\lambda$-sons of $\nu$ are given by the following table, where $\mathbf{a}_k \ldots \mathbf{a}_1\mathbf{a}_0 \rightleftharpoons [\nu]$ and $\mathbf{b}_k \ldots \mathbf{b}_0 \rightleftharpoons [[\mathbf{a}_k \ldots \mathbf{a}_0]\ominus\mathbf{1}]$, and in lines **9** and **10**, **a** is in $\{\mathbf{2} \ldots \mathbf{d}\}$.*

(24)

| $v$ | $range$ | $son$ | $metalliccode$ | **ref**. |
|---|---|---|---|---|
| **b1**, **b2** | $1\dots p-4$ | $h$ | $\mathbf{b}_k\dots\mathbf{b}_0\mathbf{h}^+$ | 1 |
| | $p-3$ | | $\mathbf{a}_k\dots\mathbf{a}_0\mathbf{0}$ | 2 |
| **w0** | $1\dots p-4$ | $h$ | $\mathbf{b}_k\dots\mathbf{b}_0\mathbf{h}$ | 3 |
| | $p-3$ | | $\mathbf{a}_k\dots\mathbf{a}_0\mathbf{0}$ | 4 |
| | $p-2$ | | $\mathbf{a}_k\dots\mathbf{a}_0\mathbf{1}$ | 5 |
| **w1** | $1\dots p-4$ | $h$ | $\mathbf{b}_k\dots\mathbf{b}_0\mathbf{h}^+$ | 6 |
| | $p-3$ | | $\mathbf{a}_k\dots\mathbf{a}_0\mathbf{0}$ | 7 |
| | $p-2$ | | $\mathbf{a}_k\dots\mathbf{a}_0\mathbf{1}$ | 8 |
| **wa** | $1\dots p-3$ | $h$ | $\mathbf{b}_k\dots\mathbf{b}_0\mathbf{h}$ | 9 |
| | $p-2$ | | $\mathbf{a}_k\dots\mathbf{a}_0\mathbf{0}$ | 10 |

Then, the proof compares $\lambda$ and $\pi$, using the property stated by Theorem 5 that the **0**-nodes coincide with the black nodes under $\pi$.

Note that the rules can be identified by their left-hand side part, which we shall do later on.

For what is the righmost assignment we have the following rules and table, see [9] for the proof:

$$\mathbf{b1}, \mathbf{b2} \rightarrow \mathbf{w3}(\mathbf{wa})^{p-7}\mathbf{wd}.\mathbf{w0}.\mathbf{b1}, \ \mathbf{wa} \rightarrow \mathbf{w2}(\mathbf{wa})^{p-6}\mathbf{wd}.\mathbf{w0}.\mathbf{b1}, \\ \mathbf{w0} \rightarrow \mathbf{w2}(\mathbf{wa})^{p-7}\mathbf{wc}.\mathbf{w0}.\mathbf{w1}.\mathbf{b2}, \ \mathbf{w1} \rightarrow \mathbf{w3}(\mathbf{wa})^{p-7}\mathbf{wd}.\mathbf{w0}.\mathbf{w1b2}. \tag{25}$$

(26)

| $v$ | range | son | metallic code | **ref**. |
|---|---|---|---|---|
| **b1**, **b2** | $1\dots p-5$ | $h$ | $\mathbf{b}_k\dots\mathbf{b}_0\mathbf{h}^{++}$ | 1 |
| | $p-4$ | | $\mathbf{a}_k\dots\mathbf{a}_0\mathbf{0}$ | 2 |
| | $p-3$ | | $\mathbf{a}_k\dots\mathbf{a}_0\mathbf{1}$ | 3 |
| **w0** | $1\dots p-5$ | $h$ | $\mathbf{b}_k\dots\mathbf{b}_0\mathbf{h}^+$ | 4 |
| | $p-4$ | | $\mathbf{a}_k\dots\mathbf{a}_0\mathbf{0}$ | 5 |
| | $p-3$ | | $\mathbf{a}_k\dots\mathbf{a}_0\mathbf{1}$ | 6 |
| | $p-2$ | | $\mathbf{a}_k\dots\mathbf{a}_0\mathbf{2}$ | 7 |
| **w1** | $1\dots p-5$ | $h$ | $\mathbf{b}_k\dots\mathbf{b}_0\mathbf{h}^{++}$ | 8 |
| | $p-4$ | | $\mathbf{a}_k\dots\mathbf{a}_0\mathbf{0}$ | 9 |
| | $p-3$ | | $\mathbf{a}_k\dots\mathbf{a}_0\mathbf{1}$ | 10 |
| | $p-2$ | | $\mathbf{a}_k\dots\mathbf{a}_0\mathbf{2}$ | 11 |
| **wa** | $1\dots p-3$ | $h$ | $\mathbf{b}_k\dots\mathbf{b}_0\mathbf{h}^+$ | 12 |
| | $p-2$ | | $\mathbf{a}_k\dots\mathbf{a}_0\mathbf{0}$ | 13 |
| | $p-2$ | | $\mathbf{a}_k\dots\mathbf{a}_0\mathbf{1}$ | 14 |

Note that **a** is in $\{\mathbf{2}\dots\mathbf{d}\}$ in lines **12–14**.

Figures 1 and 2 illustrate the property proved in Theorem 7 for the leftmost and the rightmost assignments respectively. In the figures, the red colour is used to mark the black nodes, while the white ones have a blue and a green colour. The blue and the green colours are used to distinguish between the different kinds of white nodes which appear in Eqs. (20), (24) and (26). The blue nodes correspond to the nodes

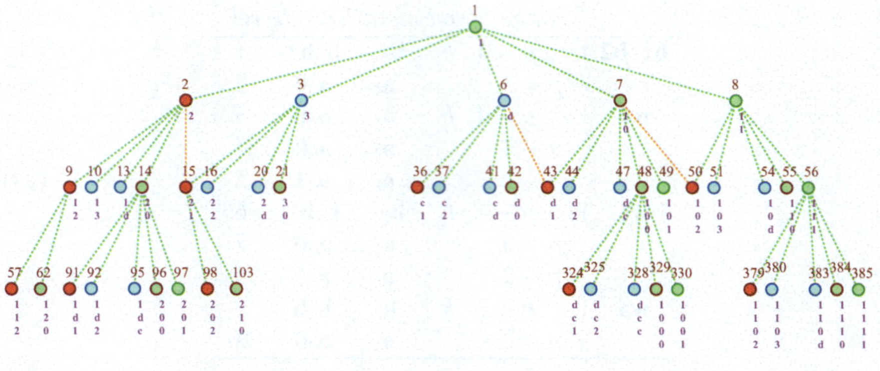

**Fig. 1** The white metallic tree. Partial representation of the first three levels of the tree when $p = 9$ with the conventions mentioned in the text

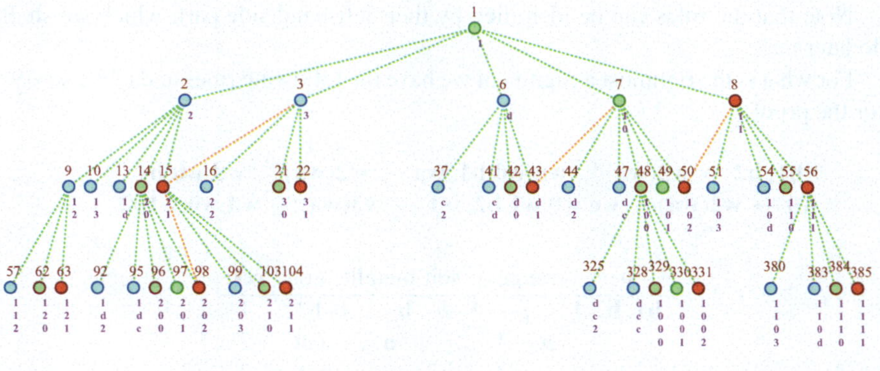

**Fig. 2** The white metallic tree under the rightmost assignment. Partial representation of the first three levels of the tree when $p = 9$ with the conventions mentioned in the text

marked by **wa**. The **w0**-nodes are indicated by a green disk with a red circle while the **w1**-nodes are indicated by a green disk with a darker green circle. The numbers in red, above the nodes, indicate the natural numbering of the tree. The metallic code is mentioned vertically, below each node.

In those illustrations, $p = 9$. However, in order to indicate the general form of the properties, in the metallic codes, **5** and **6** are replaced by **c** and **d** respectively. Indeed, it corresponds for $p = 9$ to the general values given to **c** and to **d** respectively. In order to make easier the reading of the figures, not all nodes are mentioned. We just mention those which allow us to see the application of the rules (23) and to check Eq. (24).

We defined the **0**-branch which connects the **0**-nodes we obtain which are the **0**-son of the previous one except the first one which is the root. We noted that the **0**-branch does not depend on the assignment $\alpha$ with which we equipped $\mathcal{W}$. Now, if we take a node $\nu$ whose signature is not **0**. It has a unique $\alpha$-son $\sigma$ which is a **0**-node, and we know that the position of $\sigma$ in $\mathcal{W}$ does not depend on $\alpha$. What depends on $\alpha$

is the position of $\sigma$ among the $\alpha$-sons of $\nu$. As an example, $m_{n+1}$ is the penultimate $\lambda$-son of $m_n$ while it is its ante-penultimate $\rho$-son. From what we just mentioned, we can construct a sequence $\{\varphi_n\}_{n\in\mathbb{N}}$ of nodes such that $\varphi_0 = \nu$ and $\varphi_{n+1}$ is the $\mathbf{0}$-son of $\varphi_n$ for any $n$. From Theorem 7, we know that $\varphi_{n+1}$ is always an $\alpha$-son of $\varphi_n$ and again, its position does not depend on $\alpha$. Call the sequence $\{\varphi_n\}_{n\in\mathbb{N}}$ the **0-path issued from** $\nu$. From Lemma 4,we can state:

**Theorem 8** *For any assignment $\alpha$, the $\mathbf{0}$-paths indicate the nodes in $\mathcal{W}_\alpha$ at which the application of the incrementation algorithm necessitates a carry, which produce the $\mathbf{0}$-signature of the metallic code of the node.*

## 4.4 Mid-assignments in the White Metallic Tree

Before turning to the connections between the assignments on $\mathcal{W}$ and the **nzm**-codes, we deal with a particular fixed assignment which, in some sense, synthesizes the properties we observed on the leftmost, the penultimate and the rightmost assignments. We say that an assignment is **fixed** if the black nodes are always applied the same rule and if it is the same for the white nodes. In this sub section,we consider what we call a **mid-assignment**. A mid-assignment is defined by a constant $k$ with $k \in \{2 \ldots p-4\}$ which defines the position of the black son among the sons of a node, avoiding the positions we already studied. Denote by $\mathcal{W}_{\mu,k}$ the white metallic tree equipped with such an assignment. It is illustrated by Fig. 3 in the case when $p = 7$ and $k = 4$.

The rules for the nodes are given by (27) and the sons of a node by (28). We note that the root obeys the rule **wa** of (27). From the metallic code of the leftmost son and from Lemma 4, we can see that the rule **wa** is applied until we meet the sons of the first black node on the father level. We also can check that lines **3–5** of Eq. (28) are observed.

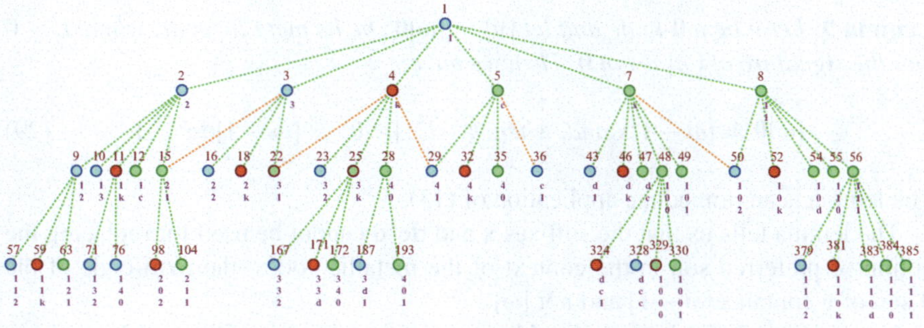


**Fig. 3** The white metallic tree. Partial representation of the first three levels of the tree when $p = 9$ with the conventions mentioned in the text

$$\begin{aligned}
&\mathbf{bk} \to \mathbf{w2}\ldots\mathbf{wk}^{-}\ldots\mathbf{bk}.\mathbf{wk}^{+}\ldots\mathbf{wd}.\mathbf{w0},\\
&\mathbf{wa} \to \mathbf{w2}\ldots\mathbf{wk}^{-}\mathbf{bk}.\mathbf{wk}^{+}\ldots\mathbf{wd}.\mathbf{w0}.\mathbf{w1},\ \text{with}\ \mathbf{0} < \mathbf{a} < \mathbf{k},\\
&\mathbf{wb} \to \mathbf{w1}\ldots\mathbf{wk}^{-}\mathbf{bk}.\mathbf{wk}^{+}\ldots\mathbf{wd}.\mathbf{w0},\ \text{with}\ \mathbf{k} < \mathbf{b} \leq \mathbf{d},\\
&\mathbf{w0} \to \mathbf{w1}\ldots\mathbf{wk}^{-}\mathbf{bk}.\mathbf{wk}^{+}\ldots\mathbf{wc}.\mathbf{w0}.\mathbf{w1}.
\end{aligned} \tag{27}$$

(28)

| $\nu$ | range | son | **nzm**-code | **ref**. |
|---|---|---|---|---|
| **bk** | $1\ldots p-4$ | $j$ | $\mathbf{a}_h\ldots\mathbf{a}_1\mathbf{a}_0^{-}\mathbf{j}^{+}$ | 1 |
| | $p-3$ | | $\mathbf{a}_k\ldots\mathbf{a}_1\mathbf{a}_0\mathbf{0}$ | 2 |
| **wa** | $1\ldots p-4$ | $j$ | $\mathbf{a}_h\ldots\mathbf{a}_1\mathbf{a}_0^{-}\mathbf{j}^{+}$ | 3 |
| | $p-3$ | | $\mathbf{a}_k\ldots\mathbf{a}_1\mathbf{a}_0\mathbf{0}$ | 4 |
| | $p-2$ | | $\mathbf{a}_k\ldots\mathbf{a}_1\mathbf{a}_0\mathbf{1}$ | 5 |
| **wb** | $1\ldots p-3$ | $j$ | $\mathbf{a}_h\ldots\mathbf{a}_1\mathbf{a}_0^{-}\mathbf{j}$ | 6 |
| | $p-2$ | | $\mathbf{a}_k\ldots\mathbf{a}_1\mathbf{a}_0\mathbf{0}$ | 7 |
| **w0** | $1\ldots p-4$ | $j$ | $\mathbf{a}_h\ldots\mathbf{a}_1\mathbf{a}_0^{-}\mathbf{j}$ | 8 |
| | $p-3$ | | $\mathbf{a}_k\ldots\mathbf{a}_1\mathbf{a}_0\mathbf{0}$ | 9 |
| | $p-2$ | | $\mathbf{a}_k\ldots\mathbf{a}_1\mathbf{a}_0\mathbf{1}$ | 10 |

The reader can find in [9] the justification of rules (27) and of Eq. (28).

## 4.5 *Assignments in* $\mathcal{W}$ *and the* **nzm**-*Codes*

As the **0**-signature has no more any meaning in **nzm**-codes, the property of the preferred son can be reformulated as follows: is there a value **a** such that each node $\nu$ has among its $\alpha$-sons exactly one of them whose **nzm**-code is $[\nu]\mathbf{a}$ for at least one assignment $\alpha$?

Before addressing that issue, note that we can easily characterise in **nzm**-terms the nodes whose signature is **0** in the metallic code. As far as the nodes do not change but their sons according to the assignment $\alpha$ we set on $\mathcal{W}$, let us still call those nodes **0**-nodes even in that context. We have:

**Lemma 9** *Let* $\nu$ *be a* **0**-*node and let* $[\nu] = [\omega]\mathbf{0}^k$ *be its metallic code, where* $k > 0$ *and the signature of* $[\omega]$ *is not* **0**. *Then we have:*

$$[\omega]\mathbf{0} = [\omega-1]\mathbf{x}\ \textit{and, when}\ k \geq 2,\ [\omega]\mathbf{0}^k = [\omega-1]\mathbf{d}\mathbf{c}^{k-2}\mathbf{d} \tag{29}$$

The lemma is an immediate application of (13).

The lemma tells us that the suffixes **x** and **dc*****d** cannot be used for replacing the notion of preferred son in the context of the metallic codes: the **nzm**-code of the **0**-son of $\nu$ contains $[\omega-1]$ and not $[\omega]$.

Now, it is not difficult to see that **11** occurs among the sons of the root **1**, whatever the assignment. Also, the first nodes on the level $\mathcal{L}_2$ are **12**, …, **1x**, **21** and **21** is the $p-2$th node. Accordingly, if **2** is $\alpha$-white, **21** occurs as its rightmost $\alpha$-son. Let us call **1-nodes** the nodes of $\mathcal{W}$ whose signature of their **nzm**-code is **1**. We may

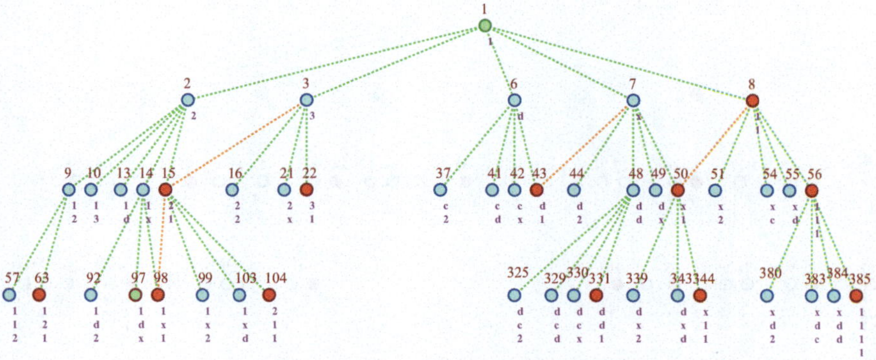


**Fig. 4** The white metallic tree and the rightmost assignment under the **nzm**-codes for the nodes. Partial representation of the first three levels of the tree when $p = 9$ with the conventions mentioned in the text

wonder what is the distribution of the **1**-nodes in $\mathcal{W}$? In fact, what we already said is a valuable hint to the solution: from Lemma 2 and its Corollary 2, we know that $[\omega - 1]\mathbf{x}\oplus\mathbf{1} = [\omega]\mathbf{1}$ and that $[\omega - 1]\mathbf{xd}^{k+1}\oplus\mathbf{1} = [\omega]\mathbf{1}^{k+2}$. This allows us to prove:

**Lemma 10** *Let $\mu$ and $\nu$ be two consecutive* **1***-nodes of the level $\mathcal{L}_n$ with $\mu < \nu$. Then $\nu - \mu = p - 3$ if and only if* **11** *is a suffix of $\nu$ and, when it is not the case, $\nu - \mu = p-2$.*

An immediate corollary is that whatever the assignment, there are infinitely many nodes $\nu$ so that **x** does not occur in their sons signatures. The detailed proof of Lemma 10 is in [9] and makes use of considerations on the decrementation algorithm on **nzm**-codes.

For any node $\nu$, call **successor** of $\nu$, denoted by $succ(\nu)$, the node whose **nzm**-code is $[\nu]_{nz}\mathbf{1}$. Lemma 10 and our study of the penultimate assignment on $\mathcal{W}$ with respect to the metallic codes suggests to state:

**Theorem 9** *Let $\mathcal{W}$ equipped with the rightmost assignment $\rho$ and consider the* **nzm***-codes of its nodes. Then, for any node $\nu$, its successor occurs among its $\rho$-sons and no other $\rho$-sons of $\nu$ is a* **1***-node, so that we can call $[\nu]_{nz}\mathbf{1}$ the* **nzm***-preferred son of $\nu$. Moreover, $\rho$ is the unique assignment $\alpha$ such that for any node, its successor occurs among its $\alpha$-sons.*

The detailed proof, see [9] is based on rules (30) and on Eq. (31) which apply to the rightmost assignment in the case of the **nzm**-codes. That proof is also based on a complete induction on the number of a node. Figure 5 illustrates $\mathcal{W}_\lambda$ while Fig. 4 illustrates $\mathcal{W}_\rho$. The colours are again those of Figs. 1 and 2. As in those latter figures, the blue and the green colours indicate an application of the rules **wa**.

$$\mathbf{b1} \rightarrow \mathbf{w2.w3}\ldots\mathbf{wd.b1},\ \mathbf{wa} \rightarrow \mathbf{w2}\ldots\mathbf{wd.wx.b1}, \tag{30}$$

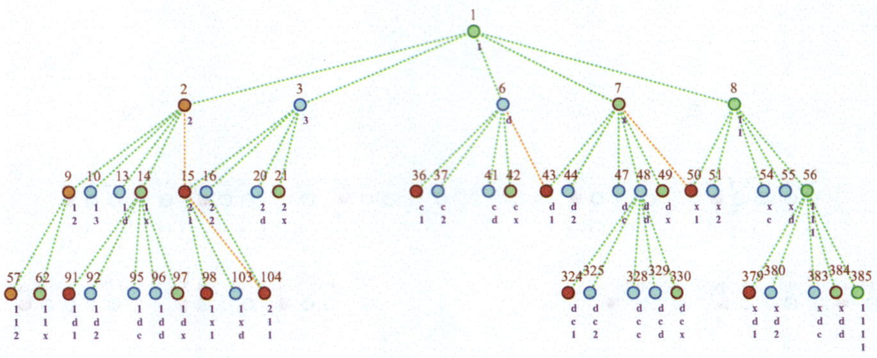

**Fig. 5** The white metallic tree and the leftmost assignment under the **nzm**-codes for the nodes. Partial representation of the first three levels of the tree when $p = 9$ with the conventions mentioned in the text

$$\begin{array}{llccc} \nu & \text{range} & \text{son} & \textbf{nzm}\text{-code} & \textbf{ref}. \\ \hline \textbf{b1} & 1\ldots p-4 & h & \mathbf{a}_k\ldots\mathbf{a}_1\mathbf{a}_0^-\mathbf{h}^+ & 1 \\ & p-3 & & \mathbf{a}_k\ldots\mathbf{a}_1\mathbf{a}_0\mathbf{1} & 2 \\ \textbf{wa} & 1\ldots p-3 & h & \mathbf{a}_k\ldots\mathbf{a}_1\mathbf{a}_0^-\mathbf{h}^+ & 3 \\ & p-2 & & \mathbf{a}_k\ldots\mathbf{a}_1\mathbf{a}_0\mathbf{1} & 4 \\ \hline \end{array} \quad (31)$$

The green colour with a red circle indicates the nodes which are before the **1**-nodes on the same level and the colour green with a green circle indicate the **1**-nodes of the rightmost branch of the tree. The black nodes lie on the leftmost branch of the tree and also in the **1**-nodes which are not on that branch. On Fig. 5, we can see that the $\lambda$-assignment does not possess the **nzm**-preferred son property. For all nodes $\nu$ except those which lie on the rightmost branch of the tree, the successor of $\nu$ is the leftmost $\lambda$-son of $\nu + 1$.

It is possible to transport the property stated in Theorem 6 and that of Theorem 8 to $\mathcal{W}_\rho$. We define the **1-branch** as the sequence of nodes $\{\omega_n\}_{n\in\mathbb{N}}$, where $\omega_0$ is the root and $\omega_{n+1}$ is the successor of $\omega_n$. The **1**-branch is the analog in the **nzm**-codes context of the **0**-branch in the context of the metallic codes. Similarly, we define the **1-paths issued from a node** $\nu$ in $\mathcal{W}_\rho$. Note the difference with the previous situation: a **0**-path is a path whose terms except the first one are sons of the previous term, whatever the assignment given to $\mathcal{W}$. In the context of the **nzm**-codes, a **1**-path is a true path in $\mathcal{W}_\rho$ and it is not a path in any other $\mathcal{W}_\alpha$ as established in the proof of Theorem 9. We can state:

**Theorem 10** *Let $\mathcal{V}$ be the set of **1**-nodes of $\mathcal{W}_\rho$, equipped with the rightmost assignment $\rho$, **1** being excepted. Define the mapping $\varphi$ from $\mathcal{V}$ onto $\mathcal{W}_\rho$ by*

$$\varphi(([\nu]_{nz}\mathbf{1})) \rightleftharpoons ([\nu]_{nz}).$$

*Define the sons of* $([\nu]_{nz}\mathbf{1})$ *as the* $\mathbf{1}$*-sons of the* $\rho$*-sons of* $([\nu]_{nz})$*. Then* $\varphi$ *defines an isomorphism between* $\mathcal{V}$ *equipped with its natural numbering and* $\mathcal{W}_\rho$ *and* $\varphi^{-1}$ *transports the* $\rho$*-assignment onto* $\mathcal{V}$*.*

# 5 Properties of the Black Metallic Tree

As defined in Sect. 2.2, the black metallic tree $\mathcal{B}$ is defined by the same rules as the white one, the difference being that the root of $\mathcal{B}$ is a black node. We know that the number of nodes on the level $n$ of $\mathcal{B}$ is $b_n$ which satisfies (4). We also know that $B_n = m_n$. Accordingly, the nodes of the rightmost branch of $\mathcal{B}$ are numbered by $m_n$, as known from Theorem 3, and we get from (20) that their **nzm**-code is $\mathbf{dc}^{n-2}\mathbf{d}$.

In [8], we proved the properties of the sons signatures of the nodes in the black metallic tree under the leftmost assignment and when the nodes are fitted with their metallic code. The properties are different from those we have noted in the white one in the similar context. In Sect. 5.1 we consider the properties of $\mathcal{B}$ when its nodes are fitted with the metallic codes. Section 5.1.1 studies the case of $\mathcal{B}_\lambda$ when $\mathcal{B}$ is constructed under the $\lambda$-assignment. Figure 6 illustrates the black metallic tree in that context for $p = 9$ as in the case of Fig. 1 to which the reader is referred for a comparison between $\mathcal{W}$ and $\mathcal{B}$. We recall the results in Sect. 5.1.1. A more detailed study of that comparison can be found in [8]. In the present section, we shall stress on the the comparison with the situation of the black metallic tree under the rightmost assignment and also, in both the leftmost and the rightmost assignments when the nodes are equipped with their **nzm**-codes. We shall write $\mathcal{B}_\lambda$, $\mathcal{B}_\rho$ for $\mathcal{B}$ equipped with the $\lambda$-, $\rho$-assignments respectively. The study of $\mathcal{B}_\lambda$ with the **nzm**-codes is dealt with in Sect. 5.2.1, while the similar study with $\mathcal{B}_\rho$ is the goal of Sect. 5.2.2.

## *5.1 The Black Metallic Tree and the Metallic Codes*

We now turn to the black metallic tree $\mathcal{B}$ and we look at properties, similar to those which hold for the white metallic tree. Some of them are still valid in that tree and we try to see the reason why for some others they are not valid. Section 5.1.1 looks at the situation for $\mathcal{B}_\lambda$, the tree $\mathcal{B}$ when it is fitted with the leftmost assignment $\lambda$. Section 5.1.2 deals with the situation for $\mathcal{B}_\rho$, the tree $\mathcal{B}$ when it is fitted with the rightmost assignment $\rho$. We recall the reader that in this subsection, we consider the metallic codes for the representations of the numbers attached to the nodes.

### 5.1.1 The Black Metallic Tree Under the Leftmost Assignment and the Metallic Codes

Figure 6 shows us that the preferred son property is not true in $\mathcal{B}_\lambda$. The leftmost son of a level, a black node, has no son whose signature is **0**. All other nodes have a son whose signature is **0**, and among them, the last node of a level has two sons whose signature is **0**, so that the leftmost assignment is not even a **0**-assignment for the leftmost one. Now, for a node $\nu$ which has a unique son whose signature is **0**, the metallic code of that node is not $[\nu]0$ but it is $[\nu - 1]0$.

Here too, call **successor** of the node $\nu$, the node whose metallic code is $[\nu]0$. We can state:

**Theorem 11** (see [8]) *In $\mathcal{B}_\lambda$, the nodes are applied the rules of (32) and the metallic codes of the $\lambda$-sons of a node $\nu$ are given by Eq. (33), the root being excepted. The root is applied the rule* **b1** $\rightarrow$ **b2w3**...**wdw0**. *For the other nodes there are two kinds of black nodes, the nodes* **b1** *and the nodes* **b0** *which follow the rules for black nodes in (32). The rule* **b1** *is also followed by node* **2**. *The nodes* **b1** *are present on the leftmost branch of $\mathcal{B}$, node* **2** *being excepted, and only on those places. The other black nodes, the leftmost one of the $\lambda$-sons of a node are* **b0** *nodes. There are two types of white nodes,* **w0**, *and* **wa** *with* **a** $>$ **0**. *The metallic codes of the sons of a node are given in Eq. (33) in terms of* $\mathbf{b}_k \ldots \mathbf{b}_0 \rightleftharpoons [\nu] - 1$. *The* **w0**-*nodes are exactly the nodes of the rightmost branch of the tree, the root being excepted.*

*The nodes of the rightmost branch of the tree being excepted, the successor of $\nu$ is the leftmost $\lambda$-son of $\nu + 1$. The tree $\mathcal{B}_\lambda$ does not observe the preferred son property. The nodes on the extremal branches of the tree being excepted but the root being included, any other node has a* **0**-*node among its $\lambda$-sons which is not its successor. The root being excepted, a node on the rightmost branch is the successor of its father*

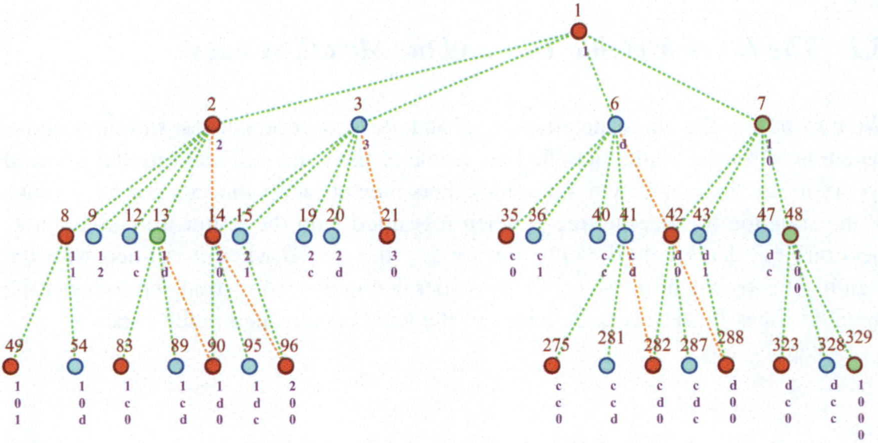


**Fig. 6** The black metallic tree with the leftmost assignment and with the metallic code of the nodes. The conventions on colours are those of Fig. 1

*which lies on the branch. Those nodes are the unique* **w0***-nodes. The root is the single node of the tree which has a preferred son. Equation (33) gives the metallic code of a node* $\nu$ *in terms of* $[\nu]$.

$$\begin{array}{ll} \mathbf{b0} \rightarrow \mathbf{b0}, \mathbf{w1}, \ldots, \mathbf{wc} & \mathbf{b1} \rightarrow \mathbf{b1}, \mathbf{w2}, \ldots, \mathbf{wd} \\ \mathbf{w0} \rightarrow \mathbf{b0}, \mathbf{w1}, \ldots, \mathbf{wc}, \mathbf{w0} & \mathbf{wa} \rightarrow \mathbf{b0}, \mathbf{w1}, \ldots, \mathbf{wd}. \end{array} \quad (32)$$

| $\nu$ | *range* | *son* | *metalliccode* | **ref.** |
|---|---|---|---|---|
| **b0** | $1 \ldots p-3$ | $h$ | $[\mathbf{b}_k \ldots \mathbf{b}_0]\mathbf{h}^-$ | 1 |
| **b1** | $1 \ldots p-3$ | $h$ | $\mathbf{10}^{k-1}\mathbf{h}$ | 2 |
| **wa** | $1 \ldots p-2$ | $h$ | $[(\mathbf{b}_k \ldots \mathbf{b}_0) - 1]\mathbf{h}^-$ | 3 |
| **w0** | $1 \ldots p-3$ | $h$ | $\mathbf{dc}^{k-1}\mathbf{h}^-$ | 4 |
| | $p-2$ | | $\mathbf{10}^{k+1}$ | 5 |

(33)

The detailed proof can be found in [9].

Note that, *par abus de langage*, we can also say for the black metallic tree equipped with the leftmost assignment that $m_{k+1}$ is the preferred son of $m_k$.

### 5.1.2 The Black Metallic Tree Under the Rightmost Assignment and the Metallic Codes

Figure 7 illustrates $\mathcal{B}_\rho$. The figure can be compared with Fig. 2. The colours indicates that the rules in the case of $\mathcal{B}_\rho$ seems to be simpler than the rules for $\mathcal{W}_\rho$, see (25). If we compare Fig. 7 with Fig. 6, we can see that the preferred son property which is not observed in $\mathcal{B}_\lambda$ as stated in Theorem 11 seems to be satisfied in $\mathcal{B}_\rho$, which is indeed the case.

In fact, we have a stronger property tightly connected with Lemma 4, reminding us what we noted in the case of the **nzm**-codes:

**Theorem 12** *Let* $\mathcal{B}_\rho$ *be the black metallic tree equipped with the rightmost assignment. Consider the metallic representations of its nodes. The rules which may be used for constructing the tree are given by (34) and the metallic codes of the* $\rho$*-sons of a node* $\nu$ *are given by Eq. (35) in terms of* $[\nu]$ *and of* $[\nu]-1$*. Equipped with the rightmost assignment,* $\mathcal{B}$ *possesses the preferred son property. But the tree does not possess that property if it is fitted with another assignment.*

$$\mathbf{b0} \rightarrow \mathbf{w1} \ldots \mathbf{wc}.\mathbf{b0}, \qquad \mathbf{wa} \rightarrow \mathbf{w1} \ldots \mathbf{wd}.\mathbf{b0}, \qquad \text{with } \mathbf{a} > \mathbf{0}. \quad (34)$$

| $\nu$ | range | son | metallic code | **ref.** |
|---|---|---|---|---|
| **b0** | $1 \ldots p-4$ | $h$ | $[\mathbf{b}_k \ldots \mathbf{b}_0]\mathbf{h}$ | 1 |
| | $p-3$ | | $\mathbf{a}_k \ldots \mathbf{a}_0\mathbf{0}$ | 2 |
| **wa** | $1 \ldots p-3$ | $h$ | $[\mathbf{b}_k \ldots \mathbf{b}_0]\mathbf{h}$ | 3 |
| | $p-2$ | | $\mathbf{a}_k \ldots \mathbf{a}_0\mathbf{0}$ | 4 |

(35)

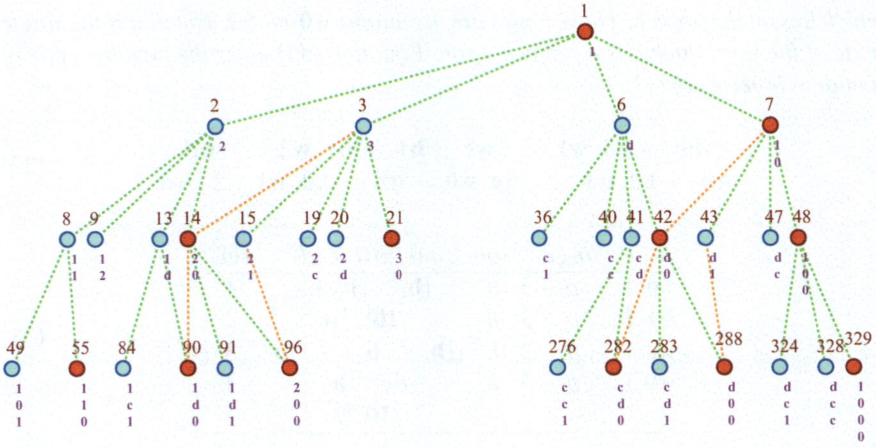


**Fig. 7** The black metallic tree with the rightmost assignment and with the metallic code of the nodes. The same convention about colours of the nodes and of the edges between nodes as in Fig. 2 is used. We can see that the preferred son property is true in the present setting

We refer the reader to [9] for the proof, notincing that Lemma 4 is also true for $\mathcal{B}$. Similarly to Theorem 10, the following result is proved in sketchilly proved in [9]:

**Theorem 13** *Let $\mathcal{V}$ be the set of $\mathbf{0}$-nodes of $\mathcal{B}_\rho$, equipped with the rightmost assignment $\rho$. Define the mapping $\varphi$ from $\mathcal{V}$ onto $\mathcal{B}_\rho$ by $\varphi(([\nu]\mathbf{0})) \rightleftharpoons ([\nu])$. Define the sons of $([\nu]\mathbf{0})$ as the $\mathbf{0}$-sons of the $\rho$-sons of $([\nu])$. Then $\varphi$ defines an isomorphism between $\mathcal{V}$ equipped with its natural numbering and $\mathcal{B}_\rho$ and $\varphi^{-1}$ transports the $\rho$-assignment onto $\mathcal{V}$.*

## 5.2 The Black Metallic Trees and the nzm-Codes

We now turn to the study of $\mathcal{B}$ when the numbers of its nodes are written as **nzm**-codes. In Sect. 5.2.1 we investigate the properties for $\mathcal{B}_\lambda$ while Sect. 5.2.2 is devoted to those of $\mathcal{B}_\rho$.

### 5.2.1 The Black Metallic Tree Under the Leftmost Assignment and the nzm-Codes

Figure 8 illustrates $\mathcal{B}_\lambda$ when the nodes are fitted with their **nzm**-codes. At first glance, whatever the digit **a**, no $\lambda$-son of a node $\nu$ has the **nzm**-code $[\nu]_{nz}\mathbf{a}$. Accordingly, the preferred son cannot be defined for $\mathcal{B}_\lambda$, a situation which reminds us that of the same tree when the metallic codes are used for the nodes.

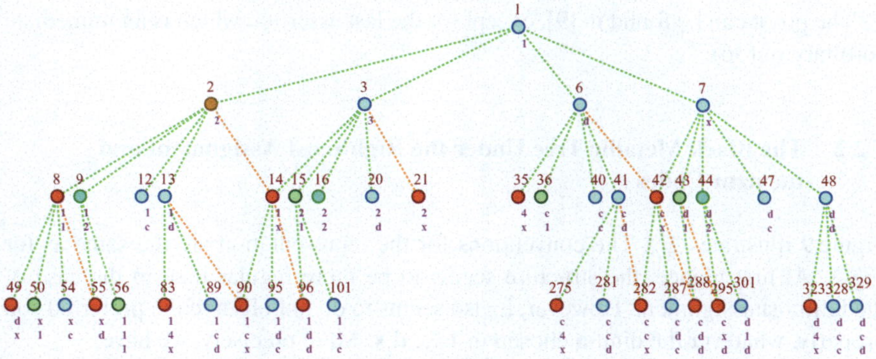


**Fig. 8** The black metallic tree still with the leftmost assignment but with the **nzm**-codes of the nodes. This time it seems that we have five types of rules for the nodes in order to define the sons signature. We can see that the preferred son property is not true in the present setting

In this situation we can state:

**Theorem 14** *In $\mathcal{B}_\lambda$, the black metallic tree dotted with the leftmost assignment, the rules giving the status and the* **nzm***-signatures of the sons of a node are given in (36), the root only being applied the rule* **r1**. *The* **nzm***-codes of the λ-sons of a node ν are given by Eq. (37), where $\mathbf{b}_k \ldots \mathbf{b}_0 \rightleftharpoons [\nu-1]$ and $\mathbf{f}_k \ldots \mathbf{f}_0 \rightleftharpoons [\nu-2]$. The tree $\mathcal{B}_\lambda$ under the leftmost assignment has no preferred son property in term of the* **nzm***-codes of its nodes. In the tree, the successor of ν is the second son of ν + 1.*

$$\begin{gathered}\mathbf{r1} \to \mathbf{b2}, \mathbf{w3}, \ldots, \mathbf{wx}, \mathbf{b2} \to \mathbf{b1}, \mathbf{w2}, \ldots, \mathbf{wd}, \mathbf{b} \to \mathbf{bx}, \mathbf{w1}, \ldots, \mathbf{wc}.\\ \mathbf{w1} \to \mathbf{bd}, \mathbf{wx}, \ldots, \mathbf{wc}, \mathbf{w2} \to \mathbf{bd}, \mathbf{w1}, \ldots, \mathbf{wd},\\ \mathbf{wa} \to \mathbf{bx}, \mathbf{w1}, \ldots, \mathbf{wd}. \text{ with } \mathbf{a} > \mathbf{2}.\end{gathered} \tag{36}$$

| ν | range | son | metallic code | **ref**. |
|---|---|---|---|---|
| **r1** | $1 \ldots p-4$ | $h$ | $\mathbf{h}^+$ | 1 |
| | $p-3$ | | $\mathbf{x}$ | 2 |
| **b2** | $1 \ldots p-3$ | $h$ | $\mathbf{1h}$ | 3 |
| **b** | 1 | | $[\mathbf{f}_k \ldots \mathbf{f}_0]\mathbf{x}$ | 4 |
| | $2 \ldots p-3$ | $h$ | $[\mathbf{b}_k \ldots \mathbf{b}_0]\mathbf{h}^-$ | 5 |
| **w1** | 1 | | $[\mathbf{f}_k \ldots \mathbf{f}_0]\mathbf{d}$ | 6 |
| | 2 | | $[\mathbf{f}_k \ldots \mathbf{f}_0]\mathbf{x}$ | 7 |
| | $3 \ldots p-2$ | $h$ | $[\mathbf{b}_k \ldots \mathbf{b}_0]\mathbf{h}^-$ | 8 |
| **w2** | 1 | | $[\mathbf{f}_k \ldots \mathbf{f}_0]\mathbf{d}$ | 9 |
| | $2 \ldots p-2$ | $h$ | $[\mathbf{b}_k \ldots \mathbf{b}_0]\mathbf{h}^-$ | 10 |
| **wa** | 1 | | $[\mathbf{f}_k \ldots \mathbf{f}_0]\mathbf{x}$ | 11 |
| | $2 \ldots p-2$ | $h$ | $[\mathbf{b}_k \ldots \mathbf{b}_0]\mathbf{h}^-$ | 12 |

(37)

The proof can be found in [9], except for the last assertion which is an immediate corollary of (36).

### 5.2.2 The Black Metallic Tree Under the Rightmost Assignment and the nzm-Codes

Figure 9 illustrates $\mathcal{B}_\rho$. The conventions for the representation are the same as for Fig. 8. At first glance, the structure seems to be more regular than in the case of the leftmost assignment. However, it also seems to do not observe the preferred son property, whatever the digit **a** chosen in **1**,...,**d**,**x**. More precisely, we have:

**Theorem 15** *Let $\mathcal{B}_\rho$ be $\mathcal{B}$, the black metallic tree, equipped with the rightmost assignment $\rho$. The rules which allow us to construct the tree under that assignment are given in (38) and the* **nzm***-codes of the $\rho$-sons of a node $\nu$ are given in (39) in terms of the* **nzm***-codes of $\nu - 1$ and of $\nu - 2$, $\mathbf{b}_k \ldots \mathbf{b}_0$ and $\mathbf{f}_k \ldots \mathbf{f}_0$ respectively. Under that assignment, the tree does not observe the preferred son property, whatever the digit chosen for that purpose. The successor of the node $\nu$ is a $\rho$-son of $\nu + 1$: its leftmost $\rho$-son or the next $\rho$-son of $\nu + 1$. No assignment allows us to establish any preferred son property on $\mathcal{B}$.*

$$\begin{array}{c}\mathbf{r1} \rightarrow \mathbf{w2}, \ldots, \mathbf{wd}, \mathbf{bx}, \mathbf{b} \rightarrow \mathbf{w1}, \ldots, \mathbf{wc}, \mathbf{bd}. \\ \mathbf{wa} \rightarrow \mathbf{w1}, \ldots, \mathbf{wd}, \mathbf{bx}, \text{ with } \mathbf{1} < \mathbf{a} < \mathbf{x}, \mathbf{w1}, \mathbf{wx} \rightarrow \mathbf{wx}, \mathbf{w1}, \ldots, \mathbf{wc}, \mathbf{bd}.\end{array} \tag{38}$$

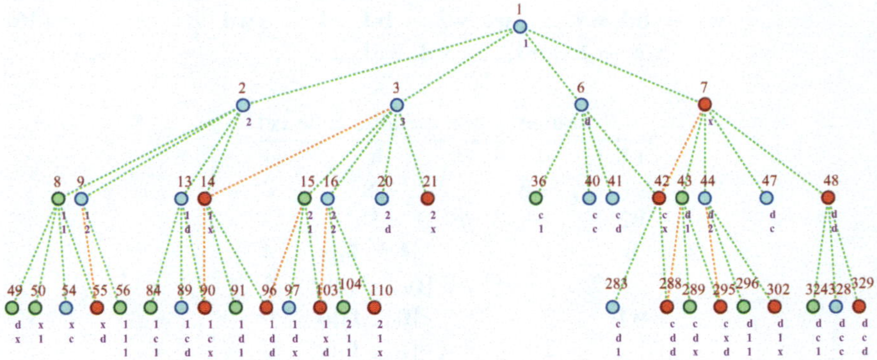

**Fig. 9** The black metallic tree still with the rightmost assignment but with the **nzm**-codes of the nodes. This time it seems that we have three types of rules for the nodes in order to define the sons signature. We can see that the preferred son property is not true in the present setting

| $\nu$ | range | | son metallic code | **ref**. |
|---|---|---|---|---|
| **r1** | $1 \ldots p-4$ | $h$ | $\mathbf{h}^+$ | 1 |
| | $p-3$ | | $\mathbf{x}$ | 2 |
| **b** | $1 \ldots p-3$ | $h$ | $[\mathbf{b}_k \ldots \mathbf{b}_0]\mathbf{h}$ | 3 |
| **w1**, **wx** | 1 | | $[\mathbf{f}_k \ldots \mathbf{f}_0]\mathbf{x}$ | 4 |
| | $2 \ldots p-2$ | $h$ | $[\mathbf{b}_k \ldots \mathbf{b}_0]\mathbf{h}^-$ | 5 |
| **wa** | $1 \ldots p-3$ | $h$ | $[\mathbf{b}_k \ldots \mathbf{b}_0]\mathbf{h}$ | 6 |
| | $p-2$ | | $[\mathbf{b}_k \ldots \mathbf{b}_0]\mathbf{x}$ | 7 |

(39)

The reader is referred to [9] for the proof which also relies on a complete induction on the number of a node and on the incrementation algorithme on **nzm**-codes.

# 6 Connection of the Metallic Trees with the Tilings $\{p, 4\}$ and $\{p+2, 3\}$ of the Hyperbolic Plane

With the previous sections, we established the properties of the metallic trees. As already mentioned in [8], the metallic trees are connected with two families of tilings of the hyperbolic plane: the tilings $\{p, 4\}$ and the tilings $\{p+2, 3\}$ with $p \geq 5$. The first family of tilings is generated by the regular convex polygon with $p$ sides and the right angle as interior angle at each vertex by reflections in its sides and, recursively, by the reflections of the images in their sides. The tilings of the second family are generated in the same way from the regular convex polygon with $p+2$ sides and with the angle $\frac{2\pi}{3}$ as interior angle at each vertex. Those angles indicate that four tiles share the same vertex in $\{p, 4\}$ and that three of them do the same in $\{p+2, 3\}$. Clearly, in these recursive constructions, infinitely many tiles are obtained twice. There is another way to generate those tilings which rely on the metallic trees, providing an injective construction. Figure 10 illustrates the considered tilings in the case when $p = 7$ and Fig. 13 illustrates the role of the metallic trees in the same tilings.

In Sect. 6.1, we define the regions of the tilings which are associated with the metallic trees and in Sect. 6.2, we explain the correspondence between the trees and the regions. In Sect. 6.3 we look carefully at the case of the white metallic tree and in Sect. 6.4, we study the case of the black one.

## *6.1 Sectors and Strips in the Tilings $\{p, 4\}$ and $\{p+2, 3\}$ of the Hyperbolic Plane*

The metallic trees are associated with two kinds of regions of the considered tilings. The sub section is devoted to the definition of those regions.

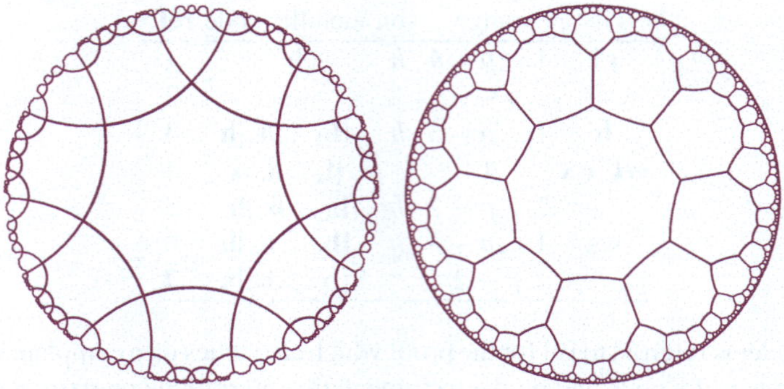

**Fig. 10** The tilings generated by the white metallic tree with $p = 7$. To left, the the tiling $\{7, 4\}$ to right, the tiling $\{9, 3\}$

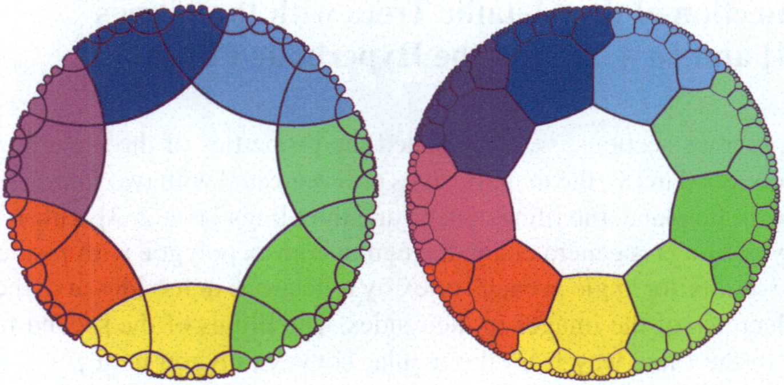

**Fig. 11** The sectors around the central tile fixed once and for all

The regions addressed by the white metallic tree is called a **sector**. The sectors are illustrated by Fig. 11 which shows us that their definitions are different according to the family of tiling we consider.

First, we define the **sectors**.

In $\{p, 4\}$ a sector of the tiling is defined by two rays $u$ and $v$ issued from a vertex $V$ of a tile $T$, $u$ and $v$ being supported by the sides of $T$ which meet at $V$. The sector defined by $u$ and $v$ is the set of tiles whose center is contained in the right angle defined by those rays. The left hand-side picture of Fig. 11 illustrates how $p$ sectors can be displayed around a once and for all fixed tile which we call the **central tile**, say $T_0$. The sectors and the central tile cover the hyperbolic plane with no hole and their interiors do not intersect.

The right hand-side picture of the figure illustrates the same display of sectors around $T_0$ in the tiling $\{p + 2, 3\}$ with, this time, $p + 2$ sectors around the central tile. In $\{p + 2, 3\}$, sectors are defined as follows. A sector is again defined by two rays $u$ and $v$. Consider a tile $T$, a vertex $V$ of $T$. Two sides of $T$, say **a** and **b** meet at $V$ where a third side **c** abut. Then, $u$ and $v$ are issued from the midpoint of **c**, $u$ and $v$ passing through the midpoints of **a** and **b** respectively. The sector defined by $u$

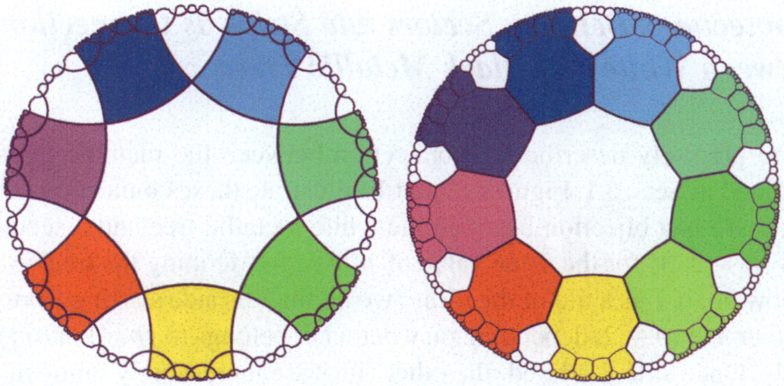

**Fig. 12** The strips around the central tile fixed once and for all

and $v$ is the set of tiles whose center lies in the acute angle defined by $u$ and $v$. The $p+2$ sectors around $T_0$ in $\{p+2, 3\}$ and $T_0$ cover the hyperbolic plane with no hole and their interiors do not intersect.

In both tilings, the tile $T$ we above considered to define a sector is called the **head** of the sector or, also, its **leading tile**.

Presently, let us define the **strips** in those tilings.

In $\{p, 4\}$, a strip is defined by two rays $u$ and $v$ together with a side **a** of a tile $T$, $u$ and $v$ being issued from the ends of **a** and being supported by the sides of $T$ which meet **a**. The left hand-side of Fig. 12 illustrates the strip in $\{p, 4\}$. In the tiling, the strip is the set of tiles whose centre lies in the intersection of the three closed half-planes defined by $u$, $v$ and **a** which contain **a** and the rays. We can see in the figure that a strip is, in some sense, smaller than a sector. As the figure points at that, the $p$ strips displayed around the central tile and $T_0$ itself do not cover the hyperbolic plane. As can be seen on the figure, in between two strips associated by two consecutive sides of $T_0$, there is a sector.

In $\{p+2, 3\}$, a strip is also defined by two rays $u$ and $v$ together with a side **a** of a tile $T$. Let **b** and **c** be the sides of $T$ which share a vertex with **a**. Then, $u$, $v$ is the ray issued from the foot of the perpendicular to **a** issued from the midpoint of **b**, **c** respectively which pass through the midpoint by which it is defined, also see the right hand-side picture of Fig. 14. In the tiling, the strip is the set of tiles whose centre lies in the intersection of the three closed half-planes defined by $u$, $v$ and the line supporting **a** which contains that side and the rays. On the right-hand side picture of Fig. 12, we can see that the strips around $T_0$ together with that tile do not cover the hyperbolic plane. Applying the definition of a sector in that context, we can see that in between the strips defined by two consecutive sides of $T_0$, there is a sector.

Here too, in both tilings, the tile $T$ we considered for defining the strip is called the **head** of the strip or also, its **leading tile**.

### 6.2 Connections Between Sectors and Strips as Connections Between White and Black Metallic Trees

It is time to precisely describe the connection between the metallic trees and the regions defined in Sect. 6.1. Figures 13 and 14 illustrate these connections. As shown in [3, 10], there is a bijection between the white metallic tree and a sector of both $\{p, 4\}$ and $\{p + 2, 3\}$ for the same value of $p$ used for defining the tree.

From now on, if $T$ is a tile of the tiling, we number its side starting from **1** up to $h$ with $h = p$ or $h = p + 2$, depending on whether $T$ belongs to $\{p, 4\}$ or to $\{p + 2, 3\}$ respectively. Once side **1** is fixed, the other sides are increasingly numbered from 1 while counterclockwise turning around the tile starting from side **1**. Denote by $(T)_i$, with $i \in \{1 \ldots h\}$ the tile which shares the side **i** of $T$ with that latter tile. In a tiling, a tile which shares a side with $T$ is called a **neighbour** of $T$.

The comparison between Fig. 11 and 13 allows us to better see the tree structure in a sector. The idea is to associate white nodes to the head of a sector and black nodes to the head of a strip.

First, consider the case of the tiling $\{p, 4\}$. The root of the white metallic tree is associated with the head $T$ of a sector $\mathcal{S}$. Let $u$ and $v$ be the rays defining $\mathcal{S}$. We fix number **1** in such a way that side **1** is supported by $u$, so that side **p** is supported by $v$, exchanging the names of $u$ and $v$ if necessary for the numbering of the sides of $T$. From that numbering and the definition of a sector, $(T)_1$ and $(T)_p$ are outside $\mathcal{S}$. From [3, 10], we know that the neighbours $(T)_i$ of $T$ with $i \in \{2 \ldots p - 1\}$ are in $\mathcal{S}$. We precisely associate the $\lambda$-sons of the root in the order of their numbers to the $(T)_i$'s inside $\mathcal{S}$ in the order of their numbers too. Next, consider a tile $\tau$ already associated with a node $\nu$ of $\mathcal{W}$. We number the sides of $\tau$ as already mentioned, the number **1** being given to the side shared with the tile associated to the father of $\nu$. If $\nu$ is white, we associate its $\lambda$-sons in the order of their numbers to the neighbours $(\tau)_i$ of $\tau$ with $i$ in $\{2 \ldots p - 1\}$ in that order. If $\nu$ is black, we associate its $\lambda$-sons in

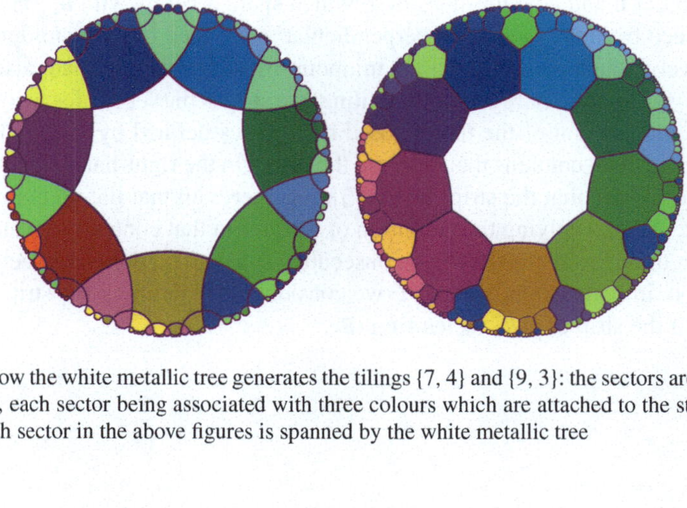

**Fig. 13** How the white metallic tree generates the tilings $\{7, 4\}$ and $\{9, 3\}$: the sectors are delimited by colours, each sector being associated with three colours which are attached to the status of the nodes. Each sector in the above figures is spanned by the white metallic tree

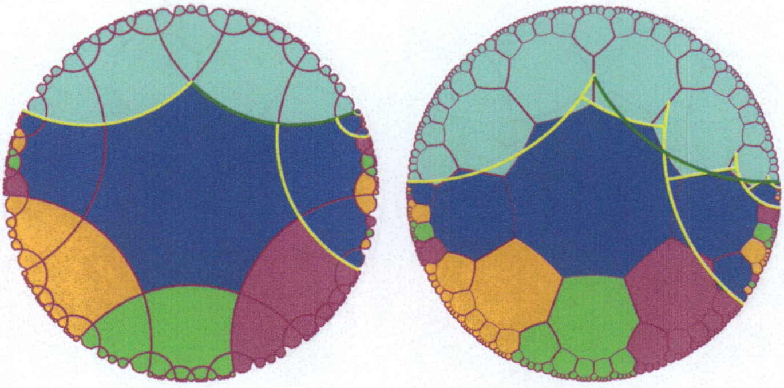

**Fig. 14** The decomposition of a sector spanned by the white metallic tree into a tile, then two copies of the same sector and a strip spanned by the black metallic tree. To left: the decomposition in the tiling $\{p, 4\}$; to right, the decomposition in the tiling $\{p+2, 3\}$. In both cases, the dark blue colour indicates the black nodes while the white ones are indicated in dark yellow, in green and in purple

the order of their numbers to the neighbours $(\tau)_j$ of $\tau$ with $j$ in $\{3...p-1\}$ in that order too. From [3, 10], it is known that this process establishes a bijection between the nodes of $\mathcal{W}$ and the tiles of $\mathcal{S}$.

Similarly, consider a strip $\mathfrak{S}$ defined by the rays $u$, $v$ and the side **a** of $T$, its leading tile. Fix **a** as side **1** of $T$ and let side **2** be supported by $u$ and side **p** be supported by $v$, exchanging the names of $u$ and $v$ if needed by the numbering of the sides of $T$. Then $(T)_1$, $(T)_2$ and $(T)_p$ are outside $\mathfrak{S}$ while the neighbours $(T)_j$ of $T$ with $j$ in $\{3...p-1\}$ are in the strip, see [3, 10]. We can repeat the above process, considering the head of $\mathfrak{S}$ as associated to the root of $\mathcal{B}$ as there are exactly $p-3$ neighbours of $T$ inside $\mathfrak{S}$. It is not difficult to prove from that that the same process as for $\mathcal{S}$ starting from the head $T$ of $\mathfrak{S}$ establishes a bijection between the nodes of $\mathcal{B}$ and the tiles of $\mathfrak{S}$. The reason is that $\mathcal{B}$ can be obtained from $\mathcal{W}_\lambda$ by removing the sub tree rooted at the rightmost son of the root of $\mathcal{W}$, and that subtree is isomorphic to $\mathcal{W}$. Now, it is proved in [3, 10], that a strip $\mathfrak{R}$ can be obtained from a sector $\mathcal{S}$ with head $T$ by removing the image of the sector defined by the sides **1** and **p** of $(T)_{p-1}$, the last neighbour of $T$ in $\mathcal{S}$, see also Fig. 14, and the head of $\mathfrak{R}$ is $T$ too.

Secondly, consider the case of the tiling $\{p+2, 3\}$. Again, we associate the root of $\mathcal{W}$ with the head $T$ of a sector $\mathcal{S}$. Let $u$ and $v$ be the rays defining $\mathcal{S}$ and let **a** be the side of another tile which meets $T$ at the vertex belonging to the consecutive sides of $T$ met by $u$ and $v$ at their midpoints. Let the side **1** met by $u$ while the side **p+2** is met by $v$, exchanging the names of $u$ and $v$ if needed in order to be coherent with the numbering of the sides of $T$. We can see that the tiles $(T)_1$, $(T)_2$ and $(T)_{p+2}$ have their centre outside $\mathcal{S}$. It is proved in [3] that the neighbours $(T)_i$ of $T$ with **i** in $\{3...p\}$ have their centre in $\mathcal{S}$. We apply the same process as in the case of the tiling $\{p, 4\}$ with this difference that to the $\lambda$-sons of a node $\nu$ associated to the tile $\tau$, we associate in the order of the numbers of the sons the neighbours $(\tau)_i$ with $i$ in $\{3...p\}$ in this order if $\nu$ is white and if $\nu$ is black, we associate the the neighbours $(\tau)_j$ with

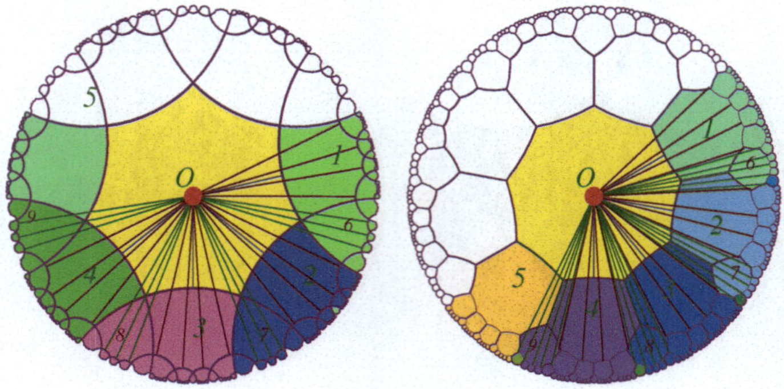


**Fig. 15** Proving the bijection theorem by estimating the distance from the centre of the central tile to the tiles on levels 1 and 2 of the metallic tree. The half-lines issued from $O$ realise the orthogonal projection of $O$ on the side of the tile indicated by the other end of the ray in the figure. The length of the corresponding segment is the distance from $O$ to the tile

$j$ in $\{4...p\}$. It is proved in [3] that the just described process establishes a bijection between $\mathcal{S}$ and the tiles of a sector in the tiling $\{p+2, 3\}$. The right hand-side of Fig. 14 illustrates the structure of the tree in $\mathcal{S}$. It also illustrates the fact that the same process establishes a bijection between $\mathcal{B}$ and the tiles of a strip $\mathfrak{S}$ in the tiling $\{p+2, 3\}$.

Now, we are in the position to sketchilly remind the proof of that property, see for instance [5, 6]. The key points are illustrated by Fig. 15, for the tilings $\{p, 4\}$, $\{p+2, 3\}$ on its left, right-hand side respectively.

The key point of the proof is that the distance from the centre of a tile $T$ to the centre $O$ of the central tile is at least $k-1$ times the half-diameter of a tile where $k$ is the level of the node which is in correspondence with the tree.

We note that the bijection of a sector both in $\{p, 4\}$ and $\{p+2, 3\}$ occurs with the same metallic tree. The difference of two sides for the regular convex polygons generating those tilings lies in the fact that as three tiles meet at a vertex instead of four of them, the number of neighbours of the head which are outside the sector is bigger in $\{p+2, 3\}$ than in $\{p, 4\}$. It is also the same situation for a strip. Let $u$, $v$ be the rays and **a** be the side of its leading tile $T$ which define a strip $\mathfrak{S}$. Take the side **1** of $T$ as **a** and number the other sides as already indicated, exchanging the names of $u$ and $v$ if needed for side **p+2** to be identified with the side which is crossed by $v$ and which shares a vertex with **a**. Then, it is not difficult to see that the centres of the neighbours $(T)_1$, $(T)_2$, $(T)_3$ and $(T)_{p+2}$ are outside $\mathfrak{S}$. The other neighbours have their centres inside $\mathfrak{S}$ and there are $p-3$ of them which explains the bijection with $\mathcal{B}$.

We close this sub section by reminding something we already mentioned in [8]. Indeed, we indicated there a property mentioned too in [4]: a sector $\mathcal{S}$ can be split into a sequence $\{\mathfrak{S}_n\}_{n\in\mathbb{N}}$. The first term of the sequence is the strip $\mathfrak{S}_0$ whose head is the head too of $\mathcal{S}$. Note that according to our conventions, the side **1** of $T$ as the head of $\mathfrak{S}_0$ is the side **p** of $T$ as the head of $\mathcal{S}$. The ray $u_0$ defining $\mathfrak{S}_0$ is the ray $u$ defining

$\mathcal{S}$. The ray $v_0$ defining $\mathfrak{S}_0$ passes through the midpoint of the side **p** of $T$ as head of $\mathfrak{S}_0$. We take this occasion to note that the same side of a tile may receive different numbers depending on the context which defines the choice of the side **1** which may differ from one situation to another one. The head of $\mathfrak{S}_{n+1}$ is the neighbour $(\tau_n)_p$ of the head $\tau_n$ of $\mathfrak{S}_n$. The ray $u_{n+1}$ which defines $\mathfrak{S}_{n+1}$ is the ray $v_n$ which defines $\mathfrak{S}_n$, and the ray $v_{n+1}$ defining $\mathfrak{S}_{n+1}$ passes through the midpoint of the side **p+2** of $(\tau_n)_p$, the side **1** of that neighbour being the side it shares with $\tau_n$. The construction of the first elements of that sequence is illustrated by Fig. 14, in its left-, right-hand side parts for the tiling $\{p, 4\}$, $\{p+2, 3\}$ respectively.

The following Sects. 6.3 and 6.4 study the applications of the numbering and their representations to location problems of the tiles in a sector and in a strip. Such problems are at the basis of an implementation of cellular automata in the settings of those hyperbolic tilings.

## 6.3 *The Case of the White Metallic Tree*

As explained in Sect. 6.2, the white metallic tree is connected with the tilings $\{p, 4\}$ and $\{p+2, 3\}$ of the hyperbolic plane with $p \geq 5$. Recall that Fig. 10 illustrates the tilings $\{7, 4\}$, $\{9, 3\}$, left-, right-hand side respectively, associated to $p = 7$. In the present sub-section, we take use of the studies of Sects. 4 and 5 in order to solve two location problems of the tiles in a sector of those tilings. The first problem which we address is to find an algorithm computing the path from a tile to the head of a sector. The problem is addressed by Sect. 6.3.1. The second problem is to compute the codes of the neighbours of a tile, which is solved in Sect. 6.3.2.

### 6.3.1 Algorithm for the Path from a Tile to the Head of a Sector

In [8], we provided an algorithm to compute the path from a tile $\tau$ of a sector $\mathcal{S}$ to the head of $\mathcal{S}$ which was based on the metallic code of $\tau$, and $\mathcal{W}$ was supposed to be fitted with the leftmost assignment. Here, we revisit the algorithm, assuming that $\mathcal{W}$ is fitted with the rightmost assignment. In [8] two algorithms were provided, the first one reading the digits of $[\tau]$ from the lowest to the highest and the second one performs the same in the reverse order. In that second algorithm two paths are constructed, one to right, the second to left and, eventually, the expected path is the to left one. The second algorithm of [8] has a decisive advantage: its complexity is linear in the size of the metallic code of $\tau$. Accordingly, we provide a similar algorithm based on the metallic codes as codes for the nodes of $\mathcal{W}_\rho$.

To that purpose, let us have a look on Fig. 2 which we reproduce as Fig. 16 for the convenience of the reader. Note that for a node $\nu$ of the level $n$ such that $\nu < m_n$, the metallic code has $n$ digits and when, on the same level, $\nu \geq m_n$, the metallic code has $n+1$ digits. In that latter case, the highest digit is **1**. If we look at the highest digit of the metallic codes of the nodes on level 2, we note it is **a** for the last two sons

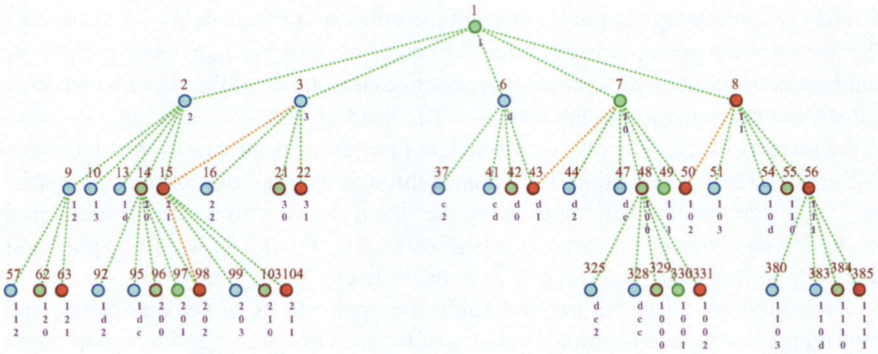


**Fig. 16** The white metallic tree under the rightmost assignment. We are here interested in the metallic codes

of the node **a** and for all sons of the node (**a**) + 1, its last two sons being excepted. At this level, there is an exception when **a** is **1**: the last three sons of **10**, the sons of **11** and those of **2**, its last two sons being excepted. If we look at the nodes of level 3, the digit $\mathbf{a}_1$ of the metallic code of $\nu$ is most often connected with the last digit of the code of $\nu - 1$ where $\nu$ is the father of $\nu$.

More generally. Assume that $\nu = \mathbf{a}_k \ldots \mathbf{a}_1 \mathbf{a}_0$. Equation (26) shows that, most often, the metallic code of its sons but the last two ones are based on $[\nu - 1]$. In particular, the digit whose index is 1 is the last digit of $[\nu - 1]$. Accordingly, if we know the path from the head of a sector down to $\nu$, we know that the node whose metallic code is $[\nu]\mathbf{b}_h \ldots \mathbf{b}_0$ with $\mathbf{b}_i$ in $\{\mathbf{0}, \mathbf{1} \ldots \mathbf{d}\}$ is either in the sub tree issued from $[\nu \mathbf{b}_h]$ or $[([\nu \mathbf{b}_h])+1]$ at a time when we know $[\nu]\mathbf{b}_h$ without knowing the digits $\mathbf{b}_i$ with $i < h$. In [9], we explicitly give the algorithm which allows us to compute the path from the head of $\mathcal{S}$. Presently, we simply outline the main lines of the algorithm. The path is given as a table whose length is that of the metallic code of $\nu$. Each entry of the table contains two pieces of information: the indication of the son $\sigma$ of a node $\nu$ as the rank of $\sigma$ among the sons of $\nu$, where rank 1 is that of the leftmost son; the indication of the status of $\nu$.

The path is computed with the help of two lists which we call the **left, right-hand side path** respectively. If the highest digit is **1**, we need to know the next one: if the next digit is **0**, then the left hand-side path goes through the preferred son and the right hand-side path goes through the rightmost son. If the next digit is **1** or greater than **1**, the left hand-side path goes through the rightmost son and the right hand-side path goes through the leftmost son of node **1**. Then, we argue on the remaining digits by a decreasing induction.

During that induction, it is assumed that when we examine the current digit **a**, the left hand-side $\text{list}_\ell$ path goes through a node $\nu$ and the right hand-side path $\text{list}_r$ goes through $\nu + 1$. The last registered digit **b** occurs in the signature of $\nu$. Let **a** be the digit we examine: it is the signature of a son of $\nu$ or of $\nu + 1$. If $\nu$ is white and if **b** = **0**, then if **a** = **2**, $\text{list}_\ell$ goes on the rightmost branch of the tree rooted at $\nu$

and list$_r$ goes through the leftmost branch of the tree rooted at $\nu + 1$ so that if $\sigma$ is the new end of list$_\ell$, $\sigma + 1$ is that of list$_r$. The same situation occurs for $\nu$ if its signature is not **0** and if **a** = **1**. In the other cases, the number to be remembered is (**a**)$-1$ for the edge going from the father of $\nu$ to $\nu$ for the left hand-side path and it is (**a**) for the edge to $\nu + 1$ for $\nu + 1$ stored in the right hand-side path. Now, if the node to be remembered is the son of the last node of the left hand-side path, the right hand-side path is identified with the left hand-side one by taking its values. If the node to be remembered is the son of the last node stored in the right hand-side path, the same is performed by exchanging the roles of the paths. All these points are easily implemented as parts of the algorithm, see [9].

When the coordinates of the nodes of $\mathcal{W}_\rho$ are the **nzm**-codes, it is possible to devise a simpler algorithm as two rules only manage the distribution of the statuses among the nodes: see [9] for the detailed algorithm. Here we give its main lines. Let us denote by $\nu$ the node such that $[\nu]_{nz} \rightleftharpoons \mathbf{a}_k\ldots\mathbf{a}_1\mathbf{a}_0$. Again, we construct two paths, left-, right-hand side paths respectively. Let us denote by $\mu$ the node which is currently at the end of the left-hand side path. By construction, $\mu + 1$ is the end of the right-hand side path. Let **a** be the current digit under examination when the left-hand side path reaches $\mu$. Let be **v** that which comes after **a** in $\mu_{nz}$. If $\mathbf{v} \neq \mathbf{1}$, the node whose coordinates is $\mathbf{a}_k\ldots\mathbf{av}$ is not a son of $\mu$ but a son of $\mu + 1$. We identify the current left-hand side path with the right-hand side one and we extend the left-hand side path to the son $\sigma$ of $\mu + 1$ whose signature is $\mathbf{v}\ominus\mathbf{1}$. The right-hand side path is extended to $\sigma + 1$. If **v** = **1**, we extend the left-hand side path to $\sigma$, this time the rightmost son of $\mu$, so that $\mu + 1$, to which we extend the right-hand side path is the leftmost son of $\mu + 1$ whose signature is **2**. The result is in the left-hand side path. Note that the updating of the left-hand side path can be performed by keeping the node of the latest update so that the total process is linear in the length of $[\mu]_{nz}$.

### 6.3.2 The Codes for the Neighbours of a Tile in a White Metallic Tree

In the present sub subsection, we turn to another problem. Knowing the coordinate $[\nu]$ or $[\nu]_{nz}$ of a tile, how to get the coordinates of the same type for its neighbours? The answer is given by Table 2 for the metallic codes and by Table 1 for the **nzm**-ones. Both tables consider $\mathcal{W}_\rho$, i.e. the metallic tree equipped with the rightmost assignment. The tables give the neighbours both in $\{p, 4\}$ and $\{p + 2, 3\}$.

Table 1 is shorter as there are only two rules for $\mathcal{W}_\rho$ with the **nzm**-codes.

Its construction is easy from Eq. (31) which gives the neighbours $(\nu)_j$ for $j \in \{3\ldots p\}$ for a white node and $j \in \{3\ldots p-1\}$ for a black one. In both cases, $(\nu)_1$ is the father and $(\nu)_2$ is the rightmost son of $\nu{-}1$ whose **nzm**-codes can easily be derived from Eq. (31). In the case of the black node, $(\nu)_p$ is $(\nu)_1 + 1$, the node which lies just after the father of $\nu$ on the level of $(\nu)_1$. The part of the tables devoted to $\{p + 2, 3\}$ involves two specific neighbours: $\nu{-}1$ and $\nu + 1$. For a white node they are $(\nu)_2$ and $(\nu)_{p+2}$ respectively. For a black node, they are $(\nu)_2$ and $(\nu)_{p+1}$ respectively as far as in that case $(\nu)_{p+2}$ is $(\nu)_1 + 1$.

**Table 1** Table of the neighbours of $\nu$ in $\mathcal{W}_\rho$ from $[\nu]_{nz}$: to left, in the tiling $\{p,4\}$, to right, in the tiling $\{p+2,3\}$. In the table, the black son of $\omega$ is denoted by $(\omega)_b$. Let $\mathbf{a}_k \ldots \mathbf{a}_0 \rightleftharpoons [\nu]_{nz}$ and let $\mu$ be defined by $[\mu]_{nz} = \mathbf{a}_k \ldots \mathbf{a}_1$

in $\{p,4\}$

**w-node**

| **rep.** | tile | **nzm**-code |
|---|---|---|
| **1** | $(\nu)_1$ | $[\mu-1]$ |
| **2** | $(\nu-1)_b$ | $[\nu-1]\mathbf{1}$ |
| **j** | **w**-sons | $[\nu-1]\mathbf{i}$ |
| **p** | $(\nu)_b$ | $[\nu]\mathbf{1}$ |

with $j \in \{3..p\text{-}1\}$, $i = j-1$

**b-node**

| **rep.** | tile | **nzm**-code |
|---|---|---|
| **1** | $(\nu)_1$ | $[\mu]$ |
| **2** | $(\nu-1)_b$ | $[(\nu-1]\mathbf{1}$ |
| **j** | **w**-sons | $[\nu-1]\mathbf{i}$ |
| **p−1** | $(\nu)_b$ | $[\nu]\mathbf{1}$ |
| **p** | $(\nu)_1+1$ | $[\mu+1]$ |

with $j \in \{3..p\text{-}2\}$, $i = j-1$

in $\{p+2,3\}$

**w-node**

| **rep.** | tile | **nzm**-code |
|---|---|---|
| **1** | $(\nu)_1$ | $[\mu-1]$ |
| **2** | $\nu-1$ | $[\nu-1]$ |
| **3** | $(\nu-1)_b$ | $[\nu-1]\mathbf{1}$ |
| **j** | **w**-sons | $[\nu-1]\mathbf{i}$ |
| **p+1** | $(\nu)_b$ | $[\nu]\mathbf{1}$ |
| **p+2** | $(\nu)_1+1$ | $[\nu+1]$ |

with $j \in \{4..p\}$, $i = j-2$

**b-node**

| **rep.** | tile | **nzm**-code |
|---|---|---|
| **1** | $(\nu)_1$ | $[\mu]\ominus\mathbf{1}$ |
| **2** | $\nu-1$ | $[\nu-1]$ |
| **3** | $(\nu-1)_b$ | $[\nu-1]\mathbf{1}$ |
| **j** | **w**-sons | $[\nu-1]\mathbf{i}$ |
| **p** | $(\nu)_b$ | $[\nu]\mathbf{1}$ |
| **p+1** | $\nu+1$ | $[\nu+1]$ |
| **p+2** | $(\nu_1)+1$ | $[\mu+1]$ |

$j \in \{4..p\text{-}1\}$, $i = j-2$

**Table 2** Table of the neighbours of $\nu$in $\mathcal{W}_\rho$: to left, in the tiling $\{p, 4\}$, to right, in the tiling $\{p + 2, 3\}$. In the table, the black son of $\omega$ is denoted by $(\omega)_b$, its preferred son by $(\omega)_\pi$. In the table, m-code stands for metallic code. If $\mathbf{a}_k \ldots \mathbf{a}_0 \rightleftharpoons [\nu]$, then $[\nu] \rightleftharpoons \mathbf{a}_k \ldots \mathbf{a}_1$

in $\{p,4\}$ / in $\{p+2,3\}$

**wa-node**

| rep. | tile | m-code |
|---|---|---|
| 1 | $(\nu)_1$ | $[\mu{-}1]$ |
| 2 | $(\nu{-}1)_b$ | $[\nu{-}1]$**1** |
| j | w-sons | $[\nu{-}1]$**i** |
| p−1 | $(\nu)_\pi$ | $[\nu]$**0** |
| p | $(\nu)_b$ | $[\nu]$**1** |

with $j \in \{3..p\text{-}2\}$, $i = j{-}1$

| rep. | tile | m-code |
|---|---|---|
| 1 | $(\nu)_1$ | $[\mu{-}1]$ |
| 2 | $\nu{-}1$ | $[\nu{-}1]$ |
| 3 | $(\nu{-}1)_b$ | $[\nu{-}1]$**1** |
| j | w-sons | $[\nu{-}1]$**i** |
| p | $(\nu)_\pi$ | $[\nu]$**0** |
| p+1 | $(\nu)_b$ | $[\nu]$**1** |
| p+2 | $\nu{+}1$ | $[\nu{+}1]$ |

with $j \in \{4..p\text{-}1\}$, $i = j{-}2$

**w0-node**

| rep. | tile | m-code |
|---|---|---|
| p−2 | $(\nu)_\pi$ | $[\nu]$**0** |
| p−1 | $(\nu)_{p-1}$ | $[\nu]$**1** |
| p | $(\nu)_b$ | $[\nu]$**2** |

| rep. | tile | m-code |
|---|---|---|
| p−1 | $(\nu)_\pi$ | $[\nu]$**0** |
| p | $(\nu)_{p-1}$ | $[\nu]$**1** |
| p+1 | $(\nu)_b$ | $[\nu]$**2** |

**w1-node**

| rep. | tile | me-code |
|---|---|---|
| 1 | $(\nu)_1$ | $[\mu{-}1]$ |
| 2 | $(\nu{-}1)_b$ | $[\nu{-}1]$**2** |
| j | w-sons | $[\nu{-}1]$**i** |
| p−2 | $(\nu)_\pi$ | $[\nu]$**0** |
| p−1 | $(\nu)_{p-1}$ | $[\nu]$**1** |
| p | $(\nu)_b$ | $[\nu]$**2** |

with $j \in \{3..p\text{-}3\}$, $i = j$

| rep. | tile | m-code |
|---|---|---|
| 1 | $(\nu)_1$ | $[\mu{-}1]$ |
| 2 | $\nu{-}1$ | $[\nu{-}1]$ |
| 3 | $(\nu{-}1)_b$ | $[\nu{-}1]$**2** |
| j | w-sons | $[\nu{-}1]$**i** |
| p−1 | $(\nu)_\pi$ | $[\nu]$**0** |
| p | $(\nu)_p$ | $[\nu]$**1** |
| p+1 | $(\nu)_b$ | $[\nu]$**2** |
| p+2 | $\nu{+}1$ | $[\nu{+}1]$ |

with $j \in \{4..p\text{-}2\}$, $i = j{-}1$

**b-node**

| | | |
|---|---|---|
| p | $(\nu)_1{+}1$ | $[\mu]$ |

| | | |
|---|---|---|
| p+1 | $\nu{+}1$ | $[\nu{+}1]$ |
| p+2 | $(\nu)_1{+}1$ | $[\mu]$ |

Table 2 displays more cases as far as there are more rules for $\mathcal{W}_\rho$ when the metallic codes are used. In order to reduce the number of repetitions, the lines for **w0**-nodes again takes the lines for **wa**-nodes except for the lines **p-2**, **p-1** and **p** in the case of $\{p, 4\}$ and the lines **p-1**, **p** and **p+1** in the case of $\{p + 2, 3\}$. The same thing was done in the case of the **b**-nodes which takes the lines of the **w1**-nodes except the line **p** for $\{p, 4\}$ and the lines **p+1** and **p+2** for $\{p + 2, 3\}$. The table rewrites the exceptional lines accordingly. The reason of the changes is clear. The codes for the first sons of a **wa**-node and for a **w0**-one are built on the same way from the metallic code of the node. For a **b**-node, a similar remark is relevant.

## *6.4 The Case of the Black Metallic Tree*

In this section, we consider the same problems for the black metallic tree. Section 6.4.1 deals with the path from a node to the root of $\mathcal{B}_\rho$, while Sect. 6.4.2 computes the codes of the neighbours of a tile in a strip.

### 6.4.1 Algorithm for the Path from a Tile to the Head of a Strip

In [9], the reader will find a detailed algorithm yielding the path from a node $\nu$ in $\mathcal{B}_\rho$ to the root of the tree.

The data of the algorithm is the metallic code of $\nu$. As in th case of $\mathcal{W}$, the algorithm constructs two lists during its computation whose elements are the status of the current node and its rank among the sons of its father, the leftmost son having rank 1.

The algorithm is very similar to the one whose main lines are given in Sect. 6.3.2. Its justification is straightforward. Outside the case when the current digit **a** is **0**, the path necessarily goes within the tree rooted at the node whose signature is **a**+1 and it is the **a**th node among the sons of its father. When **a** = **0**, except the case when the path has to move to right, it is needed to go on on both paths at a distance 1 from each other as at that moment, the next digit is not known.

In [9], the reader can find a similar algorithm when the coordinates of the nodes of $\mathcal{B}_\rho$ are given through their **nzm**-code. However, the structure of that latter algorithm is more complex than the one we sketchilly described for the case of $\mathcal{W}_\rho$ when the nodes are fitted their **nzm**-codes.

The reason is not only the fact this time we have three rules for the nodes instead of two ones for the white tree, it is also due to the fact that the occurrence of the pattern **dc*** **d** entails that when the pattern is followed by a digit **a** with **a** < **c**, the appropriate node is in the node pointed at by the right hand-side path: it follows from Eq. (39).

**Table 3** Table of the metallic codes of the neighbours of a tile $\nu$ in both tilings $\{p, 4\}$ and $\{p+2, 3\}$ in $\mathcal{B}_\rho$. We assume that $\mathbf{a}_k \ldots \mathbf{a}_1 \mathbf{a}_0 \rightleftharpoons [\nu]$, that $(\nu)_i$ indicates the neighbour $i$ and that $\mu$ is defined by $[\mu]_{nz} \rightleftharpoons \mathbf{a}_k \ldots \mathbf{a}_1$

**wa-node**

in $\{p, 4\}$

| rep. | tile | m-code |
|---|---|---|
| 1 | $(\nu)_1$ | $[\mu+1]$ |
| 2 | $(\nu-1)_p$ | $[\nu-1]$**0** |
| j | w-sons | $[\nu-1]$**i** |
| p | b-son | $[\nu]$**0** |

with $i \in \{3..p-1\}, i = j-2$

in $\{p+2, 3\}$

| rep. | tile | m-code |
|---|---|---|
| 1 | $(\nu)_1$ | $[\mu+1]$ |
| 2 | $\nu-1$ | $[\nu-1]$ |
| 3 | $(\nu\text{-}1)_{p+1}$ | $[\nu-1]$**0** |
| j | w-sons | $[\nu-1]$**i** |
| p+1 | b-son | $[\nu]$**0** |
| p+2 | $\nu+1$ | $[\nu+1]$ |

with $j \in \{4..p\}, i = j-3$

**b0-node**

in $\{p, 4\}$

| rep. | tile | m-code |
|---|---|---|
| 1 | $(\nu)_1$ | $[\mu]$ |
| 2 | $(\nu-1)_p$ | $[\nu-1]$**0** |
| j | w-sons | $[\nu-1]$**i** |
| p | b-son | $[\nu]$**0** |

with $j \in \{3..p-1\}, i = j-2$

in $\{p+2, 3\}$

| rep. | tile | m-code |
|---|---|---|
| 1 | $(\nu)_1$ | $[\mu]$ |
| 2 | $\nu-1$ | $[\nu-1]$ |
| 3 | $(\nu\text{-}1)_{p+1}$ | $[\nu-1]$**0** |
| j | w-sons | $[\nu-1]$**i** |
| p | b-son | $[\nu]$**0** |
| p+1 | $\nu+1$ | $[\nu+1]$ |
| p+2 | $(\nu)_1+1$ | $[(\mathbf{a}_k..\mathbf{a}_1)+1]$ |

with $j \in \{4..p-1\}, i = j-3$

### 6.4.2 The Codes for the Neighbours of a Tile in a Black Metallic Tree

We now turn to the computation of the coordinates of the neighbours of a tile $\nu$ which lies in a strip $\mathfrak{S}$ in bijection with $\mathcal{B}_\rho$. We first study that computation when it is based on the metallic code $[\nu]$ of $\nu$. The computation should be easier as there are two rules only for the sons of a node.

Table 3 follows immediately from the examination of rules (34) and Eq. (35). The additional neighbours of $\nu$ in $\{p+2, 3\}$ are the nodes $\nu-1$ and $\nu+1$ which are on the same level as $\nu$ in $\mathcal{B}_\rho$. This introduce a small change in the numbering of the sons of the node compared with their numbering in $\{p, 4\}$.

Table 4 shows the **nzm**-codes of the neighbours $(\nu)_i$ of $\nu$ in $\mathcal{B}_\rho$. The table follows from the rules (38) and (39). It should be remarked that the **nzm**-codes of the neighbours of a **wx**-node are very similar to those of a **w1**-one. The difference is in computation of the **nzm**-code of the father. For the **wx**-node, the **nzm**-code of the father is given by $[(\mathbf{a}_k \ldots \mathbf{a}_1)+2]_{nz}$ and not by $[(\mathbf{a}_k \ldots \mathbf{a}_1)+1]_{nz}$ as it the case for a **w1**-node.

**Table 4** Table of the **nzm**-codes of the neighbours of a tile $\nu$ in both tilings $\{p, 4\}$ and $\{p+2, 3\}$ for $\mathcal{B}_\rho$. We assume that $\mathbf{a}_k \ldots \mathbf{a}_1 \mathbf{a}_0 \rightleftharpoons [\nu]_{nz}$, that $[\mu]nz =\rightleftharpoons \mathbf{a}_k \ldots \mathbf{a}_1$ and that $(\nu)_i$ indicates the neighbour $i$. When references of neighbours for a type of nodes are missing, they have to be seen in the same column at the previous type and, again at the previous one if they are still missing

in $\{p, 4\}$ / in $\{p+2, 3\}$

**wa-node**

| rep. | tile | **nzm**-code |
|---|---|---|
| 1 | $(\nu)_1$ | $[\mu+1]_{nz}$ |
| 2 | $(\nu-1)_p$ | $[\nu-2]_{nz}\mathbf{x}$ |
| j | **w**-sons | $[\nu-1]_{nz}\mathbf{i}$ |
| p | **b**-son | $[(\nu-1)]_{nz}\mathbf{x}$ |

with $i \in \{3..p\text{-}1\}, i = j-2$

| rep. | tile | **nzm**-code |
|---|---|---|
| 1 | $(\nu)_1$ | $[\mu+1]_{nz}$ |
| 2 | $\nu-1$ | $[\nu-1]_{nz}$ |
| 3 | $(\nu\text{-}1)_{p+1}$ | $[\nu-2]_{nz}\mathbf{x}$ |
| j | **w**-sons | $[\nu-1]_{nz}\mathbf{i}$ |
| p+1 | **b**-son | $[(\nu-1)]_{nz}\mathbf{x}$ |
| p+2 | $\nu+1$ | $[\nu+1]_{nz}$ |

with $j \in \{4..p\}, i = j-3$

**w1-node**

| rep. | tile | **nzm**-code |
|---|---|---|
| 2 | $(\nu-1)_p$ | $[\nu-2]_{nz}\mathbf{d}$ |
| 3 | $(\nu)_3$ | $[\nu-2]_{nz}\mathbf{x}$ |
| j | **w**-sons | $[\nu-1]_{nz}\mathbf{i}$ |
| p | **b**-son | $[(\nu-1)]_{nz}\mathbf{d}$ |

with $j \in \{4..p\text{-}1\}, i = j-3$

| rep. | tile | **nzm**-code |
|---|---|---|
| 3 | $(\nu-1)_p$ | $[\nu-2]_{nz}\mathbf{d}$ |
| 4 | $(\nu)_3$ | $[\nu-2]_{nz}\mathbf{x}$ |
| j | **w**-sons | $[\nu-1]_{nz}\mathbf{i}$ |
| p+1 | **b**-son | $[(\nu-1)]_{nz}\mathbf{d}$ |

with $i \in \{5..p\}, j = i-4$

**wx-node**

| rep. | tile | **nzm**-code |
|---|---|---|
| 1 | $(\nu)_1$ | $[\mu+2]_{nz}$ |

| rep. | tile | **nzm**-code |
|---|---|---|
| 1 | $(\nu)_1$ | $[\mu+2]_{nz}$ |

**bd,bx-nodes**

| rep. | tile | **nzm**-code |
|---|---|---|
| 1 | $(\nu)_1$ | $[\mu+1]_{nz}$ |
| 2 | $(\nu-1)_p$ | $[\nu-2]_{nz}\mathbf{x}$ |
| j | **w**-sons | $[\nu-1]_{nz}\mathbf{i}$ |
| p | $(\nu)_p$ | $[\nu+1]_{nz}$ |

with $j \in \{3..p\text{-}1\}, i = j-2$

| rep. | tile | **nzm**-code |
|---|---|---|
| 1 | $(\nu)_1$ | $[\mu+1]_{nz}$ |
| 2 | $\nu-1$ | $[\nu-1]_{nz}$ |
| 3 | $(\nu-1)_p$ | $[\nu-2]_{nz}\mathbf{x}$ |
| j | **w**-sons | $[\nu-1]_{nz}\mathbf{i}$ |
| p+1 | $(\nu)_{p+1}$ | $[\nu+1]_{nz}$ |
| p+2 | $(\nu)_1+1$ | $[\mu+2]_{nz}$ |

with $j \in \{4..p\}, i = j-3$

## 7 Conclusion

We can conclude the paper with several remarks.

The present paper gathers in a synthetic way the researches of the papers [7–9].

It is interesting to remark that in [9], a result solves the generalization of a question raised in [2] in the case of the Fibonacci tree. The result is stated in Theorem 7. However, the statement of Theorem 7 deals with metallic codes while in [2] the question about assignments dealt with the Fibonacci representation of the natural numbers. It was noted there that there were assignments for which the preferred property in the frame of Fibonacci representations no more holds. The question raised in that paper is still open. It is interesting to note that the algorithms considered in the paper and developed in [8, 9] are linear in the size of the code which is used.

The author does not pretend to have solved any possible problems in these settings. The paper is simply a walk in those tilings.

## References


1. Iwamoto C, Andou T, Morita K, Imai K (2002) Computational complexity in the hyperbolic plane. Lecture notes in computer science, vol 2420. Proceedings of MFCS'2002, pp 365–374
2. Margenstern M (2000) New tools for cellular automata of the hyperbolic plane. J Univ Comput Sci 6(12):1226–1252
3. Margenstern M (2007) Cellular automata in hyperbolic spaces, vol I. In: Adamatzky A (ed) Theory, collection: advances in unconventional computing and cellular automata. Old City Publishing, Philadelphia, 424p
4. Margenstern M (2008) Cellular automata in hyperbolic spaces, vol II. In: Adamatzky A (ed) Implementation and computations, Collection: advances in unconventional computing and cellular automata. Old City Publishing, Philadelphia, 360p
5. Automates cellulaires hyperboliques universels (2015) Parties 1 & 2. Techniques et sciences informatiques 34(3):233–310
6. Margenstern M (2015) About embedded quarters and points at infinity in the hyperbolic plane, 17pp. arXiv:1507.08495 [cs.CG]
7. Margenstern M (2019) About Fibonacci trees - I -, 17pp. arXiv:1904.12135 [cs.DM]
8. Margenstern M (2019) About Fibonacci trees - II - : generalized Fibonacci trees, 35pp. arXiv:1907.04677 [cs.DM.]
9. Margenstern M (2019) About Fibonacci trees III: multiple Fibonacci trees, 35pp. arXiv:1909.01893 [cs.DM.]
10. Margenstern M, Skordev G (2003) Fibonacci type coding for the regular rectangular tilings of the hyperbolic plane. J Univ Comput Sci 9(5):398–422
11. Smullyan R (1992) Gödel's incompleteness theorems. Oxford University Press, Oxford

# From Additive Flowers to Additive Automata Networks

**Enrico Formenti, Christophe Papazian, Adrien Richard, and Pierre-Alain Scribot**

**Abstract** This paper surveys some old results about linear shift registers and restates them in the context of additive automata networks. The addition of new results allows an almost complete description of the dynamical behavior of additive automata networks. The computational complexity aspects of deciding such behaviors are also discussed.

## 1 Introduction

The motivation for studying the dynamics of (finite) automata networks (AN) has multiple sources. Indeed, they are simple formal models well-suited for many practical applications. Our main motivation comes from the strict connection existing between AN and gene networks.

The understanding of gene networks in different contexts and different scales is a long standing problem in biology. Many different formal approaches have been pursued to understand the overall dynamical behavior of these networks, their structure, persistence, etc. This paper follows the vein initiated by Kauffman and Thomas in the Seventies [6, 14]. They modeled a gene network by a network of $n$ finite automata. Each automaton $i \in \{1, \ldots, n\}$ has a state chosen from a finite set of states $S$ and a local function $f_i : S^n \to S$ which updates this state according to the state of the $n$ automata. A function $\theta$ from $\mathbb{N}$ to the subsets of $\{1, \ldots, n\}$, called *updating scheme*, indicates the automata of the network that will be updated at a given time. Then,

E. Formenti (✉) · C. Papazian · A. Richard · P.-A. Scribot
I3S, CNRS, Université Côte d'Azur, Nice, France
e-mail: enrico.formenti@unice.fr

C. Papazian
e-mail: christophe.papazian@unice.fr

A. Richard
e-mail: adrien.richard@unice.fr

P.-A. Scribot
e-mail: scribot@unice.fr

 
A. Adamatzky (ed.), *Automata and Complexity*, Emergence, Complexity and Computation 42, https://doi.org/10.1007/978-3-030-92551-2_18

if $x = (x_1, \dots, x_n) \in S^n$ is the global state of the system at time $t$, the global state of the system at time $t+1$ is $y = (y_1, \dots, y_n) \in S^n$ where $y_i = f_i(x)$ if $i \in \theta(t)$ and $y_i = x_i$ otherwise. In this paper, we will focus on the most classical updating scheme, the *parallel* one, i.e. for all $t \in \mathbb{N}$, $\theta(t) = \{1, \dots, n\}$.

The *interaction graph* IG is a convenient way to represent the interactions among the automata of the network. In many cases, it conveys lot of informations about the overall dynamics in a much compact form. Indeed, the size of the IG is exponentially smaller than the graph describing the dynamics. This IG is the digraph $(V, E)$ where $V = \{1, \dots, n\}$ and an edge $(j, i) \in V \times V$ iff the local function $f_i$ depends on input $j$, that is, if there are two global states $x, y \in S^n$ that only differ in $x_j \neq y_j$ such that $f_i(x) \neq f_i(y)$.

Roughly speaking, the IG describes the impact that each automaton has on the others in matter of state change. Studying the relations between structural properties of IG and the dynamics of AN based on them is a challenging research direction which have already given raise to a prolific literature (see [4, 12] for reviews).

In many cases (especially in applications), one is more interested in a concise description of the dynamics of a network and not to precise details. For instance, one can be only interested to the graph of dynamics up to isomorphism, or in the *orbit description* $D \subseteq \mathbb{N} \times \mathbb{N}$ where $(k, p) \in D$ means that there are $k$ orbits of period $p$. For a comprehensive example, refer to Sect. 5.1.

From a topological point of view, the dynamics of AN is easy. Indeed, they are finite systems and, hence, their asymptotic behavior is ultimately periodic. However, the detailed description of the dynamics is computationally difficult even for small systems (see for instance [8, 15]). In [2], it is proved that many interesting questions about AN dynamics are at least PSPACE-complete.

The difficulties of the computational task of describing the dynamics of AN calls for a better understanding of their intrinsic and structural properties. A promising research direction in this sense consists in decomposing the problem into smaller ones, find solutions for them and then use this knowledge to obtain a complete theory for all networks.

In [9], one can find the starting blocks of this program. Indeed, the idea is to start with an IG which is made of a single cycle. The simple structure allows a very precise analysis of the dynamics. Next, one can consider IGs made by two cycles which have just one vertex in common (see Fig. 1). At this point, things are already pretty complex and one has to start compromises in order to decrease the complexity. The choice made in [9] is to restrict to boolean networks and to local functions made of $\wedge$ or $\vee$ operators only. Under these constraints the dynamics has been exhaustively described.

The next step in the program is therefore to complete the study of double cycles by allowing the local rule of the common vertex to be a XOR (i.e. turning the network into and additive network). Indeed, this is the starting point of the present research.

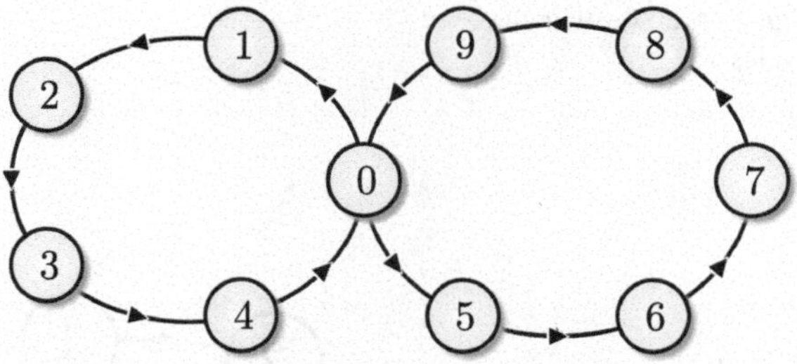


**Fig. 1** Interaction graph made of two cycles having a single common vertex

## 2 Additive Flowers

The natural research direction to follow after the study of double cycles networks was to consider the remaining boolean functions of two variables for the common vertex. The *exclusive or* $\oplus$ is the most representative of this group.

Before going on we make an assumption which we will adopt in rest of the paper: a local function $f_i$ of a vertex $i$ is a linear function i.e. $S = \mathbb{Z}_p$ for some $p \geq 1$ and

$$\forall x \in \mathbb{Z}_p^n, \qquad f_i(x) = \sum_{j=1}^{n} \lambda_{ij} x_j \mod p$$

for some coefficients $\lambda_{i1}, \ldots, \lambda_{in} \in \mathbb{Z}_p$.

Hence, the dynamics of the network is simply given by the successive iterations of the *global* function

$$f : \mathbb{Z}_p^n \to \mathbb{Z}_p^n, \qquad f(x) = (f_1(x), \ldots, f_n(x)).$$

In the following, we identify the network with its global function, saying that $f$ an *additive* AN with $n$ components over $\mathbb{Z}_p$. Such an additive AN can be conveniently expressed using the matrix $M \in \mathbb{Z}_p^{n \times n}$ in which $m_{ij} = \lambda_{ij}$ since then $f(x) = Mx$ for all $x \in \mathbb{Z}_p^n$. Note that the IG of $f$ can be simply deduced from the coefficients $\lambda_{ij}$, since it has an arc $(j, i)$ if and only if $\lambda_{ij} \neq 0$.

Generalizing cycles and double cycles, we say that a digraph is a *flower* if it contains a vertex $i$, called *control vertex*, such that $i$ meets every cycle of $G$ and all the vertices distinct from $i$ are of in-degree one. An additive AN $f$ with $n$ components over $\mathbb{Z}_p$ is an *additive flower* if its interaction graph $G$ is a flower and, for each arc $(j, i)$ of $G$ where $i$ is of in-degree one, we have $f_i(x) = x_j$ for all $x \in \mathbb{Z}_p^n$. See Fig. 2 for an example of flower with three cycles and twelve vertices.

The following example lead us to discover the general method for describing the dynamics of additive flowers.

**Example 1** For some fixed $n \in \mathbb{N}$, let $f^{(n)}$ be the additive AN with $n$ components over $\mathbb{Z}_2$ defined by $f_1^{(n)}(x) = x_1 \oplus x_n$ and $f_{i+1}^{(n)}(x) = x_i$ for all $i \in [1, n-1]$. The IG of $f^{(n)}$ consists of a cycle of length $n$ plus a loop on vertex 1, thus $f^{(n)}$ is an

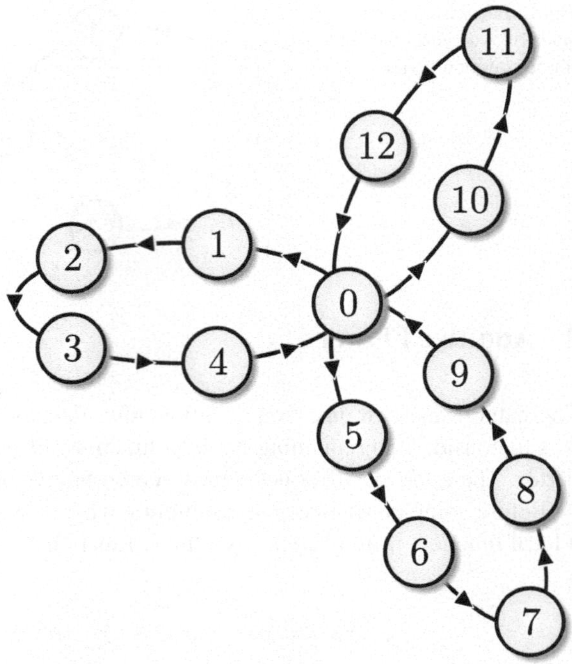


**Fig. 2** A flower with three cycles

additive flower. Let $p_n$ denote the maximal period of an orbit of $f^{(n)}$. The first terms of the sequence $\{p_n\}_{n\in\mathbb{N}}$ are

$$1, 3, 7, 15, 21, 63, 127, 63, 73, 889.$$

This is sequence *A*046932 in [10] and what is funny it is that it is connected to a question given to the Olympiads of Mathematics in 1993 [11, Problem 6]. It is not difficult to see that $P(X) = X^n + X^{n-1} + 1$ is the characteristic polynomial associated with the adjacency matrix of the IG of $f^{(n)}$. As a consequence of Theorems 2 and 4 proved later, we decompose the dynamics into simpler ones. For instance, the dynamics of $f^{(5)}$ is the Cartesian product of the dynamics of $f^{(2)}$ and $f^{(3)}$ since $P(X) = X^5 + X^4 + 1 = (X^2 + X + 1) \cdot (X^3 + X + 1) \mod 2$. Figure 4 shows the dynamics of $f^{(2)}$, $f^{(3)}$ and $f^{(5)}$ (Fig. 3).

## 2.1 Transforming Flowers

When flowers start having more than two cycles and when states of the automata are no more booleans, the dynamics becomes richer and it is difficult to analyze

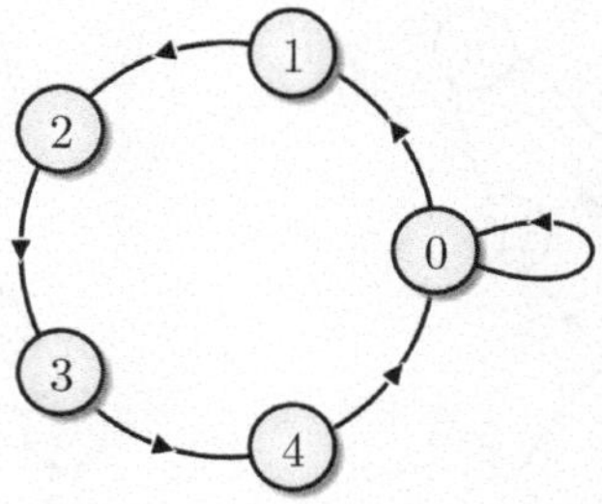


**Fig. 3** The interaction graph of $f^{(5)}$ (see Example 1)

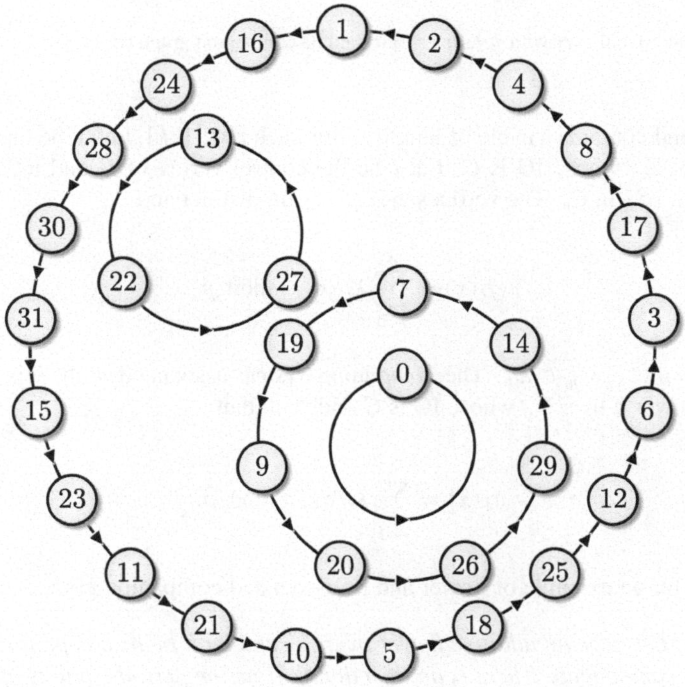

**Fig. 4** Phase space of $f^{(5)}$ (see Example 1). Here, the label of the vertex is the decimal representation of the global state of $f^{(5)}$

because of the interactions between the length of the cycles, their number and the more complex operations performed by the control vertex.

In this section, we are going to transform a generic additive flower with $k$ cycles into a suitable additive network. The idea is that the new network will have almost identical dynamics but it will allows an easy algebraic analysis.

Let $G$ be a flower with $k$ cycles $C_1, \dots, C_k$, of length $\ell_1 \leq \ell_2 \leq \cdots \leq \ell_k$. The *companion* graph $C$ of $G$ has vertex set $[1, \ell_k]$, and contains an arc from $i$ to $i+1$ for all $i \in [1, \ell_k]$ and an arc from $i$ to 1 for all $i \in \{\ell_1, \dots, \ell_k\}$. Thus $C$ has at most

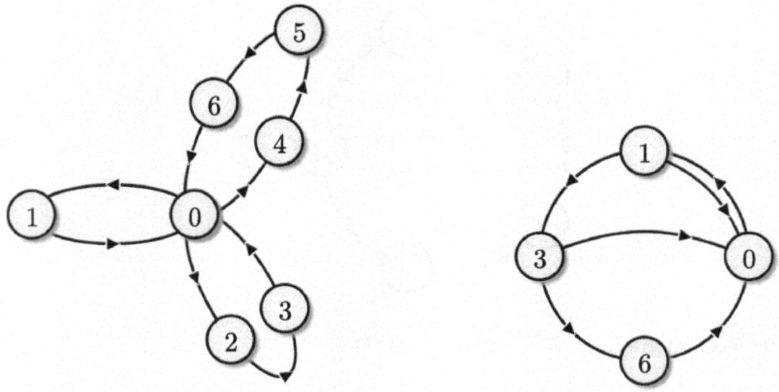


**Fig. 5** An additive flower with 7 vertices (left) and its companion graph (right)

$k$ cycles, and contains a cycle of length $\ell_r$ for each $r \in [1, k]$. Let $f$ be the additive flower over $\mathbb{Z}_p$ whose IG is $G$. Let $i$ be the control vertex of $G$ and let $j_r$ be the in-neighbor of $i$ in $C_r$. The vertices $j_1, \ldots, j_k$ are distinct and

$$f_i(x) = \sum_{r=1}^{k} \lambda_{ij_r} x_{j_r} \mod p \tag{1}$$

for some $\lambda_{ij_1}, \ldots \lambda_{ij_k} \in \mathbb{Z}_p$. The *companion system* associated with $f$ is then the additive flower $h$ over $\mathbb{Z}_p$ whose IG is $C$ and such that

$$h_1(x) = \sum_{r=1}^{k} \lambda_{ij_r} x_{\ell_r} \mod p.$$

See Fig. 5 for an example of flower and its associated companion graph.

**Lemma 1** *Let $f$ be an additive flower over $\mathbb{Z}_p$ and let $h$ be its companion system, say with $n$ components. There is an injection $\psi$ from the periodic points of $f$ to $\mathbb{Z}_p^n$ such that $\psi(f(x)) = h(\psi(x))$ for all periodic points $x$ of $f$, and if $p$ is a prime then $\psi$ is a bijection.*

***Proof*** Let $G$ be the IG of $f$ and suppose that $G$ has $k$ cycles $C_1, \ldots, C_k$, of length $\ell_1 \leq \ell_2 \leq \cdots \leq \ell_k$, so $\ell_k = n$. Let $i$ be the control vertex of $G$ and let $i_1, \ldots, i_n$ be the vertices of $C_k$ in order, starting from $i_1 = i$. For each $r \in [1, k]$, let $j_r$ the in-neighbor of $i$ in $C_r$; in particular, $j_k = i_n$. Then (1) holds for some some $\lambda_{ij_1}, \ldots \lambda_{ij_k} \in \mathbb{Z}_p$.

For $d \in [1, n]$, we denote by $J_d$ the set of vertices $j$ in $G$ such that the path from $i$ to $j$ in $G$ is of length $d - 1$. So $J_1, \ldots, J_d$ is a partition of the vertex set of $G$. Note also that $J_1 = \{i\}$ and $j_r \in J_{\ell_r}$ for all $r \in [1, k]$. Suppose that $f$ has $m$ components, and let $X$ be the set of $x \in \mathbb{Z}_p^m$ such that $x_j = x_{i_d}$ for all $d \in [1, n]$ and $j \in J_d$. Let $\phi$ be the map from $X$ to $\mathbb{Z}_p^n$ defined by $\phi(x) = (x_{i_1}, \ldots, x_{i_n})$. One easily check that $\phi$ is a bijection.

Let $x \in X$ and let us prove that $\phi(f(x)) = h(\phi(x))$. For all $d \in [2, n]$,

$$\phi(f(x))_d = f_{i_d}(x) = x_{i_{d-1}} = \phi(x)_{d-1} = h_d(\phi(x)) = h(\phi(x))_d.$$

It remains to prove that $\phi(f(x))_1 = (h(\phi(x))_1$, that is $f_i(x) = h_1(\phi(x))$. For all $r \in [1, k]$ we have $j_r \in J_{\ell_r}$, thus $x_{j_r} = x_{i_{\ell_r}}$, and we deduce that

$$f_i(x) = \sum_{r=1}^{k} \lambda_{ij_r} x_{j_r} \mod p \tag{2}$$

$$= \sum_{r=1}^{k} \lambda_{ij_r} x_{i_{\ell_r}} \mod p \tag{3}$$

$$= \sum_{r=1}^{k} \lambda_{ij_r} \phi(x)_{\ell_r} \mod p \quad = \quad h_1(\phi(x)). \tag{4}$$

Let $\Omega$ be the set of periodic points of $f$, and let us prove that $\Omega \subseteq X$. For that we show, by induction on $d$ from 1 to $n$, that $x_j = x_{i_d}$ for all $x \in \Omega$ and $j \in J_d$. This is obvious for $d = 1$ since $J_1 = \{i\}$ and $i_1 = i$. Suppose that $d > 1$ and let $x \in \Omega$. Then there is $y \in \Omega$ such that $f(y) = x$. Hence, for all $j \in I_d$, we have $x_j = f_j(y) = y_{j'}$ where $j'$ is the in-neighbor of $j$ in $G$, and $x_{i_d} = f_{i_d}(y) = y_{i_{d-1}}$ since $i_{d-1}$ is the in-neighbor of $i_d$ in $G$. Since $j' \in J_{d-1}$ and $y$ is a periodic point, we have $y_{j'} = y_{i_{d-1}}$ by induction, and we deduce that $x_j = x_{i_d}$. This completes the induction and thus $\Omega \subseteq X$.

So the restriction $\psi$ of $\phi$ to $\Omega$ has the property of the statement. To complete the proof, suppose that $p$ is a prime, and let us prove that $\Omega = X$. For that, it is sufficient to prove that $h$ is bijection. Given $x \in \mathbb{Z}_p^n$, let $y \in \mathbb{Z}_p^n$ be defined by $y_j = x_{j+1}$ for all $j \in [1, n-1]$ and

$$y_n = \left( x_1 - \sum_{r=1}^{k-1} \lambda_{ij_r} x_{\ell_{r+1}} \right) / \lambda_{ij_k},$$

which is well defined since $p$ is a prime. One easily checks that $h(y) = x$ and thus $h$ is indeed a bijection. □

The previous lemma highlights the importance of companion graphs in the study of the dynamics of additive flowers. This is the matter of the next section.

## 3 Companion Graph and Companion Matrix

This section focuses on a generalization of companion graphs in which edges are labeled with coefficients from a Galois field $\langle \mathbb{F}_p, \oplus, \odot \rangle$ ($p \in \mathbb{N}$, a prime number) and their associated matrices i.e. the companion matrices. Indeed, more notation is needed.

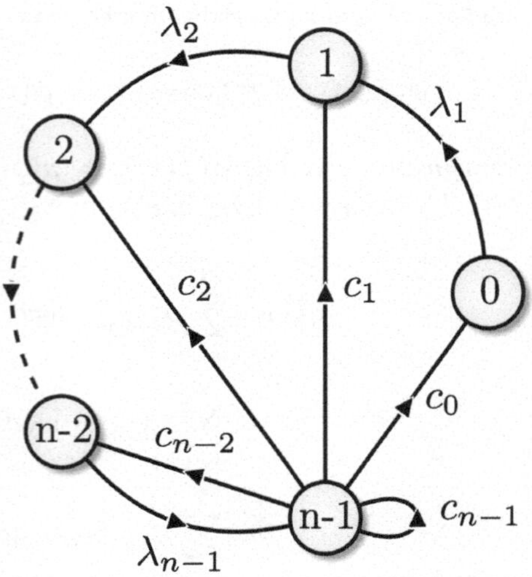


**Fig. 6** A generic companion graph with $n$ vertices

Denote $\mathbb{F}_p[X]$ the set of polynomials with coefficients in $\mathbb{F}_p$ and $deg(\mathcal{P})$ is the degree of the polynomial $\mathcal{P}$. The operations $\oplus$ and $\odot$ can be naturally extended to $\mathbb{F}_p[X]$. A polynomial $\mathcal{P} \in \mathbb{F}_p[X]$ is *irreducible* if there is no polynomials $\mathcal{A}, \mathcal{B} \in \mathbb{F}_p[X]$ such that $\mathcal{P} = \mathcal{A} \odot \mathcal{B}$ with $deg(\mathcal{A}) > 0$ and $deg(\mathcal{B}) > 0$.

As $\mathbb{F}_p[X]$ is a unique factorization domain, it is also possible to extend the usual *mod* operation in $\mathbb{F}_p[X]$. Indeed, $\mathcal{M} \mod \mathcal{P}$ is the unique polynomial $\mathcal{R}$ such that there exists $\mathcal{Q} \in \mathbb{F}_p[X]$ with $\mathcal{M} = \mathcal{Q} \odot \mathcal{P} \oplus \mathcal{R}$ and $deg(\mathcal{R}) < deg(\mathcal{P})$. Given a polynomial $\mathcal{P} \in \mathbb{F}_p[X]$, denote $\mathbb{F}_p[X]/\mathcal{P}$ the set of classes given by the equivalence relation of *having the same rest mod* $\mathcal{P}$. Clearly, the operations $\oplus$ and $\odot$ can be extended to $\mathbb{F}_p[X]/\mathcal{P}$ in a natural way.

The matrix (with coefficients in $\mathbb{F}_p$) associated to a generic companion graph with $n$ vertices (see Fig. 6) has the following structure

$$M_n = \begin{bmatrix} 0 & 0 & \cdots & 0 & c_0 \\ \lambda_1 & 0 & \cdots & 0 & c_1 \\ 0 & \lambda_2 & \cdots & 0 & c_2 \\ \vdots & & \ddots & \vdots & \vdots \\ 0 & 0 & \cdots & \lambda_{n-1} & c_{n-1} \end{bmatrix} \tag{5}$$

and it is called *companion matrix*. Indeed, in order to simplify calculations, it more interesting to work with the graph (see Fig. 7) represented by the following matrix

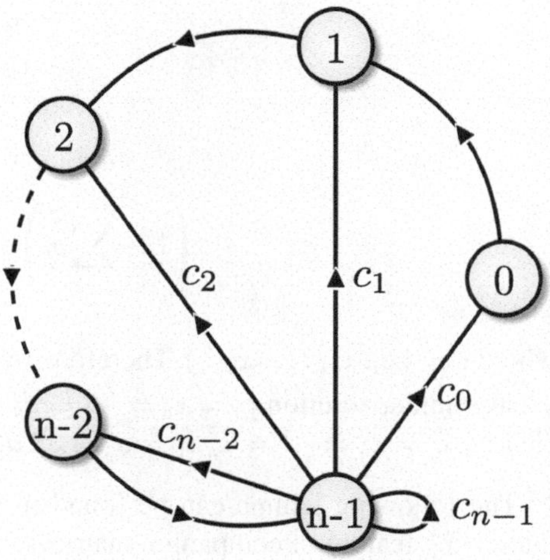


**Fig. 7** The generic companion graph with $n$ vertices used in this paper. Edges labelled with 1 are just drawn without label

$$M_n = \begin{bmatrix} 0 & 0 & \cdots & 0 & c_0 \\ 1 & 0 & \cdots & 0 & c_1 \\ 0 & 1 & \cdots & 0 & c_2 \\ \vdots & & \ddots & \vdots & \vdots \\ 0 & 0 & \cdots & 1 & c_{n-1} \end{bmatrix} \tag{6}$$

Remark that the dynamics of the new system is perfectly equivalent to the former one. Indeed, it is enough to set $\lambda'_i = 1$ and $c'_i = \left(\prod_{k=1}^{i} \lambda_k\right)$ for $i \in \{1, \ldots, n-1\}$.
**Notation.** From now on, when no confusion is possible, we will call *companion graph* the simplified version given in Fig. 7. Correspondingly, the matrix (6) will be called the *companion matrix*. Moreover, the field operations $\oplus, \odot$ over $\mathbb{F}_p$ will be simply written as $+$ and $\cdot$, respectively.

A first interesting property of companion graphs is that it allows a simple characterization of the fixed points and their number. Indeed, we have the following result.

**Proposition 1** *For $n, p \in \mathbb{N}$, let $M_n$ be a companion graph over $\mathbb{F}_p$. Then, it has*

- $\vec{0}$ *as the unique fixed point if* $\sum_{i=0}^{n-1} c_i \neq 1_{\mathbb{F}_p}$
- $p$ *fixed points if* $\sum_{i=0}^{n-1} c_i = 1_{\mathbb{F}_p}$.

*Moreover all fixed points are* uniform *i.e. they have the form $\vec{a} = (a, a, \ldots, a)$ for $a \in \mathbb{F}_p$.* □

***Proof*** Consider a companion matrix $M_n$ as in Eq. (6). By definition, we have a fixed point if $(M^T - I) \cdot \vec{x} = \vec{0}$. If we explicit this equality we have the following system of equations

$$\begin{aligned} -x_0 + x_1 &= 0 \\ -x_1 + x_2 &= 0 \\ \cdots \quad &= \cdots \\ -x_{n-2} + x_{n-1} &= 0 \\ \left(1 - \sum_{i=1}^{n-1} c_i\right) \cdot x_{n-1} &= 0 \end{aligned} \tag{7}$$

where $\vec{x} = (x_0, x_1, \ldots, x_{n-1})$. Therefore, by (7), we have that if $\sum_{i=1}^{n-1} c_i \neq 1_{\mathbb{F}_p}$, there exists a unique solution $x_0 = x_1 = \cdots = x_{n-1} = 0$. However, if $\sum_{i=1}^{n-1} c_i = 1_{\mathbb{F}_p}$, then, clearly, $x_0 = x_1 = \cdots = x_{n-1} = a$ for all $a \in \mathbb{F}_p$ is a solution. □

The following lemma can be found in any algebra textbook. It shows another interesting feature of companion matrices.

**Lemma 2** *The characteristic polynomial $\mathcal{P}_{M_n}$ of $M_n$ coincides with the minimal polynomial of $M_n$ and has the following expression*

$$\mathcal{P}_{M_n}(X) = X^k - \sum_{i=0}^{k-1} c_i \cdot X^i .$$

**Remark 1** $M_n$ is bijective if and only if $\det(M_n) = \mathcal{P}_{M_n}(0) = c_0 \neq 0$. In other words, $M_n$ is bijective if and only if $X \nmid \mathcal{P}_{M_n}(X)$.

Recall that two discrete dynamical systems $f : X \to X$ and $g : Y \to Y$ are *topologically conjugated* if there exists an homeomorphism $\phi : X \to Y$ such that $\phi \circ f = g \circ \phi$. In other words, two (topologically) conjugated systems have the same type of dynamics and topological conjugacy is an equivalence relation on dynamical systems whenever $X = Y$. Finally, remark that if $X$ is finite then $\phi$ is an isomorphism.

Define the natural bijection $\phi$ between $\mathbb{F}_p^n$ and $\mathbb{F}_p[X]/\mathcal{P}_{M_n}$ as follows

$$\forall v_0, v_1, \ldots, v_{n-1} \in \mathbb{F}_p,\ \phi(v_0, v_1, \ldots, v_{n-1}) = \sum_{i=0}^{n-1} v_i \cdot X^i .$$

In $\mathbb{F}_p[X]/\mathcal{P}_{M_n}$, the *shift map* $\sigma(\mathcal{H}) = X \cdot \mathcal{H}$ for $\mathcal{H} \in \mathbb{F}_p[X]/\mathcal{P}_{M_n}$ will play a prominent role in the sequel as it is highlighted by the following lemma.

**Lemma 3** *$M_n$ is topologically conjugated to the shift map over $\mathbb{F}_p[X]/\mathcal{P}_{M_n}$.*

***Proof*** Let $v \in \mathbb{F}_p{}^n$ and consider the bijection $\phi$ as defined above.

$$\sigma(\phi(v)) = X \cdot \sum_{i=0}^{n-1} v_i X^i = \sum_{i=0}^{n-1} v_i X^{i+1} = v_{n-1} X^n + \sum_{i=0}^{n-2} v_i X^{i+1}$$

$$
\begin{aligned}
\phi(M_n \cdot v) &= \phi((c_0 v_{n-1}, v_0 + c_1 v_{n-1}, v_1 + c_2 v_{n-1}, \ldots, v_{n-2} + c_{n-1} v_{n-1})) \\
&= c_0 v_{n-1} + (v_0 + c_1 v_{n-1})X + (v_1 + c_2 v_{n-1})X^2 + \cdots \\
&\qquad + (v_{n-2} + c_{n-1} v_{n-1})X^{n-1} \\
&= v_{n-1} \sum_{i=0}^{n-1} c_i X^i + \sum_{i=0}^{n-2} v_i X^{i+1} .
\end{aligned}
$$

Remark that $\mathcal{P}_{M_n} = X^n - \sum_{i=0}^{n-1} c_i X^i$, thus in $\mathbb{F}_p[X]/\mathcal{P}_{M_n}$ we have $X^n = \sum_{i=0}^{n-1} c_i X^i$ and then $\sigma \circ \phi(v) = \phi(M_n \cdot v)$. □

The following is a standard result that can be found in any algebra textbook. We give a proof here for completeness sake and for illustrating some mechanisms that will be used in the sequel.

**Theorem 1** (Chinese remainder theorem) *For all* $\mathcal{U} \in \mathbb{F}_p[X]$ *such that* $\mathcal{U} = \mathcal{U}_1\mathcal{U}_2 \ldots \mathcal{U}_m$ *and* $\gcd(\mathcal{U}_i, \mathcal{U}_j) = 1$ *for all* $i \neq j \in [\![1, m]\!]$, *it holds*

$$
\mathbb{F}_p[X]/\mathcal{U} \cong \mathbb{F}_p[X]/\mathcal{U}_1 \times \mathbb{F}_p[X]/\mathcal{U}_2 \times \cdots \times \mathbb{F}_p[X]/\mathcal{U}_m .
$$

***Proof*** We prove the case $m = 2$, the other cases can be obtained by finite induction on $m$. Given $\mathcal{H} \in \mathbb{F}_p[X]$ such that $\mathcal{H} = \mathcal{U}\mathcal{V}$ with $\gcd(\mathcal{U}, \mathcal{V} = 1_{\mathbb{F}_p}$, consider the ring homomorphism

$$
\begin{aligned}
h\colon \mathbb{F}_p[X]/\mathcal{H} &\to \mathbb{F}_p[X]/\mathcal{U} \times \mathbb{F}_p[X]/\mathcal{V} \\
\mathcal{W} &\mapsto (\mathcal{W} \bmod \mathcal{U}\ ,\ \mathcal{W} \bmod \mathcal{V})
\end{aligned}
$$

If $h(\mathcal{W}) = (0, 0)$, then $\mathcal{W}$ is divisible by both $\mathcal{U}$ and $\mathcal{V}$. Since $\gcd(\mathcal{U}, \mathcal{V} = 1$, it follows that $W(X) = 0 \bmod \mathcal{H}$ and therefore $\ker(h) = \{0\}$. We conclude that, in this case, $h$ is injective.[1] Moreover,

$$
\begin{aligned}
|\mathbb{F}_p[X]/\mathcal{H}| &= q^{\deg(\mathcal{H})} = q^{\deg(\mathcal{U})} \times q^{\deg(\mathcal{V})} \\
&= |\mathbb{F}_p[X]/\mathcal{U} \times \mathbb{F}_p[X]/\mathcal{V}|
\end{aligned}
$$

and thus $h$ is an isomorphism. □

By combining Lemma 3 and Theorem 1, one obtains

**Theorem 2** *For all* $\mathcal{U} \in \mathbb{F}_p[X]$ *such that* $\mathcal{U} = \mathcal{U}_1\mathcal{U}_2 \ldots \mathcal{U}_m$ *and* $\gcd(\mathcal{U}_i, \mathcal{U}_j) = 1$ *for all* $i \neq j \in [\![1, m]\!]$, *the shift map* $\sigma$ *over* $\mathbb{F}_p[X]/\mathcal{U}$ *is topologically conjugated to the dynamics of* $\prod_{i=1}^{m} \sigma_i$, *where* $\sigma_i$ *is the shift map over* $\mathbb{F}_p[X]/\mathcal{U}_i$.

In other words, Theorem 2 stresses that understanding the dynamics of $M_n$ pass through the complete understanding of the dynamics of the shift map over $\mathbb{F}_p[X]/\mathcal{H}$ for some generic $\mathcal{H} \in \mathbb{F}_p[X]$.

---

[1] Recall that $h$ injective means $h(\mathcal{U}) = h(\mathcal{V}) \Rightarrow \mathcal{U} = \mathcal{V}$. But since $h$ is a homomorphism it comes $h(\mathcal{U}) - h(\mathcal{V}) = 0 \iff h(\mathcal{U} - \mathcal{V}) = 0 \iff (\mathcal{U} - \mathcal{V}) \in \ker(h)$, hence $\ker(h) = \{0\} \iff h$ injective.

Recall that $M_n$ is bijective if and only if $X \nmid \mathcal{P}_{M_n}$. Therefore, according to Theorem 2, the study of the dynamics of the shift map over $\mathbb{F}_p[X]/\mathcal{H}$ can be split into two parts: the nilpotent part and the bijective one.

## 3.1 Nilpotent Dynamics

Given $\mathcal{U} \in \mathbb{F}_p[X]$, let $\mu_{\mathcal{U}}$ be the largest integer $i$ such that $X^i \mid \mathcal{U}$. The following theorem describes completely the dynamics associated to $X^{\mu_{\mathcal{U}}}$ and relies on the definition of $\sigma$ and the fact that $|\mathbb{F}_p| = p$.

**Theorem 3** *Let $\xi$ be a positive integer. Then, the shift map $\sigma$ over $\mathbb{F}_p[X]/X^\xi$ has the following behavior*

1. $\sigma(\vec{0}) = \vec{0}$;
2. $\forall \mathcal{U} \in \mathbb{F}_p[X]/X^\xi \setminus \{\vec{0}\}$, $\sigma^{\xi-\mu_{\mathcal{U}}}(\mathcal{U}) = \vec{0} \mod X^\xi$;
3. *if* $\mathcal{U}(X) = \sum_{i=1}^{\xi-1} \lambda_i \cdot X^i$ *then*

$$\sigma^{-1}(\mathcal{U}) = \left\{ \mathcal{V}_c \mid \forall c \in \mathbb{F}_p,\ \mathcal{V}_c(X) = c + \sum_{i=1}^{\xi-1} \lambda_i \cdot X^{i-1} \right\} .$$

   *In particular, $\mathcal{U}$ has $p$ preimages by $\sigma$.*
4. *If $\mu(\mathcal{U}) = 0$ then $\mathcal{U}$ has no preimages by $\sigma$.*

In other words, Theorem 3 tells that the graph describing the dynamics of $\sigma$ over $\mathbb{F}_p[X]/X^\xi$ is a $p$-tree with the exception that the root has $p - 1$ other predecessors and is also its own predecessor. See Fig. 8 for an illustrative example.

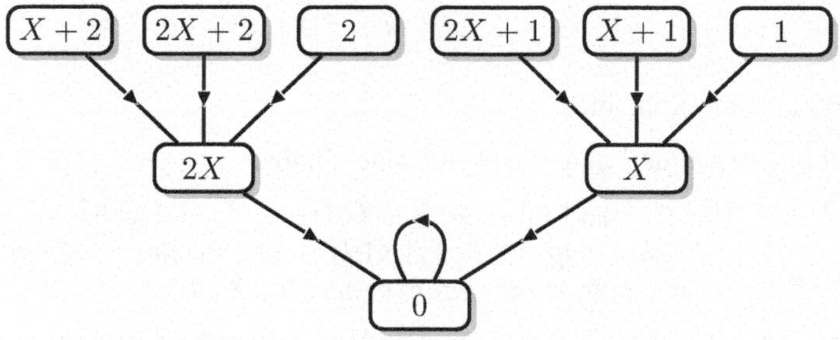


**Fig. 8** Example of transient behavior : each vertex represents a polynomial in $\mathbb{F}_3/X^2$. There is an arc from $\mathcal{U}$ to $\mathcal{V}$ if $\mathcal{V} = \sigma(\mathcal{U}) \mod X^2$

## 3.2 Bijective Dynamics

In this section we consider a companion graph $M_n$ and we assume that it is bijective over $\mathbb{F}_p$ i.e. $X \nmid \mathcal{P}_{M_n}(X)$. Remark that all the results of this section are essentially contained in [1]. We report them here in order to give a point of view contextualized to the domain of automata networks. We also provide some more explicit formulas.

**Notation.** For $\mathcal{H} \in \mathbb{F}_p[X]$, the orbit size of $\mathcal{U} \in \mathbb{F}_p[X]/\mathcal{H}$ is denoted $\theta_{\mathcal{U}}^{\mathcal{H}}$. It is the smallest positive integer $k$ such that $X^k \cdot \mathcal{U} = \mathcal{U}$ or zero if it does not exist.

Obviously, for all $\mathcal{U} \in \mathbb{F}_p[X]/\mathcal{H}$, $\theta_{\mathcal{U}}^{\mathcal{H}} = \theta_{X \cdot \mathcal{U}}^{\mathcal{H}}$.

**Theorem 4** *If $\mathcal{P} \in \mathbb{F}_p[X]$ is irreducible then all non-zero $\mathcal{U} \in \mathbb{F}_p[X]/\mathcal{P}$ have period $\theta_1^{\mathcal{P}}$. Moreover, $\theta_1^{\mathcal{P}} \mid p^d - 1$, where $d = deg(\mathcal{P})$.*

***Proof*** Since $\mathcal{P}$ is irreducible over $\mathbb{F}_p[X]$ then $\mathbb{F}_p[X]/\mathcal{P}$ is a finite field (see [7, Theorem 1.61 p. 25]). Hence, for any non-zero $u \in \mathbb{F}_p^n, \mathcal{U} = \phi(u)$ has a multiplicative inverse $\mathcal{U}^{-1}$. It follows that $\phi(M_n^t(u)) = \phi(u)$ iff $X^t \cdot \mathcal{U} = \mathcal{U}$ iff $X^t = 1$. Moreover, since the order of the multiplicative group is $p^n - 1$ then, by Lagrange's theorem, $\theta_1^{\mathcal{P}} \mid p^n - 1$. □

A consequence of the Theorem 4 is that if $\mathcal{P} \in \mathbb{F}_p[X]$ is irreducible, then $p \nmid \theta_1^{\mathcal{P}}$ (but $p - 1$ does).

**Lemma 4** *Let $\mathcal{P} \in \mathbb{F}_p[X]$. Let $\mathcal{U} \in \mathbb{F}_p[X]/\mathcal{P}$, with $\mathcal{U} = \mathcal{D} \cdot \mathcal{V}$ such that $\mathcal{D} = \gcd(\mathcal{P}, \mathcal{U})$. Then, $\theta_{\mathcal{U}}^{\mathcal{P}} = \theta_{\mathcal{D}}^{\mathcal{P}}$.*

***Proof*** As $\gcd(\mathcal{P}, \mathcal{V}) = 1$, $\mathcal{V}$ has an inverse in $\mathbb{F}_p[X]/\mathcal{P}$. Hence, the following formula is true in $\mathbb{F}_p[X]/\mathcal{P}$:

$$\forall t \in \mathbb{N} \ \left(X^t \mathcal{U} = \mathcal{U} \iff X^t \mathcal{D} = \mathcal{D}\right).$$

□

**Theorem 5** *If $\mathcal{P} = \mathcal{Q}^k$ with $\mathcal{Q} \in \mathbb{F}_p[X]$ irreducible, then*

1. $\theta_1^{\mathcal{P}} = \theta_1^{\mathcal{Q}} \cdot p^{\lceil \log_p k \rceil}$;
2. *for all* $\mathcal{U} \in \mathbb{F}_p[X]/\mathcal{P}$, *it holds* $\theta_{\mathcal{U}}^{\mathcal{P}} = \theta_{\mathcal{Q}^s}^{\mathcal{P}} = \theta_1^{\mathcal{Q}^{k-s}}$, *where* $s = \max\{n \in [\![0, k]\!], \mathcal{Q}^n \mid \mathcal{U}\}$.

***Proof*** 1. Let $t = \lceil \log_p k \rceil = \min\{r \in \mathbb{N}, p^r \geq k\}$. Then,

$$(X^{\theta_1^{\mathcal{Q}}} - 1)^k \text{ divides } (X^{\theta_1^{\mathcal{Q}}} - 1)^{p^t} = X^{\theta_1^{\mathcal{Q}} p^t} - 1 \ . \tag{8}$$

It is clear that $\mathcal{Q} \mid X^{\theta_1^{\mathcal{Q}}} - 1$ and hence $\mathcal{Q}^k \mid (X^{\theta_1^{\mathcal{Q}}} - 1)^k$. By (8), we conclude that $\mathcal{Q}^k \mid X^{\theta_1^{\mathcal{Q}} p^t} - 1$ and that

$$\theta_1^{\mathcal{P}} \mid \theta_1^{\mathcal{Q}} p^t \ . \tag{9}$$

By definition we have $X^{\theta_1^{\mathcal{P}}} = 1 \mod \mathcal{P}$ and hence $X^{\theta_1^{\mathcal{P}}} = 1 \mod \mathcal{Q}$. This implies that $\theta_1^{\mathcal{Q}} \mid \theta_1^{\mathcal{P}}$. Therefore, by (9), it must be $\theta_1^{\mathcal{P}} = \theta_1^{\mathcal{Q}} \cdot p^{t'}$ with $t' \leq t$. Assume $t' < t$, so $p^{t'} < k$ i.e. $p^{t'} + 1 \leq k$. Since $\mathcal{Q} \mid X^{\theta_1^{\mathcal{Q}}} - 1$ and $\mathcal{P} = \mathcal{Q}^k \mid X^{\theta_1^{\mathcal{P}}} - 1 = X^{\theta_1^{\mathcal{Q}} \cdot p^{t'}} - 1 = (X^{\theta_1^{\mathcal{Q}}} - 1)^{p^{t'}}$, we have $\mathcal{Q}^{k+1} \mid (X^{\theta_1^{\mathcal{Q}}} - 1)^{(1+p^{t'})}$ and hence $\mathcal{Q}^{k+1} \mid (X^{\theta_1^{\mathcal{Q}}} - 1)^k$. Remark that $X \nmid X^{\theta_1^{\mathcal{Q}}} - 1$, while its derivative $\theta_1^{\mathcal{Q}} \cdot X^{\theta_1^{\mathcal{Q}}}$ has no other factor than $X$. Hence $\gcd(X^{\theta_1^{\mathcal{Q}}} - 1, \theta_1^{\mathcal{Q}} \cdot X^{\theta_1^{\mathcal{Q}}}) = 1$. We conclude that $X^{\theta_1^{\mathcal{Q}}} - 1$ is square-free. In particular, we have $\mathcal{Q} \mid X^{\theta_1^{\mathcal{Q}}} - 1$ but $\mathcal{Q}^2 \nmid X^{\theta_1^{\mathcal{Q}}} - 1$. Since $\mathcal{Q}$ is irreducible, it is impossible that $\mathcal{Q}^{k+1}$ divides $(X^{\theta_1^{\mathcal{Q}}} - 1)^k$. Hence $t' = t$.

2. Let $\mathcal{U} \in \mathbb{F}_p[X]/\mathcal{Q}^k$ and $s \in [\![0, k]\!]$ such that $\mathcal{U} = \mathcal{V}\mathcal{Q}^s$ with $\gcd(\mathcal{Q}, \mathcal{V}) = 1$. By Lemma 4, $\theta_{\mathcal{U}}^{\mathcal{P}} = \theta_{\mathcal{Q}^s}^{\mathcal{P}}$. So the following formula is true in $\mathbb{F}_p[X]$:

$$\forall t \in \mathbb{N} \left(X^t \mathcal{Q}^s = \mathcal{Q}^s \mod \mathcal{Q}^k \iff X^t = 1 \mod \mathcal{Q}^{(k-s)}\right).$$

Hence, $\theta_{\mathcal{Q}^s}^{\mathcal{P}} = \theta_1^{\mathcal{Q}^{(s-k)}}$. □

**Theorem 6** *If* $\mathcal{P} = \prod_{i=0}^{k} \mathcal{P}_i^{m_i}$ *with*

1. $\forall i \in [0, k], \mathcal{P}_i \in \mathbb{F}_p[X]$
2. $\forall i, j \in [0, k]$ $i \neq j$ *implies* $\gcd(\mathcal{P}_i, \mathcal{P}_j) = 1$,

*then* $\theta_1^{\mathcal{P}} = \operatorname{lcm}(\theta_1^{\mathcal{P}_i^{m_i}})$.

***Proof*** By using Theorem 2, one finds

$$\begin{aligned}
\theta_1^{\mathcal{P}} &= \min\left\{t \in \mathbb{N}^+, X^t = 1 \mod \prod_{i=0}^{k} \mathcal{P}_i^{m_i}\right\} \\
&= \min\left\{t \in \mathbb{N}^+, X^t = 1 \mod \mathcal{P}_0^{m_0} \wedge \cdots \wedge X^t = 1 \mod \mathcal{P}_k^{m_k}\right\} \\
&= \operatorname{lcm}(\theta_1^{\mathcal{P}_i^{m_i}}) .
\end{aligned}$$

□

The following results allow to complete the description of the dynamics providing an explicit account for the multiplicities of cycles.

**Notation.** Given $\mathcal{P} = \mathcal{P}_{M_n} \in \mathbb{F}_p[X]$ and $\mathcal{Q} \mid \mathcal{P}$, $\#\theta_{\mathcal{Q}}^{\mathcal{P}}$ denotes the number of cycles of $M_n$ of period $\theta_{\mathcal{Q}}^{\mathcal{P}}$ for the set $S_{\mathcal{Q}}^{\mathcal{P}} = \{\mathcal{R} \mid \gcd(\mathcal{R}, \mathcal{P}) = \mathcal{Q}\}$. By Lemma 4, all the elements of this subset have the same period $\theta_{\mathcal{Q}}^{\mathcal{P}}$.

**Theorem 7** *Let* $\mathcal{P} = \mathcal{P}_{M_n} \in \mathbb{F}_p[X]$. *Then, the following cases hold*

1. *if* $\mathcal{P} = \mathcal{Q}^m$ *with* $\mathcal{Q}$ *irreducible,* $d = deg(\mathcal{Q})$ *and* $n = deg(\mathcal{P})$ *(so* $n = dm$*) then for all* $k \in [1, m-1]$ *one has*

$$\#\theta_{\mathcal{Q}^{m-k}}^{\mathcal{P}} = \frac{(p^{(k-1)d - \lceil \log_p(k) \rceil})(p^d - 1)}{\theta_1^{\mathcal{Q}}},$$

2. *if* $\mathcal{P} = \mathcal{Q}\mathcal{R}$ *with* $\gcd(\mathcal{Q}, \mathcal{R}) = 1$, $\mathcal{A} \mid \mathcal{Q}$ *and* $\mathcal{B} \mid \mathcal{R}$ *then*

$$\#\theta^{\mathcal{P}}_{\mathcal{A}\mathcal{B}} = \#\theta^{\mathcal{Q}}_{\mathcal{A}} \cdot \#\theta^{\mathcal{R}}_{\mathcal{B}} \ .$$

***Proof*** 1. The size of the set $S^{\mathcal{P}}_{Q^{m-k}}$ is $p^{kd} - p^{(k-1)d} = p^{(k-1)d}(p^d - 1)$ as the number of elements that are multiple of $Q^{m-k}$ is $p^{kd}$ and the number of elements that are multiple of $Q^{m-k+1}$ is $p^{(k-1)d}$.
The size of the cycles in $S^{\mathcal{P}}_{Q^{m-k}}$ is $\theta^{\mathcal{P}}_{Q^{m-k}} = \theta^{\mathcal{Q}^k}_1 = \theta^{\mathcal{Q}}_1 \cdot p^{\lceil \log_p(k) \rceil}$ by Theorem 5.
The number of cycles is obtained by dividing the number of elements by the size of the cycles.
2. The result is a direct consequence of the fact that $\mathbb{F}_p[X]/\mathcal{P}$ is isomorphic to $\mathbb{F}_p[X]/\mathcal{Q} \times \mathbb{F}_p[X]/\mathcal{R}$ (see Theorem 2). □

**Corollary 1** *If* $\mathcal{P} = \mathcal{P}_{M_n} \in \mathbb{F}_p[X]$ *is irreducible of degree n, then* $\theta^{\mathcal{P}}_1 \mid (p^n - 1)$.

***Proof*** From Theorem 7, we know that in this case, the integer $\#\theta^{\mathcal{P}}_1$ is equal to $\frac{(p^n-1)}{\theta^{\mathcal{P}}_1}$. □

Corollary 1 is pretty important from the practical point of view since it provides an algorithmic speed-up to find $\theta^{\mathcal{P}}_1$ knowing the factorization of $(p^n - 1)$.

## 4 Extension to Rings

The results of the previous sections cannot be extended to rings like $\mathbb{F}_{p^n}$ (for a prime $p \in \mathbb{N}$ and $n \in \mathbb{N}^\star$). Indeed, polynomials over $\mathbb{F}_{p^n}$ do not have a unique factorization. In [3], they provide the following interesting identity over $\mathbb{F}_{p^n}$

$$(X^m + p^{n-1})(X^m + p) = X^m(X^m + p^{n-1} + p)$$

and conclude that for any $m \in \mathbb{N}$, there exists in $\mathbb{F}_{p^n}$ a product of at most $n$ irreducible that is also representable as a product of more than $m$ irreducible.

**Lemma 5** ([3]) *Let* $\mathcal{P} \in \mathbb{F}_{p^n}[X]$ *for some prime* $p \in \mathbb{N}$ *and* $n \in \mathbb{N}^\star$. *If* $\mathcal{P}$ *is irreducible in* $\mathbb{F}_p[X]$, *then it is irreducible in* $\mathbb{F}_{p^n}[X]$.

The above lemma grants that if we have an irreducible polynomial in $\mathbb{F}_p[X]$, then it still irreducible in $\mathbb{F}_{p^n}[X]$. The converse is false. Indeed, $X^4 + 2$ is irreducible in $\mathbb{F}_4$ but in $\mathbb{F}_2$ it is equivalent to $X^4$. However, at least for the polynomials described in Lemma 5, one can apply Theorem 4 and perfectly describe the dynamics. Moreover, remark that Theorem 1 can be extended to the ring $\mathbb{F}_q[X]$ for a generic integer $q > 1$ and this helps in further extending the analysis of the dynamics.

## 5 Extension to General Additive Networks

The previous sections report a general framework to describe the dynamics of peculiar additive networks like additive flowers or companion graphs. This section extends the framework to general additive networks. The following result shows that the dynamics of a general additive network can be decomposed in the Cartesian product of the dynamics of a certain number of companion graphs.

**Theorem 8** *Consider an additive network $M_n$ over $\mathbb{F}_p$, then*

1. *there exist $n_0, n_1, \ldots, n_k \in \mathbb{N}$ such that $n = \sum_{i=0}^{k} n_i$;*
2. *there exists companion graphs $C_{n_0}, C_{n_1}, \ldots, C_{n_k}$;*

*such that the dynamics of $M_n$ is topologically conjugated with the dynamics of*

$$C_{n_0} \times C_{n_1} \times \cdots \times C_{n_k} .$$

***Proof*** Reducing $M_n$ to its Frobenius normal form implies that

1. there exist $n_0, n_1, \ldots, n_k \in \mathbb{N}$ such that $n = \sum_{i=0}^{k} n_i$;
2. there exists companion matrices $C_{n_i}$ of size $n_i$ for $i = 0, \ldots, k$;
3. there exists an invertible matrix $P$ of size $n$;

such that $M_n = P \cdot C \cdot P^{-1}$ where

$$C = \begin{bmatrix} C_{n_0} & \mathbb{O} & \ldots & \mathbb{O} \\ \mathbb{O} & C_{n_1} & \ddots & \vdots \\ \vdots & \ddots & \ddots & \mathbb{O} \\ \mathbb{O} & \ldots & \mathbb{O} & C_{n_k} \end{bmatrix}$$

and $\mathbb{O}$ are matrices of suitable size with all entries fixed at 0. This implies that $M_n$ and $C_n$ are topologically conjugated. Moreover, remark that for all $n \in \mathbb{N}$ we have

$$C^n = \begin{bmatrix} C^n_{n_0} & \mathbb{O} & \ldots & \mathbb{O} \\ \mathbb{O} & C^n_{n_1} & \ddots & \vdots \\ \vdots & \ddots & \ddots & \mathbb{O} \\ \mathbb{O} & \ldots & \mathbb{O} & C^n_{n_k} \end{bmatrix}$$

that is to say the dynamics of $C^n$ is the cartesian product of the dynamics of the companion matrices $C_{n_i}$. □

From the proof of the previous result, it is clear that in order to find the decomposition of a generic additive network one has to pass through the Frobenius normal form of the matrix associated with the network. Recently, Arne Storjohann has proposed a $O(n^3)$ algorithm for finding the Frobenius normal form of a square matrix of size

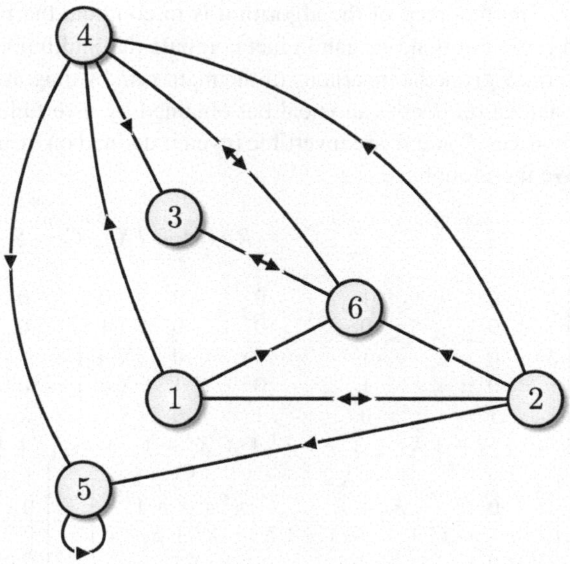


**Fig. 9** The boolean network used in Sect. 5.1

$n$ [13]. Faster algorithms exist for sparse or special matrices [16]. A nearly linear speedup (trading processors for time) has been obtained for parallel version of these algorithms [5].

However, one can remark that most of these algorithms are made in two steps. First, the Smith normal form $S$ is computed and then the Frobenius normal form is deduced from $S$. Indeed, the elements $s_{i,i}$ of $S$ which are different from 1 are exactly the characteristic polynomials of the companion matrices $C_j$ of the Frobenius normal form.

## 5.1 A Complete Example

In this section we are going to develop a complete example in order to illustrate the use of the results found so far. Consider the boolean additive network with 6 nodes which has the interaction graph as depicted in Fig. 9.

The network can be represented by its adjacency matrix:

$$\begin{bmatrix} 0 & 1 & 0 & 0 & 0 & 0 \\ 1 & 0 & 0 & 0 & 0 & 0 \\ 0 & 0 & 0 & 1 & 0 & 1 \\ 1 & 1 & 0 & 0 & 0 & 1 \\ 0 & 1 & 0 & 1 & 1 & 0 \\ 1 & 1 & 1 & 1 & 0 & 0 \end{bmatrix}$$

The first step of the algorithm is to compute the Frobenius normal form of the matrix. For that, we can in fact compute the Smith normal form. The Smith normal form $S$ gives the invariants of the matrix $M$ by diagonalizing $M - I \cdot X$. This diagonalization is not canonical but obtained by invertible transformations. Hence, the matrices $R$ and $C$ are invertible (even if defined on the ring $\mathbb{F}_p[X]$ that is not a field). We therefore have

$$R \cdot (M - IX) \cdot C = S$$

$$\begin{bmatrix} 0 & 1 & 0 & 0 & 0 & 0 \\ 0 & 0 & 0 & 0 & 1 & 0 \\ 0 & 1 & 0 & 0 & X+1 & 1 \\ 0 & 1 & 0 & 1 & X+1 & 0 \\ 1 & 0 & 1 & 0 & X+1 & X \\ X^2+X+1 & X^2+X+1 & X^2+X & X^2+1 & 0 & X^3+X^2 \end{bmatrix} \cdot \begin{bmatrix} X & 1 & 0 & 0 & 0 & 0 \\ 1 & X & 0 & 0 & 0 & 0 \\ 0 & 0 & X & 1 & 0 & 1 \\ 1 & 1 & 0 & X & 0 & 1 \\ 0 & 1 & 0 & 1 & X+1 & 0 \\ 1 & 1 & 1 & 1 & 0 & X \end{bmatrix} \cdot$$

$$\cdot \begin{bmatrix} 1 & X & 0 & X & X^3+X^2 & X^3+X^2+X \\ 0 & 1 & 0 & 1 & X^2+X & X^2+X+1 \\ 0 & 0 & 1 & X & X^3+X^2+X+1 & X^3+X^2+X+1 \\ 0 & 0 & 0 & 1 & X^2+1 & X^2 \\ 0 & 0 & 0 & 0 & 1 & 1 \\ 0 & 0 & 0 & 0 & 0 & 1 \end{bmatrix} = \begin{bmatrix} 1 & 0 & 0 & 0 & 0 & 0 \\ 0 & 1 & 0 & 0 & 0 & 0 \\ 0 & 0 & 1 & 0 & 0 & 0 \\ 0 & 0 & 0 & 1 & 0 & 0 \\ 0 & 0 & 0 & 0 & X^2+1 & 0 \\ 0 & 0 & 0 & 0 & 0 & X^4+X^3+X+1 \end{bmatrix}$$

Thee Smith normal form gives immediately the Frobenius normal form as two companion matrices block: one for $X^2 + 1$, the other for $X^4 + X^3 + X + 1$. Hence there exists an invertible matrix $A$ such that

$$M = A \cdot F \cdot A^{-1}$$

$$\begin{bmatrix} 0&1&0&0&0&0 \\ 1&0&0&0&0&0 \\ 0&0&0&1&0&1 \\ 1&1&0&0&0&1 \\ 0&1&0&1&1&0 \\ 1&1&1&1&0&0 \end{bmatrix} = \begin{bmatrix} 1&1&1&0&1&0 \\ 1&1&0&1&0&1 \\ 0&0&1&0&1&1 \\ 0&0&0&1&1&1 \\ 0&1&0&0&0&1 \\ 0&0&0&0&0&1 \end{bmatrix} \cdot \begin{bmatrix} 0&1&0&0&0&0 \\ 1&0&0&0&0&0 \\ 0&0&0&0&0&1 \\ 0&0&1&0&0&1 \\ 0&0&0&1&0&0 \\ 0&0&0&0&1&1 \end{bmatrix} \cdot \begin{bmatrix} 1&0&1&0&1&0 \\ 0&0&0&0&1&1 \\ 1&1&0&1&0&0 \\ 1&1&1&0&0&0 \\ 1&1&1&1&0&1 \\ 0&0&0&0&0&1 \end{bmatrix}$$

Note that the matrix $A$ can be computed from the inverse of matrix $R$. This inverse can easily be computed while applying the algorithm for the Smith normal form $S$.

Inspecting each companion submatrix. The first matrix is the companion matrix of characteristic polynomial $X^2 + 1 = (X + 1)^2$. By Theorems 5 and 7, we find one orbit of size 2 and two orbits of size 1. The second matrix is the companion matrix of characteristic polynomial $X^4 + X^3 + X + 1 = (X + 1)^2 \cdot (X^2 + X + 1)$. We already computed the dynamics of $(X + 1)^2$. As $X^2 + X + 1$ is irreducible, by Theorem 4, we find a unique orbit of maximum length that must divide $2^2 - 1 = 3$. As the order of $X$ is not 1, 3 is the only option. Hence, there is one orbit of size 3 and one of size 1.

Finally, Theorem 8 tells that to obtain the overall dynamics one has to take the Cartesian product. The following table gives the orbit complete description:

| First block | Second Block | | Network |
|---|---|---|---|
| $(X+1)^2$ | $(X+1)^2$ | $X^2+X+1$ | |
| (2, 1) | (2, 1) | (1, 1) | (4, 1) |
| (1, 2) | (1, 2) | | (4, 2) |
| | | | (1, 4) |
| | | (1, 3) | (4, 3) |
| | | | (4, 6) |
| | | | (1, 12) |

## 6 Extension to Affine Networks

A Boolean network $f : \mathbb{Z}_2^n \to \mathbb{Z}_2^n$ is *affine* if, for all $i \in [1, n]$, we have

$$\forall x \in \mathbb{Z}_2^n, \qquad f_i(x) = \alpha_i + \sum_{i=j}^{n} \lambda_{ij} x_j \mod p$$

for some $\alpha_i, \lambda_{i1}, \ldots \lambda_{in} \in \mathbb{Z}_2^n$. Given such an affine Boolean network $f$, let $f^+ : \mathbb{Z}_p^{n+1} \to \mathbb{Z}_p^{n+1}$ be defined as follows:

$$\forall i \in [1, n],\ \forall x \in \mathbb{Z}_2^n, \quad f_i^+(x) = \alpha_i x_{n+1} + \sum_{i=j}^{n} \lambda_{ij} x_j \mod p, \quad f_{n+1}^+(x) = x_{n+1}.$$

We then define $f^- : \mathbb{Z}_2^n \to \mathbb{Z}_2^n$ as follows:

$$\forall i \in [1, n],\ \forall x \in \mathbb{Z}_2^n, \qquad f_i^-(x) = f_i(x) - \alpha_i.$$

Remark that both $f^+$ and $f^-$ are additive Boolean networks and they completely determine the dynamics of the $f$. Indeed, for all $x \in \mathbb{Z}_2^n$ we have $f^+(x, 1) = (f(x), 1)$ and $f^+(x, 0) = (f^-(x), 0)$. Hence $(x, 1)$ is a periodic point of $f^+$ of period $p$ if and only if $x$ is a periodic point of $f$ of period $p$, and $(x, 0)$ is a periodic point of $f^+$ period $p$ if and only if $x$ is a periodic point of $f^-$ of period $p$. Therefore, we have the following.

**Theorem 9** *Let $f$ be an affine Boolean network, and let $f^+$, $f^-$ the be associated additive Boolean networks. The number of periodic points of $f$ of period $p$ is the number of periodic points of $f^+$ of period $p$ minus the number of periodic points of $f^-$ of period $p$.*

## 7 Conclusions

This paper presents a general framework for precisely describing the dynamics of the class of additive networks ( which additive flower are a peculiar subclass). Remark that these results are not new. Indeed, they were discovered and rediscovered several times (at least four plus this one) by different researchers in different contexts (mainly in the field of Linear Shift Registers theory). However, most of the times results were not presented in a unified way or proofs were not very formal in some parts. Moreover, no remarks about computational complexity were faced. This paper tries to fill these gaps.

Future research directions are already sketched in the paper. Indeed, Sect. 4 proposes to extend the results to rings such as $\mathbb{Z}_{p^k}$ for some prime $p$ and some integer $k \geq 1$. A few results in this direction are provided but the main questions are still open.

Another interesting research direction consists in quantifying possible dynamics. For instance, assume that one knows that some phenomenon has fixed points dynamics and that this fixed point is unique. How many additive networks of size $n$ have this kind of dynamics? More in general, we know that topological equivalence is an equivalence relation over the set of possible dynamics. Can we quantify the size of its classes as a function of the number of nodes? This question has simple answer for the case of unique fixed point dynamics but how about more complex situations?

Once one has established the previous issues, the next natural step is to try to add some nonlinearity in the networks and see how and at which extent it influences the dynamical behavior. Theorem 9 provides a first step in this direction but alphabets of larger sizes call for more complicated formulations. We are currently investigating these questions.

## References

1. Elspas B (1959) The theory of autonomous linear sequential networks. IEEE Trans Circuits Syst 6:45–60
2. Formenti E, Manzoni L, Porreca AE (2014) Fixed points and attractors of reaction systems. In: Beckmann A, Csuhaj-Varjú E, Meer K (eds) Language, Life, Limits - 10th conference on computability in Europe, CiE 2014, Budapest, Hungary, June 23–27, 2014. Proceedings, volume 8493 of Lecture notes in computer science. Springer, pp 194–203
3. Frei C, Frisch S (2011) Non-unique factorization of polynomials over residue class rings of the integers. Comm Algebra 30:1482–1490
4. Gadouleau M (2020) On the influence of the interaction graph on a finite dynamical system. Nat Comput 19(1):15–28
5. Jäger G, Wagner C (2009) Efficient parallelizations of hermite and smith normal form algorithms. Parallel Comput 35(6):345–357
6. Kauffman SA (1969) Metabolic stability and epigenesis in randomly constructed genetic nets. J Theor Biol 22:437–467
7. Lidl R, Niederreiter H (1986) Introduction to finite fields and their applications. Cambridge University Press, Cambridge

8. Milano M, Roli A (2000) Solving the satisfiability problem through boolean networks. In: Lamma E, Mello P (eds) AI*IA 99: Advances in Artificial Intelligence, vol 1792. Lecture Notes in Computer Science. Springer, Berlin, pp 72–83
9. Noual M (2012) Updating automata networks. PhD thesis, Ecole Normale Supérieure de Lyon
10. OEIS (2008) Sequence A046932
11. International Math Olympiads (1993) Problem 6
12. Richard A (2019) Positive and negative cycles in boolean networks. J Theor Biol 463:67–76
13. Storjohann A (1998) An $O(n^3)$ algorithm for frobenius normal form. In: International symposium and algebraic computation (ISSAC'98). ACM Press, pp 101–104
14. Thomas R (1973) Boolean formalization of genetic control circuits. J Theor Biol 42:563–585
15. Tosic PT (2005) On complexity of counting fixed points in certain classes of graph automata. In: Electronic colloquium on computational complexity (ECCC), (051)
16. Villard G (200) Computing the Frobenius normal form of a sparse matrix. In: The third international workshop on computer algebra in scientific computing. Springer, pp 395–407

# Domino Problem for Pretty Low Complexity Subshifts

**Jarkko Kari**

**Abstract** Given a set $P$ of allowed $n \times m$ rectangular patterns of colors, a coloring of the grid $\mathbb{Z}^2$ is called valid if every $n \times m$ pattern in the coloring is in $P$. It is known that if the number of allowed $n \times m$ patterns is at most $nm$ and if there exists a valid coloring of $\mathbb{Z}^2$ then there exists a valid periodic coloring, and consequently there is an algorithm to determine for given $nm$ patterns if they admit any valid coloring. If the number of allowed patterns is higher the situation changes: We prove that for every $\varepsilon > 0$ it is undecidable for given dimensions $n, m$ and a given set of at most $(1+\varepsilon)nm$ patterns of size $n \times m$ whether they admit any valid coloring. In other words, the upper bound $nm$ is multiplicatively optimal for the number of allowed patterns that guarantees decidability and periodicity. The undecidability result that we prove is actually slightly better: it holds in the square case $n = m$, and instead of bound $(1+\varepsilon)n^2$ we can use $n^2 + f(n)n$ for any computable function $f : \mathbb{N} \longrightarrow \mathbb{N}$ that is not bounded above by a constant.

## 1 Introduction

The domino problem is a decision problem asked by H. Wang in terms of Wang tiles: a finite set of unit square tiles with colored edges is given and one asks whether there exists a tiling of the plane such that everywhere the neighboring tiles have the same color in their adjacent edges [15]. In symbolic dynamics terminology the question is about emptiness of a given two-dimensional subshift of finite type. The domino problem was proved undecidable by R. Berger [1] and it has since become a fruitful source of undecidability results in tilings and cellular automata, and in multidimensional symbolic dynamics in general. A central prerequisite for the undecidability of the domino problem is the existence of aperiodic tile sets that admit valid tilings but all admitted tilings are non-periodic. This connection was already noted by H. Wang [15], and Berger's undecidability proof in [1] involves the first construction of

J. Kari (✉)
Department of Mathematics and Statistics, University of Turku, FI-20014 Turku, Finland
e-mail: jkari@utu.fi

 
A. Adamatzky (ed.), *Automata and Complexity*, Emergence, Complexity and Computation 42, https://doi.org/10.1007/978-3-030-92551-2_19

an aperiodic set of Wang tiles. It is now known that the smallest aperiodic Wang tile set contains 11 tiles [8].

Discrete complex systems involve large numbers of simple entities with complicated collective behavior. Classical examples include automata networks and cellular automata [6, 7]. Also Wang tiles provide an ideal formalism to investigate how local interactions may either lead to global structures and order or to complexity and chaos. If the number of valid local patterns of tiles is sufficiently low then the global tiling is forced to be periodic while more local variations lead to more freedom to avoid such global regularities. To quantify the threshold of this transition we consider the following formulation of the domino problem. Let $P \subseteq A^D$ be a given set of allowed patterns of rectangular shape $D = \{1, \ldots n\} \times \{1, \ldots m\}$ over a finite set $A$ of colors. We call colorings $c \in A^{\mathbb{Z}^2}$ of the infinite grid configurations and say that a configuration $c$ is valid if every $n \times m$ rectangle that appears in $c$ is in $P$. The domino problem now asks if given $P$ admits any valid configurations. The domino problem is undecidable even for fixed $n = m = 2$.

If the number $|P|$ of allowed patterns is very small then any admitted configuration must be periodic, that is, invariant under some translation of the grid. This is certainly true if $|P| \leq mn/2$, half of the size of the rectangle [5], and the famous Nivat's conjecture claims that $|P| \leq mn$ is enough to guarantee the periodicity of any admitted configuration [12]. While Nivat's conjecture is still unproved, we do know that the low complexity assumption $|P| \leq mn$ is enough to guarantee that $P$ admits a periodic configuration if it admits a configuration at all [10]. In particular, this implies by H. Wang's argument in [15] that there is an algorithm to determine if any given set $P$ of at most $mn$ rectangular patterns of size $n \times m$ admits a valid configuration.

But what might happen if the number of allowed patterns is increased to $mn + 1$; does the domino problem remain decidable and are there any aperiodic systems? What about with $mn + k$ patterns for some constant $k$? There certainly are aperiodic systems for some values $m$, $n$ and $k$ but what is the smallest $k$ admitting an aperiodic system, and is there some constant $k$ such that the domino problem is undecidable? These questions remain open, but they highlight the quest to find the boundary of undecidability in terms of the number of allowed patterns.

In this work we show that for any Wang tile set $T$ we can compute constants $k$ and $N$ such that for all $n \geq N$ and $m \geq 2$ one can effectively construct $nm + k(n + m)$ allowed $n \times m$ patterns that are "equivalent" to $T$ (Theorem 2). In particular, if $f : \mathbb{N} \longrightarrow \mathbb{N}$ is any computable function not in $\mathcal{O}(1)$, the domino problem is undecidable in the pretty low complexity setting where one is given at most $n^2 + f(n)n$ allowed patterns of size $n \times n$ (Corollary 2). Choosing $f(n) \leq \varepsilon n$ then proves that the bound $mn$ of [10] that guarantees decidability can not be multiplicatively improved to $(1 + \varepsilon)nm$ for any $\varepsilon > 0$.

The paper is organized as follows. In Sect. 2 we give more precise formalisms of the concepts involved. Section 3 details the technical construction leading to the main result (Theorem 2). We finish with Sect. 4 that contains corollaries of the main result and discusses open questions.

## 2 Definitions

We denote $[\![n, m]\!] = \{n, n+1, \ldots, m\}$ for integers $n \leq m$, and for any positive integer $n$ we set $[\![n]\!] = [\![0, n-1]\!]$. We index the columns and rows of the $n \times m$ rectangle $[\![n]\!] \times [\![m]\!]$ by $0, \ldots, n-1$ and $0, \ldots, m-1$, respectively. The $n \times m$ rectangle at position $\mathbf{u} \in \mathbb{Z}^2$ of the two-dimensional grid is $\mathbf{u} + [\![n]\!] \times [\![m]\!] \subseteq \mathbb{Z}^2$.

**Basic Concepts**

Let $A$ be a finite set of colors. A two-dimensional *configuration* is an assignment $c : \mathbb{Z}^2 \longrightarrow A$ of colors to the cells of the infinite two-dimensional grid. For any configuration $c \in A^{\mathbb{Z}^2}$ and any cell $\mathbf{u} \in \mathbb{Z}^2$ we may denote $c(\mathbf{u})$, the symbol that $c$ has in cell $\mathbf{u}$, also by $c_{\mathbf{u}}$ to avoid clutter from parentheses. For any $\mathbf{t} \in \mathbb{Z}^2$ we denote by $\tau^{\mathbf{t}}$ the *translation* that shifts a configuration $c$ so that cell $\mathbf{t}$ moves to the origin, that is, $\tau^{\mathbf{t}}(c)_{\mathbf{u}} = c_{\mathbf{u}+\mathbf{t}}$ for all $\mathbf{u} \in \mathbb{Z}^2$. We say that $c$ is *periodic* if $\tau^{\mathbf{t}}(c) = c$ for some non-zero $\mathbf{t} \in \mathbb{Z}^2$. In this case $\mathbf{t}$ is a *vector of periodicity* and $c$ is also termed $\mathbf{t}$-*periodic*. If there are two linearly independent vectors of periodicity then $c$ is called *two-periodic*. A two-periodic $c \in A^{\mathbb{Z}^2}$ has automatically horizontal and vertical vectors of periodicity $(k, 0)$ and $(0, k)$ for some $k > 0$.

Let $D \subseteq \mathbb{Z}^2$ be a finite set of cells, a *shape*. A $D$-*pattern* is an assignment $p \in A^D$ of symbols in shape $D$. A *(finite) pattern* is a $D$-pattern for some finite $D$. We call $D$ the *domain* of the pattern. We say that a finite pattern $p$ of shape $D$ *appears* in configuration $c$ if for some $\mathbf{t} \in \mathbb{Z}^2$ we have $\tau^{\mathbf{t}}(c)|_D = p$. We also say that $c$ *contains* pattern $p$. For a fixed $D$, the set of $D$-patterns that appear in a configuration $c$ is denoted by $\mathcal{L}_D(c)$. We denote by $\mathcal{L}(c)$ the set of all finite patterns that appear in $c$, i.e., the union of $\mathcal{L}_D(c)$ over all finite $D \subseteq \mathbb{Z}^2$.

Let $p \in A^D$ be a finite pattern of shape $D$. The set $[p] = \{c \in A^{\mathbb{Z}^2} \mid c|_D = p\}$ of configurations that have $p$ in domain $D$ is called the *cylinder* determined by $p$. The collection of cylinders $[p]$ over all finite patterns $p$ is a base of a compact topology on $A^{\mathbb{Z}^2}$, the *prodiscrete* topology. See, for example, the first few pages of [4] for details. The topology is equivalently defined by a metric on $A^{\mathbb{Z}^2}$ where two configurations are close to each other if they agree with each other on a large region around cell $\mathbf{0}$. Cylinders are clopen in the topology: they are both open and closed.

Convergence of a sequence $c^{(1)}, c^{(2)}, \ldots$ of configurations to a configuration $c$ in the topology has the following simple meaning: For every cell $\mathbf{u} \in \mathbb{Z}^2$ we must have $c^{(i)}_{\mathbf{u}} = c_{\mathbf{u}}$ for all sufficiently large $i$. Compactness of space $A^{\mathbb{Z}^2}$ then means that every sequence has a converging subsequence.

**Subshifts**

A subset $X$ of $A^{\mathbb{Z}^2}$ is called a *subshift* if it is closed in the topology and closed under translations, i.e., $\tau^{\mathbf{t}}(X) = X$ for every translation $\tau^{\mathbf{t}}$. By a compactness argument one has that every configuration $c$ that is not in $X$ contains a finite pattern $p$ that prevents it from being in $X$: no configuration that contains $p$ is in $X$. We can then as well define subshifts using forbidden patterns: given a set $F$ of finite patterns we define

$$\mathcal{X}_F = \{c \in A^{\mathbb{Z}^2} \mid \mathcal{L}(c) \cap F = \emptyset\},$$

the set of configurations that do not contain any of the patterns in $F$. Set $\mathcal{X}_F$ is a subshift, and every subshift is $\mathcal{X}_F$ for some $F$. If $X = \mathcal{X}_F$ for some finite $F$ then $X$ is a *subshift of finite type* (SFT). For a subshift $X \subseteq A^{\mathbb{Z}^2}$ we denote by $\mathcal{L}_D(X) = \bigcup_{c \in X} \mathcal{L}_D(c)$ and $\mathcal{L}(X) = \bigcup_{c \in X} \mathcal{L}(c)$ the sets of $D$-patterns and all finite patterns that appear in elements of $X$, respectively. Set $\mathcal{L}(X)$ is called the *language* of the subshift.

The *orbit* of configuration $c$ is the set $\mathcal{O}(c) = \{\tau^{\mathbf{t}}(c) \mid \mathbf{t} \in \mathbb{Z}^2\}$ that contains all translates of $c$. The *orbit closure* $\overline{\mathcal{O}(c)}$ of $c$ is the topological closure of the orbit $\mathcal{O}(c)$. It is a subshift, and clearly it is the intersection of all subshifts that contain $c$. In terms of finite patters, $c' \in \overline{\mathcal{O}(c)}$ if and only if $\mathcal{L}(c') \subseteq \mathcal{L}(c)$.

A configuration $c$ is called *uniformly recurrent* if for every $c' \in \overline{\mathcal{O}(c)}$ we have $\overline{\mathcal{O}(c')} = \overline{\mathcal{O}(c)}$ so that all configurations in $\overline{\mathcal{O}(c)}$ have the same finite patterns. This is equivalent to $\overline{\mathcal{O}(c)}$ being a *minimal subshift* in the sense that it has no proper non-empty subshifts inside it. A classical result by Birkhoff [2] implies that every non-empty subshift contains a minimal subshift, so there is a uniformly recurrent configuration in every non-empty subshift.

Subshifts of finite type can as well be defined in terms of *allowed patterns*. To do so we fix a finite domain $D \subseteq \mathbb{Z}^2$, and take a set $P \subseteq A^D$ of allowed patterns with domain $D$. Forbidding all other $D$-patterns yields the SFT

$$\mathcal{V}(P) = \mathcal{X}_{A^D \setminus P} = \{c \in A^{\mathbb{Z}^2} \mid \mathcal{L}_D(c) \subseteq P\},$$

the set of configurations whose $D$-patterns are among $P$. We call elements of $\mathcal{V}(P)$ *valid configurations* admitted by $P$.

We call an SFT *aperiodic* if it is non-empty but does not contain any periodic configurations. It is significant that aperiodic SFTs exist [1]. It is also worth noting that a two-dimensional SFT that contains a periodic configuration must also contain a two-periodic configuration [13].

**The Domino Problem**

The *domino problem* (aka the tiling problem) is the decision problem in two dimensions that asks whether a given set $P \subseteq A^D$ of allowed patterns admits any valid configurations, that is, whether the SFT $\mathcal{V}(P)$ is non-empty. The domino problem is undecidable [1], but if $\mathcal{V}(P)$ is a priori known not to be aperiodic then the domino problem can be decided [15]. Indeed, there are semi-algorithms for the cases when $\mathcal{V}(P)$ is empty and when $\mathcal{V}(P)$ contains a two-periodic configuration: only on aperiodic SFTs both semi-algorithms fail to give an answer. Thus if there were no aperiodic SFTs then the domino problem would be decidable.

**Wang Tiles**

Two-dimensional SFTs are commonly studied in terms of *Wang tiles*, and the first aperiodic SFTs were constructed and the undecidability of the domino problem was originality proved in the Wang tile formalism. A Wang tile is a unit square tile with colored edges, represented as a 4-tuple

$$a = (a_{\uparrow}, a_{\rightarrow}, a_{\downarrow}, a_{\leftarrow}) \in C^4$$

**Fig. 1** A Wang tile $a = (a_\uparrow, a_\rightarrow, a_\downarrow, a_\leftarrow)$

of colors of the north, the east, the south and the west edges of the tile, respectively, where $C$ is a set of colors. (See Fig. 1.) A Wang tile set $T$ is a finite set of Wang tiles. Wang tile set $T$ defines a subshift of $T^{\mathbb{Z}^2}$, where forbidden patterns are all the dominoes of two tiles that do not have the same color on their abutting edges. We say that a configuration $c \in T^{\mathbb{Z}^2}$ is correctly tiled at position $(i, j) \in \mathbb{Z}^2$ if $c(i, j)$ matches with its four neighbors on the abutting edges so that

$$\begin{aligned} c(i, j)_\uparrow &= c(i, j+1)_\downarrow, \\ c(i, j)_\downarrow &= c(i, j-1)_\uparrow, \\ c(i, j)_\rightarrow &= c(i+1, j)_\leftarrow \quad \text{and} \\ c(i, j)_\leftarrow &= c(i-1, j)_\rightarrow. \end{aligned}$$

Otherwise there is a tiling error at position $(i, j)$. We let

$$\mathcal{V}(T) = \{c \in T^{\mathbb{Z}^2} \mid c \text{ is correctly tiled at every position } \mathbf{u} \in \mathbb{Z}^2\}$$

be the set of valid tilings by tile set $T$. Clearly $\mathcal{V}(T)$ is an SFT, and in fact any given set $P \subseteq A^D$ of allowed patterns can be effectively converted into an equivalent Wang tile set $T$ so that $\mathcal{V}(T)$ and $\mathcal{V}(P)$ are conjugate, i.e., homeomorphic under a translation invariant homeomorphism. In this sense Wang tiles capture the entire complexity of two-dimensional subshifts of finite type. Note that we use the same notation $\mathcal{V}(T)$ and $\mathcal{V}(P)$ for the sets of valid tilings by a Wang tile set $T$ and of valid configurations under allowed patterns $P$, respectively. This should not cause any confusion since it is always clear from the context whether we are talking about Wang tiles or allowed patterns.

The *cartesian product* $T_1 \times T_2 \subseteq (C_1 \times C_2)^4$ of Wang tile sets $T_1 \subseteq C_1^4$ and $T_2 \subseteq C_2^4$ is the Wang tile set that contains for all $(a_\uparrow, a_\rightarrow, a_\downarrow, a_\leftarrow) \in T_1$ and $(b_\uparrow, b_\rightarrow, b_\downarrow, b_\leftarrow) \in T_2$ the tile $((a_\uparrow, b_\uparrow), (a_\rightarrow, b_\rightarrow), (a_\downarrow, b_\downarrow), (a_\leftarrow, b_\leftarrow))$. The "sandwich" tiles in $T_1 \times T_2$ have hence two layers that tile the plane independently according to $T_1$ and $T_2$, respectively.

The results reported here are based on Berger's theorem, stating in the Wang tile formalism the existence of aperiodic SFTs and the undecidability of the domino problem.

**Theorem 1** (R. Berger [1])

*(a) There exists a Wang tile set $T$ that is aperiodic, that is, such that $\mathcal{V}(T)$ is non-empty but does not contain any periodic configurations.*

*(b) It is undecidable to determine for a given Wang tile set $T$ whether $\mathcal{V}(T)$ is empty or not.*

**Pattern Complexity**

The *pattern complexity* of a configuration $c$ with respect to a shape $D$ is the number of $D$-patterns that $c$ contains, that is, the cardinality of $\mathcal{L}_D(c)$. A sufficiently low pattern complexity forces global structure in a configuration. A relevant threshold happens when the pattern complexity is at most $|D|$, the number of cells in shape $D$. Hence we say that $c$ has *low complexity* with respect to shape $D$ if

$$|\mathcal{L}_D(c)| \leq |D|.$$

We call $c$ a *low complexity configuration* if it has low complexity with respect to some finite shape $D$.

There are non-periodic configurations that have low complexity [3], but none is known that would have low complexity with respect to a rectangle $D$, or even with respect to a convex shape $D$. This is the famous Nivat's conjecture from 1997.

**Conjecture 1** (*M. Nivat* [12]) A two-dimensional configuration $c$ satisfying $|\mathcal{L}_D(c)| \leq |D|$ for some rectangle $D = [\![n]\!] \times [\![m]\!]$ is periodic.

Note that $|\mathcal{L}_D(c)| \leq \frac{1}{2}|D|$ for some rectangle $D$ is known to imply that $c$ is periodic [5]. Also $|\mathcal{L}_D(c)| \leq |D|$ for infinitely many different size rectangles $D$ guarantees periodicity [11]. Moreover, if $c$ is uniformly recurrent and satisfies $|\mathcal{L}_D(c)| \leq |D|$ with respect to a rectangle (or any other convex shape) $D$ then $c$ is periodic [10]. See [9] for a review of the algebraic approach that was used to obtain the latter two of these results.

Because the orbit closure of any configuration contains a uniformly recurrent configuration, if a subshift $X$ contains a configuration $c$ satisfying $|\mathcal{L}_D(c)| \leq |D|$ with respect to some convex shape $D$ then $X$ contains also a periodic configuration. In particular, for a rectangle $D = [\![n]\!] \times [\![m]\!]$, if a set $P \subseteq A^D$ of allowed patterns is such that $|P| \leq |D|$ then $\mathcal{V}(P)$ is either empty or contains a periodic configuration. In such a low complexity setting we therefore have an algorithm to solve the domino problem. In this work we are interested in the case when the number of allowed patterns is slightly larger than $|D|$.

In summary, for a given rectangle $D$ and a set $P \subseteq A^D$ of patterns of shape $D$, it is undecidable whether $\mathcal{V}(P)$ is empty, but in the low complexity setting $|P| \leq |D|$ the question is decidable. And there is $P \subseteq A^D$ such $\mathcal{V}(P)$ is aperiodic, but there is no aperiodic $\mathcal{V}(P)$ in the low complexity case $|P| \leq |D|$. This naturally raises the question of the pretty low complexity setting where $|P|$ is slightly greater than $|D|$, say $|P| = |D| + 1$.

## 3 Recoding Wang Tiles

In this section an arbitrary Wang tile set $T$ is converted into a pretty small set $P$ of binary rectangular allowed patters that is equivalent to $T$ in the sense that $P$ admits a (periodic) configuration if and only if $T$ admits a (resp. periodic) configuration.

Configurations that $P$ admits have bits 1 sparsely positioned to that each bit 1 represents a single Wang tile of a valid tiling, and the relative positions of bits 1 uniquely identify the corresponding Wang tiles. Allowed patterns in $P$ are so restricted that only matching Wang tiles are allowed next to each other.

So let $T$ be a given finite set of Wang tiles. We first modify the set to make sure that no tile matches itself as its neighbor. This is easy to establish by making two copies of $T$ and forcing the copies be used alternatingly on even and odd cells. More precisely, we replace $T$ by the cartesian product $T \times \{\text{EVEN}, \text{ODD}\}$ where EVEN has color 0 on its north and east sides and color 1 on south and west, while in ODD the colors are reversed. The EVEN/ODD -components of tiles form an infinite checkerboard tiling of the plane. The new tile set admits a (periodic) tiling if and only if $T$ admits a (periodic, resp.) tiling.

From now on we can hence assume that no tile of $T$ matches in color with itself. Let $t = |T|$ be the number of tiles, and denote

$$S = \{2^j - 1 \mid j = 0, 1, \ldots, t-1\}$$

and $s = 2^{t-1}$. The set $S \subseteq [\![s]\!]$ has the property that for $a, b \in S, a \neq b$, the difference $a - b$ uniquely identifies both $a$ and $b$. The proof of this fact is easy.

**Lemma 1** *For $a_1, a_2, b_1, b_2 \in S$, if $a_1 - b_1 = a_2 - b_2 \neq 0$ then $a_1 = a_2$ and $b_1 = b_2$.* □

Fix a bijection $\alpha : T \longrightarrow S$. In our coding tile $t$ will be represented as horizontal sequence of $s$ bits where bit number $\alpha(t)$ is set to 1 and all other bits are 0's.

Choose $N = 3s$ and fix $m \geq 2$ and $n \geq N$, the dimensions of the rectangular patterns considered, and define

$$D = [\![n]\!] \times [\![m]\!].$$

Denote $n' = n - s$ and $m' = m - 1$. In our coding of a Wang tiling we paste to position $(i \cdot n', j \cdot m')$ the bit sequence representing the Wang tile in position $(i, j)$. A configuration $c \in T^{\mathbb{Z}^2}$ is then represented as a binary configuration $\beta(c) \in \{0, 1\}^{\mathbb{Z}^2}$ where for all $(i, j) \in \mathbb{Z}^2$, tile $c(i, j)$ contributes symbol 1 in position $(in' + \alpha(c(i, j)), jm')$. All positions without a contribution from any tile of $c$ have value 0.

In $\beta(c)$ all symbols 1 appear in the intersections of vertical strips

$$V_i = (in' + [\![s]\!]) \times \mathbb{Z}$$

and horizontal strips

$$H_j = \mathbb{Z} \times \{jm'\},$$

for $i, j \in \mathbb{Z}$. There is exactly one symbol 1 in each intersection $I_{i,j} = V_i \cap H_j$, representing the Wang tile in position $(i, j)$. See Fig. 2 for an illustration.

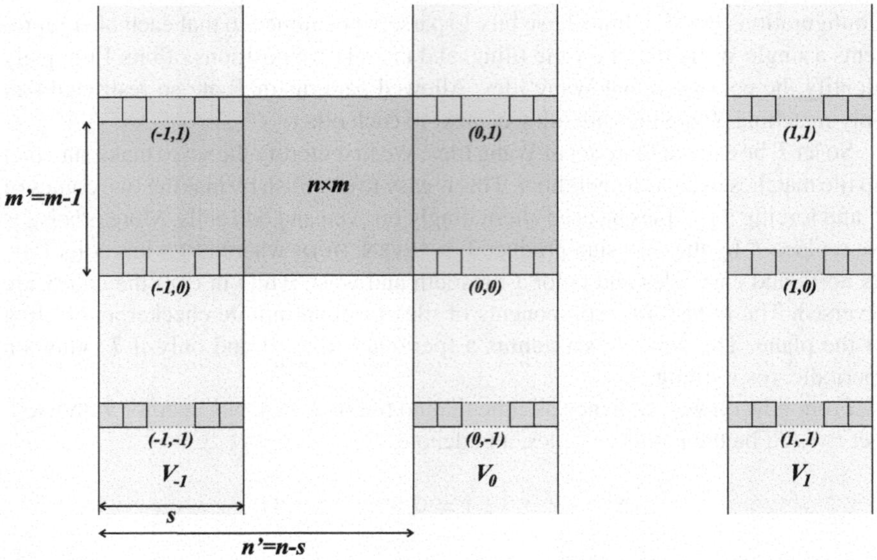


**Fig. 2** The positioning of the horizontal $s$-bit encodings of Wang tiles in coding $\beta$. The given coordinates indicate the positions in the Wang tiling that are encoded in the corresponding bit sequences. A sample rectangle of size $n \times m$ is depicted in dark shading. Three consecutive vertical $V_i$ strips are highlighted

Let us first count all rectangular $n \times m$ patterns that may appear in $\beta(c)$ for some $c \in T^{\mathbb{Z}^2}$, that is, find an upper bound on the cardinality of the set

$$Q = \bigcup_{c \in T^{\mathbb{Z}^2}} \mathcal{L}_D(\beta(c)).$$

As $n = n' + s$ we have that for all $j \in \mathbb{Z}$ there is $i \in \mathbb{Z}$ such that $in' + [\![s]\!] \subseteq j + [\![n]\!]$, that is, every $n \times m$ rectangle on the grid fully intercepts one of the vertical strips $V_i$. Analogously, the rectangle intercepts a horizontal strip $H_j$ and hence some $I_{i,j}$ is fully contained in the rectangle. This implies that every pattern in $Q$ contains at least one symbol 1. On the other hand, $n \leq 2n' - s$ so that an $n \times m$ rectangle can not intersect with more than two strips $V_i$, and analogously it cannot intersect more than two horizontal strips $H_j$. This means that there are at most four symbols 1 in each pattern of $Q$.

Let $p \in Q$ and let $c \in T^{\mathbb{Z}^2}$ be such that $p \in \mathcal{L}_D(\beta(c))$. Let $E = \mathbf{u} + D$ be a rectangle containing pattern $p$ in $\beta(c)$. We have the following four possibilities.

- Suppose that $E$ has a non-empty intersection with two consecutive vertical strips $V_i$ and $V_{i+1}$ and with two consecutive horizontal strips $H_j$ and $H_{j+1}$. Rectangle $E$ can be positioned in at most $2s$ positions relative to these strips, and there at most $t^4$ choices of the Wang tiles encoded in the intersections of the two horizontal and two vertical strips. This means that there are at most $2st^4$ patterns $p$ that can be extracted this way.

- Suppose that $E$ has non-empty intersection with two consecutive vertical strips $V_i$ and $V_{i+1}$ and with only one horizontal strip $H_j$. There are at most $2sm$ ways to position the rectangle and at most $t^2$ choices for the two tiles encoded within the block. There are hence at most $2smt^2$ patterns $p$ of this type.
- Symmetrically, if $E$ has non-empty intersection with two consecutive horizontal strips $H_j$ and $H_{j+1}$ and with only one vertical strip $V_j$ then the number of extracted patterns is bounded by $nt^2$.
- Finally, if $E$ only intersects a single vertical and horizontal strip then $E$ contains a single symbol 1. There are at most $nm$ positions for this 1 inside the $n \times m$ rectangle.

Adding up the four cases above gives the upper bound

$$nm + 2st^4 + 2st^2m + t^2n \leq nm + k(n + m)$$

for the cardinality of $Q$, where we can choose $k = 2st^4$, assuming $t \geq 1$. This choice of $k$ works by a direct calculation due to $t^2 \geq 1$, $st^2 \geq 1$ and $n \geq 2$: subtracting the left-hand-side from the right-hand-side yields

$$2st^4(n + m) - (2st^4 + 2st^2m + t^2n) = 2st^2(t^2 - 1)m + t^2(2st^2 - 1)(n - 1) - t^2 \geq 0.$$

Note that constant $k = 2st^4 = |T|^4 \cdot 2^{|T|}$ does not depend on $n$ or $m$ but only on the number of tiles in $T$. Note also that the patterns in $Q$ can be effectively constructed. We have established the following result.

**Lemma 2** *The number of different $n \times m$ patterns that appear in $\beta(c)$ over all $c \in T^{\mathbb{Z}^2}$ is at most $nm + k(n + m)$ for $k = |T|^4 \cdot 2^{|T|}$. These patterns can be effectively constructed for given $T$.* □

**Remark** A smaller constant $k$ can be obtained by using a more succinct representation $\alpha$ of tiles $T$ as numbers. One just needs to encode tiles as natural numbers whose differences $a - b$ identify uniquely $a$ and $b$, so that Lemma 1 is satisfied. Instead of the exponentially growing sequence of representatives $0, 1, 3, 7, 15, \ldots$ that we use here one can use, for example, numbers of the Mian-Chowla sequence $1, 2, 4, 8, 13, 21, 31, \ldots$ (sequence A005282 in [14]) that only grows polynomially. Then constant $k$ will be bounded by a polynomial of $|T|$.

Let us further limit the allowed patters by removing from $Q$ patterns that contain two 1's whose relative positions indicate neighboring Wang tiles whose colors do not match. More precisely, let $p \in Q$.

(H) Suppose $p$ contains on some row two symbols 1, in columns $i$ and $j$, for $i < j$. In order for $p$ to appear in $\beta(c)$ for some valid tiling $c$ we necessarily must have that the two symbols 1 are the contributions of two matching horizontally neighboring tiles in $c$, so that $i = k + \alpha(a)$ and $j = k + n' + \alpha(b)$ for some integer $k$ and tiles $a, b \in T$ such that the east color of $a$ is the same as the west color of $b$. Hence we remove $p$ from $Q$ if no matching $a, b$ exist such that $j - i = n' + \alpha(b) - \alpha(a)$.

(V) Suppose $p$ contains symbol 1 in some column $i$ of the bottom row and some column $j$ of the top row where $i - s < j < i + s$. Now $p$ can appear in $\beta(c)$ only if the two symbols 1 are the contributions of two vertically neighboring tiles in $c$, so that $i = k + \alpha(a)$ and $j = k + \alpha(b)$ for some integer $k$ and tiles $a, b \in T$ such that the north color of $a$ is the same as the south color of $b$. We remove $p$ from $Q$ if no matching $a, b$ exist such that $j - i = \alpha(b) - \alpha(a)$.

Let $P$ be the set of patterns in $Q$ that are not removed by the conditions (H) and (V) above. Set $P$ can be effectively constructed and, since $P \subseteq Q$, the upper bound $nm + k(n + m)$ from Lemma 2 holds on its cardinality.

Let us next prove that allowing the patterns in $P$ admits precisely the configurations $\beta(c)$ and all their translates, for all $c \in T^{\mathbb{Z}^2}$ that are valid Wang tilings.

**Lemma 3** *With the notations above,*

$$\mathcal{V}(P) = \{\tau^{\mathbf{t}}(\beta(c)) \mid \mathbf{t} \in \mathbb{Z}^2 \text{ and } c \in \mathcal{V}(T)\}.$$

***Proof*** By the definition of $P$ it is clear that for every valid tiling $c \in \mathcal{V}(T)$ the encoded configuration $\beta(c)$ only contains allowed patterns in $P$. Hence the inclusion "$\supseteq$" holds.

To prove the converse inclusion, consider an arbitrary configuration $e \in \mathcal{V}(P)$, that is, $e \in \{0, 1\}^{\mathbb{Z}^2}$ such that $\mathcal{L}_D(e) \subseteq P$. Every pattern in $P$ contains symbol 1 so configuration $e$ must contain symbol 1 in every $m \times n$ block.

Let us denote, for any $x, y, z, w \in T$, by Bin $\binom{z\ w}{x\ y}$ the $m \times n$ binary pattern with exactly four 1's, two of which are on the bottom row in columns $\alpha(x)$ and $n' + \alpha(y)$, and two are on the topmost row in columns $\alpha(z)$ and $n' + \alpha(w)$. In other words, the bit sequences that encode tiles $x$, $y$, $z$ and $w$ are in the four corners of the pattern, as in the dark grey block in Fig. 2. Let us call Bin $\binom{z\ w}{x\ y}$ a *standard block* if the Wang tiles $x, y, z, w$ match each other in colors as a $2 \times 2$ pattern with $x, y, z$ and $w$ at the lower left, lower right, upper left and upper right position of the $2 \times 2$ pattern, respectively.

Consider now any occurrence of symbol 1 in $e$, that is, $\mathbf{u} \in \mathbb{Z}^2$ such that $e(\mathbf{u}) = 1$. Let us prove that there is a standard block Bin $\binom{z\ w}{x\ y}$ in $e$ with this occurrence of 1 representing Wang tile $x$. Let $p = \tau^{\mathbf{u}}(e)_{|D}$ be the $m \times n$ pattern with lower left corner at cell $\mathbf{u}$, so there is symbol 1 at the lower left corner of $p$. By the definition of $P$, pattern $p$ appears in $\beta(f)$ for some $f \in T^{\mathbb{Z}^2}$. The structure of $\beta(f)$ implies that there is another symbol 1 in pattern $p$ on the same horizontal row, say $i$ position to the right of the lower left corner. By condition (H) above, $i = n' + \alpha(y) - \alpha(x)$ for some tiles $x, y \in T$ such that the east color of $x$ is the same as the west color of $y$. Because no tile in $T$ matches with itself in color, we have $x \neq y$ and hence $x$ and $y$ are unique by Lemma 1.

Let $\mathbf{v} = \mathbf{u} - (\alpha(x), 0)$, and extract the $n \times m$ pattern $q = \tau^{\mathbf{v}}(e)_{|D}$ located $\alpha(x)$ positions to the left of $p$ in $e$. Pattern $q$ contains symbol 1 on the bottom row at columns $\alpha(x)$ and $n' + \alpha(y)$. Pattern $q$ appears in $\beta(f')$ for some $f' \in T^{\mathbb{Z}^2}$ and therefore, due to the structure of encoded configurations, $q$ must be $\mathrm{Bin}\,(\begin{smallmatrix} z & w \\ x & y \end{smallmatrix})$ for some $z, w \in T$. Conditions (H) and (V) then ensure that $x$, $y$, $z$ and $w$ match in color with each other to form a valid $2 \times 2$ pattern of Wang tiles, so $q = \mathrm{Bin}\,(\begin{smallmatrix} z & w \\ x & y \end{smallmatrix})$ is a standard block.

We have seen that any occurrence of 1 in $e$ represents a Wang tile $x$ in the lower left corner of some standard block $q = \mathrm{Bin}\,(\begin{smallmatrix} z & w \\ x & y \end{smallmatrix})$ in $e$. With an analogous reasoning we see that the same occurrence of bit 1 is also encoding the tile $z'$ at the upper left corner of a standard block $q' = \mathrm{Bin}\,(\begin{smallmatrix} z' & w' \\ x' & y' \end{smallmatrix})$ in $e$. In $q'$ the symbol 1 that represents the Wang tile $w'$ at the upper right corner is the same as the one that represents tile $y$ at the lower right corner in $q$, so that $\alpha(x) - \alpha(y) = \alpha(z') - \alpha(w')$. By Lemma 1 the tiles are unique so that $x = z'$ and $y = w'$. (See Fig. 3 for an illustration.) We have that $q' = \tau^{\mathbf{v}'}(e)_{|D}$ for $\mathbf{v}' = \mathbf{v} - (0, m')$.

Analogously, symbol 1 in position $\mathbf{u}$ is also in the lower right and upper right corners of standard blocks in $e$ that overlap with $q$ and $q'$ in two encoded Wang tiles that by Lemma 1 are uniquely identified as $x$ and $z$ and as $x'$ and $z' = x$, respectively. So also the $m \times n$ blocks in $e$ with lower left corners at cells $\mathbf{v} - (n', 0)$ and $\mathbf{v} - (n', m')$ are standard blocks.

As cell $\mathbf{u}$ is any position containing bit 1 in configuration $e$, we can repeat the reasoning on the other corners of the standard blocks. By easy induction we see that $\tau^{\mathbf{v}_{i,j}}(e)_{|D}$ is a standard block for all $i, j \in \mathbb{Z}$ where $\mathbf{v}_{i,j} = \mathbf{v} + (in', jm')$. We now take $c \in T^{\mathbb{Z}^2}$ such that $c(i, j)$ is the Wang tile encoded in $e$ position $\mathbf{v}_{i,j}$, for all $i, j \in \mathbb{Z}$, that is, the unique $t \in T$ such that $\tau^{\mathbf{v}_{i,j}}(e)(\alpha(t)) = 1$. Clearly $\tau^{\mathbf{v}}(e) = \beta(c)$. Because standard blocks correspond to correctly tiled $2 \times 2$ blocks of Wang tiles we have that $c \in \mathcal{V}(T)$. □

Now we obtain the main technical contribution.

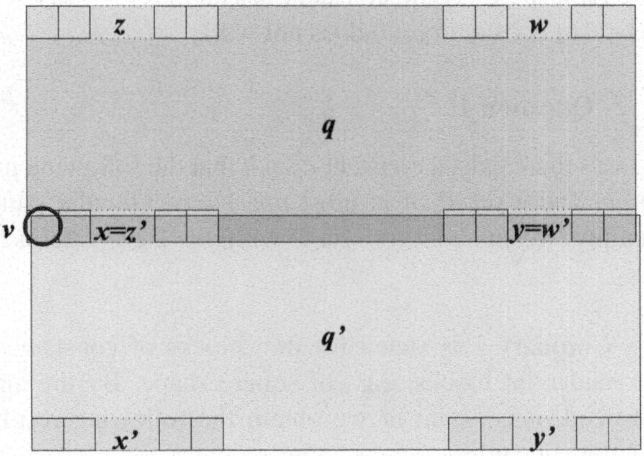


**Fig. 3** Two standard blocks sharing an encoded tile at their lower left and upper left corners, respectively. The positions of the blocks are uniquely identified by their common row, as discussed in the proof of Lemma 3. The circled cell is the position $\mathbf{v}$ in configuration $e$ in that proof

**Theorem 2** *Let $T$ be a given Wang tile set. One can effectively find positive integers $N$ and $k$ such that for any $n \geq N$ and $m \geq 2$ one can effectively construct a set $P$ of binary rectangular patterns of size $n \times m$ such that the cardinality of $P$ is at most $nm + k(n+m)$ and $\mathcal{V}(P)$ contains a (periodic) tiling if and only if $\mathcal{V}(T)$ contains a (periodic, resp.) configuration.*

***Proof*** We first construct an equivalent tile set $T'$ where no tile matches in color with itself, as shown in the beginning of the section. We then set $t = |T'|, s = 2^{t-1}, N = 3s$ and $k = 2st^4$. Let $n \geq N$ and $m \geq 2$ be arbitrary, and let us construct $P$ as above. By Lemma 2 set $P$ contains at most $nm + k(n+m)$ patterns. By Lemma 3 we have that $\mathcal{V}(P) = \emptyset$ if and only if $\mathcal{V}(T) = \emptyset$. Encoding $\beta$ maps periodic configurations to periodic configurations so also by Lemma 3 there is a periodic configuration in $\mathcal{V}(P)$ if and only if there is a periodic configuration in $\mathcal{V}(T)$. □

## 4 Conclusions

With Theorem 2 we can port results from Wang tiles to pretty low complexity SFTs.

**Corollary 1** *Let $f : \mathbb{N} \longrightarrow \mathbb{N}$ be a computable function, $f \notin \mathcal{O}(1)$. The following problem is undecidable for any fixed $m \geq 2$: Given $n$ and a set $P$ of at most $nm + f(n)n$ binary rectangular patterns of size $n \times m$, is $\mathcal{V}(P)$ empty?*

***Proof*** We many-one reduce the domino problem. Let $T$ be any given set of Wang tiles. Compute constants $N$ and $k$ of Theorem 2. For $n = N, N+1, N+2, \ldots$ compute $f(n)$ until number $n \geq N$ is found such that $f(n) \geq k + km/n$. Because $f \notin \mathcal{O}(1)$ such $n$ exists. Using Theorem 2 construct a set $P$ of at most $nm + k(n+m) \leq nm + f(n)n$ binary patterns of size $n \times m$. By Theorem 2 tiles $T$ admit a valid tiling if and only if $\mathcal{V}(P)$ is non-empty. □

Corollary 1 naturally raises the question whether the additive term $f(n)n$ can be replaced by some constant, or can at least $f(n)$ in it be replaced by a constant. By [10] constant $c = 0$ does not work.

**? Question 1**

Does there exist a constant $c$ such that the following problem is undecidable: Given $n, m$ and a set $P$ of at most $nm + c$ rectangular patterns of size $n \times m$, is $\mathcal{V}(P)$ empty? What about for sets $P$ with $nm + cn$ patterns?

Corollary 1 is stated for thin blocks of constant height $m$. It is also worth to consider fat blocks, e.g., of square shape. By the analogous proof, using $m = n$ instead of constant $m$ we obtain the following result where the additive term is almost linear in $n$.

**Corollary 2** *Let $f : \mathbb{N} \longrightarrow \mathbb{N}$ be a computable function, $f \notin \mathcal{O}(1)$. The following problem is undecidable: Given $n$ and a set $P$ of at most $n^2 + f(n)n$ binary square patterns of size $n \times n$, is $\mathcal{V}(P)$ empty?*

***Proof*** We proceed as in the proof of Corollary 1, except that we choose $n$ such that $f(n) \geq 2k$. By Theorem 2 we can effectively construct a set $P$ of at most $n^2 + k(n+n) \leq n^2 + f(n)n$ binary patterns of size $n \times n$ such that $\mathcal{V}(P)$ is non-empty if and only if $T$ admits a valid tiling. □

In particular, for any real number $\varepsilon > 0$ it is undecidable if a given set $P$ of at most $(1+\varepsilon)n^2$ square patterns of size $n \times n$ admit a valid configuration.

As usual, undecidability comes together with aperiodicity. We obtain pretty low complexity aperiodic SFTs.

**Corollary 3** *Let $f : \mathbb{N} \longrightarrow \mathbb{N}$ be a function, $f \notin \mathcal{O}(1)$. There exists $n$ and an aperiodic SFT $\mathcal{V}(P)$ where $P$ consists of at most $n^2 + f(n)n$ binary square patterns of size $n \times n$. Also, for every fixed height $m \geq 2$, there exists width $n$ and an aperiodic SFT $\mathcal{V}(P')$ where $P'$ consists of at most $nm + f(n)n$ binary rectangular patterns of size $n \times m$.*

***Proof*** Let $T$ be an aperiodic Wang tile set. Let $N$ and $k$ be as in Theorem 2, and let $n \in \mathbb{N}$ be such that $f(n) \geq 2k$. By Theorem 2 there is a collection $P$ of at most $n^2 + k(n+n) \leq n^2 + f(n)n$ binary $n \times n$ patterns such that $\mathcal{V}(P)$ is aperiodic. For fixed $m$, choosing $n$ such that $f(n) \geq k + km/n$ gives $P'$ in the second claim. □

There naturally exists a constant $c$ and a collection of $nm + c$ allowed patterns of some size $n \times m$ that defines an aperiodic SFT. But what might be the smallest such $c$? Again, by [10] constant $c = 0$ does not work.

*Question 1* Question 2 What is the smallest constant $c$ such that there exists numbers $n$ and $m$ and a set $P$ of $nm + c$ rectangular patterns of size $n \times m$ such that $\mathcal{V}(P)$ is aperiodic? Is constant $c = 1$ possible?

**Acknowledgements** The research was supported by the Academy of Finland grant 296018.

## References

1. Berger R (1966) The undecidability of the domino problem. Memoirs of the American Mathematical Society. American Mathematical Society
2. Birkhoff GD (1912) Quelques théorèmes sur le mouvement des systèmes dynamiques. Bulletin de la Société Mathématique de France 40:305–323
3. Cassaigne J (1999) Subword complexity and periodicity in two or more dimensions. In: Rozenberg G, Thomas W (eds) Developments in language theory, DLT 1999 proceedings. World Scientific, pp 14–21
4. Ceccherini-Silberstein T, Coornaert M (2010) Cellular automata and groups. Springer monographs in mathematics. Springer, Berlin Heidelberg

5. Cyr V, Kra B (2015) Nonexpansive $\mathbb{Z}^2$-subdynamics and Nivat's Conjecture. Trans Am Math Soc 367(9):6487–6537
6. Goles E, Martínez S (1990) Neural and automata networks: dynamical behavior and applications. Kluwer Academic Publishers, USA
7. Goles E, Martínez S (1998) Cellular automata and complex systems. Nonlinear phenomena and complex systems. Springer, Netherlands
8. Jeandel E, Rao M (2015) An aperiodic set of 11 wang tiles. arXiv:1506.06492
9. Kari J (2019) Low-complexity tilings of the plane. Descriptional complexity of formal systems - 21st IFIP WG 1.02 international conference, DCFS 2019 proceedings, vol 11612. Lecture notes in computer science. Springer, pp 35–45
10. Kari J, Moutot E (2020) Decidability and periodicity of low complexity tilings. In: 37th international symposium on theoretical aspects of computer science, STACS 2020 proceedings, vol 154. LIPIcs. Schloss Dagstuhl - Leibniz-Zentrum für Informatik, pp 14:1–14:12
11. Kari J, Szabados M (2015) An algebraic geometric approach to Nivat's cnjecture. In: Automata, languages, and programming - 42nd international colloquium, ICALP 2015 proceedings, Part II, vol 9135. Lecture notes in computer science. Springer, pp 273–285
12. Nivat M (1997) Keynote address at the 25th anniversary of EATCS, during ICALP 1997, Bologna
13. Robinson RM (1971) Undecidability and nonperiodicity for tilings of the plane. Inventiones mathematicae 12(3):177–209
14. Sloane NJA (2020) The on-line encyclopedia of integer sequences. http://oeis.org
15. Wang H (1961) Proving theorems by pattern recognition - II. Bell Syst Tech J 40(1):1–41

# Reversible Elementary Triangular Partitioned Cellular Automata and Their Complex Behavior

**Kenichi Morita**

**Abstract** We investigate a special type of cellular automata called elementary triangular partitioned cellular automata (ETPCAs) in which various interesting phenomena emerge. They are quite simple, since each of their local functions is specified by only four local transition rules. There are 256 ETPCAs in total, and 36 among them are reversible. Despite their extreme simplicity, some reversible ETPCAs show quite complex behavior, and they even have universal computing capability. In this survey, based on the author's past results, we discuss how complex phenomena appear in these ETPCAs, and how high-order functions, such as reversible logic gates, are realized combining useful phenomena.

**Keywords** Reversible cellular automaton · Elementary triangular partitioned cellular automaton · Reversible computing · Complex phenomena · Glider · Reversible logic gate

## 1 Introduction

Complex phenomena can emerge even in a system composed of very simple elements. This fact also holds when we further add the "reversibility constraint" to the elements. In this survey, we discuss how complex phenomena emerge from a simple reversible microscopic law, and how these phenomena are utilized to obtain higher-order functions such as computing. Since reversibility is one of the fundamental laws of nature, it is important to study this problem. Here, we investigate it using the framework of *reversible elementary triangular partitioned cellular automata*.

A three-neighbor triangular cellular automaton (TCA) is one whose cell is triangular and communicates with its three edge-adjacent cells. So far, there have been several studies on TCAs, though they are not so many. Bays [2] studied a kind of

Currently, Professor Emeritus of Hiroshima University.

K. Morita (✉)
Hiroshima University, Higashi-Hiroshima 739-8527, Japan
e-mail: km@hiroshima-u.ac.jp

 
A. Adamatzky (ed.), *Automata and Complexity*, Emergence, Complexity and Computation 42, https://doi.org/10.1007/978-3-030-92551-2_20

TCAs with a similar type of local functions as that of Game-of-Life CA [6, 7]. Gajardo and Goles [5] proposed a three-state TCA, which is defined on a hexagonal lattice, and proved its computational universality. Imai and Morita [8] studied a reversible TCA, and showed that there is an eight-state universal reversible one. An advantage of using TCAs is that their local function can be simpler than those of CAs on a square lattice, since each cell has only three neighbor cells. Hence, it is suited for studying how complex phenomena emerge from a simple local function.

A three-neighbor triangular partitioned cellular automaton (TPCA) is a CA where cells are triangular, and each cell has three parts. A TPCA is called an *elementary* TPCA (ETPCA), if it is rotation-symmetric, and each part of a cell has only two states 0 and 1 (the state 1 is also called a particle). The class of ETPCAs is one of the simplest subclasses of two-dimensional CAs. This is because the local function of each ETPCA is described by only four local transition rules. There are 256 ETPCAs in total as in the case of one-dimensional elementary CAs (ECAs) [16]. It is known that there are 36 reversible ETPCAs among them. Among 36 reversible ones, there are nine conservative ETPCAs in which the total number of symbol 1's in a configuration is conserved throughout its evolution process. This property is a similar notion to that of the conservation law of mass or energy in physics.

A particular conservative ETPCA No. 0157, where 0157 is an identification number in the class of ETPCAs, was first investigated in [8], and its computational universality was shown. In [9], it was shown that conservative ETPCA 0137 is also computationally universal. In [13], a nonconservative reversible ETPCA 0347 is studied. We think ETPCA 0347 is the most interesting one in the class of 256 ETPCAs, though not all ETPCAs have been fully investigated yet. In spite of its simplicity the local function, ETPCA 0347 shows quite interesting behavior like the Game-of-Life CA. In particular, a glider, which is a space-moving pattern, and glider guns exist in ETPCA 0347. Using gliders to represent signals, computational universality of ETPCA 0347 was also proved.

In this chapter, we discuss how interesting phenomena emerge in reversible ETPCAs 0347, 0157 and 0137, how such phenomena are utilized to realize higher-order functions such as reversible logic elements, and how reversible computers can be built from these logic elements. In Sect. 2, after giving definitions on ETPCAs, we classify 256 ETPCAs by introducing three kinds of dualities. In Sect. 3, a nonconservative reversible ETPCA No. 0347 is investigated. In this ETPCA, three kinds of patterns exist. They are periodic patterns, a space-moving pattern, and expanding patterns. Among them, a space-moving pattern called a *glider* is interesting and useful. Here, how to control its flight is explained. By this, a glider can be used as a signal for composing a reversible computer. In Sect. 4, conservative ETPCAs 0157 and 0137 are investigated. There are many space-moving patterns in both of these ETPCAs. Though behavior of these patterns is complex and seems interesting, their periods are vary large. Therefore, it is difficult to control the flying directions and the timings of the space-moving patterns. Instead, a one-particle pattern, which can move along a transmission line, is used to represent a signal for computing. Since a universal reversible logic gate is realizable, computational universality of ETPCAs 0157 and 0137 is derived.

# 2 Elementary Triangular Partitioned Cellular Automata

We give several definitions on elementary triangular partitioned cellular automata (ETPCAs). We also define three kinds of duality among them, by which 256 ETPCAs are classified into 82 equivalence classes.

## 2.1 *Triangular Partitioned Cellular Automata (TPCAs)*

A three-neighbor *triangular partitioned cellular automaton* (TPCA) is a CA defined on the cellular space shown in Fig. 1. In a TPCA, each cell has three parts, and the next state of a cell is determined by the states of the adjacent parts of the three neighbor cells as shown in Fig. 2.

All the cells of a TPCA are identical copies of a finite state machine, and each cell has three parts, i.e., the left, downward, and right parts, whose state sets are $L$, $D$ and $R$, respectively. However, the directions of the cells are not the same, i.e., there are up-triangle cells, and down-triangle cells.

We now place cells of a TPCA on $\mathbb{Z}^2$ as shown in Fig. 3. We assume that if the coordinates of an up-triangle cell is $(x, y)$, then $x + y$ must be even. It should be noted, if we define an TPCA on $\mathbb{Z}^2$, there arises a problem that the neighborhood

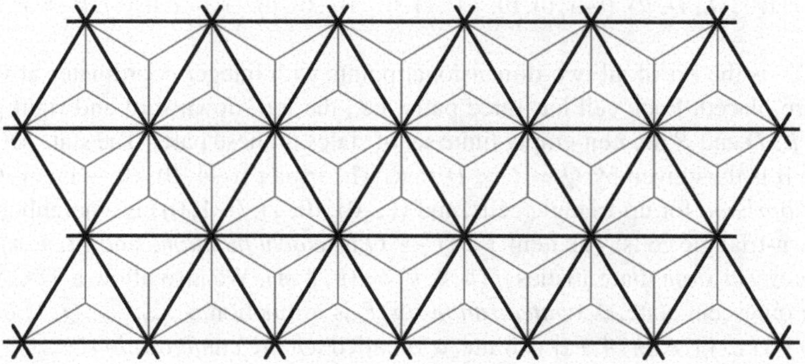

**Fig. 1** Cellular space of a three-neighbor TPCA

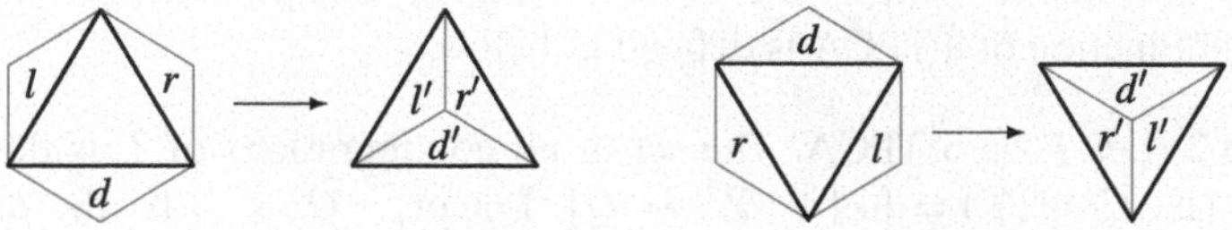


**(a)** **(b)**

**Fig. 2** Pictorial representations of the local transition rule $f(l, d, r) = (l', d', r')$, where $(l, d, r)$, $(l', d', r') \in L \times D \times R$. They are **a** for up-triangle cells, and **b** for down-triangle cells

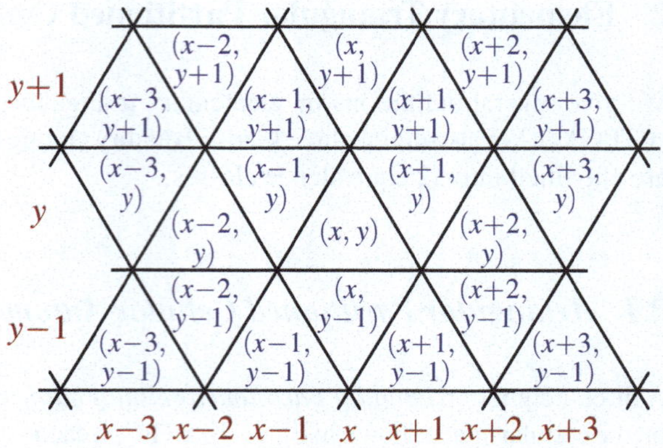


**Fig. 3** The $x$-$y$ coordinates in the cellular space of TPCA

is slightly non-uniform. Namely, for an up-triangle cell, its neighbors are the west, south and east adjacent cells (Fig. 2a), while for a down-triangle cell, its neighbors are the east, north and west adjacent cells (Fig. 2b). Though such non-uniformity is dissolved by defining a TPCA on a Cayley graph, here we define a TPCA on $\mathbb{Z}^2$.

**Definition 1** A *deterministic triangular partitioned cellular automaton (TPCA)* is a system defined by

$$T = (\mathbb{Z}^2, (L, D, R), ((-1, 0), (0, -1), (1, 0)), ((1, 0), (0, 1), (-1, 0)), f, (\#, \#, \#)).$$

Here, $\mathbb{Z}^2$ is the set of all two-dimensional points with integer coordinates at which cells are placed. Each cell has three parts, i.e., the left, downward and right parts, where $L$, $D$ and $R$ are non-empty finite set of states of these parts. The state set $Q$ of each cell is thus given by $Q = L \times D \times R$. The triplet $((-1, 0), (0, -1), (1, 0))$ is a *neighborhood* for up-triangle cells, and $((1, 0), (0, 1), (-1, 0))$ is a neighborhood for down-triangle cells. The item $f : Q \to Q$ is a *local function*, and $(\#, \#, \#) \in Q$ is a *quiescent state* that satisfies $f(\#, \#, \#) = (\#, \#, \#)$. We also allow a TPCA that has no quiescent state. A *configuration* of $T$ is a function $\alpha : \mathbb{Z}^2 \to Q$. If the set $\{(x, y) \mid \alpha(x, y) \neq (\#, \#, \#)\}$ is finite, $\alpha$ is called a *finite configuration*.

If $f(l, d, r) = (l', d', r')$ holds for $(l, d, r), (l', d', r') \in Q$, then this relation is called a *local transition rule* of the TPCA $T$ (Fig. 2). The global function induced by the local function of a TPCA is defined as below.

**Definition 2** Let $T$ be a TPCA. The set of all configurations of $T$ is denoted by $\mathrm{Conf}(T)$, i.e., $\mathrm{Conf}(T) = \{\alpha \mid \alpha : \mathbb{Z}^2 \to Q\}$. Let $\mathrm{pr}_L : Q \to L$ be the *projection function* such that $\mathrm{pr}_L(l, d, r) = l$ for all $(l, d, r) \in Q$. The projection functions $\mathrm{pr}_D : Q \to D$ and $\mathrm{pr}_R : Q \to R$ are also defined similarly. The *global function* $F : \mathrm{Conf}(T) \to \mathrm{Conf}(T)$ of $T$ is defined as the one that satisfies the following condition.

$\forall \alpha \in \mathrm{Conf}(T), \forall (x, y) \in \mathbb{Z}^2$ :

$$F(\alpha)(x, y) = \begin{cases} f(\mathrm{pr}_L(\alpha(x-1, y)), \mathrm{pr}_D(\alpha(x, y-1)), \mathrm{pr}_R(\alpha(x+1, y))) \\ \qquad \text{if } x + y \text{ is even} \\ f(\mathrm{pr}_L(\alpha(x+1, y)), \mathrm{pr}_D(\alpha(x, y+1)), \mathrm{pr}_R(\alpha(x-1, y))) \\ \qquad \text{if } x + y \text{ is odd} \end{cases}$$

From this definition, we can see that the next state of the up-triangle cell is determined by the present states of the left part of the west neighbor cell, the downward part of the south neighbor cell, and the right part of the east neighbor cell. On the other hand, the next state of the down-triangle cell is determined by the present states of the left part of the east neighbor cell, the downward part of the north neighbor cell, and the right part of the west neighbor cell. Therefore, for a local transition rule $f(l, d, r) = (l', d', r')$, there are two kinds of pictorial representations as shown in Fig. 2a, b. Namely, Fig. 2a is for up-triangle cells, while Fig. 2b is for down-triangle cells.

In [14], it is shown injectivity of the global function is equivalent to that of the local function in one-dimensional PCAs. The following lemma is proved in a similar manner as this.

**Lemma 1** ([14]) *Let $T$ be a TPCA, $f$ be its local function, and $F$ be its global function. Then, $F$ is injective if and only if $f$ is injective.*

A reversible TPCA is defined as follows.

**Definition 3** Let $T$ be a TPCA. The TPCA $T$ is called *reversible* if its local (or equivalently global) function is injective.

We define the notion of rotation-symmetry for a TPCA as follows.

**Definition 4** Let

$$T = (\mathbb{Z}^2, (L, D, R), ((-1, 0), (0, -1), (1, 0)), ((1, 0), (0, 1), (-1, 0)), f, (\#, \#, \#))$$

be a TPCA. The TPCA $T$ is called *rotation-symmetric* (or *isotropic*) if the conditions (1) and (2) holds.

(1) $L = D = R$

(2) $\forall (l, d, r), (l', d', r') \in L \times D \times R : \; f(l, d, r) = (l', d', r') \Rightarrow f(d, r, l) = (d', r', l')$

## 2.2 Elementary TPCAs (ETPCAs)

**Definition 5** Let

$$T = (\mathbb{Z}^2, (L, D, R), ((-1, 0), (0, -1), (1, 0)), ((1, 0), (0, 1), (-1, 0)), f)$$

be a TPCA. It is called an *elementary triangular partitioned cellular automaton* (ETPCA), if $L = D = R = \{0, 1\}$, and it is rotation-symmetric.

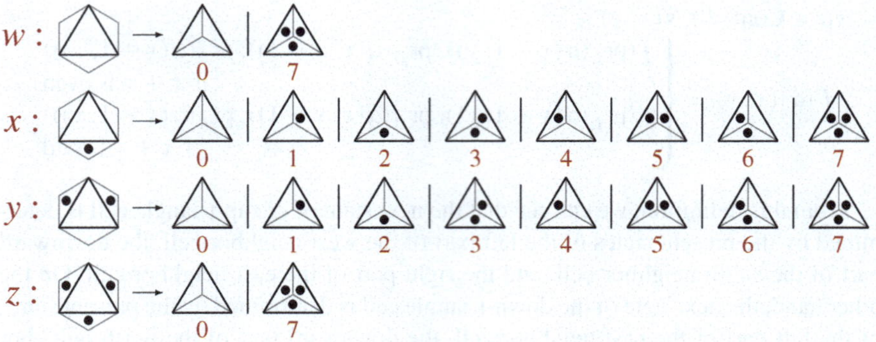


**Fig. 4** Representing an ETPCA by a four-digit number $wxyz$, where $w, z \in \{0, 7\}$ and $x, y \in \{0, 1, \ldots, 7\}$. Vertical bars indicate alternatives of a right-hand side of a rule

The set of states of a cell of an ETPCA is $L \times D \times R = \{0, 1\}^3$, and thus a cell has eight states. When drawing figures of $T$'s local transition rules and configurations, we indicate the states 0 and 1 of each part by a blank and a particle (i.e., ●), respectively.

Since ETPCA is rotation-symmetric, and each part of a cell has the state set $\{0, 1\}$, its local function is defined by only four local transition rules. Hence, an ETPCA can be specified by a four-digit number $wxyz$, as shown in Fig. 4, where $w, z \in \{0, 7\}$ and $x, y \in \{0, 1, \ldots, 7\}$. Thus, there are 256 ETPCAs. Note that $w$ and $z$ must be 0 or 7 because an ETPCA is deterministic and rotation-symmetric. The ETPCA with the number $wxyz$ is denoted by $T_{wxyz}$.

A *reversible ETPCA* is an ETPCA whose local function is injective (Definition 3). Thus, it is easy to show the following.

**Lemma 2** *Let $T_{wxyz}$ be an ETPCA. It is reversible if and only if the following conditions (1) and (2) hold.*
(1) $(w, z) \in \{(0, 7), (7, 0)\}$
(2) $(x, y) \in \{1, 2, 4\} \times \{3, 5, 6\} \cup \{3, 5, 6\} \times \{1, 2, 4\}$

A conservative ETPCA is one such that the total number of particles (i.e., ●'s) is conserved in each local transition rule. Hence, it is defined as follows.

**Definition 6** Let $T_{wxyz}$ be an ETPCA. It is called a *conservative ETPCA* if the following condition holds.

$$w = 0 \wedge x \in \{1, 2, 4\} \wedge y \in \{3, 5, 6\} \wedge z = 7$$

From Lemma 2 and Definition 6, it is clear the following holds.

**Lemma 3** *Let $T$ be an ETPCA. If $T$ is conservative, then it is reversible.*

We can see that there are 36 reversible ETPCAs (by Lemma 2), and among them there are nine conservative ones (by Definition 6). Hence, in ETPCAs, conservative ones are a subclass of reversible ones.

## 2.3 *Dualities in ETPCAs*

As seen above, there are 256 ETPCAs. However, there are some "equivalent" ETPCAs, and thus the number of essentially different ETPCAs is much smaller. Here, we introduce three kinds of dualities, and classify the ETPCAs based on them [12].

### 2.3.1 Duality Under Reflection

**Definition 7** Let $T$ and $\hat{T}$ be ETPCAs, and $f$ and $\hat{f}$ be their local functions. We say $T$ and $\hat{T}$ are *dual under reflection*, if the following holds, and it is written as $T \underset{\text{refl}}{\longleftrightarrow} \hat{T}$.

$$\forall (l, d, r), (l', d', r') \in \{0, 1\}^3 : \; f(l, d, r) = (l', d', r') \;\Leftrightarrow\; \hat{f}(r, d, l) = (r', d', l')$$

By this definition, we can see that the local transition rules of $\hat{T}$ are the mirror images of those of $T$. Therefore, an evolution process of $T$'s configurations is simulated in a straightforward manner by the mirror images of the $T$'s configurations in $\hat{T}$. For example, $T_{0137} \underset{\text{refl}}{\longleftrightarrow} T_{0467}$ holds (Fig. 5), and examples of their evolution processes are shown in Fig. 6.

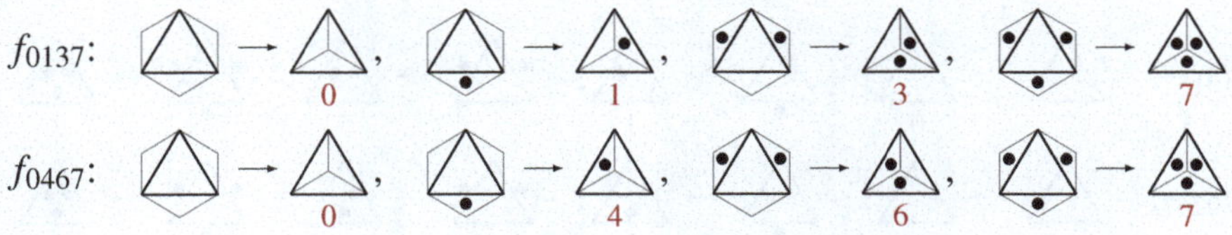


**Fig. 5** The local functions of $T_{0137}$ and $T_{0467}$ that are dual under reflection

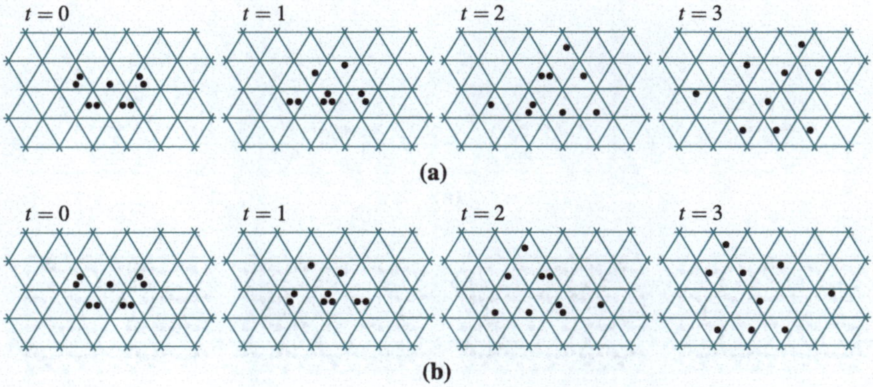


**Fig. 6** Examples of evolution processes in **a** $T_{0137}$, and **b** $T_{0467}$

### 2.3.2 Duality Under Complementation

For $x \in \{0, 1\}$, let $\overline{x}$ denote $1 - x$, i.e., the complement of $x$.

**Definition 8** Let $T$ and $\overline{T}$ be ETPCAs, and $f$ and $\overline{f}$ be their local functions. We say $T$ and $\overline{T}$ are *dual under complementation*, if the following holds, and it is written as $T \underset{\text{comp}}{\longleftrightarrow} \overline{T}$.

$$\forall (l, d, r), (l', d', r') \in \{0, 1\}^3 : \ f(l, d, r) = (l', d', r') \ \Leftrightarrow \ \overline{f}(\overline{l}, \overline{d}, \overline{r}) = (\overline{l'}, \overline{d'}, \overline{r'})$$

By this, we can see that the local transition rules of $\overline{T}$ are the 0–1 exchange (i.e., taking their complements) of those of $T$. Therefore, an evolution process of $T$'s configurations is simulated in a straightforward manner by the complemented images of the $T$'s configurations in $\overline{T}$. For example, $T_{0157} \underset{\text{comp}}{\longleftrightarrow} T_{0267}$ holds (Fig. 7), and examples of their evolution processes are shown in Fig. 8.

### 2.3.3 Duality Under Odd-Step Complementation

**Definition 9** Let $T$ be an ETPCA such that its local function $f$ satisfies the following.
(1) $\forall (l, d, r), (l', d', r') \in \{0, 1\}^3 : \ f(l, d, r) = (l', d', r') \Rightarrow f(\overline{l}, \overline{d}, \overline{r}) = (\overline{l'}, \overline{d'}, \overline{r'})$

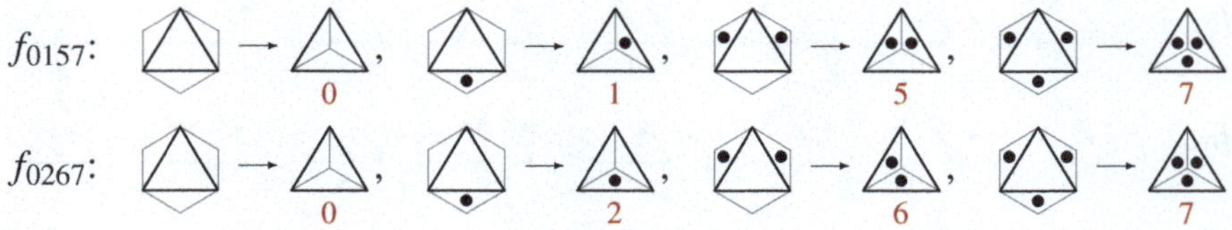


**Fig. 7** The local functions of $T_{0157}$ and $T_{0267}$ that are dual under complementation

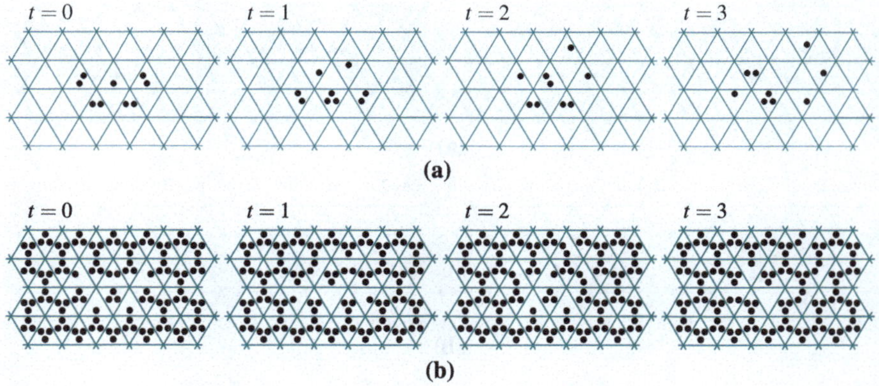


**Fig. 8** Examples of evolution processes in **a** $T_{0157}$, and **b** $T_{0267}$

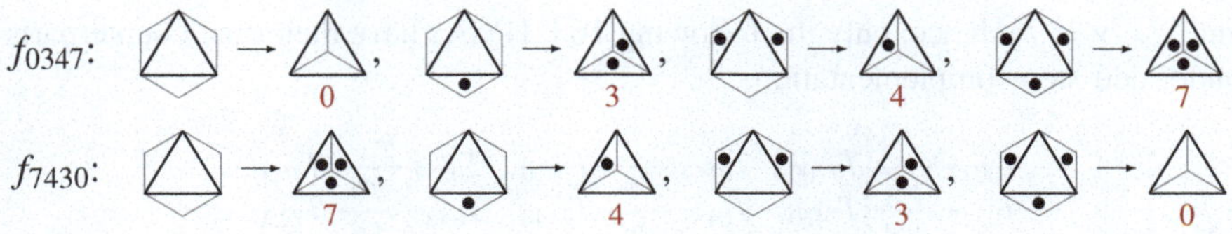


**Fig. 9** The local functions of $T_{0347}$ and $T_{7430}$, which are dual under odd-step complementation

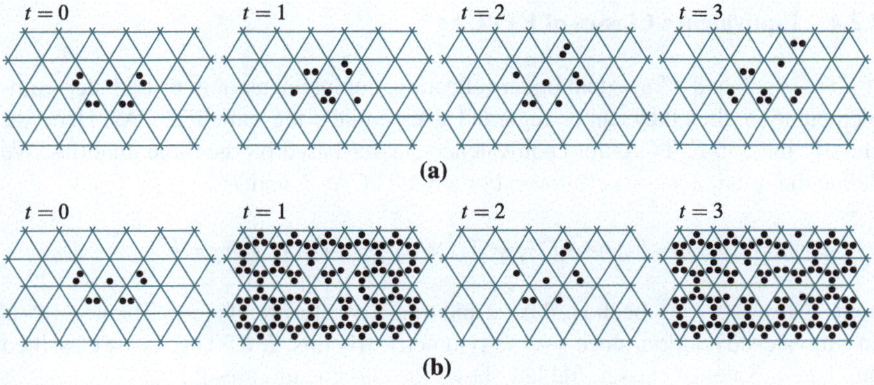


**Fig. 10** Examples of evolution processes in **a** $T_{0347}$, and **b** $T_{7430}$

Let $\tilde{T}$ be another ETPCA, and $\tilde{f}$ be its local function. We say $T$ and $\tilde{T}$ are *dual under odd-step complementation,* if the following holds, and it is written as $T \underset{\text{osc}}{\longleftrightarrow} \tilde{T}$.

$$\forall(l,d,r),(l',d',r') \in \{0,1\}^3 : \; f(l,d,r) = (l',d',r') \;\Leftrightarrow\; \tilde{f}(l,d,r) = (\overline{l'},\overline{d'},\overline{r'})$$

Since the ETPCA $T$ satisfies the condition (1), we see that for each local transition rule $f(l,d,r) = (l',d',r')$ of $T$, there are two local transition rules $\tilde{f}(l,d,r) = (\overline{l'},\overline{d'},\overline{r'})$ and $\tilde{f}(\overline{l},\overline{d},\overline{r}) = (l',d',r')$ of $\tilde{T}$ (hence $\tilde{T}$ also satisfies (1)). Let $F$ and $\tilde{F}$ be the global function of $T$ and $\tilde{T}$, respectively. If the initial configuration of $T$ is $\alpha : \mathbb{Z}^2 \to \{0,1\}^3$, then we assume $\alpha$ is also given to $\tilde{T}$ as its initial configuration. Since there is a local transition rule $\tilde{f}(l,d,r) = (\overline{l'},\overline{d'},\overline{r'})$ for each $f(l,d,r) = (l',d',r')$, the configuration $\tilde{F}(\alpha)$ is the complement of the configuration $F(\alpha)$. Furthermore, since there is a local transition rule $\tilde{f}(\overline{l},\overline{d},\overline{r}) = (l',d',r')$ for each $f(l,d,r) = (l',d',r')$, the configuration $\tilde{F}^2(\alpha)$ is the same as $F^2(\alpha)$. In this way, at an even step $\tilde{T}$ gives the same configuration as $T$, while at an odd step $\tilde{T}$ gives the complemented configuration of $T$. For example, $T_{0347} \underset{\text{osc}}{\longleftrightarrow} T_{7430}$ holds (Fig. 9). Figure 10 shows examples of their evolution processes.

Note that, in Definition 9, the ETPCA $T$ must satisfy the condition (1). Therefore, the relation $\underset{\text{osc}}{\longleftrightarrow}$ is defined on the ETPCAs of the form $T_{wxyz}$ such that $w + z = 7$

and $x + y = 7$. Hence, only the following 16 ETPCAs have their dual counterparts under odd-step complementation.

$$\begin{array}{lll} T_{0077} \underset{\text{osc}}{\longleftrightarrow} T_{7700}, & T_{0167} \underset{\text{osc}}{\longleftrightarrow} T_{7610}, & T_{0257} \underset{\text{osc}}{\longleftrightarrow} T_{7520}, \\ T_{0347} \underset{\text{osc}}{\longleftrightarrow} T_{7430}, & T_{0437} \underset{\text{osc}}{\longleftrightarrow} T_{7340}, & T_{0527} \underset{\text{osc}}{\longleftrightarrow} T_{7250}, \\ T_{0617} \underset{\text{osc}}{\longleftrightarrow} T_{7160}, & T_{0707} \underset{\text{osc}}{\longleftrightarrow} T_{7070} & \end{array}$$

### 2.3.4 Equivalence Classes of ETPCAs

If ETPCAs $T$ and $T'$ are dual under reflection, complementation, or odd-step complementation, then they can be regarded as essentially the same ETPCAs. Here, we classify the 256 ETPCAs into equivalence classes based on the three dualities. We define the relation $\longleftrightarrow$ as follows: For any ETPCAs $T$ and $T'$,

$$T \longleftrightarrow T' \;\Leftrightarrow\; (T \underset{\text{refl}}{\longleftrightarrow} T' \,\vee\, T \underset{\text{comp}}{\longleftrightarrow} T' \,\vee\, T \underset{\text{osc}}{\longleftrightarrow} T')$$

holds. Now, let $\longleftrightarrow^*$ be the reflexive and transitive closure of $\longleftrightarrow$. Then, $\longleftrightarrow^*$ is an equivalence relation, since $\longleftrightarrow$ is symmetric. By this, 256 ETPCAs are classified into 82 equivalence classes. Table 1 shows the classification result.

Wolfram classified 256 one-dimensional elementary cellular automata (ECAs) into 88 equivalence classes, which is given in the Appendix of [16]. It is based on the two dualities *reflection* and *conjugation* that correspond to reflection and complementation in ETPCAs, respectively. If we consider only these two dualities in ETPCAs, the number of equivalence classes is 88, the same as in ECAs. Note that in [16] the duality corresponding to odd-step complementation is implicitly mentioned, and if we also use it, the number of equivalence classes of ECAs becomes 82.

ECAs and ETPCAs have very different features in reversibility. In ECAs, there are six reversible ones: ECAs 15, 51, 85, 170, 204 and 240. They are grouped into two equivalence classes if we use the three dualities. We can see one class contains ECA 204 (identity), and the other contains ECA 170 (left shift). Hence, all the six are trivial ones. On the other hand, there are 36 reversible ETPCAs that are grouped into 12 equivalence classes (Table 1). Many of them are nontrivial, and as we shall see below, at least ten reversible ETPCAs in three classes are computationally universal.

**Table 1** Total numbers and numbers of equivalence classes of ETPCAs, reversible ETPCAs and conservative ETPCAs [12]

| | Total number | Equivalence classes |
|---|---|---|
| ETPCAs | 256 | 82 |
| Reversible ETPCAs | 36 | 12 |
| Conservative ETPCAs | 9 | 4 |

# 3 Complex Phenomena in Reversible ETPCA $T_{0347}$

Here, we focus on a specific nonconservative reversible ETPCA $T_{0347}$ [13]. Its local function is given in Fig. 11. Despite its extreme simplicity of the local function, it exhibits quite interesting behavior similar to the case of the Game-of-Life CA [6, 7].

Though its local function itself is very simple, it is generally hard to follow evolution processes of configurations of $T_{0347}$ by hand. So we created a simulator for it. Simulation movies can be seen in the slide file of [13]. We also created an emulator for $T_{0347}$ on the general purpose CA simulator *Golly* [15]. The file of the emulator with many examples of configurations is found in [11].

## 3.1 Patterns in $T_{0347}$

A *pattern* is a finite segment of a configuration. Patterns can be defined in any ETPCA. However, when we consider an evolution process of a finite configuration that contains pattern(s), it is convenient to restrict ETPCAs to those with an identification number of the form $0xyz$ (i.e., to those with a quiescent state). Otherwise, the initial finite configuration will become an infinite one at the next time step.

In this section, we give various interesting patterns in in $T_{0347}$. In a reversible ETPCA there are only three kinds of patterns. They are periodic patterns, space-moving patterns and expanding patterns. Note that, because of reversibility, there is no pattern that becomes a periodic (or space-moving) pattern after a positive number of transient steps.

A *periodic pattern* is one that satisfy the following condition: Starting from the finite configuration that contains one copy of it, only one copy of it appears at the same position after $p$ time steps ($p > 0$). The number $p$ is the period of the pattern. If $p = 1$, it is called a *stable pattern*.

A *block* is a stable pattern in $T_{0347}$ shown in Fig. 12. As we shall see below, combining several blocks, and appropriately colliding a space-moving pattern with them, right-turn, left-turn, backward-turn, and U-turn are realized.

A *fin* is a periodic pattern that simply rotates with period 6 (Fig. 13). Note that any pattern appearing at $t = 0, \ldots, 5$ of Fig. 13 is a fin. A *rotator* is a pattern shown in Fig. 14. Like a fin, it rotates around a point, and its period is 42. Though there are many periodic patterns in $T_{0347}$, a block, a fin and a rotator are the most useful ones.

A *space-moving pattern* is one that satisfy the following: Starting from the finite configuration that contains one copy of it, only one copy of it appears at a different

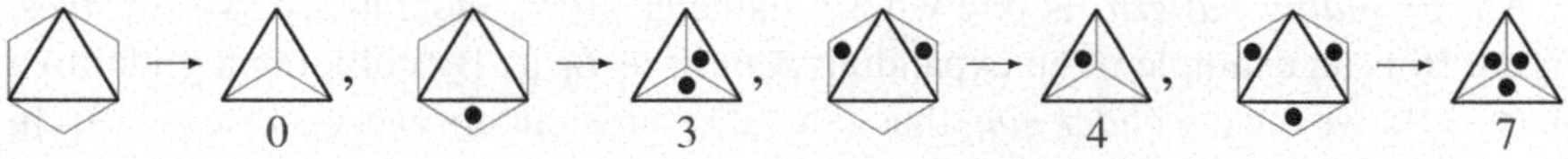


**Fig. 11** Local function of the nonconservative reversible ETPCA $T_{0347}$

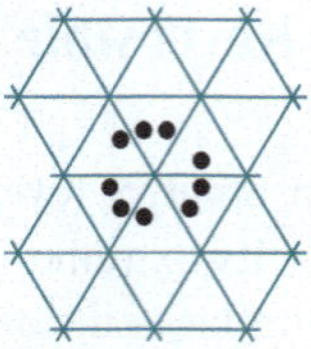

**Fig. 12** A block in $T_{0347}$. It is a stable pattern

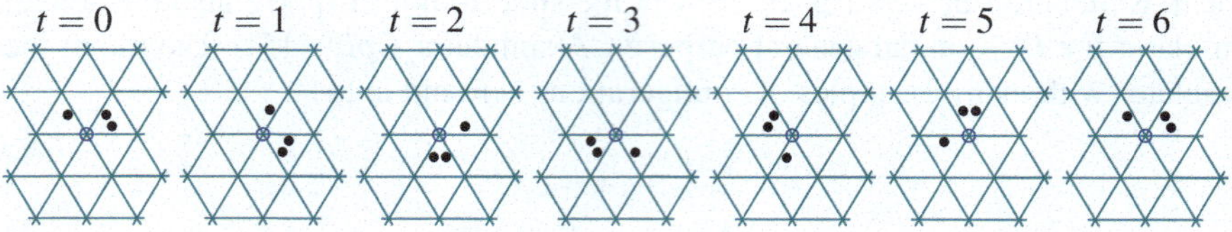


**Fig. 13** A fin. It rotates around the point ○ with period 6 in $T_{0347}$

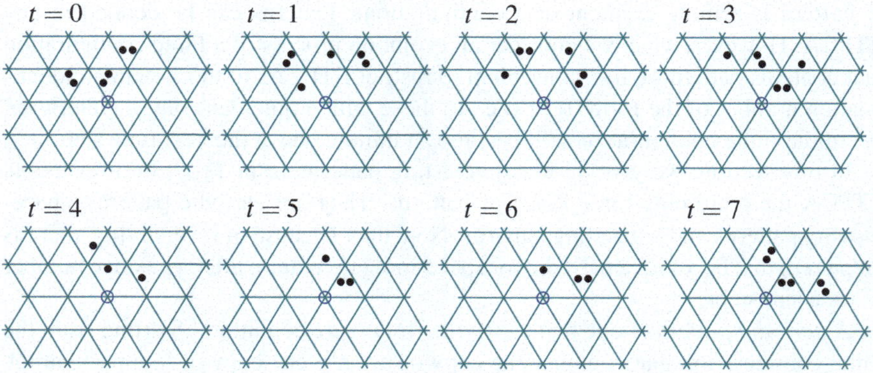


**Fig. 14** A rotator. It rotates around the point ○ with period 42 in $T_{0347}$

position after $p$ time steps ($p > 0$). The number $p$ is also called the period of the space-moving pattern. Such a patten is useful for collision-based computing [1].

Figure 15 shows a specific space-moving pattern in $T_{0347}$, which is called a *glider*. It is so named after the famous glider in the Game-of-Life [6], but it swims in the cellular space like a fish or an eel. It travels a unit distance, the side-length of a triangular cell, in 6 steps. By rotating it appropriately, it can move in any of the six directions. When constructing a computing machine in $T_{0347}$, it will be used as a signal. So far, it is not known whether there is another space-moving pattern essentially different from a glider (i.e., not composed of two or more gliders).

An *expanding pattern* is one whose diameter grows indefinitely as it evolves. Figure 16 is an example of an expanding pattern in $T_{0347}$. By colliding a glider to a fin ($t = 0$), we have a *glider gun* that generates three gliders every 24 steps [13]. In this case, the total number of particles in a configuration also grows indefinitely.

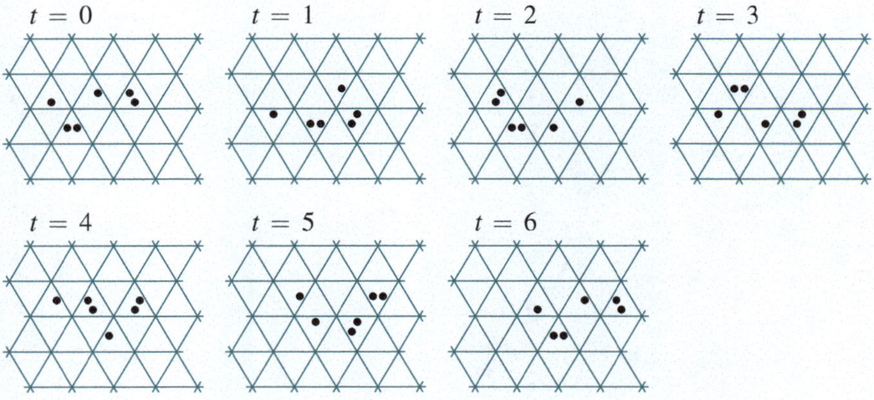


**Fig. 15** A glider. It is a space-moving pattern of period 6 in $T_{0347}$

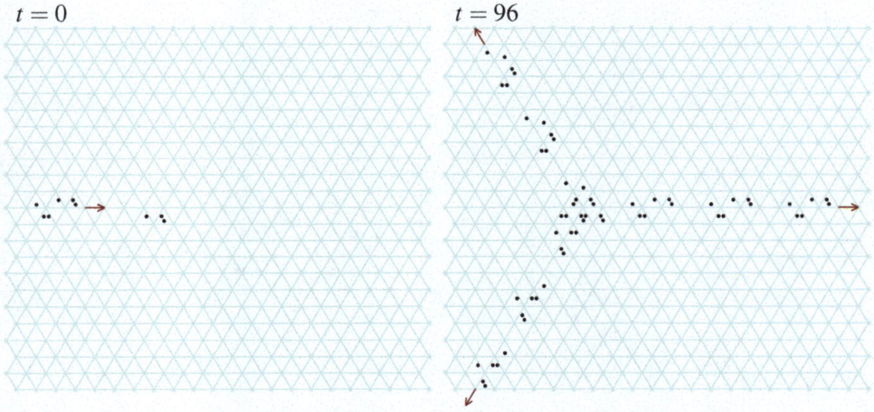


**Fig. 16** A three-way glider gun created by a collision of a glider with a fin in $T_{0347}$ [13]. It is an expanding pattern

We can show that an expanding pattern also expands to the negative time direction because of reversibility [13]. In fact, the distance between the glider and the fin at $t = 0$ in Fig. 16 grows larger and larger when we go back to the past (i.e., $t < 0$).

It is remarkable that there is a glider gun that generates gliders to the negative time direction (Fig. 17). It is also an expanding pattern. Furthermore, by appropriately combining two configurations at $t = 0$ of Fig. 16 and $t = 0$ of Fig. 17, it is possible to have a gun that generates gliders in both positive and negative time directions.

Figure 18 gives another example of an expanding pattern. If we start with a one-particle pattern, a chaotic (or disordered) pattern with many gliders is generated, and the whole pattern grows larger and larger. Also in this case, a similar chaotic pattern with gliders appears and grows indefinitely to the negative time direction [13].

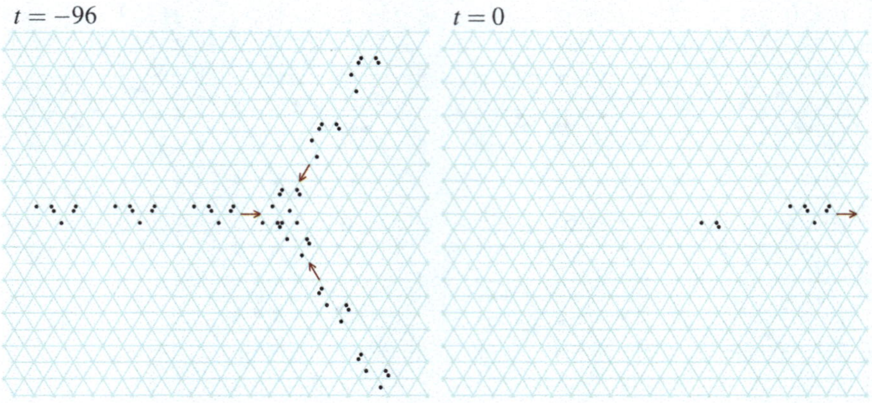


**Fig. 17** A three-way glider absorber in $T_{0347}$ [13]. It is a glider gun to the negative time direction

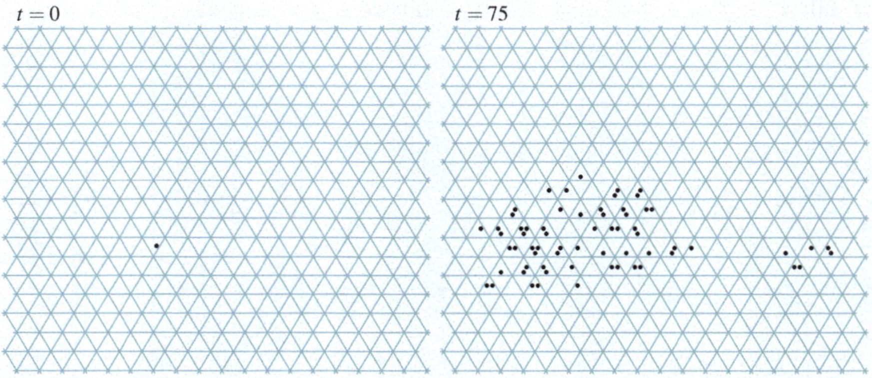


**Fig. 18** A chaotic pattern that expands indefinitely appears from the one-particle pattern in $T_{0347}$

It should be noted that a chaotic pattern often appears even from a configuration that contains only periodic patterns and gliders. Therefore, if we want to design a large pattern that perform some intended task, we should carefully do it so that the whole pattern never generates a chaotic pattern.

## 3.2 Interactions of Patterns in $T_{0347}$

It is known that various interesting phenomena emerge by interacting blocks, fins or gliders with another glider [13]. In this section, we observe what happens when we collide a glider to a sequence of blocks.

We first collide a glider with two blocks (Fig. 19a). Then, the glider is split into a rotator and a fin ($t = 56$). The fin travels around the blocks three times without

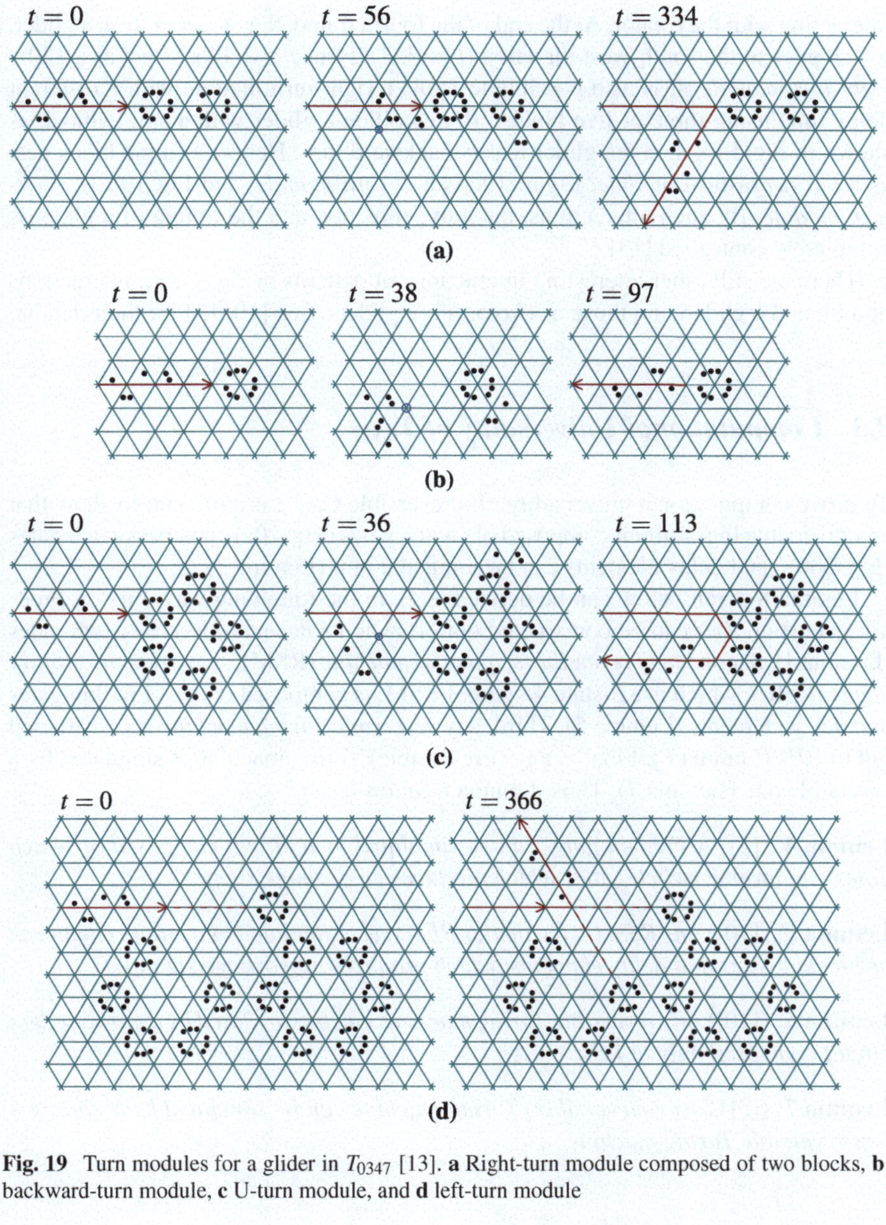


**Fig. 19** Turn modules for a glider in $T_{0347}$ [13]. **a** Right-turn module composed of two blocks, **b** backward-turn module, **c** U-turn module, and **d** left-turn module

interacting with the rotator. At the end of the fourth round, they meet to form a glider, which goes to the south-west direction ($t = 334$). Hence, two blocks act as a 120°-*right-turn module*. It is also possible to make a right-turn module with a different delay time using three or five blocks. If we collide a glider with a single block as shown in Fig. 19b, then the glider makes backward turn. Hence, a single block acts as a *backward-turn module*. Figure 19c is a *U-turn module*, and Fig. 19d is a *left-turn module*. By such interactions, the move direction and the timing of a glider is completely controlled [13].

There are still other interesting interactions of patterns in $T_{0347}$, e.g., interactions of a fin and a glider, and those of two or more gliders. See [10–13] for their details.

## 3.3 Computational Universality of $T_{0347}$

To prove computational universality of a reversible CA, it is sufficient to show that any reversible logic circuit composed of switch gates (Fig. 20a), inverse switch gates (Fig. 20b), and delay elements can be simulated in it (Lemma 8).

Lemma 8 given below can be derived, e.g., in the following way. First, a Fredkin gate (Fig. 20c) can be constructed out of switch gates and inverse switch gates (Lemma 4). Second, any *reversible sequential machine* (RSM), in particular, a *rotary element* (RE), which is a 2-state 4-symbol RSM, is composed only of Fredkin gates and delay elements (Lemma 5). Third, any *reversible Turing machine* is constructed out of REs (Lemma 6). Finally, any (irreversible) Turing machine is simulated by a reversible one (Lemma 7). Thus, Lemma 8 follows.

**Lemma 4** ([4]) *A Fredkin gate can be simulated by a circuit composed of switch gates and inverse switch gates, which produces no garbage signals.*

**Lemma 5** ([10]) *Any RSM (in particular RE) can be simulated by a circuit composed of Fredkin gates and delay elements, which produces no garbage signals.*

**Lemma 6** ([10]) *Any reversible Turing machine can be simulated by a garbage-less circuit composed only of REs.*

**Lemma 7** ([3]) *Any (irreversible) Turing machine can be simulated by a garbage-less reversible Turing machine.*

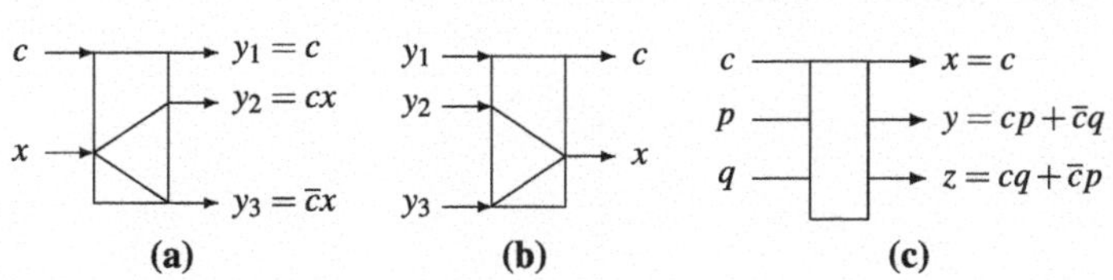


**Fig. 20** **a** Switch gate. **b** Inverse switch gate, where $c = y_1$ and $x = y_2 + y_3$ under the assumption $(y_2 \rightarrow y_1) \wedge (y_3 \rightarrow \overline{y_1})$. **c** Fredkin gate

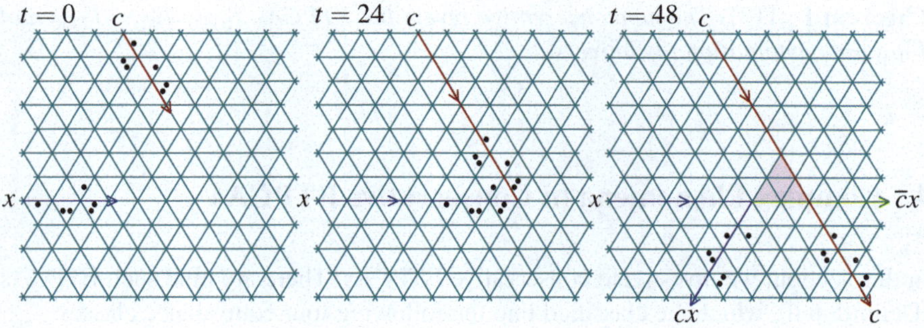


**Fig. 21** Switching operation realized by collision of two gliders in $T_{0347}$ [13]

**Lemma 8** *A reversible CA is computationally universal, if any circuit composed of switch gates, inverse switch gates, and delay elements is simulated in it.*

We now show computational universality of $T_{0347}$. It is possible to implement a switch gate and an inverse switch gate in $T_{0347}$ using gliders as signals. The switching operation $(c, x) \mapsto (c, cx, \overline{c}x)$ is realized by colliding two gliders as shown in Fig. 21. Using many turn modules to adjust the collision timing and the directions of the input gliders, we can construct a *switch gate module* as shown in Fig. 22. An inverse switch gate can be also realized in a similar manner. By above, and by the dualities among ETPCAs, we have the following.

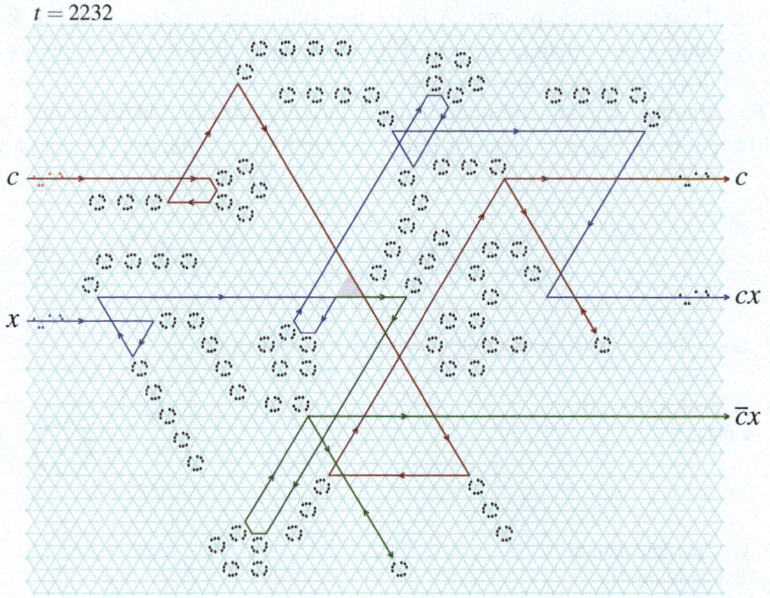


**Fig. 22** Switch gate module implemented in $T_{0347}$, whose input-output delay is 2232 [13]

**Theorem 1** ([13]) *The nonconservative reversible ETPCAs $T_{0347}$, $T_{0617}$, $T_{7430}$ and $T_{7160}$ are computationally universal.*

## 4 Complex Phenomena in Conservative ETPCAs

In this section, we investigate conservative ETPCAs. There are nine such ETPCAs (Definition 6), which are classified into the following four equivalence classes.

$$\{T_{0157}, T_{0457}, T_{0237}, T_{0267}\}, \{T_{0137}, T_{0467}\}, \{T_{0167}, T_{0437}\}, \{T_{0257}\}$$

Though the total number of particles in a configuration is conserved in them, the six ETPCAs in the first two classes still show complex behavior. In particular, they were shown to be computationally universal [8, 9]. On the other hand, those in the last two classes are trivial ETPCAs, and thus they are non-universal [9].

### *4.1 Conservative ETPCA $T_{0157}$*

Here we consider the conservative and reversible ETPCA $T_{0157}$ (Fig. 23). It was first studied in [8].

There are many periodic patterns in $T_{0157}$. Among them two are useful. They are a block and a one-particle pattern. A *block* is a stable pattern shown in Fig. 24. On the other hand, a *one-particle pattern* (Fig. 25) rotates clockwise with period 6 by the second local transition rule of Fig. 23.

In $T_{0157}$ there exist a large number of space-moving patterns. On this point, $T_{0157}$ is very different from $T_{0347}$. In fact, it is rather easy to find such patterns by watching an

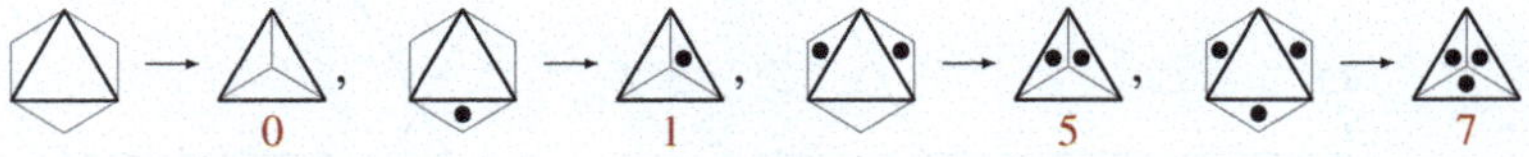


**Fig. 23** Local function of $T_{0157}$ defined by four local transition rules

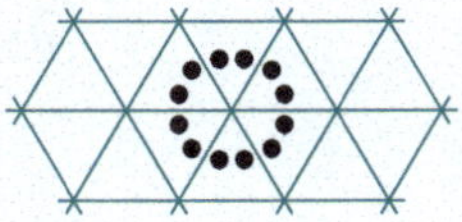

**Fig. 24** A block in $T_{0157}$. It is a stable pattern

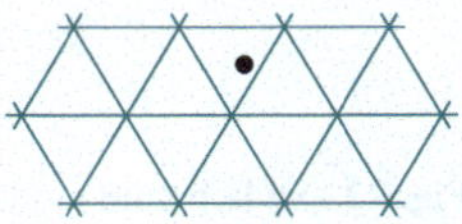

**Fig. 25** One-particle pattern in $T_{0157}$. It simply rotates with period 6

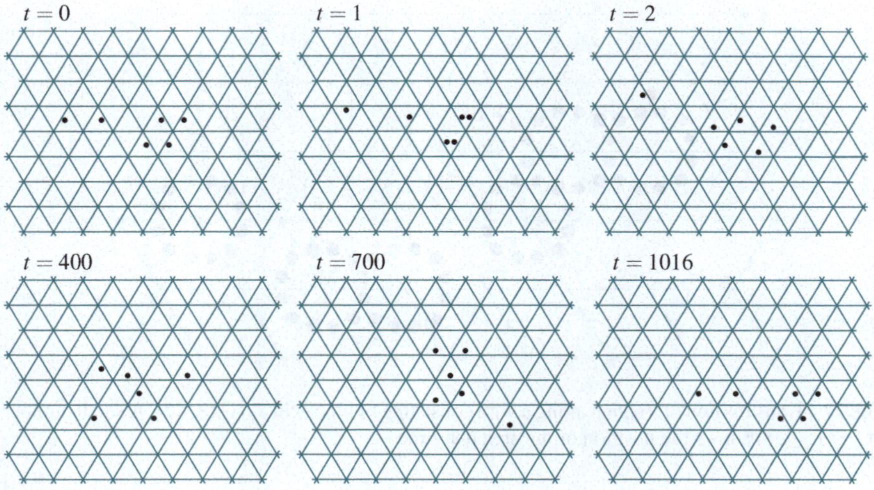


**Fig. 26** A space-moving pattern of period 1016 in $T_{0157}$

evolution process starting from a randomly given finite pattern with many particles. In many cases, some space-moving patterns and some periodic patterns will appear from it. However, periods of space-moving patterns are generally very long. Figure 26 shows an example of a space-moving pattern of period 1016. Furthermore, it exhibits complex behavior in the period. Since the period is long, it is very difficult to control a space-moving pattern, in particular to adjust its timing. So far, it is not known whether there is a space-moving pattern with a very short period (say 10 or less).

Since there are space-moving patterns in $T_{0157}$, expanding patterns also exist in it. For example, a configuration consisting of two space-moving patterns that go to different directions is an expanding pattern, though the total number of particles in a configuration is always constant.

Here we use a one-particle pattern, rather than a space-moving pattern, as a signal. A signal transmission wire, on which a signal can move, is obtained by connecting blocks as shown in Fig. 27. Wires can be bent relatively freely, and by this the timing of a signal is adjusted. In [8, 12], a module for crossing two signals in the two-dimensional space is given.

The switch gate operation $(c, x) \mapsto (c, cx, \overline{c}x)$ is realized by one cell of $T_{0157}$ as shown in Fig. 28. Likewise, the inverse switch gate operation is also realized by one cell. Complete patterns of a switch gate module, and an inverse switch gate module are given in [8, 12].

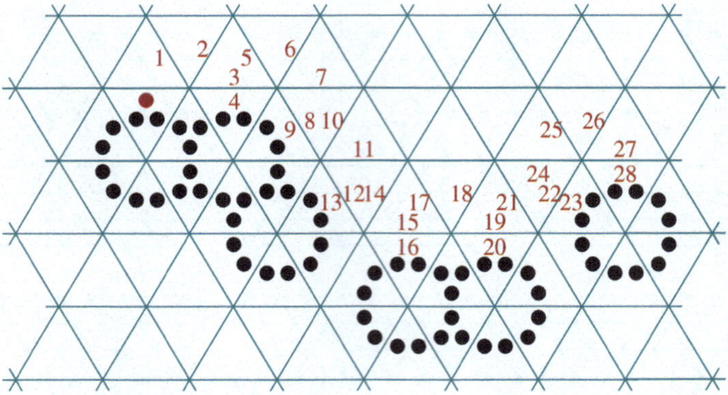


**Fig. 27** Transmission of a signal along a wire consisting of blocks in $T_{0157}$ [8, 12]. A small number $t$ ($1 \leq t \leq 28$) shows the position of the signal at time $t$

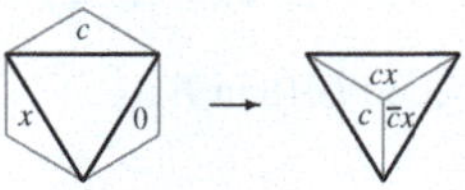


**Fig. 28** Switch gate operation realized by one cell of $T_{0157}$

## 4.2 Conservative ETPCA $T_{0137}$

Next, consider the conservative and reversible ETPCA $T_{0137}$ (Fig. 29).

Among many periodic patterns in $T_{0137}$, a block and a one-particle pattern are useful as in the case of $T_{0157}$. Figure 30 shows a *block* in $T_{0137}$. Though it is different from Fig. 24, it is stable in $T_{0137}$. The *one-particle pattern* (Fig. 25) rotates clockwise with period 6 also in $T_{0137}$.

Similar to the case of $T_{0157}$, there are many space-moving patterns in $T_{0137}$. Again, in this case, their periods are very long. Hence, it is hard to use them as signals in logic circuits. Figure 31 is an example of a space-moving pattern of period 3162.

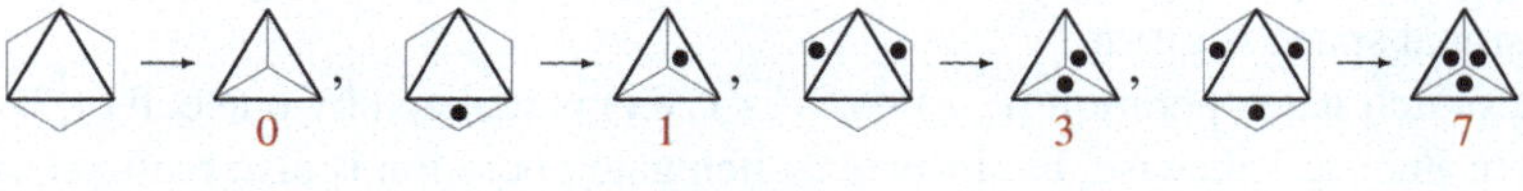


**Fig. 29** Local function of $T_{0137}$ defined by four local transition rules

**Fig. 30** A block in $T_{0137}$. It is a stable pattern

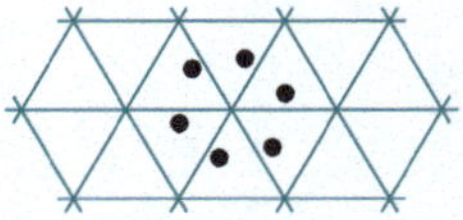

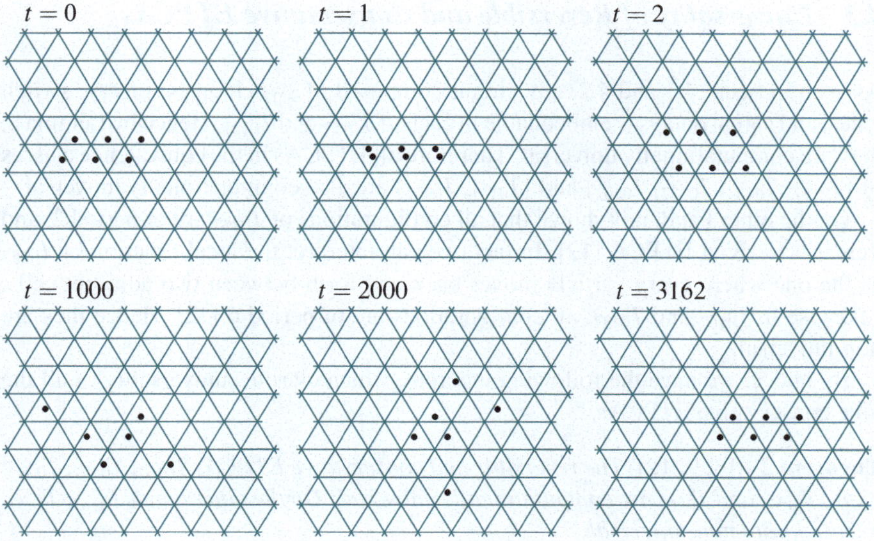


**Fig. 31** A space-moving pattern of period 3162 in $T_{0137}$

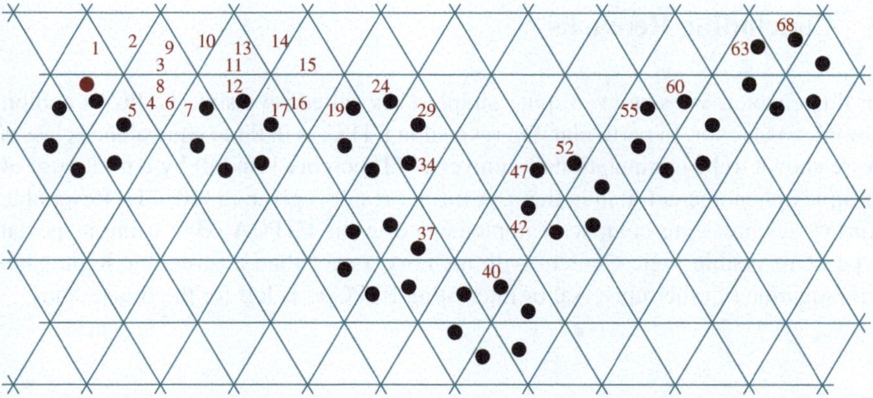


**Fig. 32** Transmission of a signal along a wire in $T_{0137}$ [9, 12]

**Fig. 33** Switch gate operation realized by one cell of $T_{0137}$

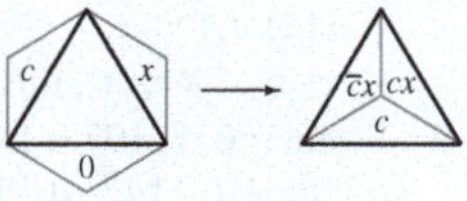


Figure 32 shows a signal transmission wire in $T_{0137}$ composed of blocks, on which a particle travels as a signal. The switch gate operation is realized by one cell of $T_{0137}$ as in Fig. 33. Using these phenomena and operations, a signal crossing module, a switch gate module and an inverse switch gate module can be constructed (see [9, 12] for the details).

### *4.3 Universality of Reversible and Conservative ETPCAs*

As seen in Sects. 4.1 and 4.2, any circuit composed of switch gates, inverse switch gates, and signal delays is embeddable in each of $T_{0157}$ and $T_{0137}$. Hence by Lemma 8, they are computationally universal. Therefore all ETPCAs in the equivalence classes of $\{T_{0157}, T_{0457}, T_{0237}, T_{0267}\}$ and $\{T_{0137}, T_{0467}\}$ are also computationally universal.

On the other hand, it is shown that all configurations of $T_{0257}$ are of period 2, and thus it is a trivial ETPCA [12]. In fact, we can interpret the local function of $T_{0257}$ as the one where every particle moves back and forth between two adjacent cells. Likewise in $T_{0167}$ and $T_{0437}$, all configurations are of period 6 [12]. Hence they are non-universal.

By above, we have the following theorem, which clarifies universality of all the nine conservative ETPCAs.

**Theorem 2** ([8, 9, 12]) *The reversible and conservative ETPCAs $T_{0157}$, $T_{0457}$, $T_{0237}$, $T_{0267}$, $T_{0137}$, and $T_{0467}$ are computationally universal. On the other hand, $T_{0167}$, $T_{0437}$, and $T_{0257}$ are non-universal.*

## 5 Concluding Remarks

In this chapter, we saw even quite simple CAs called reversible ETPCAs exhibit complex behavior. In particular, ten reversible ETPCAs in three equivalence classes were shown to be computationally universal (Theorems 1 and 2) by a tricky use of complex phenomena found in them. A further result is given in [10, 11]: Reversible Turing machines are compactly implemented in the ETPCA 0347 using a special type of reversible logic element with memory, rather than a reversible logic gate. Investigation of other universal or interesting ETPCAs is left for the future study.

## References

1. Adamatzky A (ed) (2002) Collision-based computing. Springer. https://doi.org/10.1007/978-1-4471-0129-1
2. Bays C (1994) Cellular automata in the triangular tessellation. Complex Syst 8:127–150
3. Bennett CH (1973) Logical reversibility of computation. IBM J Res Dev 17:525–532. https://doi.org/10.1147/rd.176.0525
4. Fredkin E, Toffoli T (1982) Conservative logic. Int J Theor Phys 21:219–253. https://doi.org/10.1007/BF01857727
5. Gajardo A, Goles E (2001) Universal cellular automaton over a hexagonal tiling with 3 states. Int J Algebra Comput 11:335–354. https://doi.org/10.1142/S0218196701000486
6. Gardner M (1970) Mathematical games: the fantastic combinations of John Conway's new solitaire game "life". Sci Am 223(4):120–123. https://doi.org/10.1038/scientificamerican1070-120

7. Gardner M (1971) Mathematical games: on cellular automata, self-reproduction, the Garden of Eden and the game "life". Sci Am 224(2):112–117. https://doi.org/10.1038/scientificamerican0271-112
8. Imai K, Morita K (2000) A computation-universal two-dimensional 8-state triangular reversible cellular automaton. Theor Comput Sci 231:181–191. https://doi.org/10.1016/S0304-3975(99)00099-7
9. Morita K (2016) Universality of 8-state reversible and conservative triangular partitioned cellular automata. In: ACRI 2016 El Yacoubi S, et al (eds) LNCS, vol 9863, pp 45–54. https://doi.org/10.1007/978-3-319-44365-2_5. Slides with movies of computer simulation: Hiroshima University Institutional Repository, http://ir.lib.hiroshima-u.ac.jp/00039997
10. Morita K (2017) Finding a pathway from reversible microscopic laws to reversible computers. Int J Unconv Comput 13:203–213
11. Morita K (2017) Reversible world : Data set for simulating a reversible elementary triangular partitioned cellular automaton on Golly. Hiroshima University Institutional Repository. http://ir.lib.hiroshima-u.ac.jp/00042655
12. Morita K (2017) Theory of reversible computing. Springer, Tokyo. https://doi.org/10.1007/978-4-431-56606-9
13. Morita K (2019) A universal non-conservative reversible elementary triangular partitioned cellular automaton that shows complex behavior. Nat Comput 18(3):413–428. https://doi.org/10.1007/s11047-017-9655-9. Slides with movies of computer simulation: Hiroshima University Institutional Repository, http://ir.lib.hiroshima-u.ac.jp/00039321
14. Morita K, Harao M (1989) Computation universality of one-dimensional reversible (injective) cellular automata. Trans IEICE Jpn E72:758–762
15. Trevorrow A, Rokicki T et al (2005) Golly: an open source, cross-platform application for exploring Conway's Game of Life and other cellular automata. http://golly.sourceforge.net/
16. Wolfram S (1986) Theory and applications of cellular automata. World Scientific Publishing

# Error Detection and Correction in Firing Squad Synchronization Problem

**Apostolos Kyritsis, Orestis Liolis, and Georgios Ch. Sirakoulis**

**Abstract** Firing Squad Synchronization Problem (FSSP) is one of the most well-known problem in computer science. Even though many solution algorithms have been proposed in the literature, the inherited error correction capabilities of these algorithms have been paid almost no attention. In this chapter, an error detection and correction algorithm that is utilizing the patterns that appear in Mazoyer's 8-state solution is analyzed and presented. Moreover, we investigate the patterns of 2-D Umeo's algorithm. It is proven that the 2-D FSSP solution is more tolerant to errors since more patterns can be easily found. Finally, an algorithm that is using the previous states is also presented. In particular, the usage of previous states makes the algorithm more efficient, since it is able to correct more errors. On the other hand, this algorithm is more complex than other algorithms presented in this chapter, thus it is more greedy in storage needs. The proposed error detection algorithms can protect the FSSP solution without rising the complexity of the system to excess.

## 1 Introduction

In the last decades, Cellular Automata (CA) have been proven especially effective for the solution of a vast number of various, difficult and complex mathematical and computational problems. More specifically, CA are studied in an extremely large number of scientific fields such as computational theory, mathematical and physical sciences, engineering, computer vision, robotics, theoretical biology, micro-structure modeling and many others [1, 5, 6, 8, 11, 13, 19–21, 23–25, 27, 28, 34, 36, 37, 39].

A. Kyritsis (✉) · O. Liolis · G. Ch. Sirakoulis (✉)
Department of Electrical and Computer Engineering, Democritus University of Thrace, GR–67100 Xanthi, Greece
e-mail: gsirak@ee.duth.gr

A. Kyritsis
e-mail: aposkyri2@ee.duth.gr

O. Liolis
e-mail: oliolis@ee.duth.gr



A. Adamatzky (ed.), *Automata and Complexity*, Emergence, Complexity and Computation 42, https://doi.org/10.1007/978-3-030-92551-2_21

In brief, a CA computational system consists of a grid of cells. Each CA cell can be found at one of a finite set of different states. The number of states is arbitrary and it heavily depends on the problem that someone wants to model. One of the greatest advantages of the CA models is the fact that their complex behavior of the CA emerges from cells that usually make very simple computations. In addition, CA provide great potential of inherent parallelism, since each cell is physically implemented separately but they all evolve in parallel according to the proposed corresponding CA rule.

One of the most studied CA problems is the Firing Squad Synchronization Problem (FSSP). In the literature, various solutions have been proposed [3, 7, 9, 16, 17, 29, 31, 35] for more than five (5) decades now. In addition, several practical applications of FSSP in real-life problems have been reported in literature [10, 12, 15]. Even though extended research of FSSP has been reported in literature, every work presupposes that the implemented system is protected by noise. This assumption is not realistic in all cases, since, for example, many applications of FSSP are located in communications and networks fields. In this chapter, the capabilities of error detection and correction, that are based on the patterns appearing in FSSP algorithms are studied. In addition, an algorithm that takes advantage of these capabilities is presented. The proposed algorithms are able to protect the CA from noise with low cost, due to their low complexity.

In Sect. 2 the necessary background of Cellular Automata (CA) and Firing Squad Synchronization Problem (FSSP) are presented. In Sect. 3 the error correction algorithm without using previous states is proposed. Finally, in Sect. 4 the algorithm that uses previous states is introduced, while in Sect. 5, conclusions are drawn.

## 2 Firing Squad Synchronization Problem (FSSP)

**Firing Squad Synchronization problem (FSSP)** was presented by Moore in 1962 [18]. The name of the problem arrives from "*Firing Squad*" teams. The purpose of this problem is to construct a random-length one-dimensional (1-D) CA and define the cell states and transmission rules that make the whole grid of cells "*synchronized*". The CA is triggered by one single cell. This cell is called "*General*" and when it gives the "*order*", the CA starts its operation. When the algorithm reaches its end, the CA is synchronized; namely the soldiers reach the "*Fire*" state simultaneously. This means that none of the cells has gone into this state before and every cell "*fires*" once and for all.

In the literature, many CA that solve the FSSP have been proposed. The first valid solution was presented by John McCarthy and Marvin Minsky [38]. This algorithm is implemented by the propagation of two waves at the soldiers line. The second wave is propagated three times slower than the first one. The fast wave is propagated through the grid until it reaches the last soldier. Afterwards, it is reflected until it reaches the slow wave. Then the two waves are divided to four, one slow and one fast to each direction. The same procedure is repeated until the distance of each wave is

1. When that happens, all the soldiers are ready to "*fire*". This algorithm reaches the `fire` state at $3 \times n$ time-steps, where $n$ is the number of the soldiers.

In 1962, Eiichi Goto proposed an algorithm that reaches the final state at $2 \times n - 2$ time-steps, by using thousands of different states [9]. In 1966, Waksman optimized Goto's algorithm to use only 16 states [35]. Later in 1967, Balzer introduced a further optimized solution which uses only 8 states [3]. Some years later, Mazoyer proposed two solution algorithms in 1987. The first algorithm is using 8 states while the second 6 states. Even though the second algorithm is using less states, it consists of more transition rules than the first one. This is the main reason why the 8-state algorithm has been preferred to be analyzed in the following of the chapter. Both 8 and 6 states algorithms reach `fire` state at the minimal time, which is $2 \times n - 1$. An algorithm that reaches the `fire` state in less time has not been considered easily possible, because this is the time necessary for the General's signal to reach the last soldier and return to him [16].

Mazoyer's algorithm is similar to Waksman's and Balzer's. However, the difference is established on the propagation of the waves and more specifically at the interference of the two waves. In contrast to the previous algorithms, in Mazoyer's algorithm, when the fast wave meets the slow one, only two waves are created; the first one is fast and the second one is slow. Furthermore, the soldier that is located at the meeting point of the two signals turns to be general. The soldier that became general hold this state until the `fire` state is reached, which brings the algorithm to its end. Like in the previous algorithms, the fast wave is three times faster than the slow one. An example of Mazoyer's algorithm is shown in Fig. 1. In Fig. 2 the transition table of each state is presented. It should be stated that there are not transition rules for state `X`, which is considered a boundary state and for state `F` because it is the final state of the cells, i.e. the algorithm stops when the cells reach this exact state.

In the literature, we can also find algorithms that solve the FSSP in two dimensions (2-D) [4, 14, 22, 30, 33]. In this chapter, the algorithm of Umeo, Maeda and Fujiwara, will be presented [32]. This algorithm has been published in the literature in 2002, and it has some advantages compared to the others. In this approach the cells of a $n \times m$ two-dimensional grid are divided to $n + m - 1$ sets, as Fig. 3 shows. The cells of each set are at the same state. By using this technique, the two-dimensional FSSP is equivalent with the one-dimensional and any one-dimensional solution algorithm can be used.

## 3 Error Correction Without Using Previous States

Error correction is called the procedure that detects and corrects the erroneous states created by noise in the grid without recalculating the states from the scratch. Error detection procedure has to be a universal procedure that is independent of the number of the soldiers. This can be achieved by using the features of the solving FSSP

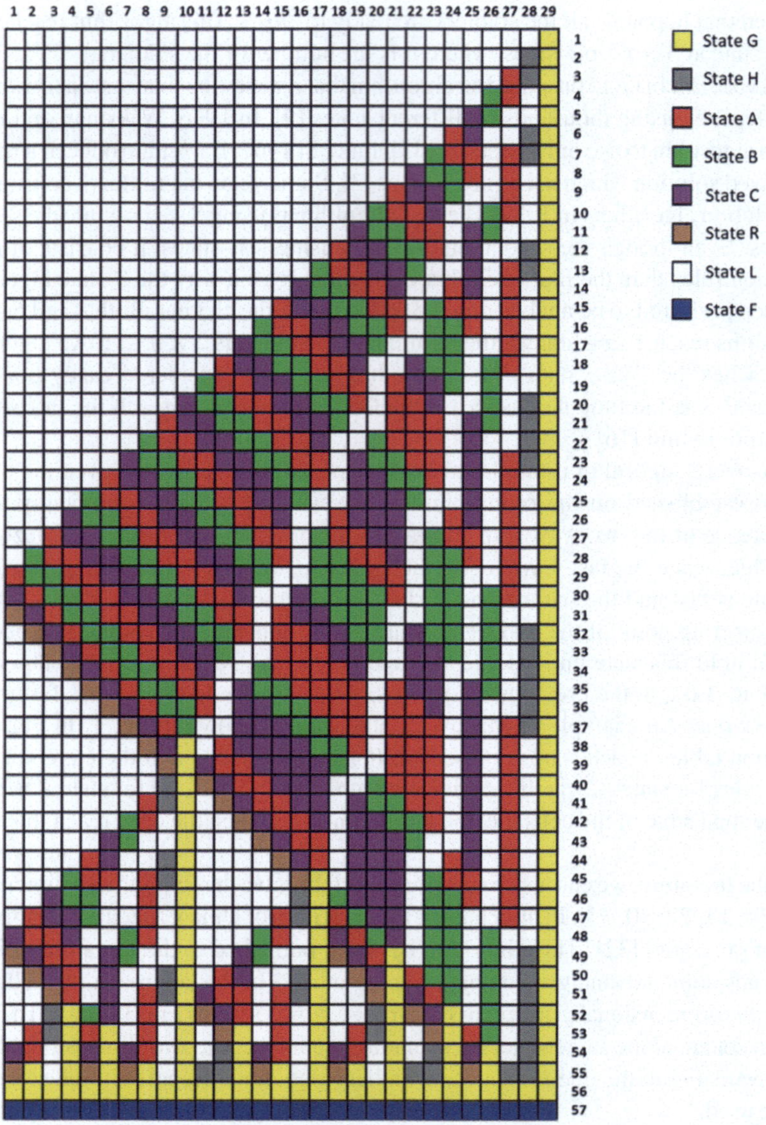


Fig. 1 Mazoyer's algorithm for $n = 29$

| A | Right Neighbor | | | | | | | |
|---|---|---|---|---|---|---|---|---|
| Left Neighbor | A | B | C | H | G | R | L | X |
| A | A | | A | | | | A | |
| B | B | | | | | | L | |
| C | C | C | | | C | | G | |
| H | | | | | | | | |
| G | | C | | | C | | | |
| R | | | | | | | | |
| L | B | | A | A | | | A | |
| X | | C | | | C | | | |

| B | Right Neighbor | | | | | | | |
|---|---|---|---|---|---|---|---|---|
| Left Neighbor | A | B | C | H | G | R | L | X |
| A | | A | A | | | | G | |
| B | B | B | | | | | B | |
| C | | C | | | | | L | |
| H | | | | | | | | |
| G | | | A | | | | | |
| R | | | | | | | | |
| L | B | B | | | | | B | |
| X | | | A | | | | | |

| C | Right Neighbor | | | | | | | |
|---|---|---|---|---|---|---|---|---|
| Left Neighbor | A | B | C | H | G | R | L | X |
| A | | | A | | | | A | |
| B | | | B | | | | G | |
| C | | C | C | | | | C | |
| H | | | | | | | | |
| G | R | | | | R | | | |
| R | R | | | | R | | | |
| L | | C | C | | | | C | |
| X | R | | | | R | | | |

| H | Right Neighbor | | | | | | | |
|---|---|---|---|---|---|---|---|---|
| Left Neighbor | A | B | C | H | G | R | L | X |
| A | | | | | L | | | |
| B | | | | | | | | |
| C | | | | | | | | |
| H | | | | | G | | | |
| G | | | | G | | | | |
| R | | | | | | | | |
| L | | | | | H | | | |
| X | | | | G | | | | |

| G | Right Neighbor | | | | | | | |
|---|---|---|---|---|---|---|---|---|
| Left Neighbor | A | B | C | H | G | R | L | X |
| A | | | | | | | G | |
| B | | | | | | | | |
| C | G | G | G | | | | G | |
| H | G | G | G | G | | G | G | G |
| G | | | | | F | | | F |
| R | G | G | G | G | | G | G | |
| L | G | G | G | G | | G | G | G |
| X | | | | | F | | | |

| R | Right Neighbor | | | | | | | |
|---|---|---|---|---|---|---|---|---|
| Left Neighbor | A | B | C | H | G | R | L | X |
| A | | | | | | | | |
| B | | | | | | | | |
| C | | | | | | | | |
| H | | | | | | | | |
| G | | | | | G | | | |
| R | | | L | | | | | |
| L | | | L | | H | | | |
| X | | | L | | G | | | |

| L | Right Neighbor | | | | | | | |
|---|---|---|---|---|---|---|---|---|
| Left Neighbor | A | B | C | H | G | R | L | X |
| A | L | L | L | | L | | | |
| B | L | L | L | L | | | L | |
| C | L | L | L | | | | L | |
| H | | | | | | | | |
| G | C | A | B | H | G | L | L | |
| R | | | | | | | | |
| L | B | C | A | A | H | L | | |
| X | C | A | B | H | G | L | | |

**Fig. 2** Mazoyer's algorithm transition rules (adopted from [16])

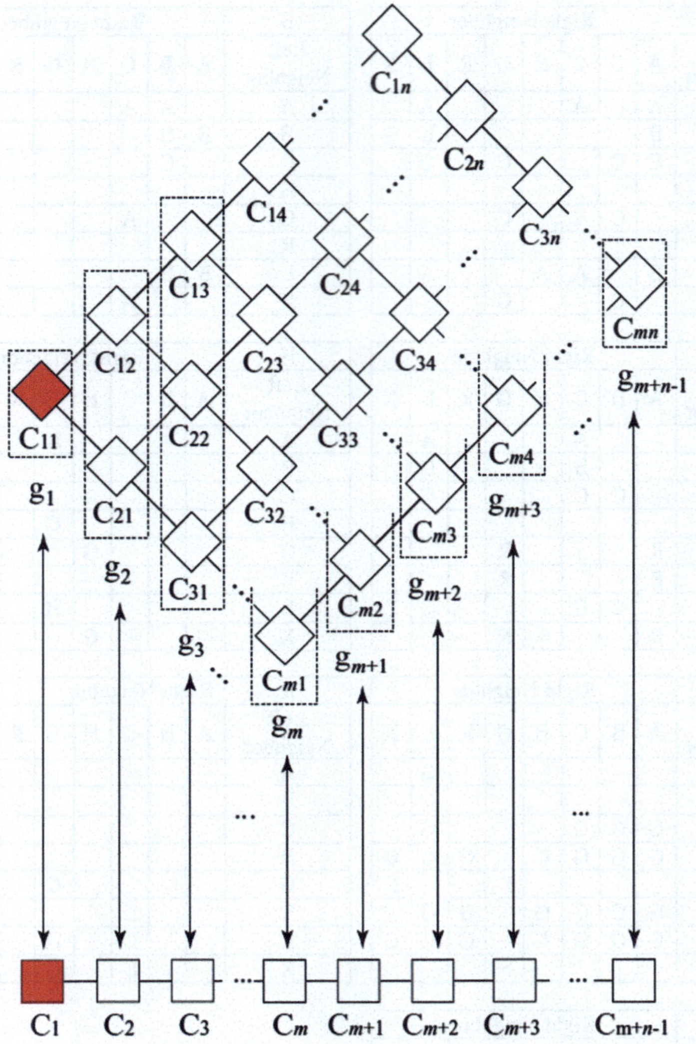


**Fig. 3** Two-dimensional CA transformation to one-dimensional, using sets of CA cells (adopted from [32])

algorithm and the patterns that appear in the solution. As it is mentioned above, the algorithm that is selected for further analysis is the 8-state Mazoyer algorithm.

## 3.1 Error Correction in One-Dimensional FSSP

At first, the capabilities of the error correction, by using a snapshot of one time-step, will be investigated. In this case, the errors that can be corrected are limited, such as the patterns that can be found in the solution are also limited.

### 3.1.1 Correction with State R

A state that can be used for error detection and correction is the state R. This state can appear only during the wave reflection. That means that the wave has reached either to the bounds of the grid or to one of the Generals. In particular, the left neighbor of this state can be only one of the following states: L, G or X. Furthermore, the L state can appear only at the left of the R state, and this pair of states changes only when it meets with either G state or X state. If any other state appears, it can be considered as error and thus it can be replaced.

### 3.1.2 Algorithmic Correction of the First and Second Wave

Another pattern can be found in the area between the first and the second wave of the General. As it is described in the previous section, the first General transmits a fast wave and then transmits waves more and more slowly. The area between the first two waves can be secured and the states of the cells between these two waves can be corrected, if this is necessary.

As depicted in Fig. 1, for the cells inside the protected area, the state of any cell changes from A to A to B to B to C to C to A, etc. That means that the General's first wave has predictable behavior. The wave is transmitted in diagonal direction at the following coordinates [0, 0], [1, 1], [2, 2], ..., $[n, n]$, where the horizontal axis is the cell's distance from the General and the vertical axis is the time. Namely, the first wave is propagating at the coordinates $[x, t]$ where $x = t$, and the value of each cell is one of the three possible states, i.e. A, B and C.

The first time that a cell of the diagonal changes to state A is at the coordinates [2, 2]. Likewise, for state B the coordinates are [3, 3] and for C the coordinates are [4, 4]. The correction algorithm can calculate the correct value by using the following equation:

$$\frac{(t-2)}{3}, \quad (t = 2, 3, 4, \ldots, n) \tag{1}$$

If the modulo of Eq. (1) equals to zero (0) then the state of the cell $x$ in time $t$, when $x = t$, is A. If the modulo equals to one (1), then the cell state is B and if it equals to two (2) then the cell state is C, respectively.

In the following step, the area between the first two waves can be defined. The distance between these two waves at each time step is shown in Table 1.

**Table 1** The resulting distance between the first two waves

| TIME | DISTANCE |
|---|---|
| 2 | 1 |
| 3 | 2 |
| 4 | 2 |
| 5 | 2 |
| 6 | 3 |
| 7 | 4 |
| 8 | 4 |
| 9 | 4 |
| 10 | 5 |
| 11 | 6 |
| 12 | 6 |
| 13 | 6 |

The aforementioned distance represents the number of the cells that are located between the two waves. The state of the most left cell can be calculated by Eq. (1) and the exact algorithmic procedure that has been already described above. After the calculation of the most left cell in the protected area, it is straightforward to proceed with the finding of the states of the rest of the cells, by just using the pattern that has been described at the beginning of this subsection, in a reversed manner.

For example, when $t = 10$, the distance between the first two waves equals to 5 and the state of the cell `10` is `C`. After defining the state of cell `10` it is easy to define the state of the corresponding cells `9`, `8`, `7` and `6` which are `B`, `B`, `A`, `A`, respectively. Moreover, the states of cell in Fig. 5 and the cells that are located at the left of cell `10` are also known. These cells are at state `L`. If one of the aforementioned cells is found at a different state from the expected one, the correction algorithm can change it to the correct one. Even though this technique is applicable at the first two waves, it can protect a considerably big area of the CA. The protected area is shown as highlighted in Fig. 4.

## 3.2 *Error Correction in Two-Dimensional FSSP*

In this subsection the ability of error detection and correction in two dimensional FSSP is analyzed. The FSSP solution algorithm that has been examined is the one introduced earlier by Umeo, Maeda and Fujiwara for two-dimensional FSSP that utilizes the 8 states algorithm of Mazoyer [32].

In contrast to one dimensional FSSP solution algorithm, in two dimensional algorithm, the corresponding patterns are more. More specifically, in two dimensional FSSP algorithm the CA cells are grouped to sets, as it was already mentioned in

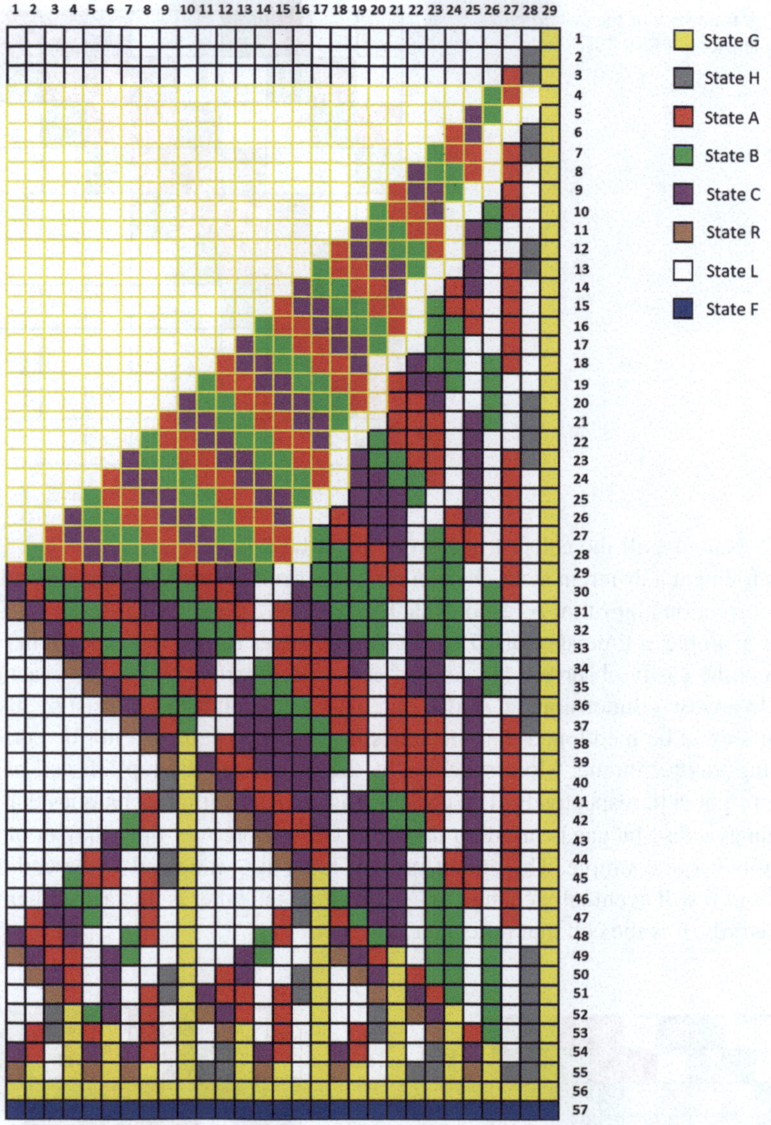


**Fig. 4** The protected area (the yellow coloured grid on the top part) between the first two waves

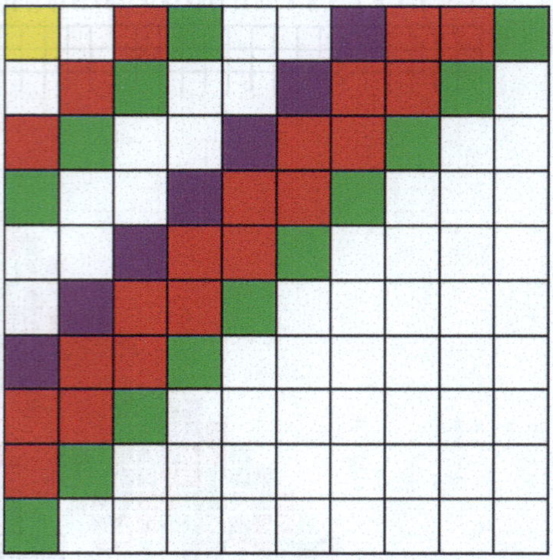

**Fig. 5** A time-shot of the two-dimensional FSSP CA

Sect. 2. Namely, all the cells of one group must be found at the same state. If one cell is found at a different state, then an error has been occurred and, as a result, the error correction algorithm is responsible to correct it.

For example, a time-shot of $10 \times 10$ CA that solves the FSSP is shown in Fig. 5. As it can be easily observed, the cells of each group correspond to the exact same state. In any two-dimensional CA, there are only two groups that consist by just one cell. It should be mentioned that the states of these two cells can not be corrected by using this technique. More specifically, these cells are the top left cell and the bottom right cell, respectively. The one-dimensional algorithm techniques that have been analyzed so far can be used for these two cells. Moreover, the behavior of these two cells is quite simple. The one of them is fixed at `G` state and at the end of the algorithm it will eventually change to `F` state while the other is the last soldier, or in other words, it is the cell that reflect the waves.

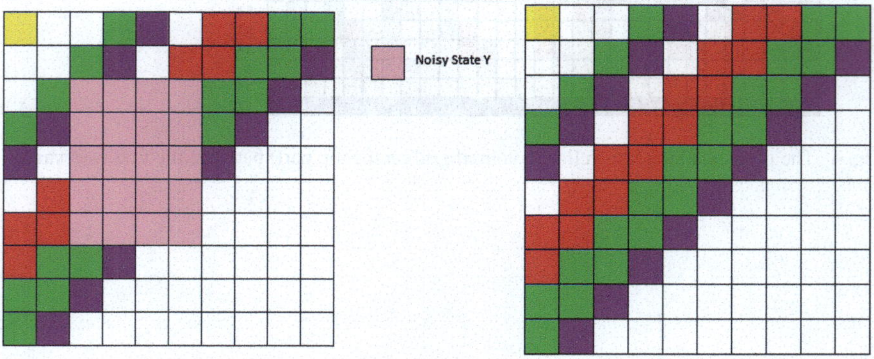


**Fig. 6** First example of error correction in two-dimensional FSSP CA

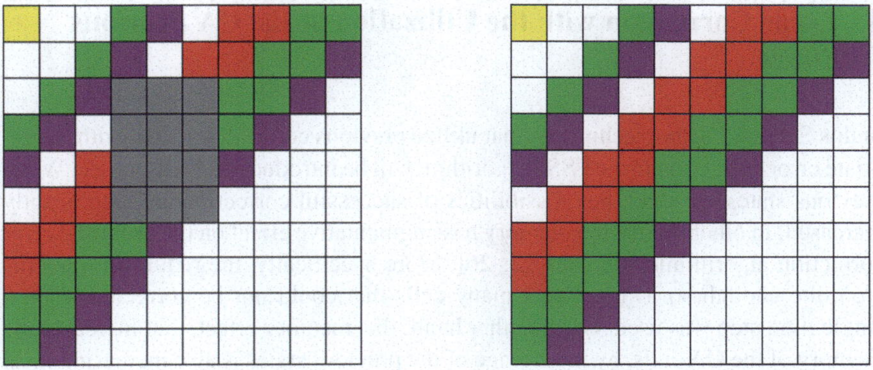

**Fig. 7** Second example of error correction in two-dimensional FSSP CA

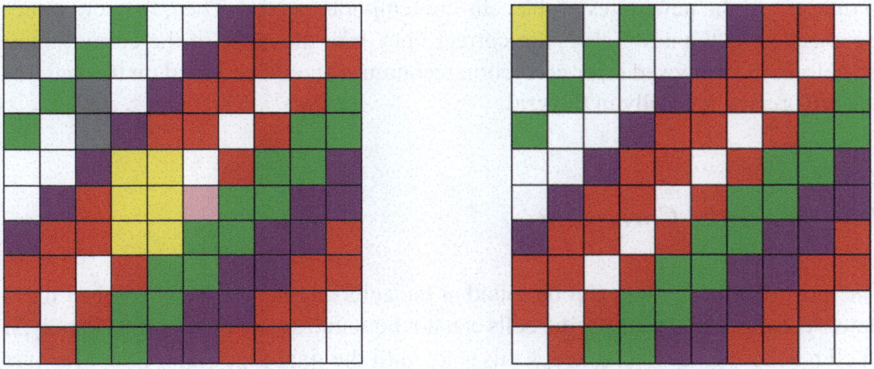

**Fig. 8** Third example of error correction in two-dimensional FSSP CA

The proposed error detection and correction algorithm pseudocode, for each time-step, is described as follows:

1. Counting the times of the appearance of each state at each group.
2. If an unknown state appear, it can be ignored.
3. The state that appears most of the times is selected for representation of each group.
4. The error correction algorithm for the one-dimensional algorithm can be used to check if the chosen state of each group is the correct one.

The following Figures, i.e. Figs. 6, 7 and 8 present some examples of the error correction algorithm application.

# 4 Error Correction with the Utilization of the CA Previous States

In this Section, further techniques that utilize previous cell states to deal with appropriate error correction in the FSSP algorithm, will be introduced and discussed. When previous states are used, the possibilities of successful correction are significantly increased. In other words, the memory has augmentative effect on the proposed error correction algorithmic approach [2, 26]. More specifically, the redundant patterns are quite enough so as to correct many cells that could not be corrected when a single time-step was used. On the other hand, this memory effect, that increases the memory of the CA cells, by the storage of the previous states is also increasing in an analogous manner the complexity of the proposed correction algorithm.

In this category of techniques, during the evolution of the CA grid from time $t$ to time $t+1$, the new states of the cells are temporary stored. Then, the new values are checked. If the new values are correct, they take the place of the current ones. As before, the proposed error correction techniques should be based on the patterns that are created naturally in the grid.

## 4.1 Generals' Correction

The most obvious pattern can be found at the cells that have already reached the `G` state. As expected, these specific cells can not be found at any other state. Namely, if a cell reaches the `G` state, it keeps this state until the final time-step of the Mazoyer algorithm, when it is changing to `F` state. So, if the state of a cell changes from `G` to another state, the algorithm takes care of this inappropriate incident and changes it back.

## 4.2 Correction During Wave's Propagation

Every General creates waves that travel through the grid until they reach either the end of the grid or until they reach another General. That means that a General sends multiple waves and every wave is slower than the previous. When a wave is reflected, then it is moving on the opposite direction until it meets with the next wave that is send by the same General.

Firstly, the area between two waves can be delimited by using two variables. The variable `start` is the position of the fast wave and variable *end* is the position of the slower wave or a general. So, the correction algorithm starts searching the grid from the opposite to the initial general edge. If the algorithm finds one of the following states `A`, `B`, `C`, `H` and `G`, the position of the current cell is stored to the variable `start`. When the algorithm finds a cell at either state `G` or state `L`, it stores the position of the previous cell to the variable *end*. For example, if $start = x$ and $k$ cells away is located a cell at L or G state, then $end = x + k - 1$. By using these two variables the area between two waves can be defined. The length of this area can change through the time.

For example, the variable $start$ for the fastest wave of the initial general is decreased by one at each time-step, i.e. the wave is moving to the left. At the same time, *end* maintains its value at some time-steps, as Fig. 1 shows.

By using these two variables and by the previous states that are stored, the new states can be verified. So, if $cell[x]$ is the current state of the cell at position $x$ and $cell'[x]$ is the temporary stored next state of the cell at position $x$, the following conditions can be applied.

For $cell[start - 1]$:

- If $cell[start]$=A then $cell'[start - 1]$=B
- If $cell[start]$=B then $cell'[start - 1]$=C
- If $cell[start]$=C then $cell'[start - 1]$=A

For $cell[start]$:

- $cell'[start] = cell[start]$

For the other cells between the two waves:

- $cell'[x] = cell[x - 1]$

## 4.3 Correction During the Wave's Impact

During the impact of a wave at either the grid's bounds or to a General, the wave's behavior is different based on the delay that is added. This impact affects the two neighbors next to the cell at position $start$, namely the $cell[start - 1]$ and $cell[start - 2]$, respectively. To be more specific, the impact is affecting the first three cells of the wave.

If an impact is occurred, the next states are defined as follows:

- If $cell[start]$=A then $cell'[start - 1]$=C and $delay = 0$. The impact is complete at the next time-step and no other additional steps are needed.
- If $cell[start]$=B then $cell'[start - 1]$=A and $delay = 1$, then an additional step is added. In this time step, the variable $start$ is recalculated and the $cell[start]$ is at state A and $cell'[start]$ should be at C state.
- If $cell[start]$=C then $cell'[start - 1]$=B and $delay = 2$, then two additional time steps are added. At the first step, $cell[start]$=B and $cell'[start]$=A. At the second step, $cell[start]$=A and $cell'[start]$=C. The variable $start$ is recalculated only at the first additional time step.

The states of the other cells between the two waves are calculated as previously.

### 4.4 Correction During the Wave's Reflection

The wave's reflection can be found at only one cell. The state of this cell is the R state. If the wave is reflected, then the value of the variable $start$ is the position of the cell that is at R state. Also, if $cell[start + 1]$=C then $cell'[start + 1]$=R and $cell'[start + 2]$=C.

The states of the other cells between the two waves are calculated as previously, until the fast wave reaches the slow wave. At this moment, a new general is created. So, if the cycle pattern of A, B, C, A is disordered and a cell at R state is appeared, then a General must be appeared in this cell at the next time-step.

## 5 Conclusions

In this chapter, several novel error correction techniques of both one-dimensional and two-dimensional FSSP solution algorithms are presented. All the proposed techniques are taking advantage of the patterns that can be found in the 8-states Mazoyer algorithm. By the combination of the proposed techniques, an error correction algorithm with great capabilities has been developed.

The algorithm is quite efficient at the one-dimensional FSSP, without using previous states; however, it can't correct every possible error that could be appeared. The states that can be observed at a time-step are not enough to extract the necessary information for the correction of every cell. This happens because most of the states can be found in many different combinations. This is normal, since the efficiency and speed of the algorithm results from complex combinations of only a few states.

If previous states are used, it has been proven that the algorithm is capable of checking and correcting the states of every cell. Future research should be focused to the improvement of the proposed algorithm by improving its efficiency and execution speed.

In the two-dimensional FSSP, the best method to detect and correct errors is by checking the states of the cells that are located at the same group, without using previous states. This method is making the error detection and correction algorithm drastically more efficient than in one-dimensional FSSP, since it can correct all cells except one, namely the last soldier.

## References

1. Adamatzky, A.: Cellular Automata: A Volume in the Encyclopedia of Complexity and Systems Science, Second Edition. Springer (2018)
2. Alonso-Sanz R (2003) Reversible cellular automata with memory: two-dimensional patterns from a single site seed. Phys D: Nonlinear Phenom 175(1):1–30

3. Balzer R (1967) An 8-state minimal time solution to the firing squad synchronization problem. Inf Control 10:22–42
4. Beyer WT (1969) Recognition of topological invariants by iterative arrays. PhD thesis, Massachusetts Institute of Technology
5. De Rango A, Furnari L, Giordano A, Senatore A, D'Ambrosio D, Spataro W, Straface S, Mendicino G (2020) OpenCAL system extension and application to the three-dimensional richards equation for unsaturated flow. Comput Math Appl
6. Dourvas N, Tsompanas MA, Sirakoulis GC, Tsalides P (2015) Hardware acceleration of cellular automata physarum polycephalum model. Parallel Proc Lett 25(01):1540006
7. Gerken HD (1987) Über synchronisations - probleme bei zellularautomaten. Diplomarbeit, Institut für Theoretische Informatik, Technische Universität Braunschweig
8. Goles E, Martínez S (1990) Neural and automata networks: dynamical behavior and applications. Kluwer Academic Publishers, Norwell
9. Goto E (1962) A minimal time solution of the firing squad problem. Dittoed course notes for Applied Mathematics 298. Harvard University, Cambridge, pp 52–59
10. Grefenstette JJ (1983) Network structure and the firing squad synchronization problem. J Comput Syst Sci 26(1):139–152
11. Karamani RE, Fyrigos IA, Tsakalos KA, Ntinas V, Tsompanas MA, Sirakoulis GC (2021) Memristive learning cellular automata for edge detection. Chaos Solitons & Fractals 145:110700
12. Liolis O, Mardiris VA, Sirakoulis GC, Karafyllidis IG (2021) Synchronization in quantum-dot cellular automata circuits and systems. IEEE Open J Nanotechnol 1:145–156
13. Lubas R, Was J, Porzycki J (2016) Cellular automata as the basis of effective and realistic agent-based models of crowd behavior. J Supercomput 72(6):2170–2196
14. Maeda M, Fujiwara N (2002) An efficient mapping scheme for embedding any one-dimensional firing squad synchronization algorithm onto two-dimensional arrays. In: International conference on cellular automata. Springer, pp 69–81
15. Mardiris VA, Liolis O, Sirakoulis GC, Karafyllidis IG (2018) Signal synchronization in large scale quantum-dot cellular automata circuits. In: Proceedings of the 14th IEEE/ACM international symposium on nanoscale architectures, NANOARCH '18. ACM, pp 153–156. https://doi.org/10.1145/3232195.3232212
16. Mazoyer J (1987) A six-state minimal time solution to the firing squad synchronization problem. Theoret Comput Sci 50:183–238
17. Moore E (1964) The firing squad synchronization problem. In: Moore E (ed) Sequential machines. Addison-Wesley, Reading, pp 213–214
18. Moore F, Langdon G (1968) A generalized firing squad problem. Inf Control 12(3):212–220. https://doi.org/10.1016/S0019-9958(68)90309-4
19. Nalpantidis L, Sirakoulis GC, Gasteratos A (2011) Non-probabilistic cellular automata-enhanced stereo vision simultaneous localization and mapping. Meas Sci Technol 22(11):114027
20. von Neumann J (1966) Theory of self-reproducing automata. University of Illinois Press, Champaign
21. Ntinas VG, Moutafis BE, Trunfio GA, Sirakoulis GC (2017) Parallel fuzzy cellular automata for data-driven simulation of wildfire spreading. J Comput Sci 21:469–485
22. Shinahr I (1974) Two-and three-dimensional firing-squad synchronization problems. Inf Control 24(2):163–180
23. Sirakoulis G, Karafyllidis I, Thanailakis A (2000) A cellular automaton model for the effects of population movement and vaccination on epidemic propagation. Ecol Modell 133(3):209–223
24. Sirakoulis G, Karafyllidis I, Thanailakis A (2005) A cellular automaton for the propagation of circular fronts and its applications. Eng Appl Art Intell 18(6):731–744
25. Sirakoulis GC, Adamatzky A (2015) Robots and lattice automata. Springer, Berlin
26. Tsiftsis A, Georgoudas IG, Sirakoulis GC (2016) Real data evaluation of a crowd supervising system for stadium evacuation and its hardware implementation. IEEE Syst J 10(2):649–660. https://doi.org/10.1109/JSYST.2014.2370455

27. Tsompanas MA, Adamatzky A, Ieropoulos I, Phillips N, Sirakoulis GC, Greenman J (2017) Cellular non-linear network model of microbial fuel cell. Biosystems 156–157:53–62
28. Tsompanas MI, Sirakoulis GC, Adamatzky A (2016) Physarum in silicon: the greek motorways study. Nat Comput 15(2):279–295
29. Umeo H (2020) How to synchronize cellular automata - recent developments -. Fund Inf 171(1–4):393–419
30. Umeo H, Kubo K (2010) A seven-state time-optimum square synchronizer. In: International conference on cellular automata. Springer, pp 219–230
31. Umeo H, Kubo K, Nomura A (2017) Smaller-state implementations of 2d fssp algorithms. In: Hung DV, Kapur D (eds) Theoretical aspects of computing - ICTAC 2017. Springer International Publishing, Cham, pp 136–152
32. Umeo H, Maeda M, Fujiwara N (2002) An efficient mapping scheme for embedding any one-dimensional firing squad synchronization algorithm onto two-dimensional arrays. In: Bandini S, Chopard B, Tomassini M (eds) Cell Automata. Springer, Berlin, pp 69–81
33. Umeo H, Nomura A (2012) A state-efficient zebra-like implementation of synchronization algorithms for 2d rectangular cellular arrays. Biomath 1(1):1209022
34. Vihas C, Georgoudas IG, Sirakoulis GC (2013) Cellular automata incorporating follow-the-leader principles to model crowd dynamics. J Cell Automata 8(5/6):333–346
35. Waksman A (1966) An optimum solution to the firing squad synchronizaton problem. Inf Control 9:66–78
36. Wolfram S (1983) Statistical mechanics of cellular automata. Rev Modern Phys 55(3):601
37. Wolfram S (1984) Cellular automata as models of complexity. Nature 311(5985):419–424
38. Wolfram S (2002) A new kind of science. Wolfram Media Inc., Champaign
39. Wolfram S (2018) Cellular automata and complexity: collected papers. CRC Press, Boca Raton

# Gardens of Eden in the Game of Life

**Ville Salo and Ilkka Törmä**

**Abstract** We prove that in the Game of Life, if the thickness-four zero-padding of a rectangular pattern is not an orphan, then the corresponding finite-support configuration is not a Garden of Eden, and that the preimage of every finite-support configuration has dense semilinear configurations. In particular finite-support Gardens of Eden are in co-NP.

## 1 Introduction

This article is dedicated to Eric Goles Chacc, and his inspiring work on all kinds of discrete complex systems, on his 70th birthday.

The Game of Life is a two-dimensional cellular automaton defined by John Conway in 1970 [6]. It consists of an infinite grid of cells, each of which is either dead or alive, and a discrete time dynamics defined by a simple local rule: a live cell survives if it has 2 or 3 live neighbors, and a dead cell becomes live if it has exactly 3 live neighbors. We study the *Gardens of Eden* of the Game of Life, that is, configurations that do not have a predecessor in the dynamics. The following theorem is our main result.

**Theorem 1** *The set of finite-support Gardens of Eden for the Game of Life is in co-NP under the encoding where the input is a rectangular pattern containing all live cells.*

Equivalently, the input can specify the values in any area of polynomial diameter, since we can always extend the domain to a rectangular one. A configuration has *finite support* if it contains a finite number of live cells. Finite-support Gardens of Eden should be contrasted with the *orphans*, which are patterns of finite domain that

V. Salo (✉) · I. Törmä
Department of Mathematics and Statistics, University of Turku, Turku, Finland
e-mail: vosalo@utu.fi

I. Törmä
e-mail: iatorm@utu.fi

 
A. Adamatzky (ed.), *Automata and Complexity*, Emergence, Complexity and Computation 42, https://doi.org/10.1007/978-3-030-92551-2_22

do not appear in the image subshift of the Game of Life. The set of orphans is in co-NP for trivial reasons.

This result is a corollary of either of the following theorems. In the first case, we also need some quantitative details of the proof.

**Theorem 2** *Let $g$ be the Game of Life cellular automaton. If $y$ is a configuration with finite support, then semilinear configurations are dense in $g^{-1}(y)$.*

**Theorem 3** *Let $g$ be the Game of Life cellular automaton, and let $P$ be a rectangular pattern. If the thickness-*4 *zero-padding of $P$ admits a preimage $Q$, then the finite-support configuration $y$ corresponding to $P$ has a preimage.*

Theorem 2 is not true if "semilinear" is replaced with "finite-support" or "co-finite-support". In Theorem 3 we do not know whether the optimal constant is 4, but it is at least 1. We state a more precise result in Theorem 5, which additionally guarantees that we need not modify the preimage of the padding of $P$ in the part that maps to $P$, to obtain a preimage for the finite-support configuration $y$. For this stronger fact, the thickness 4 is optimal, and no convex compact shape other than a rectangle can be used.

The seed of the proofs of the above theorems is the following property of the Game of Life: the set of rectangular patterns of height 2 that can be extended to a preimage of the all-0 configuration (the *trace* of the subshift of finite type $g^{-1}(0^{\mathbb{Z}^2})$) is a regular language, and all such patterns can be extended so that the preimage is vertically 3-periodic everywhere except within distance 3 of the pattern. The essential idea in the proofs of both of the theorems above is to apply this result on all sides of a rectangle containing the desired image, to change the preimage to a "better" one. Once the regularity of the trace language has been established, the periodic extension property is decidable using automata theory, and we decide both by computer. The proof did not seem to be within reach of any standard library we attempted to use, so we used pure Python instead. Our program is available in the repository [20]. We expect that some efficient enough standard libraries could solve this problem out of the box.

The technical lemmas of this paper are stated for general two-dimensional cellular automata, and the computed-assisted part can be adapted to a general cellular automaton with little work by modifying the attached program. One obtains analogues of the above theorems for any cellular automaton where the preimage of the all-zero configuration has one-sided stable/periodizable traces. Our methods do not generalize to higher-dimensional cellular automata.

This paper arose from the answer of Oscar Cunningham to the question [21] on MathOverflow, which in turn arose from the thread [15] on the ConwayLife forum. Cunningham asked whether the finite-support Gardens of Eden of the Game of Life cellular automaton form a decidable set. It follows from compactness that they are a semidecidable set for any cellular automaton (as every Garden of Eden contains an orphan), but there is no general-purpose semidecision algorithm for the other direction, in the sense that there exist two-dimensional cellular automata whose finite-support Gardens of Eden are undecidable. Our results of course solve the problem for the Game of Life.

**Corollary 1** *Finite-support Gardens of Eden for the Game of Life are a decidable set, under any natural encoding.*

The methods of this article are special cases of a more general technique we are studying with Pierre Guillon. Trace methods are a common tool in multidimensional symbolic dynamics and cellular automata theory [5, 16, 17]. See [1, 7, 19] for discussion of various aspects and generalizations of the Game of Life.

## 2 Definitions

We include $0 \in \mathbb{N}$, and unless otherwise noted, all intervals are discrete: $[a, b] = [a, b] \cap \mathbb{Z}$.

For $A$ a finite set with a special element $0 \in A$ called the *zero*, a $d$-dimensional *configuration* over $A$ is an element $x \in A^{\mathbb{Z}^d}$. The elements of $\mathbb{Z}^d$ are called *cells*, and the value of a cell $\mathbf{v}$ in $x$ is denoted $x_\mathbf{v}$. The set $A^{\mathbb{Z}^d}$ of all configurations is the *$d$-dimensional full shift*, and we give it the prodiscrete (Cantor) topology. The additive group $\mathbb{Z}^d$ acts on $A^{\mathbb{Z}^d}$ by shifts: $\sigma^\mathbf{v}(x)_\mathbf{w} = x_{\mathbf{v}+\mathbf{w}}$. A configuration $x$ is *finite-support* if $x_\mathbf{v} = 0$ for all but a finite number of cells $\mathbf{v} \in \mathbb{Z}^d$, and we denote by $\mathrm{Fin}(A)$ the set of finite-support configurations. If $X \subset A^\mathbb{Z}$ is a one-dimensional subshift, write $\mathcal{L}(X) = \{w \in A^* \mid \exists x \in X : x|_{[0,|w|-1]} = w\}$ for its *language*, the set of (finite-length) words that can be extended to infinite configurations in $X$.

A *pattern* is an element $P \in A^D$, where $D = D(P) \subset \mathbb{Z}^d$ is the *domain* of $P$, and we say $P$ is a *(finite) pattern* if it has a finite domain. If $P \in A^D$ then $\sigma^\mathbf{v}(P) \in A^{D-\mathbf{v}}$ is defined by $\sigma^\mathbf{v}(P)_\mathbf{u} = P_{\mathbf{u}+\mathbf{v}}$. If $P \in A^D$ is a pattern, write $\mathrm{pad}_0^c(P) \in A^{D+[-c,c]^d}$ for the pattern together with a zero-padding of thickness $c$ on all sides, and $\mathrm{conf}_0(P)$ is the finite-support configuration corresponding to $P$ defined by $x_\mathbf{v} = P_\mathbf{v}$ for $\mathbf{v} \in D$, and $x_\mathbf{v} = 0$ otherwise. If $P \in A^D$ and $Q \in A^E$ are two patterns, write $P \sqsubset Q$ if there exists $\mathbf{v} \in \mathbb{Z}^d$ with $D + \mathbf{v} \subset E$ and $\sigma^\mathbf{v}(Q)|_D = P$. In particular, if $E = \mathbb{Z}^d$ then $Q = x$ is a configuration, and $P \sqsubset x$ means that $P$ occurs somewhere in $x$.

A *subshift* is a topologically closed and shift-invariant set $X \subset A^{\mathbb{Z}^d}$. It follows from compactness that every subshift is defined by some set of *forbidden patterns* $F$ as

$$X = X_F = \{x \in A^{\mathbb{Z}^d} \mid \forall P \in F : P \not\sqsubset x\},$$

If $F$ can be chosen finite, then $X$ is a *shift of finite type*, or *SFT* for short.

A *cellular automaton* or *CA* is a function $f : A^{\mathbb{Z}^d} \to A^{\mathbb{Z}^d}$ that commutes with shifts and is continuous for the prodiscrete topology of $A^{\mathbb{Z}^d}$. By the Curtis-Hedlund-Lyndon theorem [11], every CA is defined by a finite *neighborhood* $N \subset \mathbb{Z}^d$ and a *local rule* $F : A^N \to A$ by $f(x)_\mathbf{v} = F(\sigma^\mathbf{v}(x)|_N)$. Usually we have $N = [-r, r]^d$ for some *radius* $r$ and think of the local rule $F : A^N \to A$ as part of the structure of the CA. If $P \in A^D$ is a finite pattern, we can apply $f$ to $P$ by defining $E = \{\mathbf{v} \in D \mid \mathbf{v} + N \subset D\}$ and $f(P) = Q \in A^E$ where $Q_\mathbf{v} = F(\sigma^\mathbf{v}(Q)|_N)$. A *Garden of Eden* for $f$ is a configuration $x \in A^{\mathbb{Z}^d}$ such that $f^{-1}(x) = \emptyset$; equivalently, $x \in A^{\mathbb{Z}^d} \setminus f(A^{\mathbb{Z}^d})$. The *finite-support Gardens of Eden* of $f$ are

$$\mathrm{FinGoE}(f) = \mathrm{Fin}(A) \setminus f(A^{\mathbb{Z}^d}).$$

An *orphan* is a finite pattern $P$ such that $P \not\sqsubset f(x)$ for all $x \in A^{\mathbb{Z}^d}$.

The images of SFTs under cellular automata are *sofic shifts*. A one-dimensional subshift is sofic iff its language is regular, iff it can be defined by a regular language of forbidden words.

The *Game of Life* is the two-dimensional cellular automaton $g : A^{\mathbb{Z}^2} \to A^{\mathbb{Z}^2}$ over $A = \{0, 1\}$ defined by

$$g(x)_{(a,b)} = \begin{cases} 1 \text{ if } x_{(a,b)} = 0 \text{ and } \sum x|_{(a,b)+B} = 3 \\ 1 \text{ if } x_{(a,b)} = 1 \text{ and } \sum x|_{(a,b)+B} \in \{3, 4\} \\ 0 \text{ otherwise,} \end{cases}$$

where $B = \{-1, 0, 1\}^2$. It has radius 1.

A configuration $x \in A^{\mathbb{Z}^d}$ is *semilinear* if for every symbol $a \in A$, the set $\{\mathbf{v} \in \mathbb{Z}^d \mid x_\mathbf{v} = a\}$ is *semilinear*, meaning it is a finite union of *linear* sets. A *linear* set is a set of the form $\mathbf{v} + \langle \mathbf{v}_1, \mathbf{v}_2, \ldots, \mathbf{v}_k \rangle$, where $\mathbf{v}, \mathbf{v}_i \in \mathbb{Z}^d$, and $\langle V \rangle$ denotes the monoid generated by $V$. It is well-known that a cellular automaton image of a semilinear configuration is effectively semilinear. This is a very robust class of sets, see [9] for characterizations.

For the purpose of complexity theory and computability theory, we fix a bijection between $B^*$ and $\mathrm{Fin}(A)$ for a fixed alphabet $B$, so that $\mathrm{FinGoE}(f)$ can be seen as a language. For $A = \{0, 1, \ldots, |A| - 1\}$, $|A| \geq 2$ and $d = 2$, we fix $B = A$ and use the encoding where $w = 0^M 1^N 0u$ with $|u| = (2M + 1)(2N + 1)$ represents the finite-support configuration $x \in A^{\mathbb{Z}^2}$ where $x_{(a,b)} = 0$ if $(a, b) \notin [-M, M] \times [-N, N]$ and $x_{(a,b)} = u_{(2M+1)a+b}$ otherwise. A similar encoding is used for finite patterns with domains of the form $[-M, M] \times [-N, N]$.

For the remainder of this article, we fix $d = 2$. Up and north are synonyms (and refer to the vector $(0, 1)$), and similarly left, right and down are synonymous with west, east and south, respectively.

We need some basic facts and definitions of symbolic dynamics, automata theory and complexity theory, some standard references are [2, 12, 13, 18]. In particular, all Boolean operations on regular languages, the concatenation operation $KL = \{uv \mid u \in K, v \in L\}$, and the equality of two given regular languages, are computable.

## 3 Traces

A pattern $P \in A^D$ is *height-n* if $D \subset \mathbb{Z} \times [0, n-1]$. We define *width-n* similarly, and $P$ is *size-n* if it is width-$n$ and height-$n$. For a set of finite patterns $F$, define $X_F^{\mathbb{N}} = \{x \in A^{\mathbb{Z}\times\mathbb{N}} \mid \forall P \in F : P \not\sqsubset x\}$ and $X_F^n = \{x \in A^{\mathbb{Z}\times[0,n-1]} \mid \forall P \in F : P \not\sqsubset x\}$. These are the upper half-planes and horizontal stripes of height $n$ where patterns of $F$ do not occur.

**Definition 1** Let $F$ be a set of forbidden patterns. Write

$$T_n(F) = \{x|_{\mathbb{Z}\times[0,n-1]} \mid x \in X_F^{\mathbb{N}}\} \subset A^{\mathbb{Z}\times[0,n-1]}$$

for the *one-sided trace* of $X_F$. Define $\pi_n : \bigcup_{[0,n-1]\subset I\subset\mathbb{Z}} A^{\mathbb{Z}\times I} \to A^{\mathbb{Z}\times[0,n-1]}$ by $\pi_n(x) = x|_{\mathbb{Z}\times[0,n-1]}$. Define

$$S_{n,\ell}(F) = \pi_n(X_F^{n+\ell}).$$

For $k \geq 0$ and $p \geq 1$ define

$$P_{n,k,p}(F) = \{x|_{\mathbb{Z}\times[0,n-1]} \mid x \in X_F^{\mathbb{N}}, \forall a \in \mathbb{Z}, b \geq n+k : x_{(a,b)} = x_{(a,b+p)}\}.$$

The trace $T_n(F)$ is the set of those height-$n$ stripes that can be extended to the entire upper half-plane without introducing a pattern of $F$, while the stripes of $S_{n,\ell}(F)$ can be extended by $\ell$ additional rows, and the stripes of $P_{n,k,p}(F)$ can be extended into upper half-planes that are vertically $p$-periodic after $k$ rows. The inclusions $P_{n,k,p}(F) \subset T_n(F) \subset S_{n,\ell}(F)$ always hold, and $T_n(F) = \bigcap_{\ell\in\mathbb{N}} S_{n,\ell}(F)$. We often identify $A^{\mathbb{Z}\times[0,n-1]}$ with $(A^{[0,n-1]})^{\mathbb{Z}}$, so that the traces can be seen as infinite sequences of finite patterns of shape $1 \times n$. We could also define two-sided traces in the analogous way, but we have omitted them from this article since they are not needed for our results, computing them is more resource-intensive, and the relationships between the relevant properties of one-sided and two-sided traces is not entirely trivial.

### 3.1 Periodizable Traces

Write $\circlearrowright$ for the operation that rotates a pattern, configuration, or every element of a set of such, 90 degrees clockwise around the origin.

**Definition 2** Let $F$ be a set of forbidden patterns of size $n+1$. We say $F$ has *one-sided periodizable traces* if there exist $k, p \in \mathbb{N}$ with $p \geq 1$ such that

$$T_n(\circlearrowright^i(F)) \subset P_{n,k,p}(\circlearrowright^i(F)) \tag{1}$$

holds for all $i \in \{0, 1, 2, 3\}$.

For $i = 0$ this property means that if an arbitrary row of height $n$ (too thin to contain forbidden patterns) can be extended upward into a half-plane that contains no forbidden patterns, then it can also be extended by $k + p$ rows so that the last $p$ rows can be repeated periodically. For other $i$, we can interpret this as a similar property for extensions to the left, downward and right. If $X \subset A^{\mathbb{Z}^2}$ is an SFT where a set of forbidden patterns $F$ of height at most $n + 1$ has been fixed, we sometimes say $X$ has one-sided periodizable traces if $F$ does.

**Lemma 1** *Let $f : A^{\mathbb{Z}^2} \to A^{\mathbb{Z}^2}$ be a cellular automaton with neighborhood $[-r, r]^2$. Let $F$ be the patterns $P \in A^{[-r,r]^2}$ that map to a nonzero symbol in the local rule of $f$. If $F$ has one-sided periodizable traces, then semilinear configurations are dense in $f^{-1}(y)$ for every $y \in \mathrm{Fin}(A)$, and $\mathrm{FinGoE}(f)$ is in co-NP.*

***Proof*** Denote $n = 2r$, and let $k, p \in \mathbb{N}$ be given by the assumption of one-sided periodizable traces. We may assume $p \geq n$. Let $y \in \mathrm{Fin}(A)$ be arbitrary. We prove that semilinear configurations are dense in $f^{-1}(y)$. If $y$ has no preimage, we are done; otherwise, take an arbitrary preimage $f(x) = y$, and let $N > p$ be such that the support of $y$ is contained in $D = [-N, N]^2$. We construct a semilinear preimage of $y$ that agrees with $x$ inside $R = [-N - r, N + r]^2$.

Consider the upper half-plane containing the top $r$ rows of $D$ and everything above them. We have $\sigma^{(0,N-r+1)}(x)|_{\mathbb{Z}\times[0,\infty)} = z \in X_F^{\mathbb{N}}$. Then the restriction to the bottom $2r$ rows of this half-plane satisfies $\sigma^{(0,N-r+1)}(x)|_{\mathbb{Z}\times[0,2r-1]} = \pi_{2r}(z) \in T_{2r}(F) = P_{2r,k,p}(F)$, where the last equality is given by (1) with $i = 0$. This means we can replace the contents of an upper half-plane in $x$ with a vertically periodic pattern, i.e. there exists $x^1 \in f^{-1}(y)$ such that $x^1|_{\mathbb{Z}\times(-\infty,N+r]} = x|_{\mathbb{Z}\times(-\infty,N+r]}$ and $x^1_{(a,b)} = x^1_{(a,b+p)}$ for all $a \in \mathbb{Z}$ and $b > N + r + k$. We can apply the exact same argument to the configuration $x^1$ below the rectangle $[-N, N]^2$, using (1) for $i = 2$, giving us $x^2 \in f^{-1}(y)$ such that $x^2|_{\mathbb{Z}\times[-N-r,N+r]} = x|_{\mathbb{Z}\times[-N-r,N+r]}$ and $x^2_{(a,b)} = x^2_{(a,b+p)}$ whenever $a \in \mathbb{Z}$ and $\min(|b|, |b + p|) > N + r + k$.

Next, we apply the same argument on the left and right borders of $D$ in $x^2$, using (1) with $i = 1, 3$ in either order. This gives us a preimage $x^3 \in f^{-1}(y)$ such that $x^3|_{[-N-r,N+r]^2} = x|_{[-N-r,N+r]^2}$ and, defining $A_{\mathrm{north}} = [-N - r, N + r] \times [N + r + k + 1, \infty)$, $A_{\mathrm{south}} = [-N - r, N + r] \times (-\infty, -N - r - k - 1]$, $A_{\mathrm{west}} = (-\infty, -N - r - k - 1] \times \mathbb{Z}$, $A_{\mathrm{east}} = [N + r + k + 1, \infty) \times \mathbb{Z}$, we have $x^3_{(a,b)} = x^3_{(a,b+p)}$ whenever $(a, b), (a, b + p) \in A_{\mathrm{north}} \cup A_{\mathrm{south}}$, and $x^3_{(a,b)} = x^3_{(a+p,b)}$ whenever $(a, b), (a + p, b) \in A_{\mathrm{west}} \cup A_{\mathrm{east}}$.

The "aperiodic region" $B = \mathbb{Z}^2 \setminus (A_{\mathrm{north}} \cup A_{\mathrm{south}} \cup A_{\mathrm{west}} \cup A_{\mathrm{east}})$ that is left after this process is infinite, so we need a final periodization step to obtain semilinearity. Since $A_{\mathrm{north}}$ is $(0, p)$-periodic, and rows of $A_{\mathrm{west}}$ and $A_{\mathrm{east}}$ are $(p, 0)$-periodic, the number of distinct height-$n$ stripes $\sigma^{(0,h)}(x)|_{\mathbb{Z}\times[0,n-1]}$ for $h > N + r + k$ is bounded by $q = p|A|^{2n(k+p)}$, where the factor $p$ comes from the phase of the period in $A_{\mathrm{north}}$ and the term $|A|^{2n(k+p)}$ comes from the two patterns of shape $n \times (k + p)$ just to the west and east sides of $A_{\mathrm{north}}$, which include the width-$k$ aperiodic regions and the repeating parts of $A_{\mathrm{west}}$ and $A_{\mathrm{east}}$.

Thus the stripes defined by some $N + r + k < h_1 < h_2 \leq N + r + k + q + 1$ are equal, and since we assumed $p \geq 2r$, we can form a new configuration $x^4 \in f^{-1}(y)$

by repeating the part between there stripes in the upper half-plane $[h_1, \infty)$. Then $x^4$ agrees with $x^3$ on $\mathbb{Z} \times (-\infty, N + r]$ and is vertically $p'$-periodic in $[h_1, \infty)$ for some $p' \leq q$. We apply the same argument to the south half-plane of $x^4$, obtaining a configuration $x^5 \in f^{-1}(y)$ with $x|_{[-N-r,N+r]^2} = x^5|_{[-N-r,N+r]^2}$ that has horizontal period $p$ outside $[-N - r - k, N + r + k] \times \mathbb{Z}$ and vertical period $q!$ outside $\mathbb{Z} \times [-q, q]$. In particular, $x^5$ is semilinear.

The co-NP claim follows from the quantitative statements above about the semilinear preimages. A semilinear preimage with the periodicity properties of $x^5$ can be summarized as a polynomial-size certificate, by giving the restriction

$$x^5|_{[-N-r-k-p,N+r+k+p]\times[-N-q-q!,N+q+q!]}.$$

One can check in polynomial time that continuing the periods does not give a forbidden pattern for $X$. Note that while $q!$ grows very fast as a function of $|A|$, $p$ and $k$, it is a constant for any fixed CA that satisfies the assumptions. This proves that the complement of FinGoE$(f)$ is in NP, as claimed. □

We include the statement that semilinear preimages imply co-NP, because this gives a very natural certificate – an actual preimage. However, the details of working computationally with semilinear configurations, while standard, are not entirely trivial, and are omitted in the above proof. For the co-NP certificate obtained from Lemma 4 the verification algorithm is much more obvious.

## 3.2 *Stable Traces*

There is no general method of computing the one-sided traces $T_n(F)$ exactly. On the other hand, the languages $\mathcal{L}(S_{n,\ell}(F))$ of the approximate traces are regular and easy (though often resource-intensive) to compute. If the sequence $(S_{n,\ell}(F))_{\ell \in \mathbb{N}}$ stabilizes after finitely many steps, then $\mathcal{L}(T_n(F))$ is also a regular language and we can analyze it using finite automata theory.

**Definition 3** If $S_{n,\ell}(\circlearrowright^i(F))) = S_{n,\ell+1}(\circlearrowright^i(F))$ for some $\ell \in \mathbb{N}$ and all $i \in \{0, 1, 2, 3\}$ and $F$ has size at most $n + 1$, then we say $F$ has *one-sided stable traces*.

The following is shown by induction, by extending a configuration legally, row by row, obtaining a valid half-plane in the limit.

**Lemma 2** *If $F$ has height at most $n + 1$ and $S_{n,\ell}(F)) = S_{n,\ell+1}(F)$, then $S_{n,\ell}(F) = S_{n,\ell+m}(F) = T_n(F)$ for all $m \geq 0$.*

We recall a basic symbolic dynamics lemma, a version of the pumping lemma.

**Lemma 3** *Let $L \subset A^*$ be a regular factor-closed language and $X \subset A^{\mathbb{Z}}$ the largest subshift with $\mathcal{L}(X) \subset L$. Then there exists $C \geq 0$ such that $uwv \in L$ and $|u|, |v| \geq C$ implies $w \in \mathcal{L}(X)$.*

***Proof*** It is easy to see that $X = \{x \in A^{\mathbb{Z}} \mid \forall k : x|_{[-k,k]} \in L\}$. Take a nondeterministic finite-state automaton (NFA) for $L$ with $n$ states all of whose states are initial and final, and pick $C = n + 1$. Suppose $uwv \in L$ and $|u|, |v| \geq C$, and pick any accepting path $P$ for $uwv$ in the automaton. By the pigeonhole principle, some state repeats in the length-$C$ prefix of $P$ corresponding to $u$, and the same is true for the suffix corresponding to $v$, so we obtain decompositions $u = u_1u_2u_3$, $v = v_1v_2v_3$ with $|u_2|, |v_2| > 0$ such that $u_2^k u_3 w v_1 v_2^k \in L$ for all $k \in \mathbb{N}$. Then ${}^{\infty}u_2u_3wv_1v_2^{\infty} \in X$, so that $w \in \mathcal{L}(X)$. □

By the proof, we can always pick $C = n + 1$ where $n$ is the number of states in any NFA accepting $L$ with all states initial and final. Often we can do better.

**Lemma 4** *Let $f : A^{\mathbb{Z}^2} \to A^{\mathbb{Z}^2}$ be a cellular automaton with neighborhood $[-r, r]^2$. Let $F$ be the patterns $P \in A^{[-r,r]^2}$ that map to a nonzero symbol in the local rule of $f$. If $F$ has one-sided stable traces, then there exists $c \in \mathbb{N}$ such that whenever $P \in A^{[0,M)\times[0,N)}$ and $\mathrm{pad}_0^c(P)$ admits a preimage $Q$, then $\mathrm{conf}_0(P)$ admits a preimage $x \in A^{\mathbb{Z}^2}$ with*

$$x|_{[-r,M+r)\times[-r,N+r)} = Q|_{[-r,M+r)\times[-r,N+r)}.$$

We use the same padding $c$ on all sides for notational convenience only. For the Game of Life this does not change anything, but for less symmetric CA and CA where the constant $C$ in the proof is not equal to 0, one may optimize this by using different paddings on all sides.

***Proof*** Denote $n = 2r$ and let $\ell \in \mathbb{N}$ be such that $S_{n,\ell}(F) = S_{n,\ell+1}(F)$, so that $S_{n,\ell}(F) = T_n(F)$ by Lemma 2. Consider the set

$$L_{n,\ell}(F) = \{Q|_{[0,k)\times[0,n)} \mid Q \in A^{[0,k)\times[0,n+\ell)}, \forall P \in F : P \not\sqsubset Q\}$$

of height-$n$ rectangular patterns that can be extended upward by $\ell$ steps into a pattern not containing any pattern from $F$ as a subpattern. We consider it as a language over the alphabet $A^n$. Since it is regular and factor-closed, and the largest subshift whose language is contained in $L_{n,\ell}(F)$ is clearly $S_{n,\ell}(F)$, by the previous lemma there exists $C \in \mathbb{N}$ such that if a word of $L_{n,\ell}(F)$ is extendable by $C$ steps in both directions, it occurs in $S_{n,\ell}(F) = T_n(F)$.

Let $a = C + \ell$ and $b = \ell$, and suppose a pattern $P \in A^{[-a,M+a)\times[-b,N+b)}$ has its support contained in $D = [0, M) \times [0, N)$ and admits a preimage. Let $x$ be a configuration with $f(x)|_{[-a,M+a)\times[-b,N+b)} = P$. Then the $r$ top rows of $D$ and the $r$ rows above them satisfy $\sigma^{(0,N-r-1)}(x)|_{[-a-r,M+a+r)\times[0,2r)} \in L_{n,\ell}(F)$, therefore $\sigma^{(0,N-r-1)}(x)|_{[-r-\ell,M+r+\ell)\times[0,2r)} \in \mathcal{L}(S_{n,\ell}(F)) = \mathcal{L}(T_n(F))$ by the assumption on $C$. Hence we can extend this pattern into a stripe of $T_n(F)$ and then extend the upper half-plane into one that maps to 0s, obtaining a configuration $x^1 \in A^{\mathbb{Z}^2}$ with $x^1|_{[-a,M+a)} = x|_{[-b,N+r)}$ and $x^1|_{\mathbb{Z}\times(N-r,\infty)}$ not containing occurrences of any pattern in $F$.

Perform the same operation symmetrically on the south border of $D$ in $x^1$ to obtain a configuration $x^2$. Then the support of $f(x^2)$ is contained in $([0, M) \times [0, N)) \cup$

$(R \times [0, N))$ where $R = (-\infty, -\ell] \cup [M + \ell, \infty)$. We can now apply the exact same argument on the west and east borders of $D$ in $x^2$ (in either order) to obtain a configuration $x^3$ such that the support of $f(x^3)$ is contained in $[0, M) \times [0, N)$ and $f(x^3)|_{[0,M)\times[0,N)} = f(x)|_{[0,M)\times[0,N)}$. Picking $c = \max(a, b) = C + \ell$, this argument shows the first claim.

For the claim that FinGoE($f$) many-one reduces to orphans in polynomial time, given an element $y$ of Fin($A$) with support contained in $[0, M) \times [0, N)$, we simply note that the pattern $y|_{[-c,M+c)\times[-c,N+c)}$ is an orphan if and only if $y \in$ FinGoE($f$). □

Since orphans (of any finite shape) are in co-NP for any cellular automaton, the previous lemma gives another verification algorithm for showing that FinGoE($f$) is co-NP. It is easier to implement as that given by Lemma 1 when the conditions of both results apply. However, the certificate is less useful: unlike in Lemma 1, the certificate does not describe an actual preimage of the finite-support configuration. Together, Lemmas 1 and 4 imply (when the assumptions hold) that whenever a pattern extends to a non-orphan with a large enough zero-padding, the corresponding finite-support configuration admits a semilinear preimage. This will be illustrated in the next section.

## 4 Application to the Game of Life

**Theorem 4** *The preimage SFT of* $0^{\mathbb{Z}^2}$ *in the Game of Life has stable and periodizable one-sided traces.*

***Proof*** Let $g : \{0, 1\}^{\mathbb{Z}^2} \to \{0, 1\}^{\mathbb{Z}^2}$ be the Game of Life. Let $F$ be the natural forbidden patterns for $g^{-1}(0^{\mathbb{Z}^2})$. Since $g$ has radius 1, we pick $n = 2$. Clearly $F = \circlearrowleft^i(F)$ for all $i$, so it is enough to find $k, \ell, p$ such that $S_{n,\ell}(F) \subset P_{n,k,p}(F)$. One can show that the choice $\ell = 4$, $k = 3$, $p = 3$ works, by computer. □

Our program in [20] verifies this result.

By using the minimal automata for the languages of the approximate traces, one can check[1] that

$$w = \begin{matrix} 1\,1\,0\,0\,1\,0\,0\,0\,0\,0\,0\,0\,0\,0\,0\,1\,0\,0\,0\,1\,0\,1\,0\,1\,0\,1\,0\,0\,0\,1\,0 \\ 0\,0\,1\,0\,0\,0\,0\,1\,1\,1\,0\,0\,1\,0\,1\,0\,1\,0\,1\,0\,0\,0\,1\,0\,0\,0\,1\,0\,0\,0 \end{matrix} \in \mathcal{L}(S_{2,3}(F)) \setminus \mathcal{L}(S_{2,4}(F)),$$

i.e. $w$ can be extended infinitely on both sides so that the resulting configuration can be continued by three rows upward without introducing a 1 in the Game of Life image, but not by four rows. Since these subshifts are not equal, $\ell = 4$ is the minimal value we can use. One can similarly show that $k = 3$, $p = 3$ are also optimal; $p = 3$

[1] Once the word $w$ is given, one can prove by hand straight form the definitions that it separates the regular languages, as a constraint-solving puzzle.

is optimal by using the word from the proof of Proposition 1 below. The optimality of $k = 3$ can be shown similarly as that of $\ell$ by studying the minimal automata.

**Lemma 5** *In the situation of the Game of Life, we have $L_{2,4}(F) = \mathcal{L}(S_{2,4}(F))$, i.e. in Lemma 3 applied to $L_{2,4}(F)$ one can pick $C = 0$.*

***Proof*** It can be checked, either with a case analysis or by computer, that $S_{2,1}(F)$ contains all words of length 6 over its alphabet. Given a word $w \in L_{2,4}(F)$, let $P$ be a height-6 rectangle whose two bottom rows form $w$ and that contains no pattern from $F$. The two rightmost columns of $P$ are in $\circlearrowright(S_{2,1}(F))$, so $P$ can be extended to the right by one column without introducing a pattern of $F$. Symmetrically, it can be extended to the left, and thus $w$ is extendable indefinitely in both directions within $L_{2,4}(F)$. □

Our program in [20] also verifies Lemma 5.

**Theorem 5** *Let g be the Game of Life. Then*

- *semilinear configurations are dense in $g^{-1}(y)$ for every finite-support configuration y,*
- *if $P \in \{0, 1\}^{[0,M)\times[0,N)}$ and $\mathrm{pad}_0^4(P)$ admits preimage $Q$, then $\mathrm{conf}_0(P)$ is not a Garden of Eden, and admits a preimage $x \in \{0, 1\}^{\mathbb{Z}^2}$ with*

$$x|_{[-1,M]\times[-1,N]} = Q|_{[-1,M]\times[-1,N]}$$

- $\mathrm{FinGoE}(g)$ *reduces in polynomial time to orphans, in particular* $\mathrm{FinGoE}(g)$ *is in co-NP.*

***Proof*** The first claim follows from Theorem 4 and Lemma 1. The second and third follow from Theorem 4 (whose proof gives stability at $S_{2,4}(F)$), and then applying the proof of Lemma 4 using the constant $C = 0$ from Lemma 5. □

The constants are not included in the statements of the lemmas for simplicity, but their values are explicitly stated in the proofs.

In the above theorem, we have a constant bound on the periods of the semilinear sets, by the proof of Lemma 1. The vertical period $q! = (p|A|^{2n(k+p)})! = 50331648!$ stated in the proof of Lemma 1 is not practically usable. By analyzing the situation a bit more carefully, we see that it is better to use separate periods of at most $p|A|^{n(k+p)} = 12288$ on the west and east borders of $A_{\text{north}}$ and $A_{\text{south}}$. This is already usable, and should be hugely improvable by further analysis.

## Example

We illustrate how to construct a semilinear preimage of a finite-support configuration from a preimage of a rectangular pattern containing its support. The following shows a pattern $P$ and its thickness-4 padding $\mathrm{pad}_0^4(P)$:

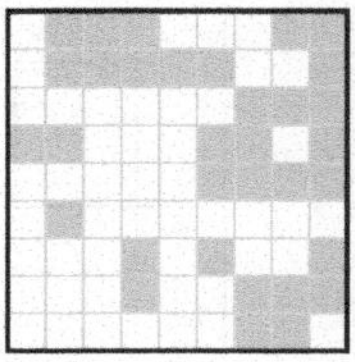

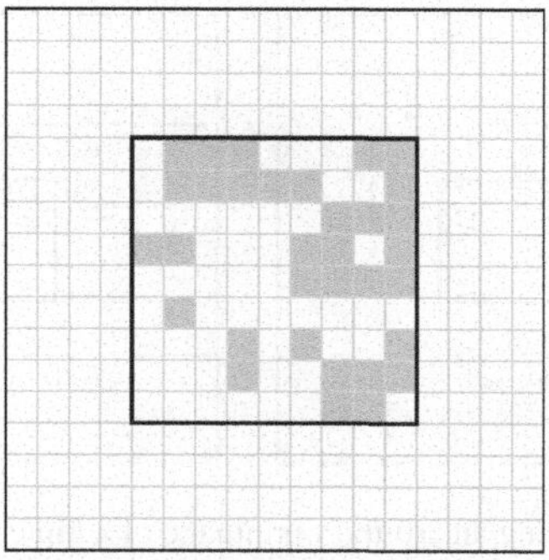

We want to know if the finite-support configuration corresponding to $P$ is in the image of the Game of Life. The first step is to find a preimage for the padded pattern, for example using a SAT solver.[2] One possible preimage (for the thickness four padding) is the following one.

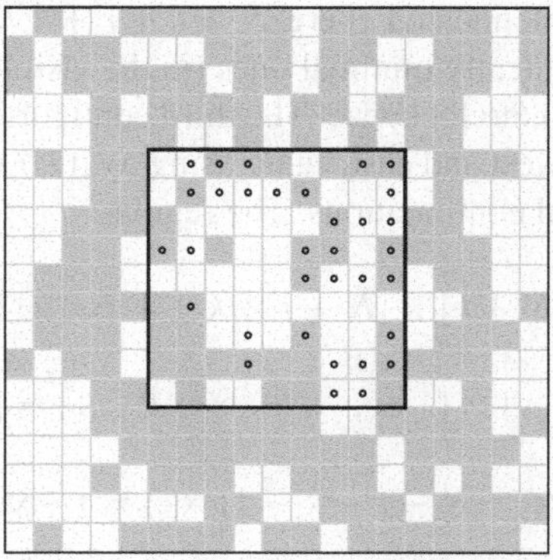

The occurrences of 1s in the image correspond to the black dots, and the 1s in the preimage are gray tiles. One can check that this preimage cannot be extended to the south without introducing a 1 in the image, can be extended to the west by one line (but not two), and to the east by two lines (but not three).

Lemma 4 implies that, since we found a preimage for the thickness-4 padding, there exists a preimage for the finite-support configuration $y = \mathrm{conf}_0(P)$. Figure 1 shows such a preimage, and the process of extracting a semilinear preimage from it: The figures denote central $[-16, 16]^2$ patterns of configurations $x, x^2, x^3$, and their common Game of Life image $y$, whose support contained in $[-N, N]^2 = [-4, 4]^2$ is $P$. These correspond to $x, x^2, x^3, y$ in the proof of Lemma 1. The 1-cells of the patterns $x, x^2, x^3$ are gray tiles, the 1-cells in $y$ are black dots, and the inner square is $[-N, N]^2$. The first configuration $x$ was sampled together with $y$, by picking 1-cells of $x$ from a Bernoulli distribution, and then adding 1-cells outside $[-N, N]^2$ until the image $y = g(x)$ outside $[-N, N]^2$ contained only 0-cells. We have $S_{2,4}(F) \subset P_{2,3,3}(F)$, and $r = 1, n = 2, \ell = 4, k = 3, p = 3$ in the notation of the proof of Lemma 1.

[2] PicoSAT [4] solves this particular instance in under a second. The solution shown here was not constructed by a SAT solver.

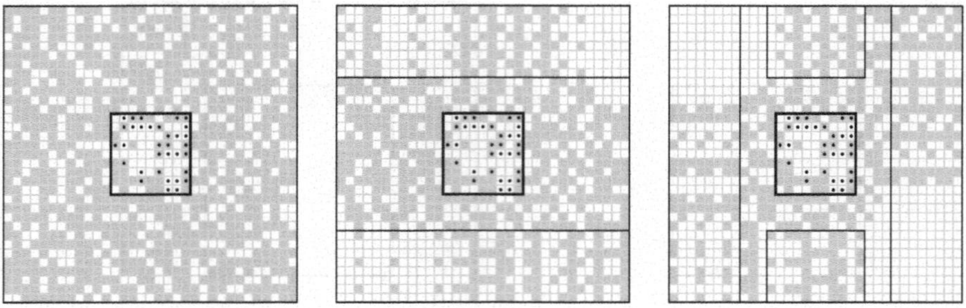

**Fig. 1** Turning an arbitrary preimage of a finite-support configuration into a semilinear one

The second configuration, $x^2$ was obtained by periodizing the contents below and above $[-N, N]^2$ using $S_{2,4}(F) \subset P_{2,3,3}(F)$, and the periodic half-planes $\mathbb{Z} \times [9, \infty)$ and $\mathbb{Z} \times (-\infty, -9]$ are delineated (here $N + k + r + 1 = 9$). The continuations picked are the lexicographically minimal ones (in the visible area) with preperiod 3 and period 3, moving counterclockwise. The third configuration $x^3$ is obtained from $x^2$ by periodizing on the west and east, again using $S_{2,4}(F) \subset P_{2,3,3}(F)$ and picking lexicographically minimal continuations. The regions

$$A_{\text{north}} = [-N - r, N + r] \times [N + r + k + 1, \infty) = [-5, 5] \times [9, \infty),$$

$$A_{\text{south}} = [-N - r, N + r] \times (-\infty, -N - r - k - 1] = [-5, 5] \times (-\infty, -9],$$

$$A_{\text{west}} = (-\infty, -N - r - k - 1] \times \mathbb{Z} = (-\infty, -9] \times \mathbb{Z},$$

$$A_{\text{east}} = [N + r + k + 1, \infty) \times \mathbb{Z} = [9, \infty) \times \mathbb{Z},$$

are delineated.

The configuration $x^4$ would be obtained by finding a repetition in the rows to the north and south of the pattern. Only the stripe at the east border of $A_{\text{north}}$ is not yet periodic, and we can simply replace the entire quadrant with 1-cells to obtain a semilinear preimage.

## 5 Questions and Additional Observations

The following proposition shows that (the first claim in) Theorem 5 is no longer true if semilinear configurations are replaced by finite-support or co-finite-support configurations. Say a configuration is *asymptotically horizontally $N$-periodic* if it is asymptotic to a configuration with period $(N, 0)$.

**Proposition 1** *Let g be the Game of Life and $N \in \mathbb{N}$. Then asymptotically horizontally N-periodic configurations are not dense in $g^{-1}(0^{\mathbb{Z}^2})$.*

Obviously the same is true for vertically periodic configurations.

***Proof*** One can verify that $g^{-1}(0^{\mathbb{Z}^2})$ contains the pattern

$$P_n = \begin{matrix} 0\,1\,0\,0\,0\,1\,0\,1\,0\,1\,0\,1\,0\,0\,0\,1\,0\,0 \\ 0\,1\,0\,1\,0\,1\,0\,0\,0\,1\,0\,0\,0\,1\,0\,0\,0\,1 \end{matrix} \begin{pmatrix} 0\,0 \\ 0\,1 \end{pmatrix}^n \begin{matrix} 0 \\ 0 \end{matrix}$$

for any $n \in \mathbb{N}$, by continuing it left and right with 0s, and then extending each column with period-3 so that the repeating pattern has an even number of 1s.

It turns out that this is the only possible continuation of the columns intersecting the pattern: At an occurrence of

$$\begin{matrix} 0\,1\,0\,0\,0\,1\,0 \\ 0\,1\,0\,1\,0\,1\,0 \end{matrix}$$

one can easily deduce that the only possible continuation downward is 0001000, and this forces the contents of the next row. The pattern is chosen so that this subpattern is seen on the first 3 iterations, after which we repeat the original pattern, which determines the entire column.

To prove the claim, pick $n > N$. Then no preimage of $0^{\mathbb{Z}^2}$ containing $P_n$ has a horizontal $N$-period. □

The pattern above was found by generalizing the shortest word separating $S_{2,3}(F)$ from $S_{2,4}(F)$. It forces the period in both directions, thus cannot appear in a preimage of $0^{\mathbb{Z}^2}$ together with its 90-degree rotation. Thus, it cannot directly be used to show that $g^{-1}(0^{\mathbb{Z}^2})$ does not have dense asymptotically doubly periodic points (when the period is not fixed). Indeed, we do not know whether asymptotically doubly periodic points are dense in $g^{-1}(0^{\mathbb{Z}^2})$.

Our first attempt at proving the decidability of the set of finite-support Gardens of Eden was to show that all finite-support configurations that have a preimage have a finite-support or co-finite-support preimage. This stays open.

**? Question**

Does every finite-support configuration in the image of the Game of Life have a finite-support preimage? A co-finite-support preimage? An asymptotically doubly periodic preimage with some fixed periods $(N, 0)$, $(0, N)$?

The first and second subquestions are equivalent, as observed by user dvgrn in [15]: in the SFT $g^{-1}(0^{\mathbb{Z}^2})$, the rectangular all-1 pattern with support $[-n-2, n+2]^2 \setminus [-n, n]^2$ can be glued to the all-0 pattern with support $[-n-c-2, n+c+2]^2 \setminus [-n-c, n+c]^2$, for any large enough $c$, and vice versa when 0 and 1 are exchanged.

Proposition 1 shows that "yes" answers in Question 5 cannot be proved by only modifying an arbitrary preimage outside a finite region. These subquestions would be solved in the negative by finding a finite-support configuration $y \in \{0, 1\}^{\mathbb{Z}^2}$ that is not a Garden of Eden but whose every preimage contains an occurrence of some $P_n$ from the proof of Proposition 1 outside the convex hull of the support of $y$.

Anecdotal evidence for Question 1 is that, as mentioned, the preimages shown for $P$ and its paddings in Example 4 were not found by a SAT solver, and in fact if the problem is fed into PicoSAT with a lexicographic order on the positions, the preimage that is found for the 4-padding of $P$ in Example 4 always has two layers of 1-cells in the two outermost layers, in other words finding a cofinite-support preimage for this particular pattern seems to be easier than finding a "bad" preimage (so we constructed one by other methods). If the problem is obfuscated (by shuffling the cell positions) before feeding it to the solver, it occasionally gives solutions that do not have 1-cells on the border of the rectangle, but this still happens very rarely.

Much of the study of the Game of Life concentrates on finite configurations. Thus, the following question seems very relevant, if the first subquestion of Question 5 has a negative answer.

**? Question**

Is it decidable whether a given finite-support configuration has a finite-support preimage?

Finite-support Gardens of Eden are $\Sigma^0_1$ (in the arithmetical hierarchy) and non-trivially decidable (by this paper), since if there is no preimage, there is an orphan. Finite-support configurations without a finite-support preimage, on the other hand, are $\Pi^0_1$, since we can prove the *no*-instances by exhibiting a preimage, but there is no obvious reason there should be a finite certificate for not having one.

In Theorem 5, the padding thickness 4 is optimal if we are not allowed to change the preimage, by positioning any word of $\mathcal{L}(S_{2,3}(F)) \setminus \mathcal{L}(S_{2,4}(F))$ on the boundary of the rectangle. We do not know if it is optimal if the preimage can be changed, i.e. we do not know whether there exist patterns $P$ such that $\mathrm{pad}^3_0(P)$ has a preimage pattern but $\mathrm{conf}_0(P)$ is a Garden of Eden. The optimal constant is at least 1: the pattern

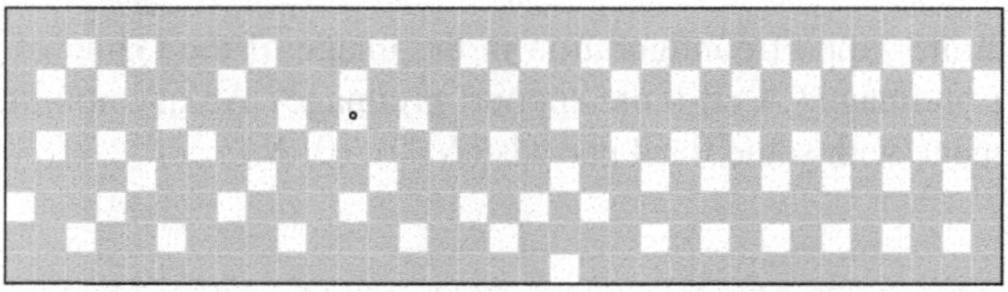

was obtained by modifying one bit in the orphan of Banks [7] (marked by a black dot). It is not an orphan, but PicoSAT reports its thickness-1 zero-padding to be an orphan.

One may ask if it is important in Theorem 5 that $P$ is rectangular. For the result that finite-support Gardens of Eden are co-NP, this is not essential: as long as the inputs of size $n$ specify values only in a polynomial-sized rectangle in $n$, to prove that the configuration $\mathrm{conf}_0(P)$ corresponding to a given pattern $P$ is not a Garden of Eden, we can extend the $P$ to a rectangular one by adding 0-cells and apply our methods. However, the second claim of the theorem fails for all convex shapes except rectangles, at large enough scales.

For a general $D \subset \mathbb{Z}^2$, we define the 0-*padding of thickness* $C$ of $P \in A^D$ as the pattern $Q$ with domain $E = (D + [-C, C]^2) \cap \mathbb{Z}^2$, where $D + [-C, C]^2$ is interpreted in $\mathbb{R}^2$ and $[-C, C] \subset \mathbb{R}$ is a continuous interval, defined by $Q|_D = P$, $Q|_{E \setminus D} = 0^{E \setminus D}$.

**Proposition 2** *Let $g$ be the Game of Life and $N \in \mathbb{N}$. Let $K \subset \mathbb{R}^2$ be a compact convex set that is the closure of its interior. If $K$ is not a rectangle aligned with the standard axes of $\mathbb{R}^2$, then for all $C \in \mathbb{R}$, for all large enough $r > 0$, the 0-padding of thickness $C$ of the all-zero pattern $P$ with shape $rK \cap \mathbb{Z}^2$ admits a preimage pattern $Q$, such that the subpattern of $Q$ that maps to $P$ does not extend to a preimage of* $\mathrm{conf}_0(P)$.

The constant $C$ can be replaced by any sublinear function of $r$.

***Proof*** Among compact convex subsets of $\mathbb{R}^2$, the axis-aligned rectangles are exactly the sets where $(a, b), (c, d) \in K$ implies $(a, d), (c, b) \in K$. Thus, suppose $K$ is not a rectangle, and pick $(a, b) \neq (c, d)$ such that w.l.o.g. $(a, d) \notin K$. By the assumption that $K$ is the closure of its interior, $(a, d)$ has positive distance to $K$, and we may assume $(a, b), (c, d)$ are interior points by moving them slightly if necessary. For any $M \in \mathbb{N}$, there then exists $r > 0$ such that the set $rK \cap \mathbb{Z}^2$ contains a translate of the square grid $[-M, M]^2$ whose convex hull in turn contains $(ra, rb)$. The same is true for $(c, d)$.

Let $P$ be the all-0 pattern of shape $rK \cap \mathbb{Z}^2$, and let $Q$ be a preimage pattern of $\mathrm{pad}_0^C(P)$ that contains a copy of the pattern $P_0$ from the proof of Proposition 1 near $(ra, rb)$ and a copy of the 90-degree rotation $\circlearrowleft(P_0)$ near $(rc, rd)$. As long as $r$ is large enough compared to $C$, such a preimage exists, since we have enough room to put these patterns near $(ra, rb)$ and $(rc, rd)$, and then we can continue them 3-periodically in both directions and fill the rest of $Q$ with 0-cells. The restriction of $Q$ to $rK \cap \mathbb{Z}^2$ does not extend to a preimage of $0^{\mathbb{Z}^2}$, since the 3-periodic patterns forced by the copies of $P_0$ would intersect near $(ra, rd)$. □

It is also natural to ask whether the complexity-theoretic aspects of our results are optimal. We have shown that for the Game of Life, the finite-support Gardens of Eden are polynomial-time reducible to orphans. For the other direction, we do not know any polynomial-time reduction, thus orphans could in principle be harder than finite-support Gardens of Eden. We conjecture that both problems are hard for co-NP, in particular are equally hard by our results.

**? Conjecture**

Finite-support Gardens of Eden and rectangular orphans for the Game of Life are co-NP-complete (under the encodings of this paper, for polynomial time many-one reductions).

The doubly periodic Gardens of Eden give another natural decision problem, and we conjecture that it is not computable. This would in particular imply that not all semilinear configurations admit semilinear preimages.

**? Conjecture**

The doubly periodic Gardens of Eden for the Game of Life are an undecidable set.

One possible method of proving these conjectures would be to encode arbitrary Wang tiles into preimages of rectangular patterns, and reduce to the co-NP-complete problem of untileability of a rectangular region [8], or to the undecidable problem of tileability of the infinite plane [3].

In the encoding we use, the input specifies the values in an area of polynomial size. One may wonder what happens if this input is given in other ways.

**? Question**

Given a finite tuple of vectors $(v_1, \ldots, v_n)$ written in binary, what is the computational complexity of checking whether the finite-support configuration with support $\{v_i \mid i = 1, \ldots, n\}$ is a Garden of Eden?

One may wonder if the Game of Life is special, or whether these results are true for a larger class of CA. The Game of Life has strong symmetry properties, so it is a natural candidate to look at, and possibly the automata-theoretic problems are particularly easy for this reason. However, as far as we know there is no reason to believe one-step properties of the Game of Life should be special among, for example, totalistic radius-1 rules with the Moore neighborhood. It would be interesting to go through a larger set of rules, and analyze the stability and periodizability properties of their traces. Of course, if the trace of a given SFT happens to be non-sofic (and in particular not stable), then our methods cannot be applied to it. Even if the trace is sofic but not stable, we do not have a general method of determining it.

We also do not know if our results apply to powers of the Game of Life. If finite-support configurations that are not Gardens of Eden for the Game of Life would always have finite-support preimages, or if semilinear such configurations would always have semilinear preimages, then we would obtain decidability of finite-support Gardens of Eden for all powers of the Game of Life. It is also unknown if all powers of the Game of Life have different sets of Gardens of Eden, i.e. whether the

Game of Life is *stable* in the terminology of [14]; according to the LifeWiki website the first 4 powers have been separated in this sense [10].

**Acknowledgements** Ville Salo supported by Academy of Finland grant 2608073211. Ilkka Törmä supported by Academy of Finland grant 295095.

## References

1. Adamatzky A (2010) Game of life cellular automata. Springer, London
2. Arora S, Barak B (2009) Computational complexity: a modern approach. Cambridge University Press
3. Berger R (1966) The undecidability of the domino problem. Mem Amer Math Soc No 66:72 pp
4. Biere A (2008) PicoSAT essentials. J Satisf Boolean Model Comput 4:75–97
5. Di Lena P (2006) Decidable properties for regular cellular automata. In: Navarro G, Bertossi L, Kohayakawa Y (eds) Fourth IFIP international conference on theoretical computer science-TCS 2006. Springer US, Boston, MA, pp 185–196
6. Gardner M (1970) Mathematical games: the fantastic combinations of John Conway's new solitaire game "Life." Sci Am 223(4):120–123
7. Gardner M (1983) Wheels, life and other mathematical amusements. W. H. Freeman and Co, San Francisco, CA
8. Garey MR, Johnson DS (1979) Computers and intractability: a guide to the theory of NP-completeness. W. H. Freeman & Co, New York, NY, USA
9. Ginsburg S, Spanier E (1966) Semigroups, presburger formulas, and languages. Pac J Math 16(2):285–296
10. Grandfather problem – LifeWiki. https://www.conwaylife.com/wiki/Grandfather_problem. Accessed: 2019-12-01
11. Hedlund GA (1969) Endomorphisms and automorphisms of the shift dynamical system. Math Syst Theory 3:320–375
12. Hopcroft JE, Motwani R, Ullman JD (2006) Introduction to automata theory, languages, and computation, 3rd edn. Addison-Wesley Longman Publishing Co Inc, Boston, MA, USA
13. Lind D, Marcus B (1995) An introduction to symbolic dynamics and coding. Cambridge University Press, Cambridge
14. Maass A (1995) On the sofic limit sets of cellular automata. Ergod Theory Dynam Syst 15(4):663–684
15. Orphan pattern / garden of eden. https://conwaylife.com/forums/viewtopic.php?f=2&t=797. Accessed: 2019-12-03
16. Pavlov R (2012) Approximating the hard square entropy constant with probabilistic methods. Ann Probab 40(6):2362–2399
17. Pavlov R, Schraudner M (2015) Classification of sofic projective subdynamics of multidimensional shifts of finite type. Trans Am Math Soc 367(5):3371–3421
18. Perrin D, Pin JÉ (2004) Infinite words: automata, semigroups, logic and games. Pure and applied mathematics. Elsevier Science, London, UK
19. Rendell P (2002) Turing universality of the game of life. In: Collision-based computing. Springer, London, pp 513–539
20. Salo V, Törmä I (2021) Game of Life traces. https://github.com/ilkka-torma/gol-traces. GitHub repository
21. van der Zypen D (2019) (https://mathoverflow.net/users/8628/dominic-van-der-zypen). Decision problems for which it is unknown whether they are decidable – MathOverflow. https://mathoverflow.net/q/345388. Accessed: 2019-11-15

# Amoebae for Clustering: A Bio-Inspired Cellular Automata Method for Data Classification

**Amaury Saint-Jore, Nazim Fatès, and Emmanuel Jeandel**

**Abstract** We present a bio-inspired mechanism for data clustering. Our method uses amoebae which evolve according to cellular automata rules: they contain the data to be processed and emit reaction-diffusion waves at random times. The waves transmit the information across the lattice and causes other amoebae to react, by being attracted or repulsed. The local reactions produce small homogeneous groups which progressively merge and realise the clustering at a larger scale. Despite the simplicity of the local rules, interesting complex behaviour occur, which make the model robust to various changes of its settings. We evaluate this prototype with a simple task: the separation of two groups of integer values distributed according to Gaussian laws.

**Keywords** Bio-inspired models · Cellular automata · Data clustering · Discrete dynamical systems · Collective intelligence

## Introduction

Biology is an inexhaustible source of inspiration for computer science. Indeed, many phenomena we observe in living organisms can be understood with simple interactions which create a self-organising process with a rich behaviour, e.g. self-reproduction [16], cellular sorting [17], etc. The model we present here aims at performing a clustering task. Various nature-inspired models have already been explored to perform this task [12, 15]. The particularity of our model is that it is inspired from the behaviour of the social amoeba *Dictyostelium discoideum*. It is a cellular automaton model, that is, space, time and states of components are discrete and the interactions are purely local. In the most simple setting of the problem, the

A. Saint-Jore · N. Fatès (✉) · E. Jeandel
Université de Lorraine, CNRS, Inria, LORIA, 54000 Nancy, France
e-mail: nazim.fates@loria.fr

A. Saint-Jore
e-mail: amaury.saint-jore@loria.fr

E. Jeandel
e-mail: emmanuel.jeandel@loria.fr

 
A. Adamatzky (ed.), *Automata and Complexity*, Emergence, Complexity and Computation 42, https://doi.org/10.1007/978-3-030-92551-2_23

model consists of a grid in which two types of data are found, initially located in random places, these items have to form two separated groups according to their types.

This clustering behaviour can be related to various segregation phenomena, such as the ones studied by Goles et al. [8]. In such models, as Schelling's model and its variations, a population separated into two types is initially randomly spread on a lattice. The agents may move to an empty location if this increases the number of agents of the same type in their neighbourhood. One observes that even if an agent is satisfied with a given minimal number of similar neighbouring agents, the systems ends up with a global segregation phenomenon.

Goles and his collaborators were also interested in the social structures which appear when applying a model of general social interactions as proposed by Sakoda [13]. They examined the robustness of the patterns which are interpreted in terms of social structures. The model implies no less than 45 different interaction rules which drive the system irreversibly to an equilibrium or a steady state.

This line of research falls into the general class of inverse problems, where cellular automata are seen as a model of computation that is used to process data. For example, one can ask a cellular automaton to decide about the type of its initial condition, for example whether it has more 0's than 1's, as in the density classification problem (see e.g. [2, 3]), or whether it has an odd number of 1's, as in the parity problem [14].

Our proposition relies on a previous discrete model inspired by the behaviour of this social amoeba [5]. This original model was used to gather thousands of particles initially scattered randomly on a lattice without any central coordination. From this original model, a variant was proposed to locate a hidden source on a lattice with the process called *infotaxis*, which is specific to the case where the detection of the source needs to be achieved with rare events [4].

Our goal now is to continue this work of exploration of the "amoebae models" and examine here whether they could be used to perform data clustering: imagine that we have set of objects initially random scattered on a lattice and we need to group this data into classes which share some similarity. The difficulty of the task lies in the propagation of information which needs to be achieved by simple cells with as few states as possible and with nearest-neighbour simple interactions.

We begin by presenting our model with formal definitions (Sect. 1). This presentation is then followed by an empirical observation of its behaviour (Sect. 2.1), the description of a quantification tool (Sect. 2.2) and a first sequence of statistical experiments (Sect. 2.3). We then briefly explore some properties of robustness of the model (Sect. 3) and say a few words on the work that remains to be tackled.

# 1 Presentation of the Model

Our model is described as a stochastic dynamical system where space, time and states are discrete. It is composed of two layers: the first layer corresponds to the environment, which propagates reaction-diffusion waves. The second layer contains the amoebae, which move and interact with the environment. Our objective is to design a system which will be as simple as possible in terms of cell states and behaviour, and for which the amoebae form groups according to the data they contain.

Space is modelled by a two-dimensional lattice with periodic boundary condition (torus), represented by $\mathcal{L} = (\mathbb{Z}/X \cdot \mathbb{Z}) \times (\mathbb{Z}/Y \cdot \mathbb{Z})$, where $X$ and $Y$ are the dimensions of the lattice. In this text, we only consider square lattices, that is, we take $X = Y = L$, where $L$ is called the *size* of the lattice.

The set of states is now the Cartesian product $Q \times D$, where $Q = \{\mathtt{N}, \mathtt{E}, \mathtt{R}\}$, with the states N, E, R which respectively correspond to the *neutral*, *excited* and *refractory* states, and the set $D$ corresponds to the data. The set $D$ is dependent on the data set on which we want to apply the clustering process; we will assume that it contains a particular value, called the *void* value, denoted by $V$, which represents the *absence of data* in a given cell.

Formally, for a cell $c \in \mathcal{L}$ and time $t \in \mathbb{N}$, we denote by $\sigma_c^t \in Q$ the state of this cell and by $\rho_c^t \in D$ its data. For the sake of brevity, we will say *state* only to refer to the first part of the cell and say *data* for the second part. Carefully note that as the cells constitute the environment, the data of a cell state $\rho_c^t$ represents the information that is being propagated, which will often differ from the data of the amoebae it contains.

The set $\mathcal{A}_c^t$ represents the amoebae contained in a cell $c$. Formally, $\mathcal{A}_c^t$ is represented by a *multi-set* of elements of $D$, that is, it may contain several elements of $D$ with an identical value. Recall that each amoeba is in charge of carrying an element of the data set that has to be clustered.

The interactions between cells follow a local rule and we use here the Moore neighbourhood, that is, a cell $c$ will only "see" its own state and the state of its 8 nearest neighbours. This set is denoted by $\mathcal{N}(c)$.

In all the experiments that follow, the initial condition of the system is that all cells are initially in the neutral state N and the elements of the data set are randomly and uniformly distributed on the grid. For the data, we simply assign to each cell a given probability $d_{\text{ini}}$ to contain a data, which is drawn at random according to a non-uniform distribution that will be described later. The probability $d_{\text{ini}}$ controls the initial *density* of data, hence the use of the letter $d$.

Informally, our model relies on three main features: (a) the amoebae have the ability to randomly trigger reaction-diffusion waves that will make other amoebae react; (b) these waves propagate without attenuation and by partially colliding and merging when they meet ; (c) when an amoeba sees such a wave in its environment, it reacts by being attracted or repulsed according to the data it sees and its own data. Let us now describe these steps more precisely.

## 1.1 The Reaction-Diffusion Waves

The rules that govern the propagation of reaction-diffusion waves are simple: (a) A neutral cell becomes excited if it has an excited neighbour; (b) an excited cell always

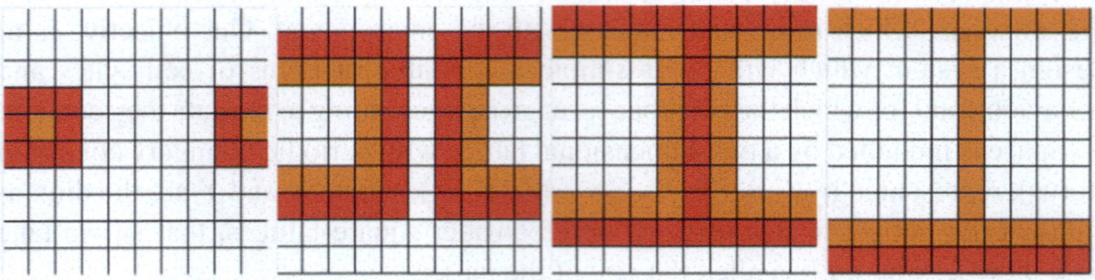

**Fig. 1** Propagation of two excitation waves: red, orange and white squares respectively correspond the excited, refractory and neutral cells (colour online)

becomes refractory at the next time step: (c) a refractory cell always become neutral at the next time step. Once a neutral cell gets excited, the excitations propagate in form of waves formed by lines of excited cells that we call *fronts*, which are followed lines of refractory cells. The role of the refractory state is to prevent the retropropagation of excitations. When two waves meet, they annihilate if the fronts propagate in opposite directions and merge if the fronts propagate in the same direction (see Fig. 1).

In addition to the possibility to being excited by its neighbours, a cell which contains at least one amoeba becomes excited with a probability $\lambda$, called the *excitation rate*. In this case, the excited cell takes the state $\mathtt{E}$, and its data $\rho_c^t$ will be chosen randomly and uniformly among the data of amoebae it contains. This choice is represented by $R[\mathcal{A}_c^t]$ where the operation of random uniform choice in a multiset $X$ is denoted by $R[X]$. We note $B(p)$ as a Bernoulli law of probability $p$.

We denote by $E_c^t = \{c' \in N(c) : \sigma_c^t = \mathtt{E}\}$ the set of cells which are in the excited state in the neighbourhood of $c$ and by $D_c^t = \{\rho_c^t : c \in E_c^t\}$ the set of data that these cells contain. It should be noted that in general the values of the data contained in a given wave are uniform, as they come from the same amoeba source. However, when two waves of excitation collide and merge, a given cell might see in its neighbour two excited cells with different data, in which case it will choose randomly and uniformly in the multiset of values it sees.

The state of a cell is updated as follows:

If $\sigma_c^t = \mathtt{N}$, then

if $\mathcal{A}_c^t \neq \emptyset$ and $B(\lambda) = 1$ then $\sigma_c^{t+1} = \mathtt{E}$ and $\rho_c^{t+1} = R[\mathcal{A}_c^t]$

else if $card(E_c^t) > 0$ then $\sigma_c^{t+1} = \mathtt{E}$ and $\rho_c^{t+1} = R[D_c^t]$

else if $\sigma_c^t = \mathtt{E}$ then $\sigma_c^{t+1} = \mathtt{R}$ and $\rho_c^{t+1} = V$

else if $\sigma_c^t = \mathtt{R}$ then $\sigma_c^{t+1} = \mathtt{N}$ and $\rho_c^{t+1} = V$

else $\sigma_c^{t+1} = \mathtt{N}$ and $\rho_c^{t+1} = V$.

We have defined our four cases: the self-excitation of a cell by an amoeba, the excitation by a neighbouring cell, the de-excitement and the return to the neutral state.

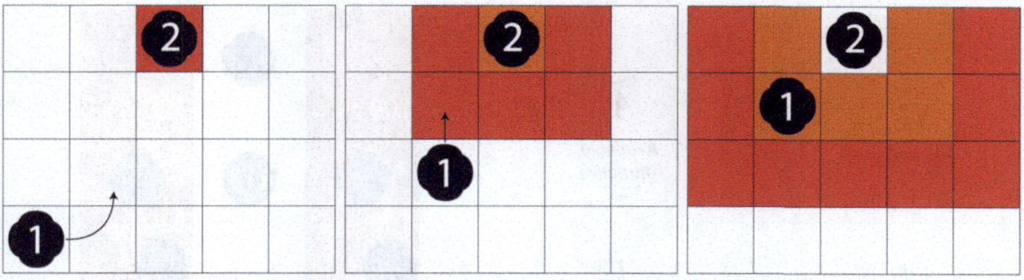


**Fig. 2** Movements of the amoebae. The cell states *excited*, *neutral* and *refractory* are represented in red, white and orange, respectively; amoebae are in black. From step 1 to step 2, Amoeba 1 moves randomly; amoeba 2 triggers a reaction-diffusion wave. From step 2 to step 3, Amoeba 1 moves by following the excitation front

## *1.2 Movements of the Amoebae*

Let us describe the movements of the amoebae, which take place in the second layer of the model (See Fig. 2). To this end, we introduce some intermediary definitions. A cell is *empty* if it does not contain any amoeba and it is *available* it is either empty or if it contains only one amoeba. In order to limit the number of amoebae contained per cell we use the following rules: (a) Amoebae can move only to the available cells of their neighbourhood. (b) If a cell contains several amoebae, only one may move to a neighbour. (c) For the sake of simplicity, there is no limit on how many amoebae an available cell can simultaneously receive.

Let us now describe how the amoebae react to the presence of excited cells in their neighbourhood. The key point of the model is that an excitation can be either attractive or repulsive for an amoeba, depending on the value of the excitation and the data of the amoeba. The use of a repulsive interaction is a novelty compared to the previous models using amoebae (see Introduction) and its purpose is to separate the groups of amoebae that will correspond to different data clusters.

The idea is that for each excited cell in the neighbourhood, we will determine whether the excitation is attractive or repulsive according to a probability law, denoted by $p_{\text{att}}$, depending on the absolute value of the difference $\delta$ between the value of the amoeba and the value of the nearby excitation. The probability to be attractive $p_{\text{att}}$ is defined with the following equation: $p_{\text{att}}(\delta) = e^{-\delta/\xi}$: the higher $\delta$ is, the lower the probability to be attractive and thus the higher the probability to be repulsed. The constant $\xi$ is set by the user of the model and depends on the data set that is used; it should be of the same order of the expected distance between the centres of the clusters of the data set.

Naturally, a problem arises when a cell contains several amoebae. In this case, we simply choose uniformly at random which of the amoebae is selected to move and we calculate $\delta$ according to the data of the selected amoeba.

For a given cell $c \in L$ which contains one amoeba or more, we define for each cell $c'$ in the neighbourhood of $c$ the quantity $\delta_{c,c'} = \left|\rho^t_{c'} - R[\mathcal{A}^t_c]\right|$. (Note that the random value $R[\mathcal{A}^t_c]$ is chosen once and not for each neighbour.) We map $\delta_{c,c'}$ to

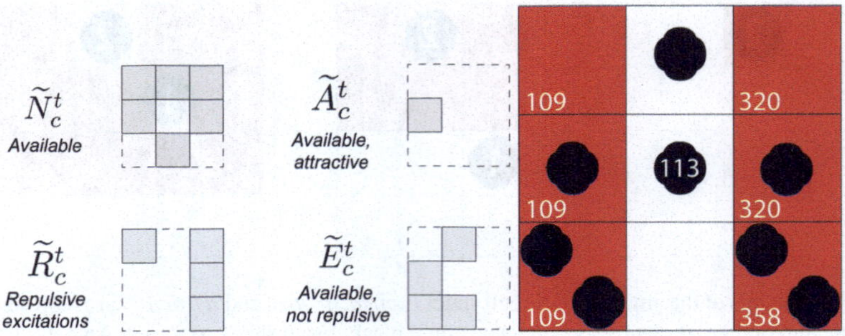


**Fig. 3** Representation of the four sets used to define the movements of an amoeba. Amoebae are represented in black and excited cells in red (colour online). The data contained in the excitations are represented in white

a value $\delta^*_{c,c'} \in \{\mathtt{A}, \mathtt{R}\}$, where $\mathtt{A}$ and $\mathtt{R}$ respectively denote an attractive or repulsive signal, where the probability to be attractive is given by $p_{\mathrm{att}}(\delta_{c,c'})$.

To express the local rules which define the movements of the amoebae, we need to define three sets. For a given cell $c$ the set $\widetilde{N}^t_c$ is the set of available neighbours: $\widetilde{N}^t_c = \{c' \in \mathcal{N}(c) : |(|\,\mathcal{A}^t_{c'}) \in \{0, 1\}\}$. The sets $\widetilde{A}^t_c$ (resp. $\widetilde{R}^t_c$) is the set of available neighbours that are attractive (resp. repulsive): $\widetilde{A}^t_c = \{c' \in \widetilde{N}^t_c : \delta^*_{c,c'} = \mathtt{A}\}$ and $\widetilde{R}^t_c = \{c' \in \widetilde{N}^t_c : \delta^*_{c,c'} = \mathtt{R}\}$. The set $\widetilde{E}^t_c$ is the complement of the previous set; it is the set of cells where an amoeba can *escape* in case of repulsion: $\widetilde{E}^t_c = \widetilde{N}^t_c \setminus \widetilde{R}^t_c$. These sets are represented on Fig. 3.

The movements will be determined by a list of rules with an imposed order of priority: an amoeba may make a random movement to an available cell according to: (a) the phenomenon of agitation (with prob. $p_{\mathrm{ag}}$), (b) the phenomenon of attraction, (c) the phenomenon of repulsion. Second, the priority is given to attraction over repulsion, that is, if an amoeba sees at least one excited cell $\mathtt{A}$, it chooses attraction, otherwise it will choose a repulsive movement.

Denoting by $\Delta^t_c$ the cell where the selected amoeba moves, this yields:

$$\Delta^t_c = \begin{cases} R[\widetilde{N}^t_c] & \text{if } B(p_{\mathrm{ag}}) = 1 \text{ and } |(|\,\widetilde{N}^t_c) > 0 \\ R[\widetilde{A}^t_c] & \text{if } \sigma^t_c = \mathtt{N} \text{ and } |(|\,\widetilde{A}^t_c) > 0 \\ R[\widetilde{E}^t_c] & \text{if } \sigma^t_c = \mathtt{N} \text{ and } |(|\,\widetilde{R}^t_c) > 0 \text{ and } |(|\,\widetilde{E}^t_c) > 0 \\ c & \text{otherwise.} \end{cases}$$

These rules represent the four possible movements: by agitation, by attraction, by repulsion and finally staying in the same cell. The positions of the amoebae are then updated synchronously for each non-empty cell $c$ with the movement from $c$ to $\Delta^t_c$.

# 2 Study of the Model

## 2.1 A First Observation of the Behaviour

Let us now observe some simulations. Figure 4 shows the clustering process on a lattice of size $L = 50$ cells. The initial condition is set with $d_{\text{ini}} = 0.1$, that is, each cell has a probability of 10% to contain a data. The value initially attributed to the amoebae is an integer drawn randomly in the interval $I = [\![0,\ 400]\!]$. The data is not uniformly distributed: its distribution is the sum of two Gaussian laws of parameters $(\mu_1, \sigma_1)$ and $(\mu_2, \sigma_2)$ where the mean values are set to $\mu_1 = 100$ and $\mu_2 = 300$ and the standard deviations are set to $\sigma_1 = \sigma_2 = 0.2$.

The parameters settings were determined empirically after a series of tests. We set the value of the emission rate of waves to $\lambda = 0.01$ and the attraction probability law, $p_{\text{att}}$, is set with $\xi = 40$. In order to introduce some "difficulty" in the clustering process, we also add some agitation, setting $p_{\text{ag}} = 0.1$, that is, the amoebae will make a random move every ten steps in average.

One can observe the following general behaviour: the amoebae start forming small groups where the type is rather homogeneous. Thanks to the randomness of the excitation waves, these groups progressively merge until they form larger groups.

At this stage two possibilities are generally observed: either the large groups directly form two well-sorted clusters, or the sorting stagnates with approximately two or three formed groups and a few isolated amoebae. In this case, the sorting process is somehow delayed due to two main reasons: (a) isolated amoebae may find themselves blocked between the different excitations, and (b) two groups of amoebae that contain amoebae of the same type might try to get closer but get repelled by another large group formed of amoebae of the opposite type. However, as one can observe on Fig. 4, the clusters are well formed after a few thousand steps and remain stable over time.

## 2.2 Quantify Sorting Progress and Efficiency

We now describe how to evaluate more quantitatively the evolution of the clustering process. We continue to consider the simple case where the system should separate the data into two clusters only. Our problem is to evaluate, for a given distribution of amoebae on the grid, how far we are from our goal. Our clustering estimator will associate to each data a type, A or B, depending on its value. For example if the data is distributed in the range of integers from 0 to 100, *type-A* data will cover the range 0 to 49 and *type-B* data will cover the rest of the interval.

With this particular setting of the problem, two main features can be isolated:

1. *homogeneity*: once the groups of connected amoebae are identified, we estimate to which extent they are mixed or not,

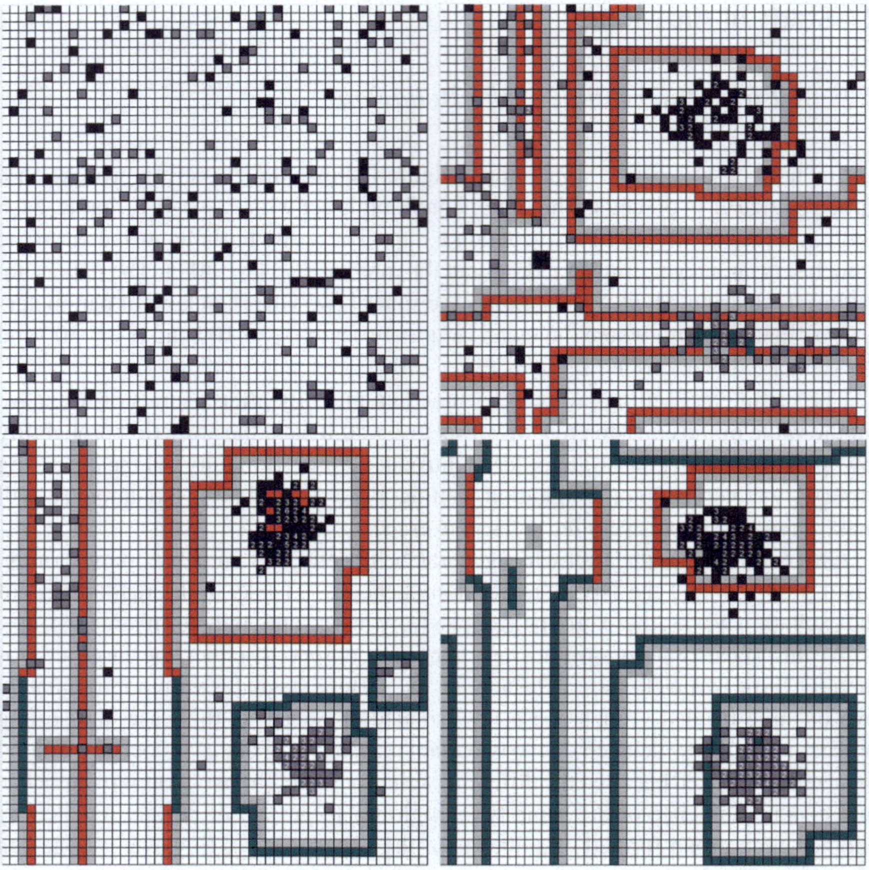

**Fig. 4** Evolution of the clustering process for times: 0, 450, 880 and 1270. The amoebae are represented in black and dark grey: this colour differentiation is only visual and indicates whether the value they contain are greater or lower than the average value 200. When several amoebae are present in the same cell, the number of amoebae is indicated by a white number (this number can me magnified in the online version). A neutral cell is white and a refractory cell in light grey. Excited cells are coloured in red and blue depending on whether the data they contain is greater or lower than the average value 200

2. *compactness*: we estimate how far the global configuration of amoebae is from forming two main groups.

For a given configuration, we thus partition the set of amoebae into connected components of amoebae, where Moore's neighbourhood is used to define the connexity property. Let us denote by $C_1, \ldots, C_k$ the components obtained, with the convention that $C_1$ and $C_2$ are respectively the largest and second largest sets (in terms of number of amoebae they contain). Given two components $i$ and $j$, we denote by $d_{\mathrm{H}}(i, j)$ the Hausdorff distance between these two components. This distance is frequently used to evaluate distances between sets of points and displays various good analytical properties [7]. We then make a partition of the set components $\{C_1, ..., C_k\}$ into two sets of components $P_1$ and $P_2$ by assigning each component $C_i$ to $P_1$ if the distance $d_{\mathrm{H}}(C_i, C_1)$ is smaller than $d_{\mathrm{H}}(C_i, C_2)$ and to $P_2$ otherwise.

To define the *homogeneity* criterion, we simply calculate the number of amoebae $n_1^A$ and $n_1^B$ (resp. $n_2^A$ and $n_2^B$) that are of respective type A and B that are contained in $P_1$ (resp. in $P_2$). The *homogeneity* of $P_m$ for $m \in \{1, 2\}$ is then set equal to $h_m = \left|n_m^A - n_m^B\right| / (n_m^A + n_m^B)$. The global homogeneity $h$ is then simply the minimum of $h_1$ and $h_2$. This form is chosen such that a homogeneous set of components will lead to $h = 1$ while a set of components with an equal number of amoebae of each type will give us $h = 0$.

Let us now examine the *compactness criterion*. We first calculate the average distance between the components and the main component they belong to, that is,

$$\tilde{d} = \frac{1}{k} \cdot \Big( \sum_{C_i \in P_1} d_{\mathrm{H}}(C_i, C_1) + \sum_{C_i \in P_2} d_{\mathrm{H}}(C_i, C_2) \Big).$$

We compare this value to the distance between the two main components of the system $d_{\mathrm{H}}(C_1, C_2)$. The compactness $\kappa$ is then defined as $\kappa = 1 - \tilde{d}/d_{\mathrm{H}}(C_1, C_2)$; a form that is chosen in order to approach 1 as $\tilde{d}$ gets smaller (while $d_{\mathrm{H}}(C_1, C_2)$ remains stable).

The combination of these two criteria will give us an estimation of how the clustering process evolves in time. Note that criteria are not defined when we have only one component and note that we considered here the case of two types, but this method can of course be extended to any number of types. It should also be emphasised that the types are only given for the analysis of the process, but that the process itself never uses this information in the local rules it applies. Let us now observe the clustering process on some simple examples.

## 2.3 Quantitative Analysis of the Behaviour

Let us now analyse the relevance of the two criteria introduced above to quantify the progress of the sorting of amoebae in the system. The two connected components of amoebae are illustrated Fig. 5 and the evolution of the temporal values of these criteria are represented on Fig. 6. The parameters are set identically to those exposed in Sect. 2.1.

Let us first put our attention on the *homogeneity* criterion. One can observe that during the first moments of the simulation, the amoebae are scattered in space and the two main sets are very heterogeneous, which leads to a low homogeneity. This continues until amoebae form larger groups of amoebae with similar data, which yields a gradual increase in the homogeneity criterion, with an additional noise due to the randomness inherent to the model. As mentioned earlier, after this stage is

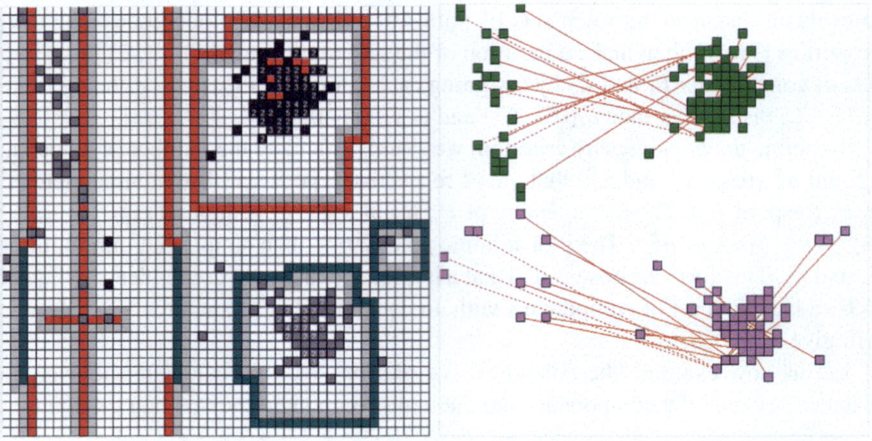

**Fig. 5** Representation of a particular configuration obtained at time 880. (left) The same conventions are applied, in particular amoebae are differentiated in black and dark grey depending on whether their value is higher or lower than the average value 200. (right) The amoebae are coloured in green or magenta according to the partition realised to calculate the homogeneity and compactness criteria (see Sect. 2.2). The red segments materialise Hausdorff distances; the dotted lines indicate that the shortest distance are calculated with periodic boundary conditions

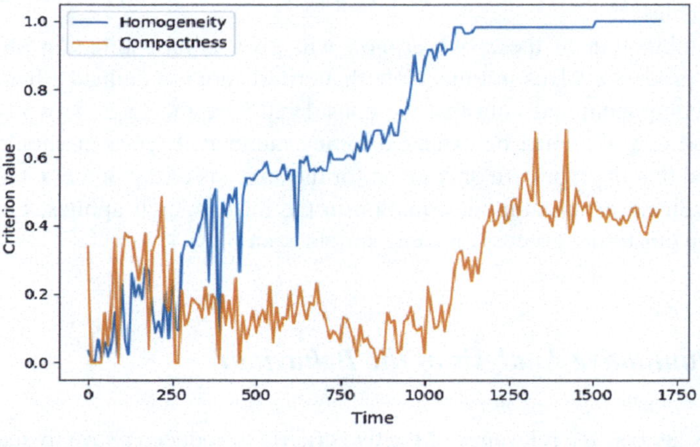


**Fig. 6** Temporal evolution of the criteria of homogeneity and compactness of a particular simulation

complete, two possibilities generally occur: we either have a rather steady increase of the criteria, which indicates that the amoebae are sorted into two groups, or the values will stagnate for a while (as shown on Fig. 5), and then find the sorted stable.

The stochastic nature of the model makes that some evolutions may last longer to reach the sorted state. For instance, when three main groups of amoebae are formed, with one group of one type and the other two of the second type, the value of homogeneity remains around 50% for a moment until the two groups of the same type can merge.

The *compactness* criterion allows us to complete our analysis of the clustering process. Recall it estimates the average relative distance of the connected components of the amoebae to the closet main component. In the beginning of the simulation, since amoebae are scattered randomly in space and these main components are not stable and the criterion is of little relevance. As one can observe, the compactness criterion becomes relevant only when larger groups of amoebae are formed, in which case its values become greater than 25%.

Note that when the clustering has occurred (that is, when we have only two groups), the homogeneity approaches 100% but the compactness stabilises at lower values (between 40 and 50%). Indeed, the rules we imposed on the movements of amoebae (agitation, limit of their number per cell) do not allow them to form perfect components.

## 3 Robustness of the Model

*Data.* Let us examine how the model responds to variations of the distribution of data. We take again the interval $I = [\![0,\ 400]\!]$ and generate the data distribution according to the sum of two Gaussian laws of parameters $(\mu_1, \sigma_1)$ and $(\mu_2, \sigma_2)$. We the standard deviation to $\sigma_1 = \sigma_2 = 0.4$.

We perform different experiments by varying $(\mu_1, \mu_2)$. The attractivity of the waves is set with $\xi = 40$ (see Sect. 1.2). We define the *convergence time* as the number of time steps that are needed to attain both a homogeneity greater than $h^* = 0.8$ and compactness of $\kappa^* = 0.25$. These values are fixed empirically by observing the situations which correspond to a good clustering quality and a relatively stable global state. Since we aim at achieving a fast convergence, we stop the simulation after 3000 time steps.

The results are presented on Fig. 7. We observe that the clustering is rapid when the two Gaussian laws are well separated $(\mu_1, \mu_2) = (100, 300)$. More surprising, we note that when the averages are made closer $(\mu_1, \mu_2) = (125, 275)$, and even though the two laws begin to overlap, the convergence time remains low and is even better than the previous case. A limit case is observed for $(\mu_1, \mu_2) = (150, 250)$ where the convergence time jumps above the limit value of 3000 steps. A visual inspection of the simulations and an observation of the evolution of the criteria shows that on the two first case, the clusters remain stable and well-sorted after the convergence time. The value of $\xi$ could be adjusted to improve these results but we choose here to keep the value constant for the sake of simplicity.

We also examined the behaviour of the model when the data is a superposition of three Gaussian laws: in this case, we observe that the clustering process occurs as expected by sorting the data into three groups, without any significant delay when $\lambda$ is set to the appropriate value.

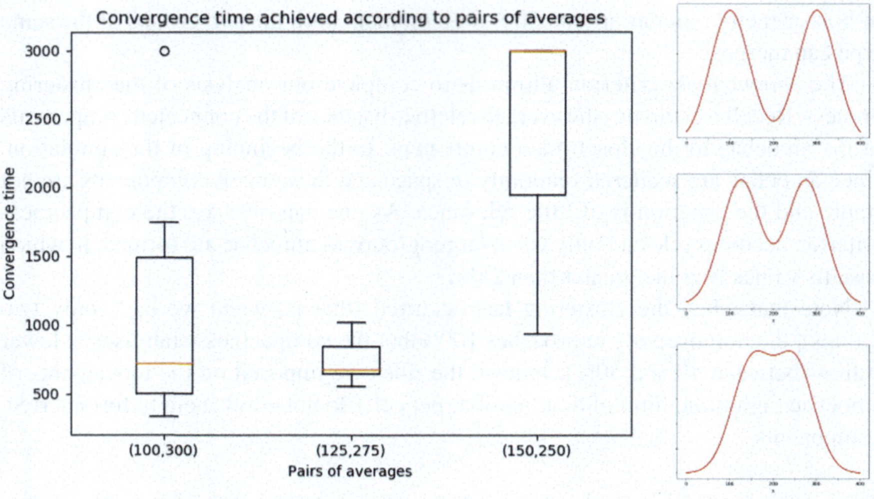


**Fig. 7** (left) Convergence time for three distribution laws which are sums of two Gaussian laws (right) Representation of the distribution laws

*Initial conditions.* One may also examine what happens if a large number of amoebae is initially introduced. A new series of simulations is carried out with the following initial condition for an initial density between 10% and 75%. We empirically observed that although the movements of the amoebae are more constrained and although the propagation of cell excitations are more difficult, the model succeeds in achieving the clustering operation in times that are comparable to those expressed above. It is only when the initial density reaches $d_{\text{ini}} = 0.75$ that strong difficulties appear and that the convergence time significantly rises. This suggest that the size of the lattice should always be set in agreement with the quantity of data that needs to be processed; with, say, at least three times as many cells as amoebae.

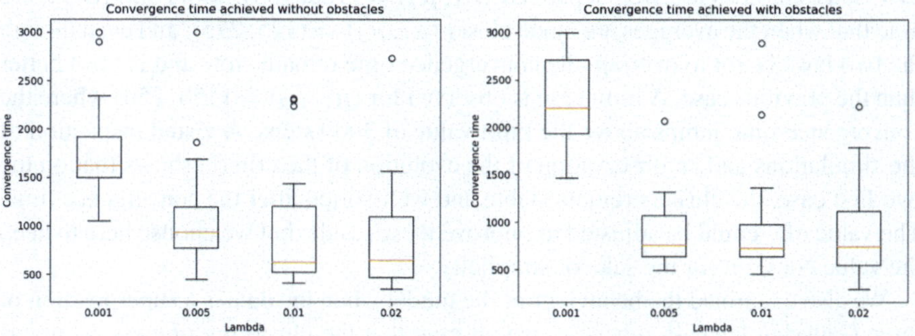


**Fig. 8** Convergence time as a function of the excitation rate λ: (left) regular lattice (right) lattice with obstacles

*Obstacles.* As mentioned earlier, one of the motivations for the use of bio-inspired models is their robustness to errors and failures in the computing medium. We thus consider the case where obstacles can be present on the lattice, that is, cells that cannot transmit any information which cannot hold any amoeba.

Simulations in such conditions are represented on Fig. 8: we consider different values of the emission rate $\lambda$ with and without obstacles. One can observe that the overall behaviour is not perturbed by the presence of obstacles, although some isolated amoebae may take a longer time to join other groups. The convergence of the amoebae can be slowed down but we empirically observed that the process always manages to achieve the clustering task.

## 4 Application to the Fisher's Iris Data Set

We now propose to test our amoebae-inspired model for classifying real-world data. As a starting point, we consider one of the most known data sets, namely the iris data set, established in 1936 by Fisher to study linear discriminant analysis techniques [6]. The data set is composed of 150 samples, divided into three groups of equal size, which correspond three species of flowers: *Iris setosa*, *Iris virginica* and *Iris versicolor*. Four features are measured for each sample: the length and the width of the sepals and length and width of petals. This dataset is known to be delicate to manipulate because its contains many samples of different class with close characteristics.

The amoebae, which encode each piece of data, thus contain the measure of the four characters, coded as they are given, that is, directly in centimetres. In order to grant each of this feature the same weight, we divide each measure value by its average value. This renormalisation ensures that all the parameters are equally taken into account but of course, one can also imagine granting various weights to the different features.

Both the amoebae and the propagated signals are thus now represented by 4-dimensional vectors. Denoting by $\rho^t_c \in \mathbb{N}^4$ the state of a cell and $\mathcal{A}^t_c \in \mathbb{N}^4$ the state of an amoeba, for a cell $c$ that contains an amoeba, the repulsive or attractive property of an excitation in a neighbouring cell $c'$ will be determined with the function: $\delta_{c,c'} = \left|\rho^t_{c'} - R[\mathcal{A}^t_c]\right|$, where $|v|$ is the Euclidean norm (in 4 dimensions) of a vector $v$.

In order to set the parameters of the model, we proceed by examining how the clustering process can be applied between *two* different classes, leaving one class aside. We set the size of the lattice to $L = 50$ and choose two classes by taking the 100 amoebae which correspond to these two classes and distribute them randomly on the lattice. For the sake of simplicity, we keep the value of the emission rate of waves to $\lambda = 0.01$ and the agitation rate to $p_{\text{ag}} = 0.1$, as seen previously. Our goal is to set $\xi$ to a value which will minimise the convergence time. Recall that $\xi$ determines the attraction-repulsion probability $p_{\text{att}}$. The convergence time will be measured, as above, for a homogeneity criterion of $h^* = 0.80$ and a compactness of $\kappa = 0.25$.

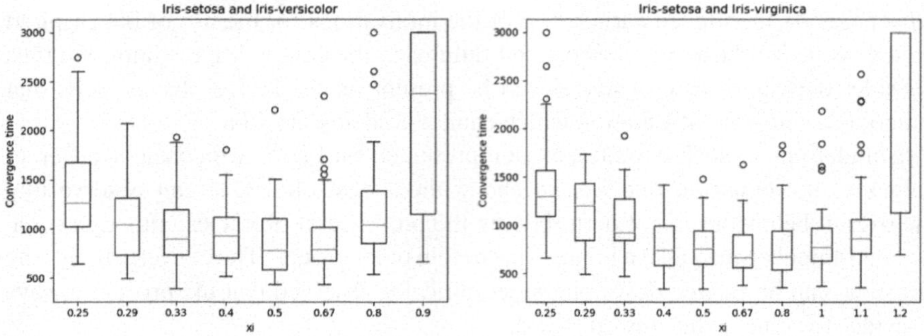


**Fig. 9** Convergence time obtained according to the value of $\xi$ for the dataset composed by the couple of Iris

The results of these experiments are shown on Fig. 9. For the two couples of data sets *Iris setosa*/*Iris versicolor* and *Iris setosa*/*Iris virginica*, we observe on the simulations that the clustering process succeeds quite rapidly and the curves expressing the variation of the convergence time with respect to $\xi$ have a regular aspect which clearly show the existence of an optimal range of settings for $\xi$. However, for the last couple *Iris versicolor*/*Iris virginica*, we could not find a suitable setting to separate these classes rapidly. This is not surprising given that these two types of flowers have close characteristics and are known to be difficult to separate. We thus don't take into account this couple for our setting of $\xi$. Observing the values of $\xi$ which minimise the convergence time for the two couples *Iris setosa*/*Iris versicolor* and *Iris setosa*/*Iris virginica* groups, we set $\xi = 0.5$, since this value is suitable for the two experiments.

We can now visually observe the clustering process on the entire Fisher Iris data set: see Fig. 10. It can be seen that, as expected, the *Iris setosa* cluster is well isolated from the others and is compact; in the great majority of cases, it is also the first group to form. For the two other classes, the separation process takes much longer but they eventually manage to form two groups that can be partially distinguished. However, this separation is not perfect as some amoebae do not succeed in reaching their class.

From the simulations, we noticed that the difficulty to finish the clustering process stems from the fact that when a group is formed, its excitations inhibit the formation of the other groups. One way of diminishing this undesirable effect would be to limit the range of propagations of the excitations. However, adding such a mechanism would increase the complexity our the rules to employ, which would go against the spirit of keeping things as simple as possible in which we have worked so far.

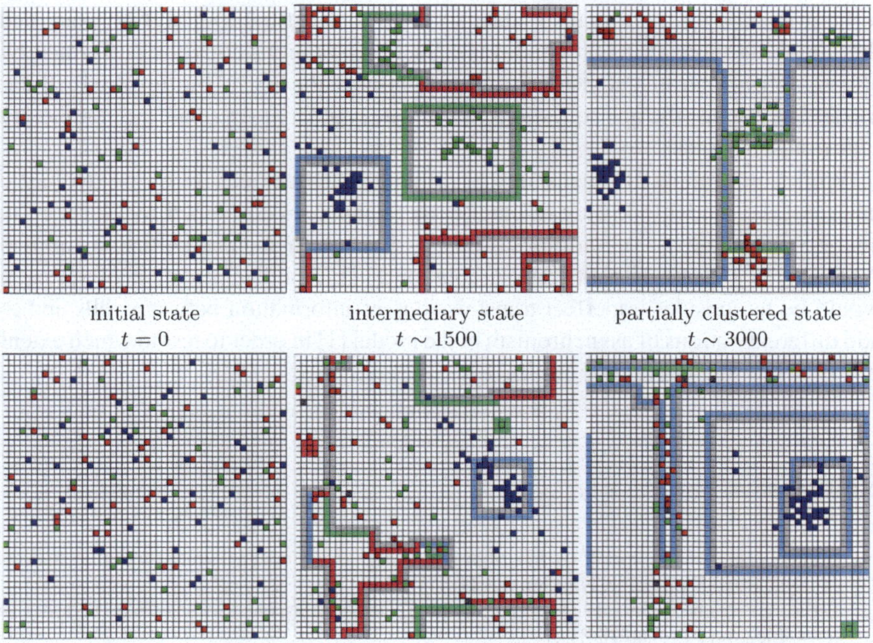


**Fig. 10** Two evolutions of the clustering process for $\xi = 0.5$. The amoebae are represented in blue, red and green, which respectively correspond to the classes *Iris setosa*, *Iris virginica* and *Iris versicolor*. When several amoebae are present in the same cell, the number of amoebae is indicated by a white number (this number can me magnified in the online version). A neutral cell is white and a refractory cell in light grey. Excited cells are light-coloured according to the type of the data they contain

# 5 Openings

These first tests are rather encouraging and show that our model remains robust to various modifications of its settings. The particularity of our bio-inspired method is its specific use of *randomness*: the moments where the amoebae emit their waves is determined randomly, and should be set in order to allow a good balance between "speaking" and "hearing". The clustering task results from a probabilistic setting, which defines a balance between attraction and repulsion when a given type of data is received. Observations show that the model easily adapts to various data sets and various initial conditions. Moreover, if some parts of the system that performs the computation are altered, this should not impede the clustering task as the behaviour results from local and simple computations only.

We presented here only a prototype tested on a rather simple clustering task with only two or three groups to form. The tests performed on the Fischer's Iris data set are quite encouraging and indicate the possibility to extend our method to a higher number of classes and a higher amount of data. The model thus now needs to be evaluated on more realistic data sets and its robustness should be examined thoroughly in more stringent conditions. Such explorations could be performed by all researchers who are interested in the exploration of the field of bio-inspired cellular models and among the open questions raised, one could try to evaluate the complexity of the clustering task.

While this complexity seems at first difficult to define because of the stochastic nature of the model, one may re-employ the techniques developed by Goles et al. analyse to complexity of various stochastic phenomena such as Diffusion-Limited Aggregation (DLA) or bootstrap percolation [11]. Another direction of research would be to consider imperfect transmissions of information between cells and to add different notions of asynchronism in the model [1] in order to test to which extent it is robust to such perturbations [9, 10].

## References


1. Bouré O, Fatès N, Chevrier V (2012) Probing robustness of cellular automata through variations of asynchronous updating. Nat Comput 11:553–564
2. Bušić A, Fatès N, Mairesse J, Marcovici I (2013) Density classification on infinite lattices and trees. Electron J Probab 18(51):1–22
3. de Oliveira PPB (2014) On density determination with cellular automata: results, constructions and directions. J Cell Autom 9(5–6):357–385
4. Fatès N (2016) Collective infotaxis with reactive amoebae: a note on a simple bio-inspired mechanism. In: Yacoubi SE, Was J, Bandini S (eds) Proceedings of ACRI 2016, vol 9863. Lecture notes in computer science. Springer, pp 157–165
5. Fatès N (2010) Solving the decentralised gathering problem with a reaction-diffusion-chemotaxis scheme. Swarm Intell 4:91–115
6. Fischer RA (1936) The use of multiple measurements in taxonomic problems. Ann Hum Genet 7:179–188
7. Fujita O (2013) Metrics based on average distance between sets. Jpn J Ind Appl Math 30:1–19
8. Goles E, Gómez L (2018) Combinatorial game associated to the one dimensional Schelling's model of social segregation. Nat Comput 17:427–436
9. Goles E, Lobos F, Montealegre P, Ruivo ELP, De Oliveira PPB (2020) Computational complexity of the stability problem for elementary cellular automata. J Cell Autom 15:261–304
10. Goles E, Montealegre P (2020) The complexity of the asynchronous prediction of the majority automata. Inf Comput 274
11. Goles E, Montealegre-Barba P, Todinca I (2013) The complexity of the bootstraping percolation and other problems. Theor Comput Sci 504:73–82
12. Handl J, Meyer B (2007) Ant-based and swarm-based clustering. Swarm Intell 1(2):95–113
13. Medina P, Goles E, Zarama R, Rica S (2017) Self-organized societies: on the Sakoda model of social interactions. Complexity, pp 427–436
14. Montalva-Medel M, de Oliveira PPB, Goles E (2018) A portfolio of classification problems by one-dimensional cellular automata, over cyclic binary configurations and parallel update. Nat Comput 17:663–671
15. Nanda SJ, Panda G (2014) A survey on nature inspired metaheuristic algorithms for partitional clustering. Swarm Evol Comput 16:1–18
16. von Neumann J, Burks AW (1966) Theory of self-reproducing automata. University of Illinois Press, Urbana, USA
17. Voß-Böhme A, Deutsch A (2010) The cellular basis of cell sorting kinetics. J Theor Biol 263(4):419–436

# Model Discovery and Discrete Inverse Problems with Cellular Automata and Boolean Networks

**Hector Zenil, Yanbo Zhang, and Narsis A. Kiani**

**Abstract** We introduce and explore some methods based on algorithmic complexity and algorithmic probability that help address the challenge of empirical causal model discovery and inverse problems. These methods, based on Algorithmic Information Dynamics (AID), are designed to describe a possible pathway from observation to causal reconstruction of the dynamics and space-time evolution of discrete systems, with consideration given to inference cost. We apply these methods to two of the most popular discrete dynamical systems, cellular automata and Boolean networks. We show that an algorithmic-probability-guided simulation of dynamic properties of these discrete systems can connect back to fundamental questions of causality and scientific discovery, whereas complexity science has traditionally tended to obfuscate such connections or obviate them with informal concepts of emergence and self-organisation. In the inverse problem of phase-space reconstruction, we consider the cost trade-off between observation and simulation in the challenge of model inference. We combine both algorithmic complexity and Bayesian methods to characterise an observation as a sequence of small computable models allowing incremental scientific model discovery, thereby providing a complexity framework that contributes to the study of causation.

**Keywords** Cellular automata · Boolean networks · Inverse problems · Causal model generation · Observability · Simulation · Scientific discovery · Causality · Algorithmic information dynamics · Mechanistic explanation

---

H. Zenil (✉)
Oxford Immune Algorithmics, Davidson House, The Forbury, Reading RG1 3EU, UK

The Alan Turing Institute, British Library, 96 Euston Rd, London NW1 2DB, UK

H. Zenil (✉) · Y. Zhang · N. A. Kiani
Unit of Computational Medicine, Algorithmic Dynamics Lab, Karolinska Institute, Visionsgatan 18, L8, 171-76 Stockholm, Sweden
e-mail: hector.zenil@cs.ox.ac.uk

Y. Zhang
School of Earth and Space Exploration, Arizona State University, 781 Terrace Mall, Tempe, AZ 85287, USA

 
A. Adamatzky (ed.), *Automata and Complexity*, Emergence, Complexity and Computation 42, https://doi.org/10.1007/978-3-030-92551-2_24

# 1 Preliminaries

In this paper, we follow an inverse-problem approach similar to the statistical-computational one introduced in [10] combined with some of the methods in [13] to reverse engineer and find underlying computational models able to approximate empirical phenomena computable or not. The main difference in this work is the introduction of a probabilistic computational approach complementing the statistical-computational one that offers more flexibility to find computational models based on sets of most probable rules thus being able to find sequential sub-models faster than waiting for the computational rule able to explain the observation in full.

To illustrate the proposed method, we use two of the most frequently used discrete dynamical system models, both equally computationally powerful as models (not as instances). On the one hand, there are cellular automata, whose evolution rules are constrained by local neighbours in space, but unlike Boolean networks are not bounded in size and can grow over time and space. On the other hand, there are Boolean networks that, once each individual network is defined, are constrained in size and forced to repeat themselves according to their Poincaré recurrence limit, though their evolutions are not bounded by local neighbour states.

## 1.1 Cellular Automata

A cellular automaton is a deterministic rewriting dynamical system that evolves in discrete time and discrete space, this latter usually a grid. It consists of a grid of cells that are locally but synchronously updated across the grid according to a global time scale and a global recursive rule governing the evolution of the state of each cell as a function of the state of neighbouring cells. While cellular automata have the same computational power as other Turing universal models and are therefore fundamentally equivalent, as they can emulate each other, one of the most salient features of cellular automata is the qualitative diversity of their space-time evolutions when exploring different rules and different initial conditions.

In 1983 [8], and later in 2002 [9], Wolfram was the first to explore CA in a systematic fashion, and conceived a simple enumeration scheme based on the rule representation. Each ECA rule can be represented by a row of all 3-tuples (the central cell and its closest neighbours), of which there are 8 cases, followed by a row of the rule assignation to either a white or a black cell. This latter row, consisting of binary digits, given that the ECA is a 2-state rule space, can be read as a binary number, e.g. 00000010, which in decimals is, e.g., rule 3 (for the depicted case). All possible mappings give a total number of $2^8 = 256$ rules.

A *cellular automaton* (CA) is a tuple $\langle S, (\mathbb{L}, +), T, f \rangle$ with a set $S$ of states, a lattice $\mathbb{L}$ with a binary operation $+$, a neighbourhood template $T$, and a local rule $f$.

The *set of states* $S$ is a finite set with elements $s$ taken from a finite alphabet $\Sigma$ with at least two elements. It is common to take an alphabet composed entirely of integers modulo $s$: $\Sigma = \mathbb{Z}_s = \{0, \ldots, s-1\}$.

An element of the lattice $i \in \mathbb{L}$ is called a cell. The lattice $\mathbb{L}$ can have $D$ dimensions and can be either infinite or finite with cyclic boundary conditions.

The *neighbourhood template* $T = \langle \eta_1, \ldots, \eta_m \rangle$ is a sequence of $\mathbb{L}$. In particular, the neighbourhood of cell $i$ is given by adding the cell $i$ to each element of the template $T$: $T = \langle i + \eta_1, \ldots, i + \eta_m \rangle$.

Each cell $i$ of the CA is in a particular state $c[i] \in S$. A *configuration* of the CA is a function $c : \mathbb{L} \to S$. The *set of all possible configurations* of the CA is defined as $S_{\mathbb{L}}$.

The *evolution of the CA* occurs in discrete time steps $t = 0, 1, 2, \ldots, n$. The transition from a configuration $c_t$ at time $t$ to the configuration $c_{(t+1)}$ at time $t+1$ is induced by applying the local rule $f$. The local rule is to be taken as a function $f : S^{|T|} \to S$ which maps the states of the neighbourhood cells of time step $t$ in the neighbourhood template $T$ to cell states of the configuration at time step $t+1$:

$$c_{t+1}[i] = f(c_t[i + \eta_1], \ldots, c_t[i + \eta_m]) \tag{1}$$

The general transition from configuration to configuration is called the *global map* and is defined as: $F : S^{\mathbb{L}} \to S^{\mathbb{L}}$.

In the following we consider only 1-dimensional (1-D) CA as introduced by Wolfram [8, 9] who named them Elementary Cellular Automata. The lattice can be either finite, i.e. $\mathbb{Z}_N$, having the length $N$, or infinite, $\mathbb{Z}$. In the 1-D case it is common to introduce the *radius* of the neighbourhood template which can be written as $\langle -r, -r+1, \ldots, r-1, r \rangle$ and has length $2r+1$ cells. With a given radius $r$ the local rule is a function $f : \mathbb{Z}_{|S|}^{|S|^{(2r+1)}} \to \mathbb{Z}_{|S|}$ with $\mathbb{Z}_{|S|}^{|S|^{(2r+1)}}$ rules. The so called Elementary Cellular Automata (ECA) with radius $r = 1$ have the neighbourhood template $\langle -1, 0, 1 \rangle$, meaning that their neighbourhoods comprise a central cell, one cell to the left of it and one to the right. The rulespace for ECA contains $2^{2^3} = 256$ rules.

## 1.2 Block Decomposition Method

In [2, 7], a measure motivated by algorithmic probability was introduced that entails exploring an enumeration of Turing machines running on a reference Universal machine that is assumed to be optimal, and whose variant specifications across several papers are widely used in other areas of the theory of computation, a machine based on Rado's so-called Busy Beaver problem [5].

Let $(n, m)$ be the space of all $n$-state $m$-symbol Turing machines, $n, m > 1$ and $s$ a sequence. Then we define CTM (standing for Coding Theorem Method) as: $CTM(n, m)(s) = \frac{|\{T \in (n,m) : T \text{ produces } s\}|}{|\{T \in (n,m)\}|}$.

What CTM does is to precompute these programs for a large set of short strings and combine them with classical information theory using what we call BDM (standing for Block Decomposition Method) in order to build a short program based on small pieces than can explain a larger piece of data for which CTM has not (yet) been calculated. CTM can also be seen as exploring all possible computable compression algorithms, each represented by a decidable (time-bounded) Turing machine from a proper universal enumeration, whose machine number is stored and allows for a decompression process, thus honouring the true spirit of Kolmogorov complexity as the ultimate data compression method, rather than privileging its much more shallow connection to statistical compression only.

CTM is motivated by the relation [7] between the algorithmic probability of an object such as a string (the frequency of production of a string from a random program) and the algorithmic complexity of the said string (the length of the shortest description under the reference machine).

Just as resource-bounded complexity compromises on resources like runtime, and MDL, MML and LZW give up on Turing completeness (and therefore on algorithmic complexity for the most part) in order to model data, and LZW is intended to capture no other feature of an algorithmic nature beyond repetitions in data, the assumptions that CTM makes are that the invariance theorem is valid and that the additive (or other) constant is small enough not to impact the ranking or the top ranking of the smallest computable models explaining data, which is precisely what CTM investigates empirically, with a view to making changes to the underlying computing models and testing against the data.

CTM is intended not just as a theoretical construct, but as a technique to be deployed in practice. CTM outperforms or complements lossless compression algorithms and other techniques such as MDL in various respects. For example, CTM can provide a finer-grained classification of short strings and is able to provide computable descriptions in the form of specific computer programs in the language of Turing machines, rather than black boxes (compressed files) with little state-to-state correspondence, and most likely unrelated to the generating process other than in terms of trivial statistical redundancy. MDL, MML and other such approaches present their ‘flaws’ as features (or even advantages), such as the claim that MML does not require Turing completeness or that MDL is computable (without ever mentioning what’s sacrificed). MDL avoids assumptions about the data-generating process.

One way to see CTM is as effectively exploring the space of all the possible computer programs able to compress a piece of data. This is in contrast to settling on a single program that one can ascertain in advance to be incapable of dealing with any algorithmic feature in data (as opposed to just statistical features, i.e. substring repetitions).

The block decomposition method (BDM) is an extension of CTM defined as $BDM(s) = \sum_i CTM(s_i) + \log(n_i)$, where each $s_i$ corresponds to a substring of $s$ for which its CTM value is known and $n_i$ is the number of times the string $s_i$ appears in $s$. A thorough discussion of BDM may be found in [11]. BDM extends the power of CTM to deal with larger strings by sticking together, using classical information theory, the estimations of complexity values of shorter substrings based on CTM under the universal model of computation. BDM is explored in [11], and among its properties is the fact that it cannot perform worse than Shannon entropy but can indeed improve upon it when combining values of algorithmic complexity for its substrings.

BDM glues together the programs producing each of the data pieces using classical information. The result is a hybrid measure that provides a candidate upper bound of algorithmic complexity.

In contrast to statistical compression algorithms, CTM can, in principle, deal with simple objects such as 123456 ... in a natural way (or $1, 2, 3, 4, 5, 6, \ldots$ seen as a data stream from a generating process) by identifying the underlying successor function in any base from observing the piece of data and extending the Turing machine rule space, as shown with 2 examples in [12]. And even for $\pi$ [11], both in principle and in practice, as suggested in [11], when normalising by highest values, Shannon entropy (and therefore LZW and cognates) will by definition retrieve maximal randomness, but for CTM and BDM, it does not have maximal randomness and actually shows a trend of decreasing values. What CTM does is find the set of computer programs that can explain segments of $\pi$, something that statistical compression cannot do.

## *1.3 Algorithmic Information Dynamics*

Algorithmic Information Dynamics (AID) [14] is an algorithmic probabilistic Bayesian framework for causal discovery and causal analysis. It enables a numerical solution to inverse problems based on or motivated by principles of algorithmic probability. AID studies dynamical systems in software space, where all possible computable models can be found or approximated under the assumption that discrete longitudinal data such as particle orbits in state and phase space can approximate continuous systems by Turing-computable means.

AID combines perturbation analysis and algorithmic information theory to guide a search for sets of models compatible with observations and to precompute and exploit those models as testable generative mechanisms and causal first principles underlying data and systems. AID is an alternative or a complement to other approaches and methods of experimental inference, such as statistical machine learning and classical information theory.

AID connects with and across other parallel fields of active research such as logical inference, causal reasoning, and neuro-symbolic computation. AID studies how candidate discrete computable equations as generating mechanisms are affected by changes in observed phenomena over time as a result of a system evolving (e.g. under the influence of noise) or being externally perturbed.

AID is related to other areas such as computational mechanics and program synthesis. However, unlike methods such as Bayesian networks, AID does not rely on graphical models or the (often inaccessible) empirical estimation of mass probability distributions. AID encompasses the foundations and methods that make the area of algorithmic information and algorithmic complexity more relevant to scientific discovery and causal analysis.

We define sensitivity as average complexity change after perturbation or "information difference," following Algorithmic Information Dynamics. Let $|S|$ denote the size of the object $S$, and not only the number of constitutive elements. Let $N$ denote the total number of constitutive elements of $S$. (Therefore $N$ may be smaller than $|S|$ , but $|S| = \mathbf{O}(f_c(N))$, where $f_c$ is a total computable function). For example, in the case of networks, $|S|$ grows as the number of all possible edges of a graph (in $\mathbf{O}(N^2)$) and $N$ is the total number of vertices.

Let $F \neq$ be a subset of the set of elements of the object $S$ to be deleted. In order to grasp such a formal measure of the algorithmic-informational contribution of each element, one may look into the difference in algorithmic complexity between the original data $S$ and the destructively perturbed data $S \setminus F$. We define the "information difference" [13] between $S$ and $S \setminus F$ as:

$$I(S, F) = K(S) - K(S \setminus F)$$

Note that the same holds analogously for constructive perturbation, i.e., the insertion of elements. When $F = \{e\}$ with $I(S, F) = I(S, e)$ and this difference assumes an absolute value:

$$|I(S, e)| = |K(S) - K(S \setminus e)| \leq \log(N) + \mathbf{O}(1),$$

we say that $e$ is a "neutral information element", since the algorithmic information contribution of the insertion of $e$ back into $S$ (or its deletion from $S$) would necessarily be of the same order as a computable reversible procedure. Furthermore, we say $e$ is a "positive information" element, if it is not neutral and $I(S, e) > 0$; and a negative information subset of elements, if it is not neutral and $I(S, e) < 0$. Thus, negative content is immediately associated with information gain from perturbations, and positive content with information loss.

In the context of causal analysis, when considering a stochastically random bit-flip being performed on a string pattern, such as 0000000... repeated $n$ times, such a perturbation will have a significant effect with respect to the length of the original shortest program $p$ consisting of a loop printing $n$ number of 0s, because the new program $p'$ would need to take into account the 1 inserted at a stochastically random position $x \leq n$. If, however, the string is algorithmically random, the effect of any stochastically random single-bit perturbation will have no effect (up to a logarithmic term of the string's algorithmic complexity, which is already maximal), because every bit is already necessary information for the string's own description. Not only for strings, but also for graphs and other multidimensional data structures, (stochastically) random single perturbations have similar worst-case effects in both simple and algorithmically random objects, amounting to up to about $\log(|S|)$. Thus, in the case of objects for which $|S| \leq N^C$, where the constant $C \geq 2$ that does not depend on $N$ (e.g. if the object is a graph), the information difference impact of moving $S$ toward or away from algorithmic randomness may be:

- of the same order (i.e., $I(S, e) = \mathbf{\Theta}(\log(N))$) as the object's algorithmic complexity $K(S)$, if, e.g., $S$ is computable and therefore logarithmically compressible in the form $K(S) \leq \log(N) + \mathbf{O}(\log(\log(N))) + \mathbf{O}(1)$;
- or strongly asymptotically dominated by the object's algorithmic complexity $K(S) \geq |S| - \mathbf{O}(1)$, as $I(S, e) = \mathbf{O}(\log(|S|)) = \mathbf{o}(|S|)$ , if, e.g., $S$ is algorithmically random and therefore incompressible (up to a constant).

In other words, at the asymptotic limit when $N \to \infty$, the ratio of worst-case information difference per bit of algorithmic complexity:

$$\frac{I(S, e)}{K(S)}$$

tends to 0 for algorithmically random objects and to a constant $\epsilon > 0$ for computable objects. The more the object's algorithmic complexity diverges from the maximal value (i.e., the larger the algorithmic randomness deficiency), the larger the relative impact of the worst case on the object's algorithmic complexity (i.e., on its optimal lossless compressibility).

In [3], we showed how the methods behind AID can help reconstruct the initial conditions and rules of several systems, including differential equations and cellular automata. This expands and takes a different direction on approximately sequential rules as probabilistic rule-based models, and hence is a hybrid approach, combining statistical inference and pure symbolic computation.

## 2 Naïve Reconstruction of Cellular Automata Rules

In [6], it was shown that this process can be generalised to any cellular automata and rule space. Once a rule space is suspected to be the rule producing the observed space-time evolution, all local rule templates should be sampled, in this case the 8 local rules of an ECA. The time to infer a global rule is a function of the number of local rules that must be observed to fill out all templates. Rule 0, for example, can be sampled from a single observation, as all others are the same (all blank or black cells go to blank cells), when assuming that the observation is complete, i.e., covers all possible states of the observed phenomena. This is, of course, rarely the case in the real world, where no terminal observation exists and new observations can vary due to limitations of the measuring device and noise. We refer to the number of ideal observations (that complete the generating rule) as the time cost to infer a rule. The example of Rule 0 may suggest a complexity measurement to guide the place and time when observations should be made to sample the space-time evolution of a certain phenomenon to maximise informativeness in the process of inference. We also consider the computational cost of observation versus simulation in model discovery in the process of rule discovery (Figs. 1, 2, 3, 4 and 5).

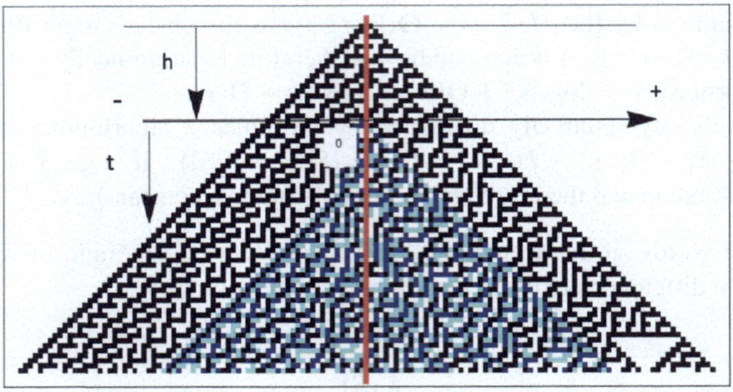


**Fig. 1** Impact of a perturbation after $h$ steps on the space-time evolution of ECA rule 30. Cells in blue are those affected. The causal effect of a perturbation at point 0 is contained within the light cone determined by the maximum propagating speed, in turn determined by number of steps. A single cell perturbation cannot propagate faster (wider on each side) than the ECA runtime. The study of the effect of a perturbation after $t$ steps requires the simulation of ECA up to runtime $t$ steps, and a cell-to-cell comparison against its original evolution is thus computationally expensive. Methods of gauging the effect of such perturbations without recourse to these 2 processes (simulation and comparison) are thus relevant

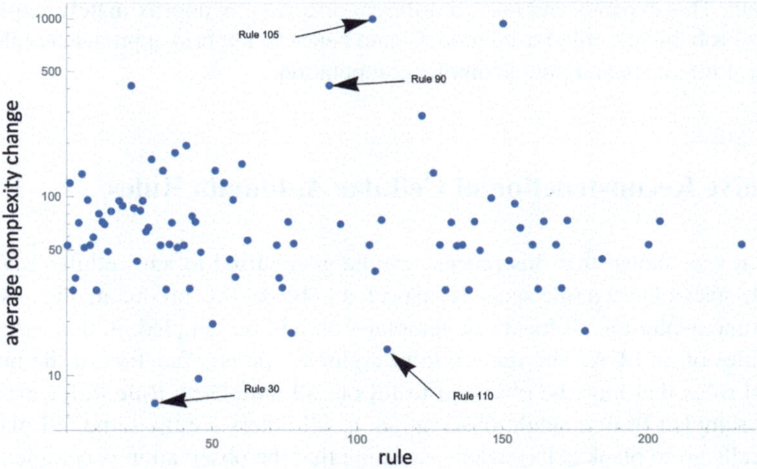


**Fig. 2** ECA rule sensitivity analysis (from simple initial condition). Illustration of complexity change is estimated by BDM (an estimation of program-length). One of the most random-looking ECA is also the least sensitive to point mutations, because the resulting space-time evolution remains in a random-looking state. This appears to be the case for complex rules like 110 as well, but not so with highly structured fractal-like rules like 105, 90 and 22, all starting from simplest single-black-cell initial condition

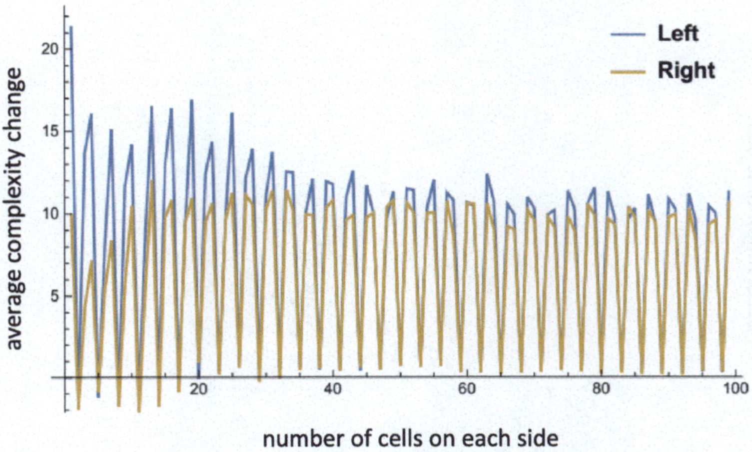


**Fig. 3** As expected, the left side of rule 30 has a greater sensitivity to perturbations than the right side, from single-black-cell initial condition

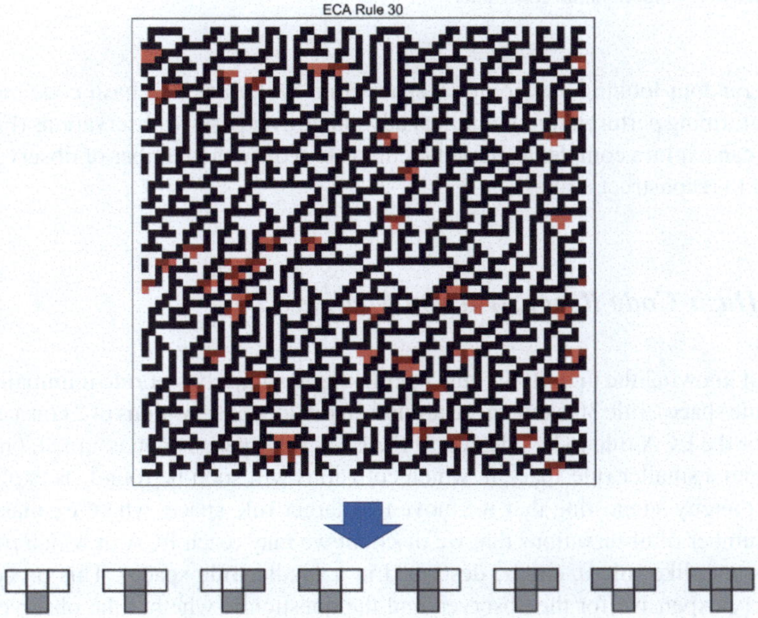


**Fig. 4** Random block observation times for reconstruction of rule 30 from random observations. The naïve reconstruction method takes random blocks from the space-time evolution of a discrete dynamical system and reconstructs the generating rules under assumptions of minimalism (no redundant local rules) and determinism (each configuration always leads to the same future state)

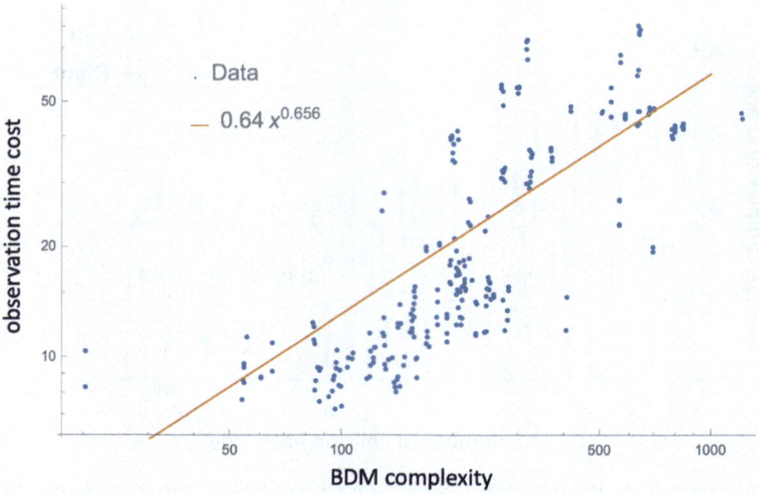


**Fig. 5** Time cost and BDM Complexity [11] comparison. Time costs of rules with their BDM Complexity for random initial conditions

For random-looking and complex space-time evolutions, the hash code method and preforming perturbations has a clear advantage over passive observations (Fig. 8).

We can explore complexity-based methods to reduce the number of observations needed to reconstruct a rule.

## 2.1 *Hash-Code Reconstruction Method*

Without knowing the underlying rule but making assumptions of rule minimality (in ECA rule space), rule 30 would need around 16 random observations of 2 consecutive rows for the ECA rule to be inferred. If the ECA rule space is not assumed, one can start from a smaller rule space in which collisions will soon be found, as explained in [6], thereby suggesting that we move to a larger rule space, where by making a fixed number of observations that we discount we may reach ECA in which rule 30 lives (some, like rule 0, can be described in a smaller rule space). This process is thus very expensive for the observer, and the question is whether the observer can guide the observations and even intervene to shorten the process.

After an initial sequence of observations, some hypotheses can be made, which can be represented by a sequence of possible candidate rules. What needs to be done is to find which hypothesis fits the data. Assuming we have a "test" $t$. If $t(observations, rule_i) = True$, then we have $p$ $(0 < p < 1)$ possibility of saying that $rule_i$ is the right hypothesis. This relationship can be explained in this hypothesis network: Fig. 6.

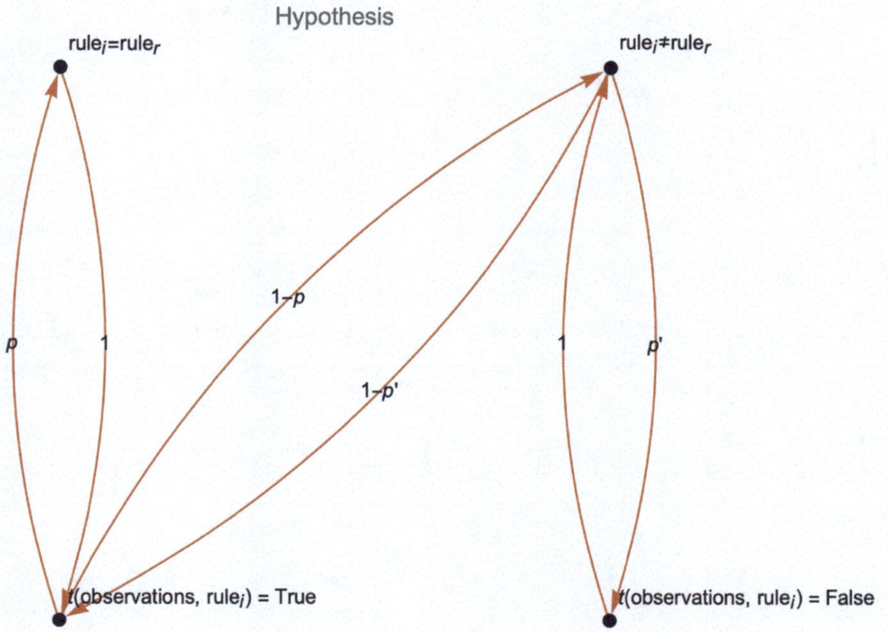


**Fig. 6** Model-driven hypothesis network

**Fig. 7** The red region (second and third rows from top to bottom) is the input $s$ of the hash function $H$, which is also the data being observed to test the hypothesis $\{r_1, r_2, \ldots, r_n\}$. The blue region (fifth row, 4 central cells) is the output of the hash function, i.e., $H(s)$. In the green region (fourth row, 6 central cells), information can have interactions which make the output more "random", so that $p'$ will be higher

The purpose is to know $p$. Based on Bayes' theorem, $p = P(r_i = .r_r|t = \text{True}) = \frac{P(r_i=r_r)P(t=\text{True}|r_i.=r_r)}{P(t=\text{True})}$. It is clear that $P(t = True|r_i = r_r) = 1$, and $P(r_i = r_r) = 1/n$. Also we find that $P(t = \text{True}) = P(r_i = r_r)P(t = \text{True}|r_i. = r_r) + P(r_i \neq r_r)P(t = \text{True}|r_i \neq r_r.) = \frac{(n-1)(1-p')}{n} + \frac{1}{n}$. So, $p = \frac{1}{n(\frac{(n-1)(1-p')}{n}+\frac{1}{n})} = \frac{1}{(n-1)(1-p')+1}$ (Figs. 7, 8 and 9).

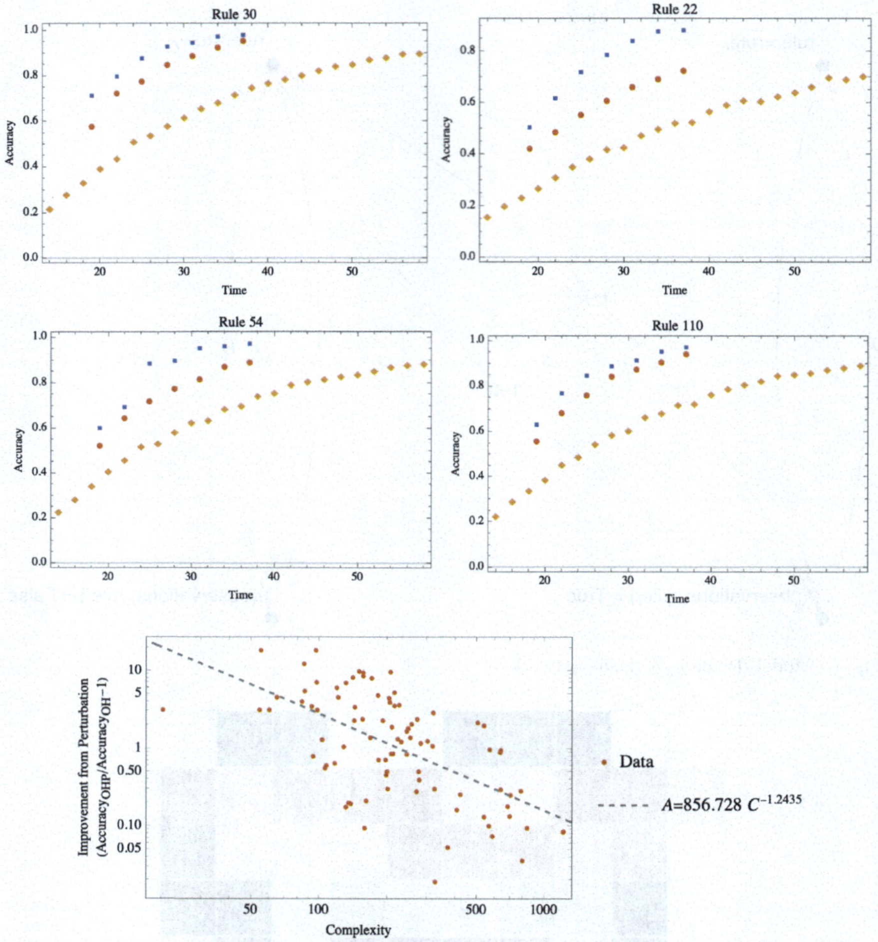


**Fig. 8** Individual cases followed by a correlation analysis between complexity and perturbation speedup across all 88 non-symmetric ECA rules. $Accuracy_{OHP}$ means observation + hash + perturbation, and $Accuracy_{OH}$ means observation + hash

To increase accuracy, we therefore need to increase $p$. According to the formula, there are two ways to increase $p$.

1. Decrease $n$;
2. Increase $p'$.

To decrease $n$, we need to observe some cells before using the hash method, so that we can make $n \sim 10$. To increase $p'$, we need to increase $P(r(r_i) = \text{False}|r_i \neq r_r.)$

If we make the requirement stronger, which means that even if $r_i$ is slightly different from $r_r$, a coding function $cod(r_i)$ would be significantly different from $cod(r_r)$, i.e., we need $cod(r)$ be a "hash function", like MD5, SHA64, etc.

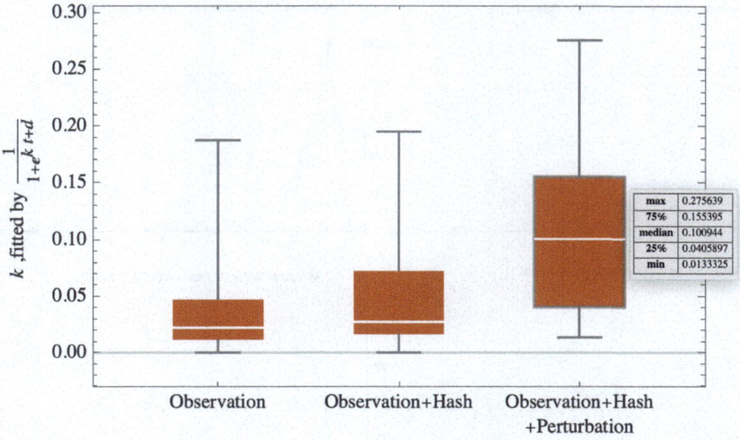


**Fig. 9** The hash approach followed by the hash approach and complexity guided perturbations leads to a significant improvement for model fitting for the same number of observations

Thus a better hash function will make $p'$ higher, so that $p$ will be higher, and the accuracy will be greater.

The reconstruction problem that we consider starts from random block observation chosen from a space-time evolution starting from a random initial condition. The problem is equivalent to a "Coupon Collector's Problem", consisting of taking sequences at random and finding how many samples are needed to obtain all block combinations.

Here we use a simplified model, similar to a mean-field model. We assume, in each measurement, that there is a possibility $p$ of getting a new kind of sequence. So the distribution of minimal measurement time $T$ is:

$$\mathcal{D}(T; p, n) = \binom{T-1}{n-1} p^n (1-p)^{T-n},$$

$$E(\mathcal{D}(T; p, n)) = n/p$$

where $n$ is the size of the rule space. For example, for ECA, $n$ equals to 8.

Here we conjectured that the expectation of $n$ is lowest only when all kinds of sequences have the same frequency of appearance. And according to the paper [1], at this condition, $E(T) = n \cdot H(n)$, where $H(n)$ is the $n$-th harmonic number, and $n$ is the number of different kinds. So $p$ has an upper limit, which means,

$$\frac{n}{p} \geq n \cdot H(n),$$

$$p \leq \frac{1}{H(n)}.$$

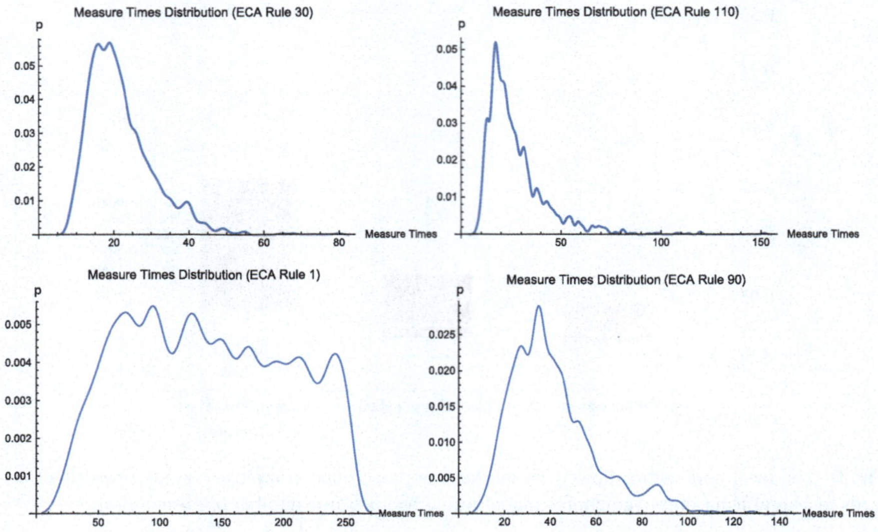


**Fig. 10** Distribution of measurement times $T$ for four typical rules. Most of them have the same pattern

For ECA, $n = 8$, so $p \leq 0.368$. Experiments show that for all ECA rules with random initial conditions, the distributions of $T$ have the same pattern for most ECA rules (see Fig. 10) that have similar patterns as $\mathcal{D}(T)$. And it also shows a good fit for most rules.

The result means that the random reconstruction can be considered an "unbalanced coupon collector's problem", which means that the frequency of appearance of different consequences is different. So a way to speed up the reconstruction procedure is to balance the frequencies; also, to make the space-time array more random, or make measurements in an area where there is greater randomness, which connects to the BDM method. Figure 11 shows the average measurement times of all ECA rules.

For random initial conditions, part of the randomness of the space-time array is owed to the initial conditions, especially for some linear rules. But previous work on random conditions has strongly inspired us to work on simple conditions. For example, we can speed up the reconstruction program by selecting specific parts of the space-time array guided by algorithmic probability.

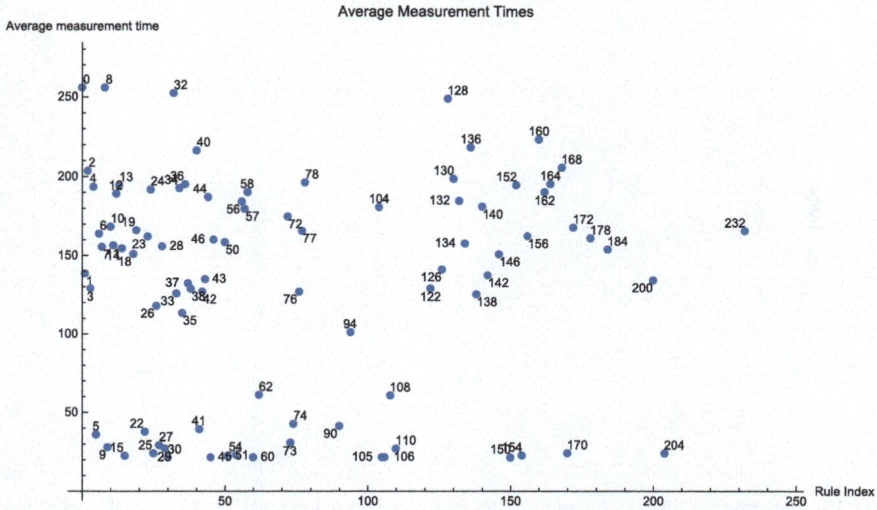


**Fig. 11** Average measurement times of ECA rules

# 3 Enumerating and Sorting Rule Neighbours

A CA consists of a rule of the following type:

$$f(x_{r_1}, x_{r_2}, \ldots, x_{r_k}),$$

where $r_i$ is the $i$-th neighbour of the cell (including the cell itself), $x$ is the colour of the space, $k$ is the number of neighbours. To reconstruct CA rules, the first thing to do is to enumerate all possible rule neighbours, and sort them in a particular way.

In order to start with a simple rule space the program should start with low $k$ and increase it when necessary. But for a certain $k$, like $k = 2$, if the maximum CA growth (commonly known as the "light speed" of the CA) is 1, then there should be three different configurations. See Fig. 12.

In Fig. 12, the first and the last one are symmetrical templates with the same complexity. But the complexity of the middle local rule may be different (lower under some boundary conditions according to BDM). An important aspect to consider is whether the template  needs to be retained, or whether it can just be combined with .

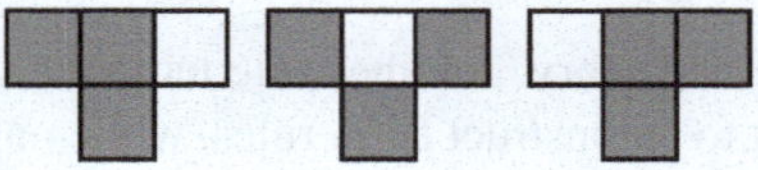

**Fig. 12** Three different rule templates that have two neighbours. The dark cells in the first row are available cells

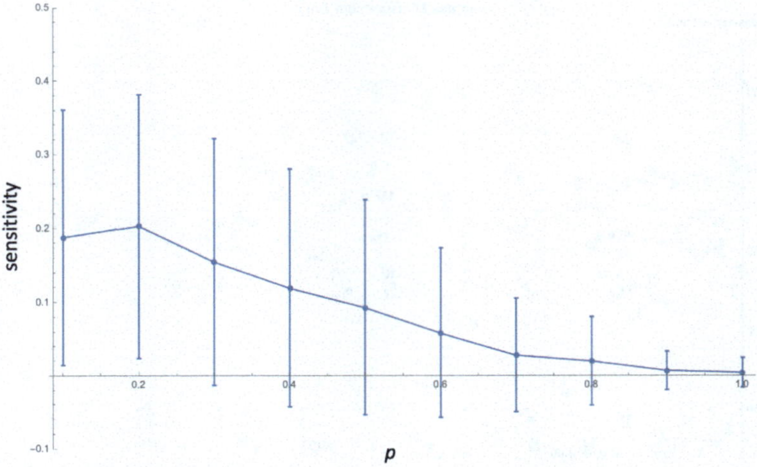


**Fig. 13** Dynamic Sensitivities of random BNs with different $p$. Let $\{seed, p, rule\}$ be the definition of a Cellular Automaton encoding Boolean Network (CABN). Changing $p$ will average number of links in BNs. It can be found that the dynamic sensitivities decrease when $p$ increases

Under boundary condition assumptions (completing the $3 \times 3$ square), BDM determines that has a complexity similar to , so can be combined with . And has a complexity similar to , so it can be combined with . So to reconstruct CA rules, experiments shown that blocks with different complexity minimise inference time and model fit. Figure 5 shows that these templates have different values of algorithmic complexity measured according to the Block Decomposition Method [11] (Fig. 13).

Template and are symmetrical templates, so in the following experiment, we follow both procedures.

In the reconstructed template program, the order is: $\rightarrow$ $\rightarrow$ $\rightarrow$ , and $\rightarrow$ $\rightarrow$ $\rightarrow$ .

The order of the templates was checked by enumerating all possible orders, and it was found that the order does not have a significant impact. It was also found that some templates should not be discarded because can strongly affect the experiments' results.

For example, when rule templates that have zero neighbours (empty templates) were discarded, meaning that the CA function is $f(x_{i-1}, x_i, x_{i+1}) = \text{Constant}(0 \text{ or } 1)$, the omission strongly affected the result. If the program observes evaluations of rule 0 without an empty template, it takes more time to get both and . This shows that the empty templates are necessary. Another rule template discarded, , would also reduce the time cost to reconstruct ECA rules, while only adding a few steps in certain cases. These changes strongly affect the results.

There are some parameters that need to be determined for this method to work. First, the minimal observation time $t_{\min}$. For example, when the program is reconstructing a rule with 2 neighbours, it needs at least $2^2 = 4$ observations, but to get enough accuracy, the $t_{\min}$ should be higher. Another parameter to be determined is the maximal observation time $t_{\max}$—to limit the time costs while still ensuring enough accuracy.

The reconstruction starts at a random condition, which means that the program chooses blocks randomly from a space-time array generated with random conditions. This is similar to the Coupon Collector's Problem, which consists of finding out how many random sequences are needed to complete all the sequences.

The formula of the distribution is too complex, especially when the appearance frequencies of the sequences are different. Here we use a simplified model, which is a kind of mean-field model. We assume that with each measurement there is a possibility $p$ of getting a new kind of sequence. So the distribution of minimal measurement time $T$ is given by:

$$\mathcal{D}(T; p, n) = \binom{T-1}{n-1} p^n (1-p)^{T-n},$$

$$E(\mathcal{D}(T; p, n)) = n/p,$$

$$\mathrm{D}(T) = \frac{n^2}{p^2} - p^{2n}\, {}_2F_1(n, n; 1; (p-1)^2),$$

where $F$ is a hypergeometric function, $n$ is the size of the rule space, for example, for ECA, and $n = 8$.

Based on the results above, we can let $t_{\min} = n/p$, and $t_{\max} = k\sqrt{\mathrm{D}(T)} \approx k t_{\min}$, which make for parameters with fewer artificial contents.

In this experiment, we use several different space-time array sizes—8, 16, 32, 64, and 128—with random initial conditions. Two different behaviours have been found; see Fig. 5. With the growth of the array size, the time cost may increase or decrease.

The reason is that for some rules their dynamics can only be inferred from the beginning of their space-time evolutions. When the program makes observations randomly, the time cost will increase, and the accuracy of reconstruction will decrease. While for other rules, like rule 30, the information about dynamics is spread over the space-time evolution. When the size increases, the time cost will actually decrease. Figure 11 shows time costs for reconstructing all ECA rules.

As shown in Fig. 14, the space-time dynamics of a BN, can be represented by the permutation of the nodes and thus its description is not captured uniquely by a single adjacency matrix, the reconstruction method has thus to have certain flexibility to capture more fundamental dynamics than those of the change of representation. This was achieved by finding a discriminator of the adjacency matrix to only take into consideration the representation that minimises the (BDM) complexity of the

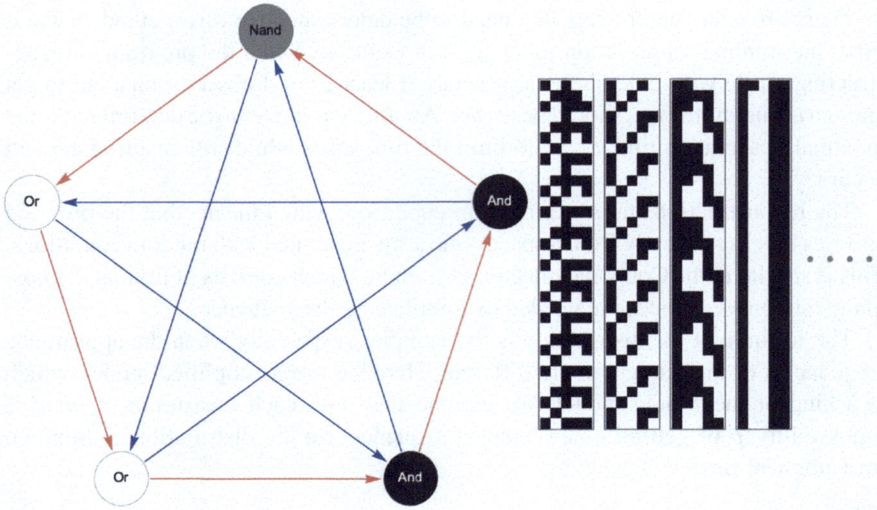


**Fig. 14** A Boolean Network (left) can be defined by three parameters $\{seed, p, rule\}$. This figure illustrates a coding system (right) to describe the dynamics of a Boolean network using a quadratic list. 0 means AND, 1 means OR, 2 means NAND, 3 means NOR. A random seed and number $p$ is used to generate a network

final encoding. As shown in Fig. 15, given that every encoding fully reconstructs the observation/instance of the dynamical system, any other representation with higher complexity would have been the result of the original complexity and an additional factor introduced by that other encoding which does not belong to the system itself and can therefore be discarded.

## 4 Boolean Networks as Inverse Problems

In Fig. 15, are shown distributions of the network in Fig. 14 (left) corresponding to each of the space-time encodings (adjacency matrices) displayed in Fig. 14 (right) and showing similarities in their own internal local distributions (groups of ECA rules used to characterise each encoding).

The distributions are sorted by the BDM, the encodings ($D_1, D_2, \ldots, D_{n!}$), where $n$ is the size of the Boolean network. $D_1$ has the lowest BDM, which also gives us the most information about the systems' dynamics. So $D_1$ was taken as representing the complexity of this BN's dynamics.

Network reconstruction is a highly investigated topic in computational biology. Once equipped with efficient tools to reconstruct a rule from partial random observations, we apply them to the challenge of reconstructing not the topology of the Boolean network but the qualitative dynamics of the network. Here, we are interested

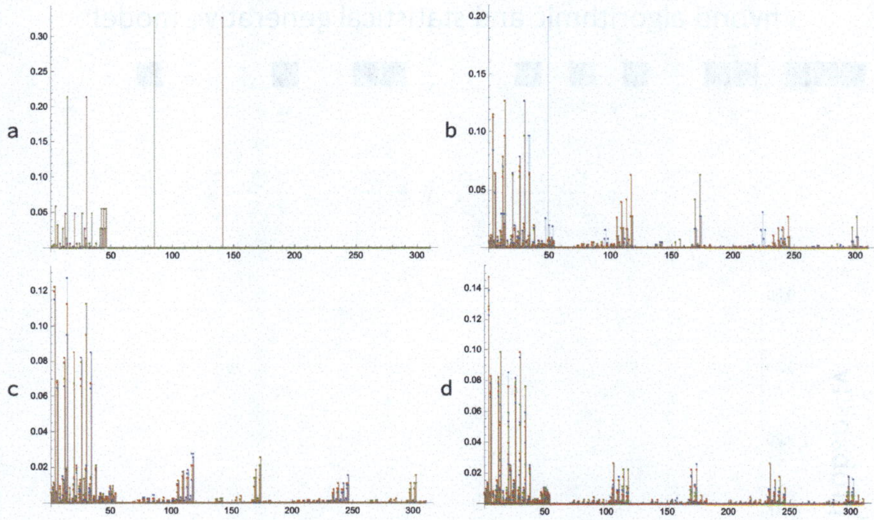


**Fig. 15** One challenge is the impact of the encoding of the dynamics of the BN on the model found explained by the ECA rules sequence distribution

in recovering/reconstructing the dynamics of a network rather than its topological features.

In Boolean networks, it is not possible to embed the nodes in a geometrical space. So if the CA space size $s$ is smaller than number $N$, there will be overlaps. But we cannot eliminate the overlaps by increasing the CA space size $s$. For most Boolean networks, cycle lengths are much shorter than the number of nodes $N$. According to Kauffman's original paper [4], $L_c \sim 0.6 \log N$, where $L_c$ is the cycle length. So for our observation, there are about $0.6N \log N$ cells that truly provide information. But when increasing CA space size $s$, the number of possible fragments $n_r$ increases exponentially, $n_r = 2^s$. In the best case, all observations give different fragments, but we still only get a very small part $\eta$ of the rule space $\eta(N) \sim (N log N)/(2^N)$. For example, $\eta(3) = 0.41, \eta(4) = 0.34$, $\eta(5) = 0.25$, etc.

If $\eta$ is too small it means we do not have enough observations to reconstruct a CA. And it can even lead to "over-fitting", which cannot help us classify BNs.

For the reasons set forth above, here we set the space size $s = 3$, which is the ECA rule space.

We can use $s \sim \log_2(N \log N)$, so that $\eta \sim 1$. For example, when $N = 5, s \sim 3$, which is an ECA, and when $N = 64$, $s \sim 8$, which means we can choose a 8-size CA to reconstruct a 64-size BN.

We took one model (ECA) to find generators explaining the behaviour of an unknown BN to which no access to its own generator is given or assumed. To this end, a probabilistic dynamic distribution of ECA rules can explain all or part of the evolution dynamics of the BN as illustrated in Fig. 16. The Bayesian pseudo-generative model is then the sequence of most probable (ECA) generative rules.

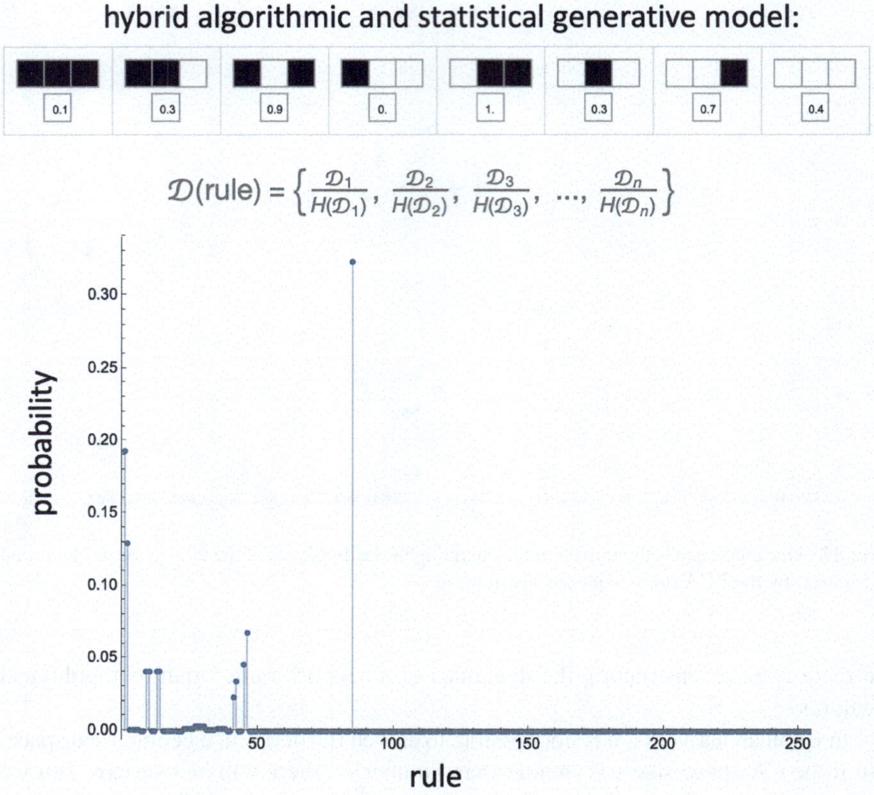


**Fig. 16** ECA rules ($x$ axis) explaining the space-time evolution of a BN

Notice that the representation of the space-time dynamics of a BN is not unique and is a function of the enumeration of the BN's nodes, and as such, the ECA rules can also permute (both in sequential order and the local rules) hence providing a rather robust description of the BN itself.

## 5 Conclusion

We have introduced the basics of an inverse-problem Bayesian framework that combines classical and algorithmic probability to find and produce generative (computable) candidate models corresponding to a set of observations compatible with the evolution of a dynamical system.

The concept was applied to the problem of inverse engineering a network and causal discovery from both an algorithmic complexity and computational cost/time perspectives suggesting trade-offs between inference time and complexity as applied to the unfolding evolution of an observed system. In other words, in inverse problems, a complexity-guided sampling method (e.g. using templates) is suggested over other sampling methods including random but also non-complexity based.

We have shown how a set of sequential cellular automata rules can describe and potentially approximate the evolution of a Boolean network providing a hybrid statistical and causal inference model that can partially provide both a statistical and rule-based description of the dynamical system.

## References


1. Dawkins B (1991) Siobhan's problem: the coupon collector revisited. Am Stat 45(1):76–82
2. Delahaye J-P, Zenil H (2012) Numerical evaluation of algorithmic complexity for short strings: a glance into the innermost structure of randomness. Appl Math Comput 219:63–77
3. Hernández-Orozco S, Zenil H, Riedel J, Uccello A, Kiani NA, Tegnér J (2020) Algorithmic probability-guided machine learning on non-differentiable spaces. Front Artif Intell
4. Kauffman S (1969) Homeostasis and differentiation in random genetic control networks. Nature 224(5215):177–178
5. Radó T (1962) On non-computable functions. Bell Syst Tech J 41(3):877–884
6. Riedel J, Zenil H (2018) Rule primality, minimal generating sets and turing-universality in the causal decomposition of elementary cellular automata. J Cell Autom 13:479–497
7. Soler-Toscano F, Zenil H, Delahaye J-P, Gauvrit N (2014) Calculating Kolmogorov complexity from the output frequency distributions of small turing machines. PLoS ONE 9(5):e96223
8. Wolfram S (1983) Statistical mechanics of cellular automata. Rev Mod Phys 55(3):601–644
9. Wolfram S (2002) A new kind of science, Wolfram research. IL, Champaign
10. Zenil H, Riedel J (2016) Asymptotic intrinsic universality and reprogrammability by behavioural emulation. In: Adamatzky A (ed) Advances in unconventional computation. Springer, pp 205–220
11. Zenil H, Hernàndez-Orozco S, Kiani NA, Soler-Toscano F, Rueda-Toicen A (2018) A decomposition method for global evaluation of shannon entropy and local estimations of algorithmic complexity. Entropy 20(8):605
12. Zenil H, Kiani NA, Zea A, Tegnér J (2019) Causal deconvolution by algorithmic generative models. Nat Mach Intell 1:58–66
13. Zenil H, Kiani NA, Marabita F, Deng Y, Elias S, Schmidt A, Ball G, Tegnér J (2019) An algorithmic information calculus for causal discovery and reprogramming systems. iScience S2589-0042(19):30270-6
14. Zenil H, Kiani NA, Abrahao FS, Tegnér J (2020) Algorithmic information dynamics. Scholarpedia 15(7):53143

# On Fungal Automata

**Andrew Adamatzky, Eric Goles, Michail-Antisthenis Tsompanas, Genaro J. Martínez, Han A. B. Wosten, and Martin Tegelaar**

**Abstract** Fungi are iniquitous creatures capable for adaptation in hush environments. Recently there is a growing that intelligence of the fungi comparable with that of slime mould and plans and that fungi sense and process information in a highly efficient way. As a first ever attempt to formalise informaiton processing in fungi we developed two cellular automaton models. 1D fungal automata and 2D fungal automata. Both model involve cellular automaton (CA) models of information dynamics on a single hypha of a fungal mycelium. Such a filament is divided in compartments (here also called cells) by septa. These septa are invaginations of the cell wall and their pores allow for flow of cytoplasm between compartments and hyphae. The septal pores of the fungal phylum of the Ascomycota can be closed by organelles called Woronin bodies. Septal closure is increased when the septa become older and when exposed to stress conditions. Thus, Woronin bodies act as informational flow valves. The 1D fungal automata is a binary state ternary neighbourhood CA, where every compartment follows one of the elementary cellular automata (ECA) rules if its pores are open and either remains in state '0' (first species of fungal automata) or its previous state (second species of fungal automata) if its pores are closed. The Woronin bodies closing the pores are also governed by ECA rules. We analyse a structure of the composition space of cell-state transition and pore-state transitions rules, complexity of fungal automata with just few Woronin bodies, and exemplify several important local events in the automaton dynamics. The 2D fungal automata is 2D CA where communication between neighbouring cells can be blocked on demand.

A. Adamatzky (✉) · M.-A. Tsompanas
Unconventional Computing Lab, UWE, Bristol, UK
e-mail: andrew.adamatzky@uwe.ac.uk

M.-A. Tsompanas
e-mail: antisthenis.tsompanas@uwe.ac.uk

E. Goles
Faculty of Engineering and Science, University of Adolfo Ibáñez, Santiago, Chile

G. J. Martínez
High School of Computer Science, National Polytechnic Institute, Mexico, Mexico

H. A. B. Wosten · M. Tegelaar
Microbiology Department, University of Utrecht, Utrecht, The Netherlands

 
A. Adamatzky (ed.), *Automata and Complexity*, Emergence, Complexity and Computation 42, https://doi.org/10.1007/978-3-030-92551-2_25

We demonstrate computational universality of the fungal automata by implementing sandpile cellular automata circuits there. We reduce the Monotone Circuit Value Problem to the Fungal Automaton Prediction Problem. We construct families of wires, cross-overs and gates to prove that the fungal automata are P-complete.

## 1 Introduction

The fungal kingdom represents organisms colonising all ecological niches [12] where they play a key role [15, 18, 30, 54]. Fungi can consist of a single cell on the one side and can form enormous underground networks on the other extreme [58]. Part of the fungi form microscopic fruit bodies, while others form fruit bodies weighting up to half a ton [19]. The underground mycelium network can be seen as a distributed communication and information processing system linking together trees, fungi and bacteria [11]. Mechanisms and dynamics of information processing in mycelium networks form an unexplored field with just a handful of papers published related to space exploration by mycelium [31, 32], patterns of electrical activity of fungi [1, 53, 57] and potential use of fungi as living electronic and computing devices [2–4]. Filamentous fungi grow by means of hyphae that grow at their tip and that branch sub-apically. Hyphae may be coenocytic or divided in compartments by septa. Filamentous fungi in the phylum *Ascomycota* have porous septa that allow for cytoplasmic streaming [38, 49]. Woronin bodies plug the pores of these septa after hyphal wounding to prevent excessive bleeding of cytoplasm [16, 36, 46, 59, 63, 65]. In addition, they plug septa of intact growing hyphae to maintain intra- and inter-hyphal heterogeneity [8, 9, 60, 60, 61].

Woronin bodies can be located in different hyphal positions (Fig. 1a). When first formed, Woronin bodies are generally localised to the apex [5, 47, 64]. Subsequently, Woronin bodies are either transported and anchored to the cell cortex (*Neurospora crassa*, *Sordaria fimicola*) or close to the septum (*Aspergillus oryzae*, *Aspergillus nidulans*, *Aspergillus fumigatus*, *Magnaporthe grisea*, *Fusarium oxysporum*, *Zymoseptoria tritici*) until they are translocated to the septal pore due to cytoplasmic flow or ATP depletion [6, 37, 46, 47, 52, 59, 60, 64, 66]. Woronin bodies that are not anchored at the cellular cortex or the septum, are located in the cytoplasm and are highly mobile (*Aspergillus fumigatus*, *Aspergillus nidulans*, *Zymoseptoria tritici*) [5, 47, 60]. Septal pore occlusion can be induced by bulk cytoplasmic flow [60] or developmental [10] and environmental cues, like puncturing of the cell wall, high temperature, carbon and nitrogen starvation, high osmolarity and low pH. Interestingly, high environmental pH reduces the proportion of occluded apical septal pores [61].

Aiming to lay a foundation of an emerging paradigm of fungal intelligence—distributed sensing and information processing in living mycelium networks—we decided to develop a formal model of mycelium and investigate a role of Woronin bodies in potential information dynamics in the mycelium. After exploring behaviour complexity of 1D fungal automata we asked ourselves on whether 2D fungal

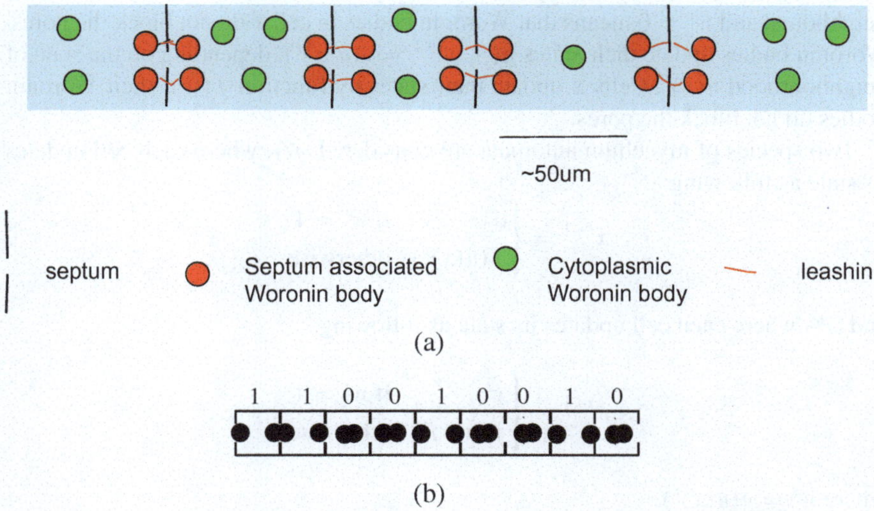


**Fig. 1** **a** A biological scheme of a fragment of a fungal hypha of an ascomycete, where we can see septa and associated Woronin bodies. **b** A scheme representing states of Woronin bodies: '0' open, '1' closed

automata are universal. We found that 'yes' they are! To demonstrate their computational universality we modify state transition rules of sand pile, or chip firing, automata [7, 14, 21, 25, 26] to allow a control for moving of sand grains, or chips, between neighbouring cells. The local control of the interactions between cells is inspired by a control of cytosol flow control in fungal hyphae [8, 9, 38, 49, 61]. Then we used developed tools of sand pile automata universality [13, 24, 27, 28, 42, 48] to show that functionally complete sets of Boolean gates can be realised in the fungal automata.

## 2 1D Fungal Automata

### 2.1 Fungal Automata $\mathcal{M}$

A fungal automaton is a one-dimensional cellular automaton with binary cell states and ternary, including central cell, cell neighbourhood, governed by two elementary cellular automata (ECA) rules, namely the cell state transition rule $f$ and the Woronin bodies adjustment rule $g$: $\mathcal{M} = \langle \mathbf{N}, u, \mathbf{Q}, f, g \rangle$. Each cell $x_i$ has a unique index $i \in \mathbf{N}$. Its state is updated from $\mathbf{Q} = \{0, 1\}$ in discrete time depending of its current state $x_i^t$, the states of its left $x_{i-1}^t$ and right neighbours $x_{i+1}^t$ and the state of cell $x$'s Woronin body $w$. Woronin bodies take states from $\mathbf{Q}$: $w^t = 1$ means Woronin bodies (Fig. 1) in cell $x$ blocks the pores and the cell has no communication with its neighbours, and $w^t = 0$ means that Woronin bodies in cell $x$ do not block the pores. Woronin bodies update their states $g(\cdot)$, $w^{t+1} = g(u(x)^t)$, depending on the state of neighbourhood $u(x)^t$. Cells $x$ update their states by function $f(\cdot)$ if their Woronin bodies do not block the pores.

Two species of mycelium automata are considered $\mathcal{M}_1$, where each cell updates its state as following:

$$x^{t+1} = \begin{cases} 0 & \text{if } w^t = 1 \\ f(u(x)^t) & \text{otherwise} \end{cases}$$

and $\mathcal{M}_2$ where each cell updates its state as following:

$$x^{t+1} = \begin{cases} x^t & \text{if } w^t = 1 \\ f(u(x)^t) & \text{otherwise} \end{cases}$$

where $w^t = g(u(x)^t)$.

State '1' in the cells of array $x$ symbolises metabolites, signals exchanged between cells. Where pores in a cell are open the cell updates its state by ECA rule $f : \{0, 1\}^3 \to \{0, 1\}$. In automaton $\mathcal{M}_1$, when Woronin bodies block the pores in a cell, the cell does not update its state and remains in the state '0' and left and right neighbours of the cells can not detect any 'cargo' in this cell. In automaton, $\mathcal{M}_2$, where Woronin bodies block the pores in a cell, the cell does not update its state and remains in its current state. In real living mycelium glucose and possibly other metabolites [8] can still cross the septum even when septa are closed by Woronin bodies, but we can ignore this fact in present abstract model.

Both species are biologically plausible and, thus, will be studied in parallel. The rules for closing and opening Woronin bodies are also ECA rules $g : \{0, 1\}^3 \to \{0, 1\}$. If $g(u(x)^t) = 0$ this means that pores are open, if $g(u(x)^t) = 1$ Woronin bodies block the pores. Examples of space-time configurations of both species of $\mathcal{M}$ are shown in Fig. 2.

## 2.2 Properties of Composition $f \circ g$

**Predecessor Sets**

Let $\mathbf{F} = \{h : \{0, 1\}^3 \to \{0, 1\}\}$ be a set of all ECA functions. Then for any composition $f \circ g$, where $f, g \in \mathbf{F}$, can be converted to a single function $h \in \mathbf{F}$. For each $h \in \mathbf{F}$ we can construct a set $\mathbf{P}(h) = \{f \circ g \in \mathbf{F} \times \mathbf{F} \mid f \circ g \to h\}$. The sets $\mathbf{P}(h)$ for each $h \in \mathbf{F}$ are available online.[1]

A size of $\mathbf{P}(h)$ for each $h$ is shown in Fig. 3c. The functions with largest size of $\mathbf{P}(h)$ are rule 0 in automaton $\mathcal{M}_1$ and rule 51 (only neighbourhood configurations (010, 011, 110, 111 are mapped into 1) in $\mathcal{M}_2$.

[1] https://figshare.com/s/b7750ee3fe6df7cbe228.

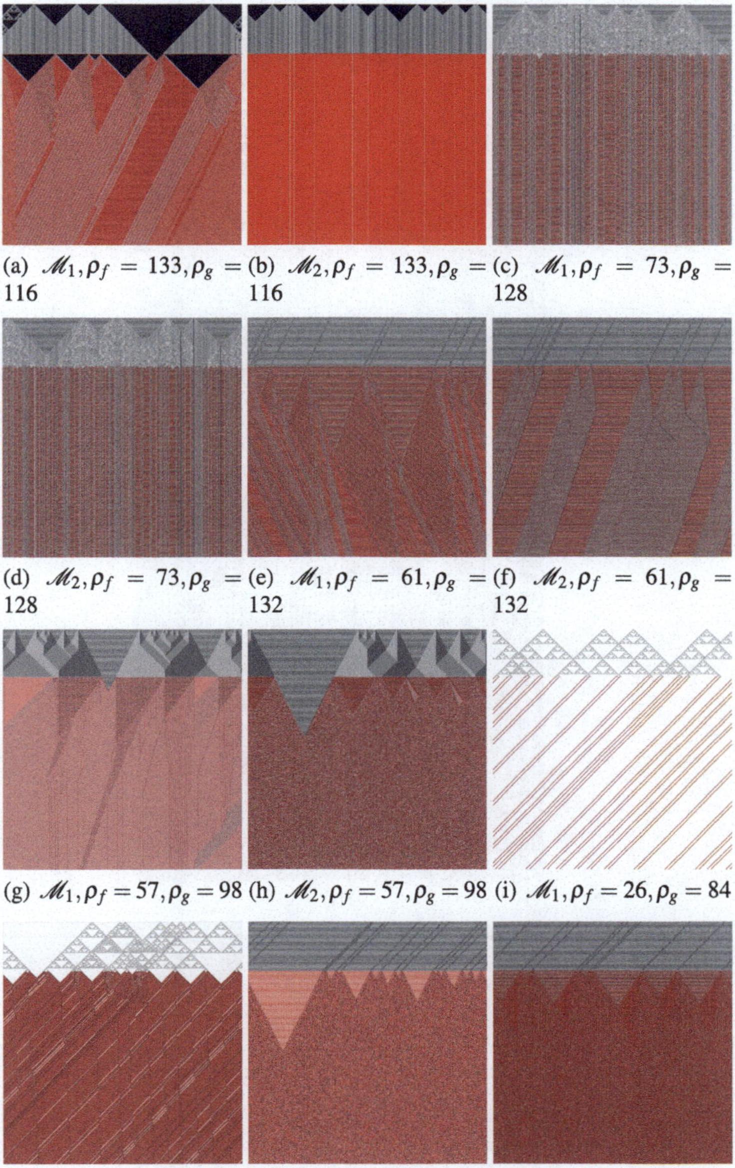

(a) $\mathscr{M}_1, \rho_f = 133, \rho_g = 116$ (b) $\mathscr{M}_2, \rho_f = 133, \rho_g = 116$ (c) $\mathscr{M}_1, \rho_f = 73, \rho_g = 128$

(d) $\mathscr{M}_2, \rho_f = 73, \rho_g = 128$ (e) $\mathscr{M}_1, \rho_f = 61, \rho_g = 132$ (f) $\mathscr{M}_2, \rho_f = 61, \rho_g = 132$

(g) $\mathscr{M}_1, \rho_f = 57, \rho_g = 98$ (h) $\mathscr{M}_2, \rho_f = 57, \rho_g = 98$ (i) $\mathscr{M}_1, \rho_f = 26, \rho_g = 84$

(j) $\mathscr{M}_2, \rho_f = 26, \rho_g = 84$ (k) $\mathscr{M}_1, \rho_f = 125, \rho_g = 105$ (l) $\mathscr{M}_2, \rho_f = 125, \rho_g = 105$

**Fig. 2** Examples of space-time dynamics of $\mathscr{M}$. The automata are $10^3$ cells each. Initial configuration is random with probability of a cell $x$ to be in state '1', $x^0 = 1$, equals 0.01. Each automaton evolved for $10^3$ iterations. Binary values of ECA rules $f$ and $g$ are shown in sub-captions. Rule $g$ is applied to every iteration starting from 200th. Cells in state '0' are white, in state '1' are black, cells with Woronin bodies blocking pores are red. Indexes of cells increase from the left to the right, iterations are increasing from the to the bottom

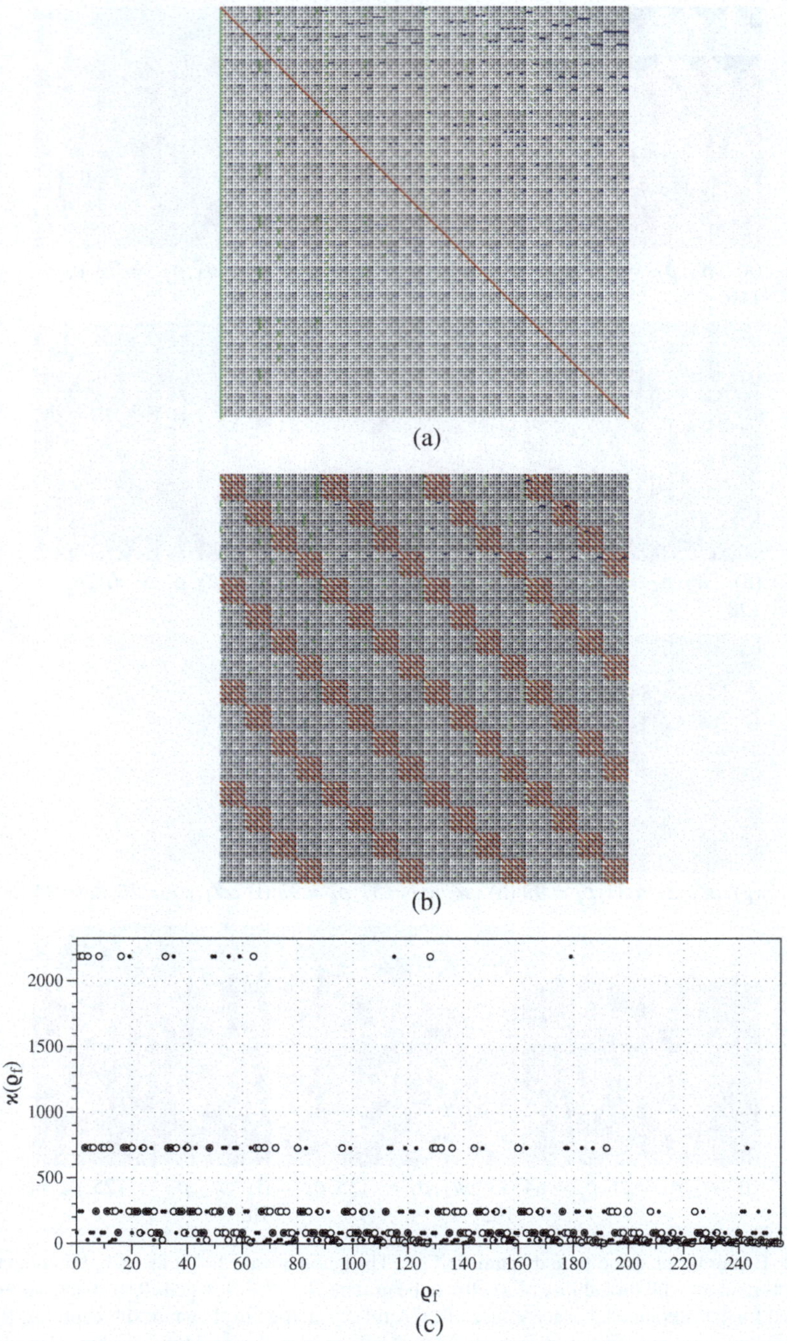


**Fig. 3** Mapping $\mathbf{F} \times \mathbf{F} \rightarrow \mathbf{F}$ for automaton $\mathscr{M}_1$ (a) and $\mathscr{M}_2$ (b) is visualised as an array of pixels, $\mathbf{P} = (p)_{0 \le \rho_f \le 255, 0 \le \rho_f \le 255}$. An entry at the intersection of any $\rho_f$ and $\rho_g$ is a coloured as follows: red if $p_{\rho_f \rho_g} = p_{\rho_g \rho_f}$, blue if $\rho_g = p_{\rho_g \rho_f}$, green if $\rho_f = p_{\rho_g \rho_f}$. (c) Sizes of $\mathbf{P}(h)$ sets for $\mathscr{M}_1$, circle, and $\mathscr{M}_2$, solid discs, are shown for every function $h$ apart of rule 0 ($\mathscr{M}_1$) and rule 51 ($\mathscr{M}_2$)

**Table 1** Characterisations of automaton mapping $\mathbf{F} \times \mathbf{F} \to \mathbf{F}$. a Size $\sigma$ of $\mathbf{P}(h)$ vs a number $\gamma$ of functions $h$ having set $\mathbf{P}(h)$ of size $\sigma$. T (b) Sizes of sets $\mathbf{P}(h)$ for rules from Wolfram class III. b Sizes of sets $\mathbf{P}(h)$ for rules from Wolfram class IV

(a) Rules per $|\mathbf{P}(h)|$

| $\sigma$ | $\gamma$ |
|---|---|
| 1 | 1 |
| 3 | 8 |
| 9 | 28 |
| 27 | 56 |
| 81 | 70 |
| 243 | 56 |
| 729 | 28 |
| 2187 | 8 |
| 6561 | 1 |

(b) $\mathscr{M}_1$: Class III rules

| Rule | $\sigma$ |
|---|---|
| 18 | 729 |
| 22, 146 | 243 |
| 30, 45, 60, 90, 105, 150 | 81 |
| 122 | 27 |
| 126 | 9 |

(c) $\mathscr{M}_1$: Class IV rules

| Rule | $\sigma$ |
|---|---|
| 41 | 243 |
| 54, 106, 110 | 81 |

(d) $\mathscr{M}_2$: Class III rules

| Rule | $\sigma$ |
|---|---|
| 18 | 729 |
| 22, 146 | 243 |
| 30, 45, 60 90, 105, 150 | 81 |
| 122 | 243 |
| 126 | 81 |

(e) $\mathscr{M}_2$: Class IV rules

| Rule | $\sigma$ |
|---|---|
| 41 | 243 |
| 54 | 729 |
| 106 | 81 |
| 110 | 27 |

Size $\sigma$ of $\mathbf{P}(h)$ vs a number $\gamma$ of functions $h$ having set $\mathbf{P}(h)$ of size $\sigma$ is shown for automata $\mathscr{M}_1$ and $\mathscr{M}_2$ in Table 1a.

With regards to Wolfram classification [68], sizes of $\mathbf{P}(h)$ for rules from Class III vary from 9 to 729 in $\mathscr{M}_1$ (Table 1b). Rule 126 would be the most difficult to obtain in $\mathscr{M}_1$ by composition two ECA rules chosen at random, it has only 9 'predecessor' $f \circ g$ pairs. Rule 18 would be the easiest, for Class III rules, to be obtained, it has 729 predecessors, in both $\mathscr{M}_1$ (Table 1b) and $\mathscr{M}_2$ (Table 1d). In $\mathscr{M}_1$, one rule, rule 41, from the class IV has 243 $f \circ g$ predecessors, and all other rules in that class have 81 (Table 1c). From Class IV rule 54 has the largest number of predecessors in $\mathscr{M}_2$, and thus can be considered as most common (Table 1d).

**Diagonals**

In automaton $\mathscr{M}_1$ for any $f \in \mathbf{F}$ $f \circ f = 0$. Assume $f : \{0, 1\}^3 \to 1$ then Woronin bodies close the pores and, thus, second application of $f$ produces state '0'. If $f : \{0, 1\}^3 \to 0$ then Woronin bodes does not close pores but yet second application of the $f$ produce state '0'.

For automaton $\mathscr{M}_2$ a structure of diagonal mapping $f \circ f \to h$, where $f, h \in \mathbf{F}$ is shown in Table 2. The set of the diagonal outputs $f \circ f$ consists of 16 rules: (0, 1, 2, 3), (16, 17, 18, 19), (32, 33, 34, 35), (48, 49, 40, 51). These set of rules can be reduced to the following rule. Let $C(x^t) = [u(x)^t = (111)] \vee [u(x)^t = (111)]$ and $B(x^t) = [u(x)^t = (011)] \vee [u(x)^t = (010)]$. Then $x^t = 1$ if $C(x)^t \vee C(x)^t \wedge B(x^t)$.

**Table 2** Diagonals of automaton $\mathscr{M}_2$

| $f \circ f$ | $f$ |
|---|---|
| 0 | 0, 1, 2, 3, 16, 17, 18, 19, 32, 33, 34, 35, 48, 49, 50, 51 |
| 1 | 128, 129, 130, 131, 144, 145, 146, 147, 160, 161, 162, 163, 176, 177, 178, 179 |
| 2 | 64, 65, 66, 67, 80, 81, 82, 83, 96, 97, 98, 99, 112, 113, 114, 115 |
| 3 | 192, 193, 194, 195, 208, 209, 210, 211, 224, 225, 226, 227, 240, 241, 242, 243 |
| 16 | 8, 9, 10, 11, 24, 25, 26, 27, 40, 41, 42, 43, 56, 57, 58, 59 |
| 17 | 136, 137, 138, 139, 152, 153, 154, 155, 168, 169, 170, 171, 184, 185, 186, 187 |
| 18 | 72, 73, 74, 75, 88, 89, 90, 91, 104, 105, 106, 107, 120, 121, 122, 123 |
| 19 | 200, 201, 202, 203, 216, 217, 218, 219, 232, 233, 234, 235, 248, 249, 250, 251 |
| 32 | 4, 5, 6, 7, 20, 21, 22, 23, 36, 37, 38, 39, 52, 53, 54, 55 |
| 33 | 132, 133, 134, 135, 148, 149, 150, 151, 164, 165, 166, 167, 180, 181, 182, 183 |
| 34 | 68, 69, 70, 71, 84, 85, 86, 87, 100, 101, 102, 103, 116, 117, 118, 119 |
| 35 | 196, 197, 198, 199, 212, 213, 214, 215, 228, 229, 230, 231, 244, 245, 246, 247 |
| 48 | 12, 13, 14, 15, 28, 29, 30, 31, 44, 45, 46, 47, 60, 61, 62, 63 |
| 49 | 140, 141, 142, 143, 156, 157, 158, 159, 172, 173, 174, 175, 188, 189, 190, 191 |
| 50 | 76, 77, 78, 79, 92, 93, 94, 95, 108, 109, 110, 111, 124, 125, 126, 127 |
| 51 | 204, 205, 206, 207, 220, 221, 222, 223, 236, 237, 238, 239, 252, 253, 254, 255 |

**Commutativity**
In automaton $\mathscr{M}_1$, for any $f, g \in \mathbf{F}$ $f \circ g \neq g \circ f$ only if $f \neq g$. In automaton $\mathscr{M}_2$ there are 32768 pairs of function which $\circ$ is commutative, their distribution visualised in red in Fig. 3b.

**Identities and Zeros**
In automaton $\mathscr{M}_1$ there are no left or right identities, neither right zeros in $\langle \mathbf{F}, \mathbf{F}, \circ \rangle$. The only left zero is the rule 0. In automaton $\mathscr{M}_2$ there are no identities or zeros at all.

**Associativity**
In automaton $\mathcal{M}_1$ there 456976 triples $\langle f, g, h \rangle$ on which operation $\circ$ is associative: $(f \circ g) \circ h = f \circ (g \circ h)$. The ratio of associative triples to the total number of triples is 0.027237892. There are 104976 associative triples in $\mathcal{M}_2$, a ratio of 0.006257057.

## 2.3 *Tuning Complexity: Rule 110*

To evaluate on how introduction of Woronin bodies could affect complexity of automaton evolution, we undertook two series of experiments. In the first series we used fungal automaton where just one cell has a Woronin body (Fig. 5). In the second series we employed fungal automaton where regularly positioned cells (but not all cells of the array) have Woronin bodies.

State transition functions $g$ of Woronin bodies were varied across the whole diapason but the state transition function $f$ of a cell was Rule 110, $\rho_f = 110$. We have chosen Rule 110 because the rule is proven to be computationally universal [17, 39], P-complete [51], the rules belong to Wolfram class IV renown for exhibiting complex and non-trivial interactions between travelling localisation [67], rich families of gliders can be produce in collisions with other gliders [43–45].

We wanted to check how an introduction of Woronin bodies affect dynamics of most complex space-time developed of Rule 110 automaton. Thus, we evolved the automata from all possible initial configurations of 8 cells placed near the end of $n =$ 1000 cells array of resting cells and allowing to evolve for 950 iterations. Lempel–Ziv complexity (compressibility) LZ was evaluated via sizes of space-time configurations saved as PNG files. This is sufficient because the 'deflation' algorithm used in PNG lossless compression [20, 33, 56] is a variation of the classical Lempel–Ziv 1977 algorithm [69]. Estimates of LZ complexity for each of 8-cell initial configurations are shown in Fig. 4a. The initial configurations with highest estimated LZ complexity are 10110001 (decimal 177), 11010001 (209), 10000011 (131), 11111011 (253), see example of space-time dynamics in Fig. 4b.

We assumed that a cell in the position $n - 100$ has a Woronin body which can be activated (Fig. 5), i.e. start updating its state by rule $f$, after 100th iteration of the automaton evolution. We then run 950 iteration of automaton evolution for each of 256 Woronin rules and estimated LZ complexity. In experiments with $\mathcal{M}_1$ we found that 128 rules, with even decimal representations, do not affect space time dynamics of evolution and 128 rules, with even decimal representations, reduce complexity of the space-time configuration. The key reasons for the complexity reduction (compare Fig. 4b, c) are cancellation of three gliders at c. 300th iteration and simplification of the behaviour of glider guns positioned at the tail of the propagating wave-front. In experiments with $\mathcal{M}_2$ 128 rules, with even decimal representations, do not change the space-time configuration of the author. Other 128 rules reduce complexity and modify space-time configuration by re-arranging the structures of glider guns and establishing one oscillators at the site surrounding position of the cell with Woronin body (Fig. 4d).

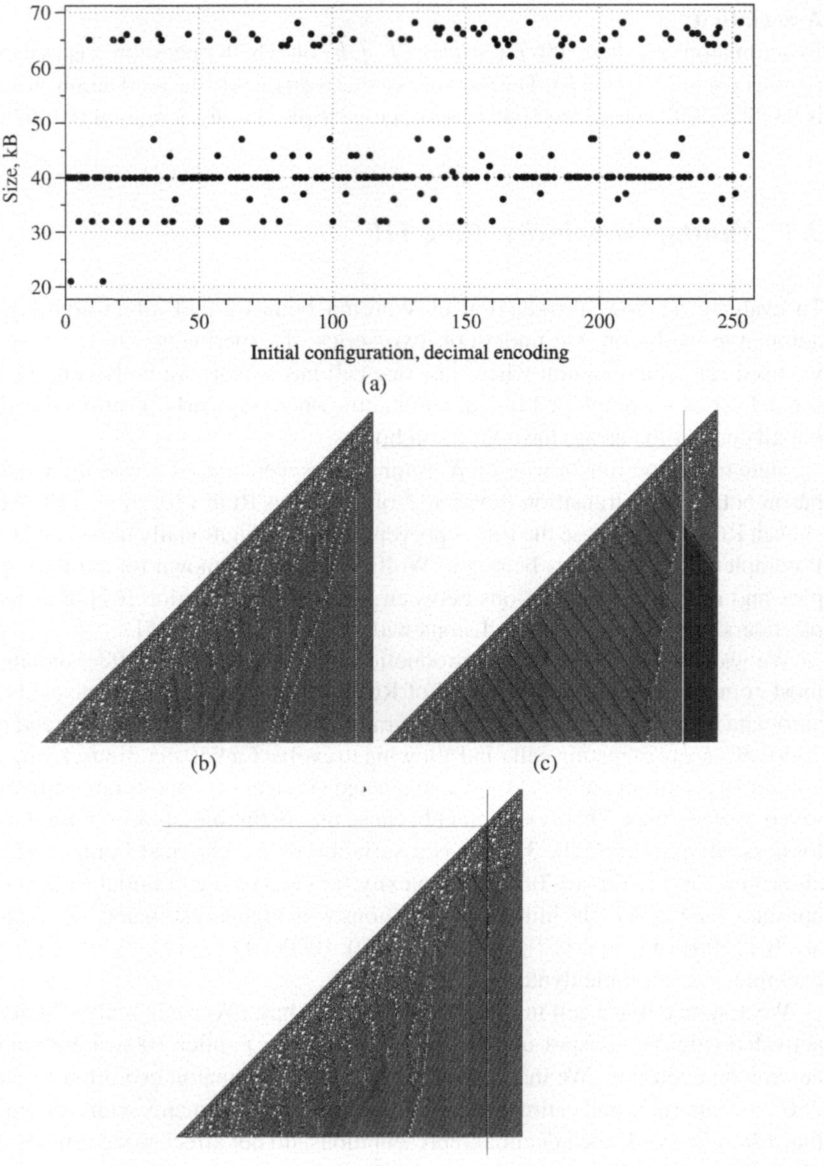


**Fig. 4** **a** Estimates of LZ complexity of space-time configurations of ECA Rule 110 without Woronin bodies. **b** A space-time configuration of ECA Rule 110 evolving from initial configuration 10110001 (177), no Woronin bodies are activated. **c** A space-time configuration of $\mathcal{M}_1$ Rule 110 evolving from initial configuration 10110001 (177), Woronin body is governed by rule 43; red lines indicate time when the body was activated and position of the cell with the body. In (bcd), a pixel in position $(i, t)$ is black if $x_i^t = 1$

**Fig. 5** Only one cell has has two Woronin bodies by which it can close itself from the other compartments

In second series of experiments we regularly positioned cells with Woronin bodies along the 1D array: every 50th cell has a Woronin body. Then we evolved fungal automata $\mathcal{M}_1$ and $\mathcal{M}_2$ from exactly the same initial random configuration with density of '1' equal to 0.3. Space-time configuration of the automaton without Woronin bodies is shown in Fig. 6a. Exemplar of space-time configurations of automata with Woronin bodies are shown in Fig. 6b–h. As seen in Fig. 7 both species of fungal automata show similar dynamics of complexity along the Woronin transition functions ordered by their decimal values. The automaton $\mathcal{M}_1$ has average LZ complexity 82.2 ($\overline{\sigma} = 24.6$) and the automaton $\mathcal{M}_2$ 78.4 ($\overline{\sigma} = 22.1$). Woronin rules $g$ which generate most LZ complex space-time configurations are $\rho_g = 133$ in $\mathcal{M})_1$ (Fig. 6b) and $\rho_g = 193$ in $\mathcal{M})_2$ (Fig. 6e). The space-time dynamics of the automaton is characterised by a substantial number of gliders guns and gliders (Fig. 6b). Functions being in the middle of the descending hierarchy of LZ complexity produce space-time configurations with declined number of travelling localisation and growing domains of homogeneous states (Fig. 6cg). Automata with Woronin functions at the bottom of the complexity hierarchy quickly (i.e. after 200–300 iterations) evolve towards stable, equilibrium states (Fig. 6dh).

## 2.4 *Local Events*

Let us consider some local events happening in the fungal automata discussed in Sect. 2.3: every 50th cell of an array has a Woronin body. **Retaining gliders.** A glider can be stopped and converted into a station localisation by a cell with Woronin body. As exemplified in Fig. 8a, the localisation travelling left was stopped from further propagation by a cell with Woronin body yet the localisation did not annihilate but remained stationary.

**Register memory.** Different substrings of input string (initial configuration) might lead to different equilibrium configurations achieved in the domains of the array separated by cells with Woronin bodies. When there is just two types of equilibrium configurations they be seen as 'bit up' and 'bit down' and therefore such fungal automaton can be used a memory register (Fig. 8b).

**Reflectors.** In many cases cells with Woronin bodies induce local domains of stationary, sometimes time oscillations, inhomogeneities which might act as reflectors for travelling localisations. An example is shown in Fig. 8c where several localisations are repeatedly bouncing between two cells with Woronin bodies.

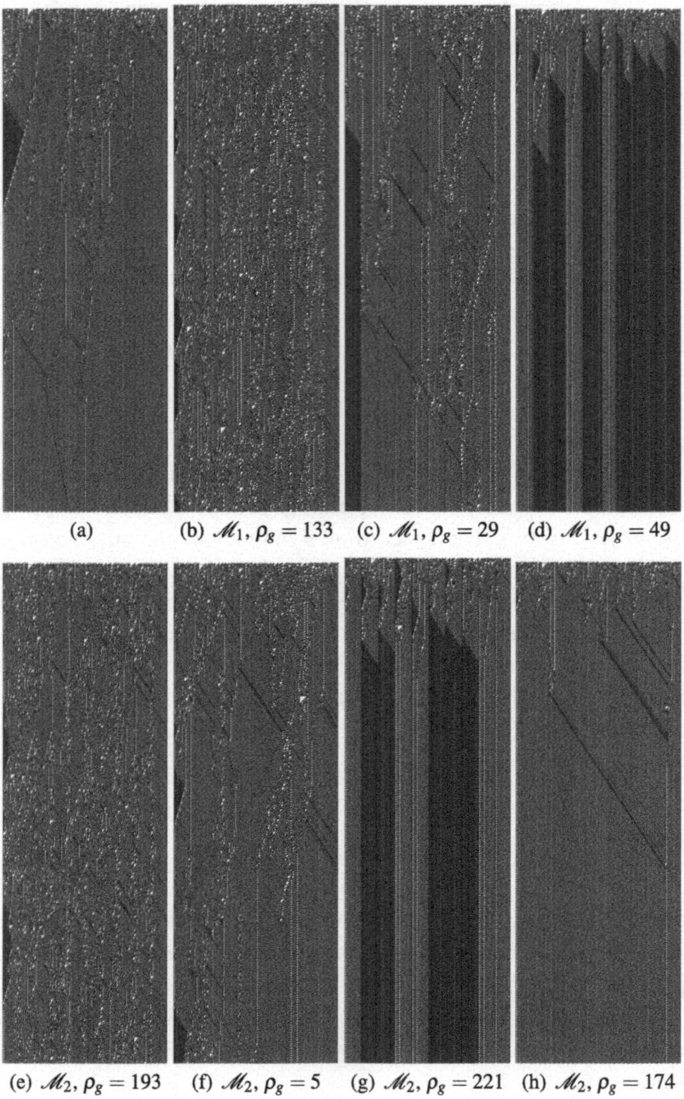


**Fig. 6** **a** ECA Rule 110, no Woronin bodies. Space-time evolution of $\mathcal{M}$ (**b**–**d**) and $\mathcal{M}$ (**e**–**h**) for Woronin rules shown in subcaption. LZ complexity of space-time configurations decreases from **b** to **d** and from **e** to **h**. Every 50th cell has a Woronin body

**Modifiers.** Cells with Woronin bodies can act as modifiers of states of gliders reflected from them and of outcomes of collision between travelling localizations. In Fig. 8d we can see how a travelling localisation is reflected from the vicinity of Woronin bodies three times: every time the state of the localisation changes. On the third reflection the localisation becomes stationary. In the fragment (Fig. 8e) of

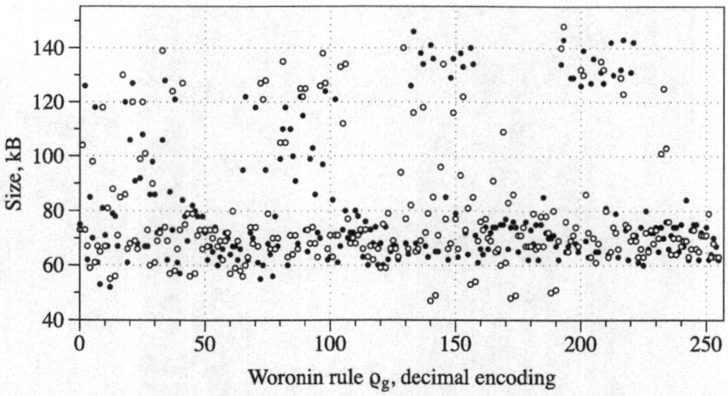


**Fig. 7** Estimations of LZ complexity of space-time, 500 cells by 500 iterations, configurations of $\mathcal{M}_1$, discs, and $\mathcal{M}_1$, circles, for all Woronin functions $g$

space-time configuration of automaton with Woronin bodies governed by $\rho_g = 201$ of the fragment we can see how two localisations got into contact with each in the vicinity of the Woronin body and an advanced structure is formed two breathing stationary localisations act as mirror, and there are streams of travelling localisations between them. A multi-step chain reaction can be observed in Fig. 8f: there are two stationary, breathing, localisations at the sites of the cells with Woronin bodies. A glider is formed on the left stationary localisation. The glider travel to the right and collide into right breather. In the result of the collision the breath undergoes structural transitions, emits a glider travelling left and transforms itself into a pair of stationary breathers. Meantime the newly born glider collided into left breather and changes its state.

# 3 2D Fungal Automata

## 3.1 Two Dimensional Fungal Automata

Filamentous fungi of the phylum Ascomycota have porous septa that allow for cytoplasmic streaming throughout hyphae and the mycelium [38, 49]. The pores of damaged hyphae will be plugged by a peroxisome-derived organelle to prevent bleeding of cytoplasm into the environment [36, 55, 63, 64]. These Woronin bodies can also plug septa of intact hyphae [9, 60]. The septal pore occlusion in these hyphae can be triggered by septal ageing and stress conditions [9, 10, 61].

A scheme of the mycelium with Woronin bodies is shown in Fig. 9. An apical compartment has one neighbouring sub-apical compartment, while a sub-apical compartment has a neighbouring compartment at both ends. Because compartments can also branch, they can have one or more additional neighbouring compartments.

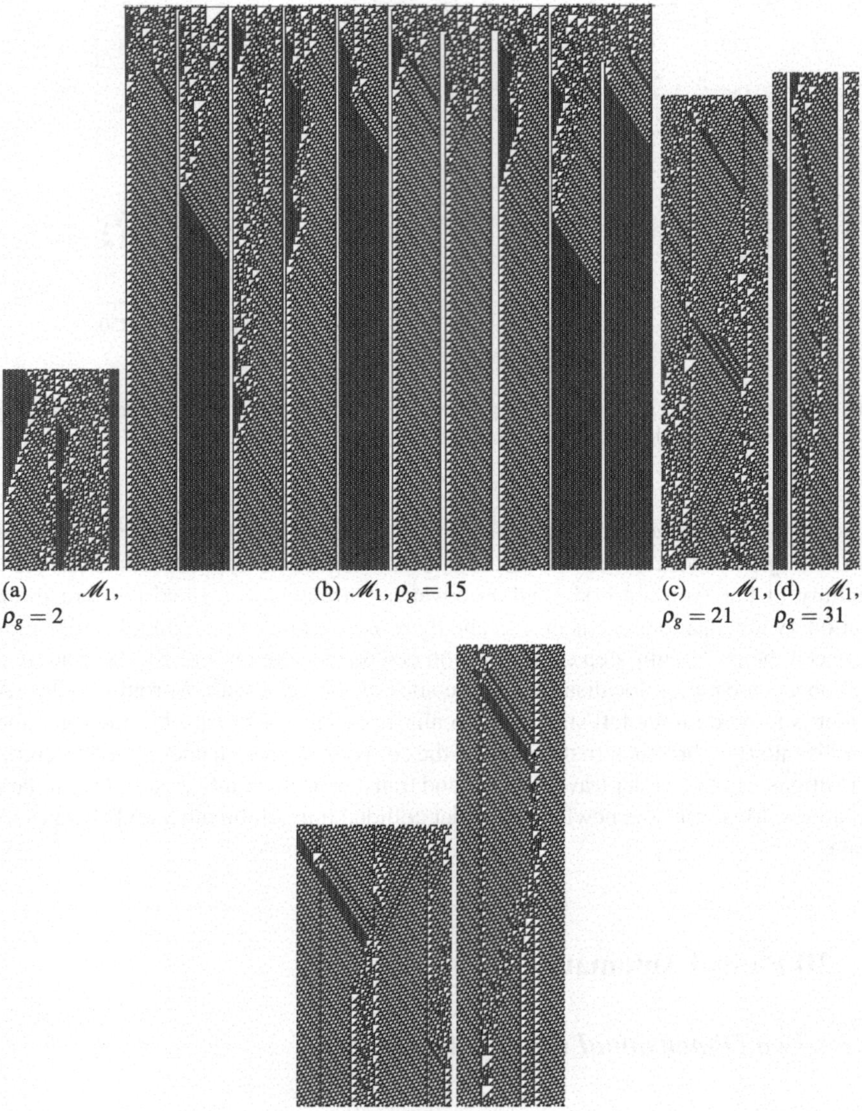

(a) $\mathcal{M}_1$, $\rho_g = 2$ (b) $\mathcal{M}_1$, $\rho_g = 15$ (c) $\mathcal{M}_1$, $\rho_g = 21$ (d) $\mathcal{M}_1$, $\rho_g = 31$

(e) $\mathcal{M}_1$, $\rho_g = 29$ (f) $\mathcal{M}_1$, $\rho_g = 201$

**Fig. 8** **a** Localisation travelling left was stopped by the Woronin body. **b** Analog of a memory register. **c** Reflections of travelling localisations from cells with Woronin bodies. **d** Modification of glider state in the vicitinity of Woronin bodies. **e** A fragment of configuration of automaton with $\rho_g = 29$, left cell states, right Woronin bodies states. **f** Enlarged sub-fragment of the fragment **d** where Wonorin body tunes the outcome of the collision. For both automata $\rho_f = 110$

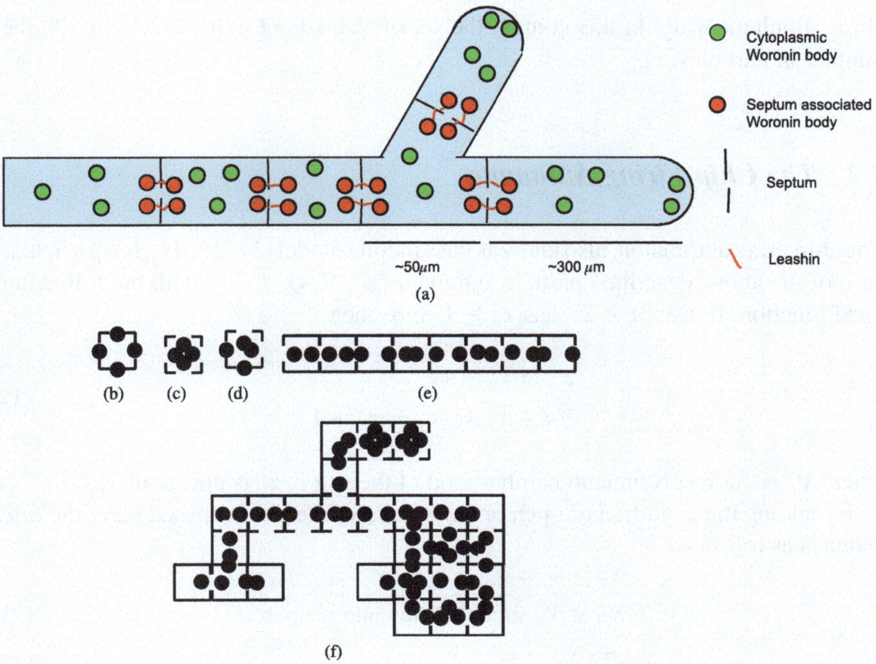


**Fig. 9** Fungal automata. **a** Biological scheme. **b–f** Abstract schemes. **b** All pores are closed, **c** All pores are open, **d** North and East pores are open, **e** one-dimensional automaton, **f** an example of an arbitrary architecture of fungal automata

Thus, the compartment with pores and Woronin bodies is a elementary unit of fungal automata, Fig. 9b–d. From these compartments one can assemble quasi-one-dimensional, Fig. 9e, and two-dimensional, Fig. 9f, structures.

In this context, let us consider a cellular automaton in the two dimensional grid $\mathbb{Z}^2$ with the von Neumann neighbourhood, with set of states $Q$ and a global function $F$. Each cell of the grid has four sides that could be open or closed. An open side means that the information (the state) of both cells is shared. If not, when the side is closed, both sites mutually ignore each other. When every side is open we have the usual cellular automata model [68]. On the other hand, if sides are open or closed in some random or periodical way we get for the same local functions different dynamic behaviours. In this paper we will consider only "uniform" ways to open-close the sides. Actually at a given step to open every vertical side (every column of the grid) or every horizontal side, rows of the array. So the fungal automata model becomes a tuple $FA = \langle \mathbb{Z}^2, \mathbf{Q}, V, F, w \rangle$, where $V$ is the von Neumann neighbourhood, $w$ is a finite word on the alphabet $H$, $V$ (horizontal, vertical). Each iteration of automaton evolution is associated with one letter of $H$, $V$.

In this work we focus on "particles" rules. That is to say at each site there are a finite amount of particles or chips, that, according to a specific rule are disseminated in the vicinity of a site. Every step is going synchronously, so each site lose and receive chips simultaneously. In this context the set of states is $\mathbf{Q} = \{0, 1, 2, ...\} \subset \mathbb{N}$, the number of particles.

## 3.2 The Chip Firing Automata

The chip firing automaton, also know as the sandpile model [27, 28, 48], is a particular case of the above described particle automata $\langle \mathbb{Z}^2, V, \mathbf{Q}, F, w \rangle$, with the following local function. If a site $v \in \mathbb{Z}^2$ has $x_v \geq 4$ chips then:

$$\begin{aligned} &x'_v = x_v - 4 \\ &\forall u \in V_v \Rightarrow x'_u = x_u + 1 \end{aligned} \tag{1}$$

where $V_v$ is the von Neumann neihborhood of the site $v$, $x'_v$ is update of $x_v$.

By adding the condition of open or close sides of the site proposed here; the rule changes as follows:

$$\begin{aligned} &x_v \geq 4 \Rightarrow x'_v = x_v - \alpha \\ &\forall u \in V_v \text{ such that the gate is open} \\ &\Rightarrow x'_u = x_u + 1 \end{aligned} \tag{2}$$

where $\alpha$ is the number of gates that are open.

When the rule is applied in parallel on every site, the new state at a site $v$ is:

$$x'_v = x_v - \alpha + \beta$$

where $\beta$ is the number of chips which the site $v$ receives from its open and firing neighbours.

If every side (columns and rows) is always open, then we have the usual chip firing automaton. When a word $w$ of open or close sides is considered, for instance $w = HHVV$, at each step we open (or close) the rows or columns of the grid periodically

$$(HHVV)^* = HHVV\ HHVV\ HHVV\ ... \tag{3}$$

## 3.3 Computational Complexity Notions

In order to study the complexity of an automaton we can analyse a power of the automaton to simulate Boolean functions, i.e., by selecting specific initial conditions and sites as inputs and outputs to determine the different Boolean functions the automaton may compute by its dynamics [29]. More Boolean functions are founded, more complex is the automaton. A similar notion, related to some prediction problems, appears in the framework of the theory of computational complexity. Essentially this is similar to trying to determine the computational time related to the size of a problem, that a Turing Machine take to solve it. In our context, let us consider the following decision or prediction problem.

**PRE**: Consider the chip firing fungal automaton $FA = \langle \mathbb{Z}^2, V, \mathbf{Q}, F, w \rangle$, an initial assignment of chips to every site , $x(0) \in \mathbf{Q}^{\mathbb{Z}^2}$, an integer number $T > 0$ of steps of the automaton and a site $v \in \mathbb{Z}^2$, such that $x_v(0) = 0$. Question: Will $x_v(t) > 0$ be for some $t \leq T$?

Of course one may give an answer by running the automaton at most $T$ steps, which can be done in a serial computer in polynomial time. But the question is a little more tricky: could we answer faster than the serial algorithm, ideally, exponentially faster in polylogarithmic time in a parallel computer with a polynomial number of processor?

To answer the questions we consider two classes of decision problems, those belonging to $P$, the class of problems solved by a polynomial algorithm, and the class $NC$, being the problems solved in a parallel computer in $O(log^q n)$ steps (polylogarithmic time). This is straightforward that $NC \subseteq P$ because any parallel algorithm solved in polylogarithmic time can be simulated efficiently in a serial computer. But the strict inclusion is a very hard open problem (like the well know $P = NP$).

An other notion from computational complexity related with the possibility that the two classes melt is P-completeness. A problem is *P-complete* if it is in the class $P$, (that is to say, there exist a polynomial algorithm to solve it) and every other problem in $P$ can be reduced, by a polynomial transformation, to it. Clearly, if one of those *P-complete* problems is in $NC$, both classes collapsed in one. So, to prove that a problem is $P - Complete$ gives us an idea of its complexity.

One well known *P-complete* problem is the *Circuit Value Problem*, i.e., the evaluation of a Boolean Circuit (Boolean function). Roughly because any polynomial problem solved in a serial computer (a Turing machine) can be represented as a Boolean circuit. On the other hand, Boolean circuits intuitively are essentially serial because in order to compute a layer of functions it is necessary to compute previous layers so in principle it is not clear how to determine the output of the circuit in parallel. Further, when the circuit is monotonous, i.e., it admits only OR and AND gates (no negations) it is also a *P-complete* problem. This is because negation can be put in the input (the two bits of the variable 0 and 1) and for the gates which are a negation, to use the De Morgan laws.

The complexity of the chip firing automata was first studied in [27, 28], where it was proved that in arbitrary graphs (in particular, non-planar ones) the chip firing automata are Turing Universal. To prove this a universal set of Boolean circuits is built by using specific automata configurations, so, also **PRE** is *P-complete*. In a similar way, but in a d-dimensional grid, $\mathbb{Z}^d$, it was proved in [48] that for $d \geq 3$ the problem is *P-complete* and the complexity, until today, remains open for a two-dimensional grid. In [24] it was proved that in a two-dimensional grid and the von Neumann and Moore neighbourhood it is not possible to cross signals by constructing wires over quiescent configurations. That can be done only for bigger neighbourhood, so, in fact, over non planar graphs.

## 3.4 Computational Complexity of the Fungal Automata

We will study the computational complexity of the Fungal Sand Pile Automaton, by proving that for the word $w = H^4V^4 = HHHHVVVV$, the Prediction Problem, **PRE**, is *P-complete*. That is to say one can not determine an exponentially faster algorithm to answer unless $NC = P$.

**Proposition 1** *For the word* $H^4V^4 = HHHHVVVV$ *the fungal chip firing automaton is P-complete.*

***Proof.*** Clearly the problem is in $P$. It suffices to run the automaton at most $T$ steps and see if the site $i$ changes, which is done in $O(T^3)$: in fact we have to compute the "cone" between site $i$, its neighbourhood at step $T-1$, $T-2$, and so on, until the initial values in the site in an square $(2T-1) \times (2T-1)$. So the number of sites is to consider is $1^2 + 3^2 + 5^2 + \cdots . + (2T-1)^2$ which is bounded by a cubic polynomial, so one may compute the state of site $i$ in $T^3$.

To establish the completeness, we will reduce the Monotone Circuit Value Problem to the Fungal Automaton Prediction Problem, **PRE**. That is to say, to establish specific automaton configurations which simulates a wire, the AND and the OR gates, as well as a CROSS- OVER. This last gadget is important to compute non-planar circuits in the two dimensional grid.

In the constructions below every cell which is not in the diagram is understood initially empty, without chips ($x_i = 0$).

To construct the wire let us first see what happen when $V^4$ is applied in the particular structure showed in Fig. 10a. The important issue is that the initial site with 4 chips, after the application of $V^4$, remains with 4 chips. Only the adjacent sites and down change their number of chips. Then one applies $H^4$ to obtain Fig. 10b which is similar to the initial configuration, shifted to the right Fig. 10c.

To implement AND gate Fig. 11 and OR gate Fig. 12 we have to connect two wires (this corresponds to a branching of mycelium). In the AND gate two single chips arrive to a central site with 2 chips, so to trigger firing, threshold 3, the signal has to arrive. With $H$ the signal continues to the right, thus the output is 1. The OR gate functions similarly to AND gate but the central site has 3 chips. There is an unwanted signal coming back signal but the computation is made to the right.

The cross-over is demonstrated in Fig. 13. Here we apply four $V$ and $H$ steps. In Fig. 13a we illustrate the crossing of a horizontal signal (by applying $H^4$). For the vertical signal the dynamics is similar but $V^4$ is applied. Figure 13b shows the case when two signals arrive at the junction at the same time.

## 3.5 Other Words of Automaton Updates

For other shorter words, like the usual chip firing (with sides always open) and the words in the set $\mathbf{B} = \{HV, H^2V^2, H^3V^3\}$ we are able to construct wires, the OR and

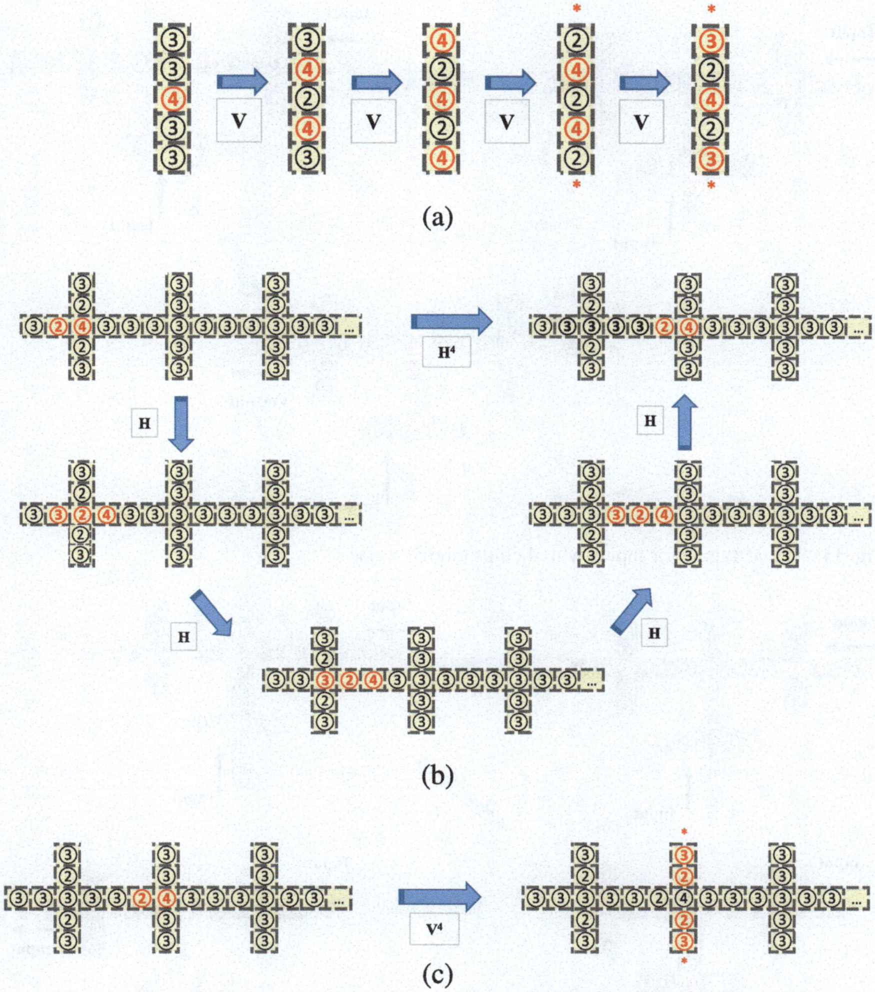


**Fig. 10** Wires. **a** Application of $V^4$ to the wire. **b** Application of $H^4$ to the wire. **c** Final state of the wire after the application of $V^4$

the AND gates, but we are unable to built a cross-over. In such cases we can only implement planar circuits with non-crossing wires.

Below we exhibit the different constructions for those words. It is important to point out that the strategy we used to built the constructions has been by taking as initial framework a quiescent configuration, i.e. a fixed point of the automaton. In [24] it has been proved that with this strategy, for a two dimensional grid with the von Neumann or Moore neighbourhoods it is impossible to cross information, i.e. to built a cross-over. It seems that is also the case for the words in the set **B**. In this

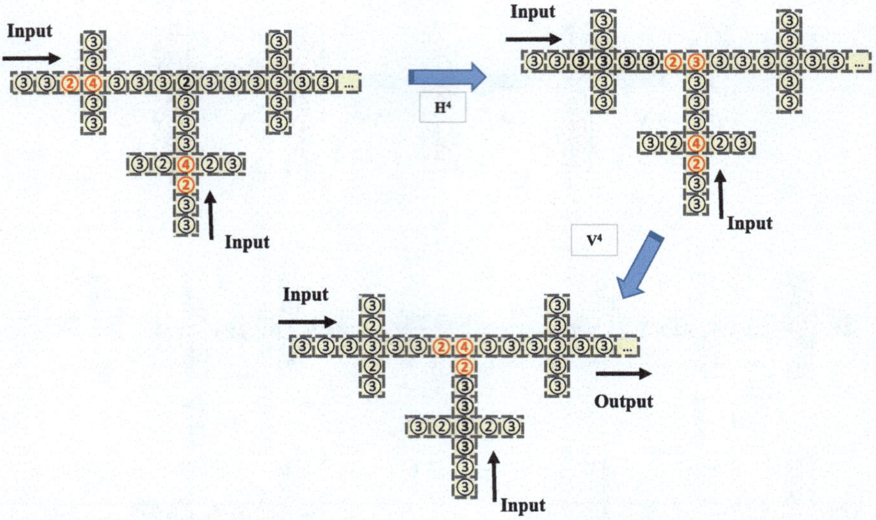


**Fig. 11** The AND gate for inputs with 4 chips (signal = 1)

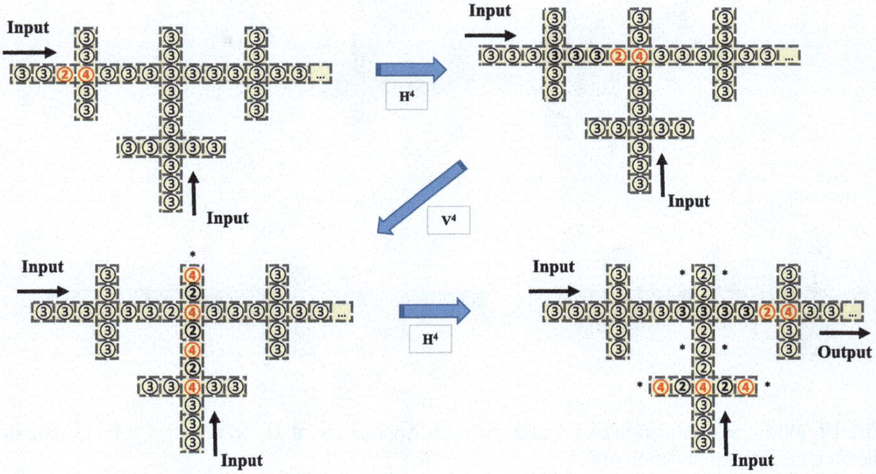


**Fig. 12** The OR gate for one firing input

sense one may say that our result is the best possible: no shortest word allows to cross information, at least following the quiescent strategy.

In previous situations for the usual chip firing automaton the constructions are given in Fig. 14. For the word $HV$: wire is shown in Fig. 15a and the AND gate in Fig. 15. For the word $HHVV$, the wire is shown in Fig. 16a, the OR gate is shown in Fig. 16a and the AND gate in Fig. 16c.

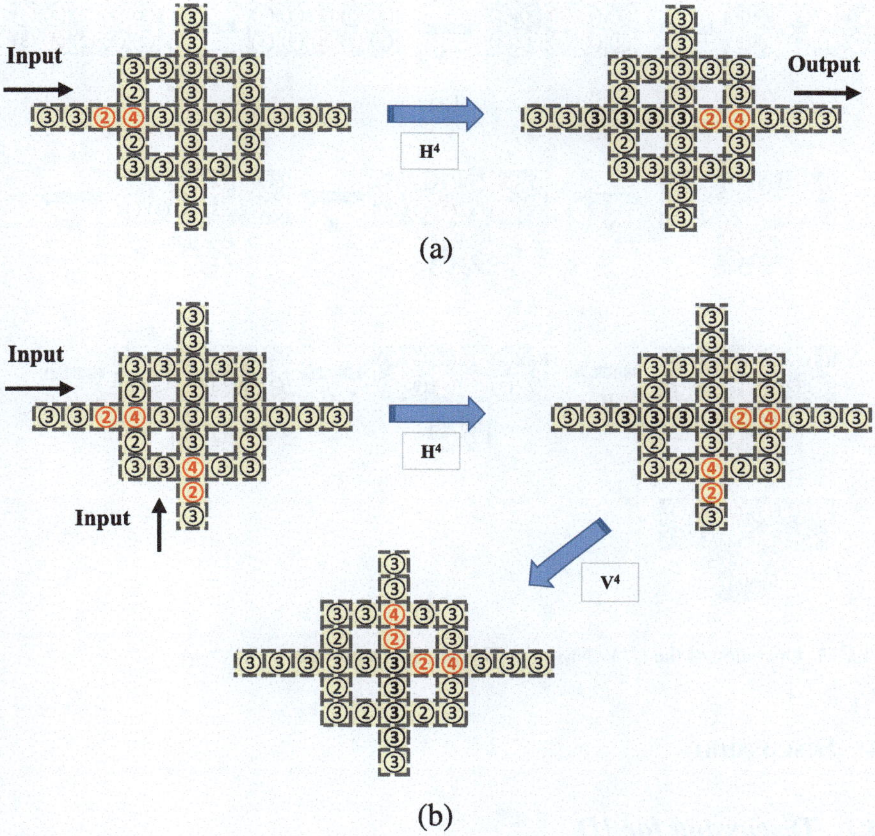


**Fig. 13** The cross-over **a** with a horizontal signal and **b** for two signals

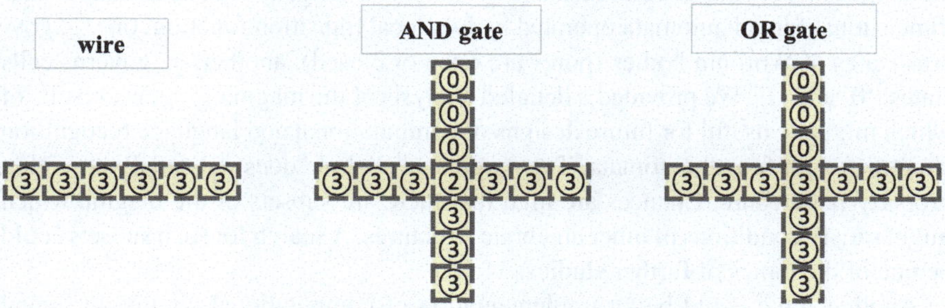


**Fig. 14** The wire and gates for the classical chip firing automaton: every side is open

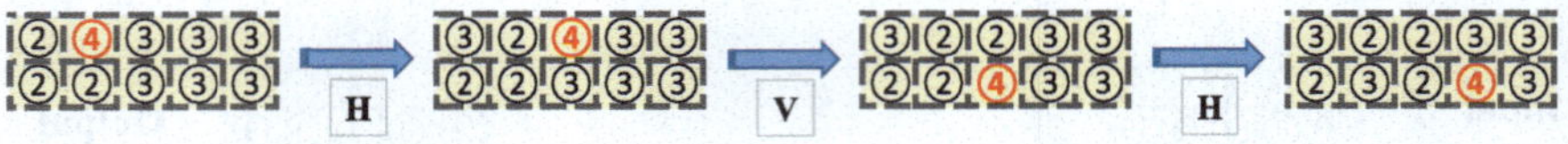


(a)

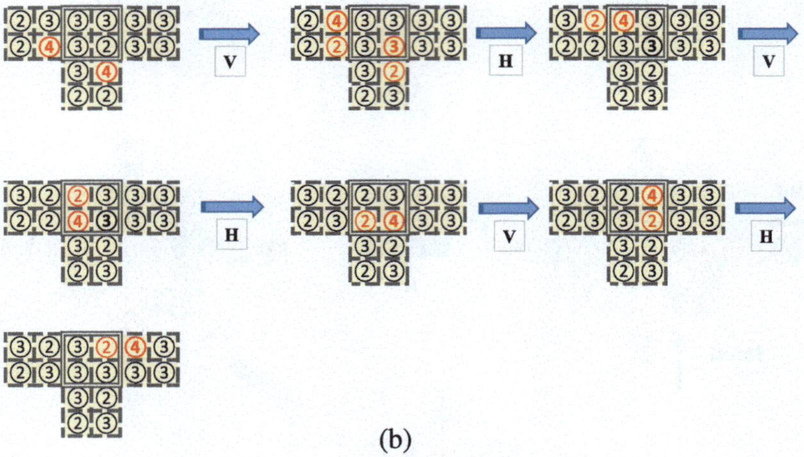


(b)

**Fig. 15** Operation of the $(HV)^*$ word. **a** The wire. **b** The OR gate

# 4 Discussion

## *4.1 Discussion for 1D*

As a first step towards formalisation of fungal intelligence we introduced one-dimensional fungal automata operated by two local transition function: one, $g$, governs states of Woronin bodies (pores are open or closed), another, $f$, governs cells states: '0' and '1'. We provided a detailed analysis of the magma $\langle f, g, \circ \rangle$, results of which might be useful for future designs of computational and language recognition structures with fungal automata. The magma as a whole does not satisfy any other property but closure. Chances are high that there are subsets of the magma which might satisfy conditions of other algebraic structures. A search for such subsets could be one of the topics of further studies.

Another topic could be an implementation of computational circuits in fungal automata. For certain combination of $f$ and $g$ we can find quite sophisticated families of stationary and travelling localisations and many outcomes of the collisions and interactions between these localisations, an illustration is shown in Fig. 17. Thus the target could be, for example, to construct a $n$-binary full adder which is as compact in space and time as possible.

The theoretical results reported show that by controlling just a few cells with Woronin bodies it is possible to drastically change dynamics of the automaton array.

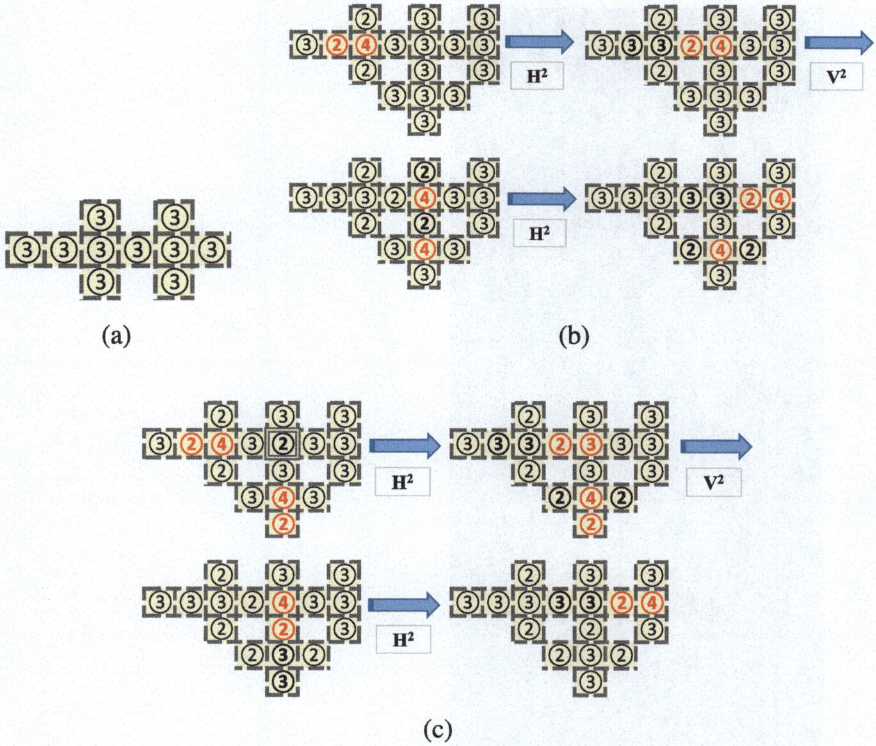


**Fig. 16** Operation of the $(HHVV)^*$ word. **a** The wires. **b** The OR gate. **c** The AND gate

Third direction of future studies could be in implemented information processing in a single hypha. In such a hypothetical experimental setup input strings will be represented by arrays of illumination and outputs could be patterns of electrical activity recorded from the mycelium hypha resting on an electrode array.

## *4.2 Discussion for 2D*

Using sandpile, or chip firing, automata we proved that Fungal Automata are computationally universal, i.e., by arranging positions of branching in mycelium it is possible to calculate any Boolean function.

The structure of Fungal Automata presented can be relaxed by consider the site firing chips only when it has as many chips as open side. In present model, since at each step there are only two sides can be open, the firing threshold is 2. In this situation, the wire AND and the OR gates can be built as in previous cases but not the cross-over. Dynamics of the wire is shown in Fig. 18, the AND gate in Fig. 19.

**Fig. 17** An example of 5-inputs-7-outputs collision in $\mathcal{M}_2$, $\rho_f = 110$, $\rho_g = 40$. Every 50th cell has a Woronin body. Cells state transitions are shown on the left, Woronin bodies state transitions on the right. A pixel in position $(i, t)$ is black if $x_i^t = 1$, left, or $w_i^t = 1$, right

**Fig. 18** Two chip-firing wire

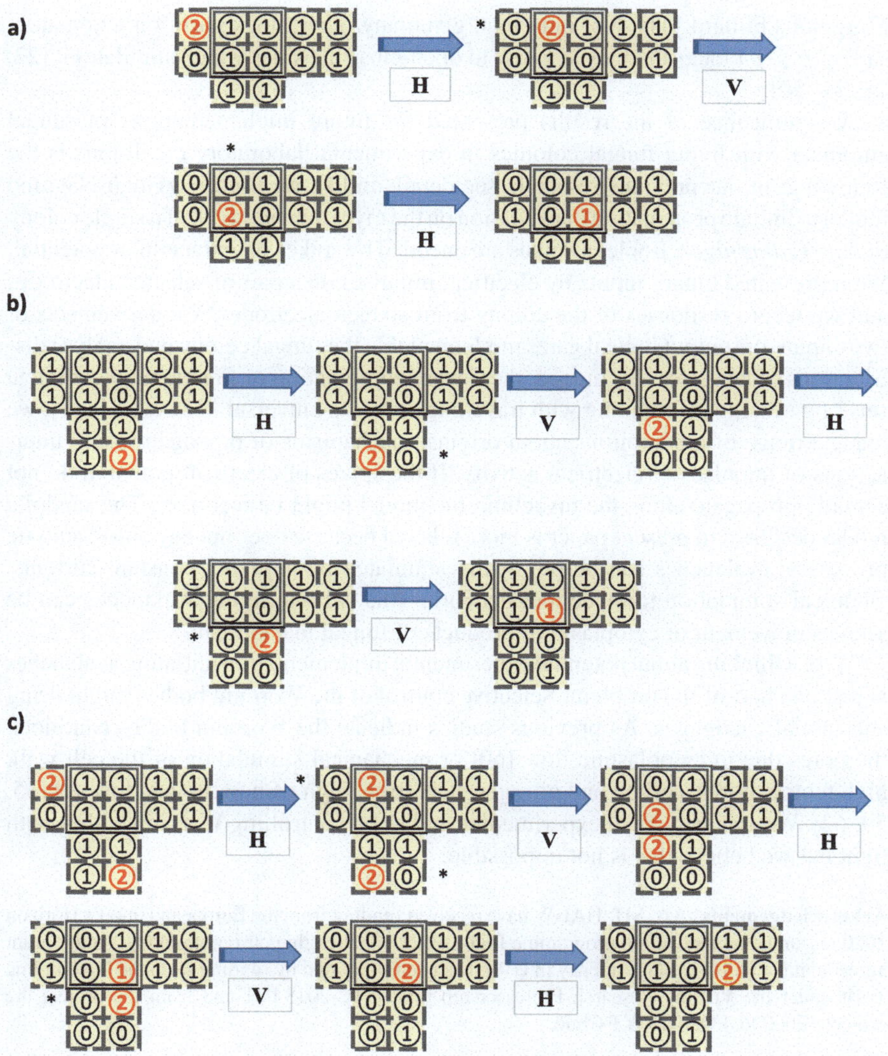


**Fig. 19** The AND gate for firing threshold 2. **a** One horizontal signal. **b** One vertical signals. **c** Two signals and one output signal

Consider the $N \times N$ grid $\{0, 1, 2, ..., N-1\} \times \{0, 1, 2, ..., N-1\}$. Another possibility to open-close sides could be the following: at even steps $t = 0, 2, ...,$ open the even rows and columns and at odd steps, $t = 1, 3, 5, ...$ open the odd rows and columns.

If we do that we simulate exactly the Margolous partitions ($2 \times 2$ blocks) [40]. This give us another way to determine the universality and, in this case, reversibility of this specific Fungal Automaton, because with this strategy one may simulate the

Margolous billiard [41]. Not only that, given any other block partition automaton, say by $p \times p$ blocks there exist a way to open-close the sides which simulates it [22, 23, 34, 50].

A significance of the results presented for future implementations of fungal automata with living fungal colonies in experimental laboratory conditions is the following: in our previous research, see details in [3], we used FitzHugh-Nagumo model to imitate propagation of excitation on the mycelium network of a single colony of *Aspergillus niger*. Boolean values are encoded by spikes of extracellular potential. We represented binary inputs by electrical impulses on a pair of selected electrodes and we record responses of the colony from sixteen electrodes. We derived sets of two-inputs-on-output logical gates implementable the fungal colony and analyse distributions of the gates [3]. Indeed, there were combination of functionally complete sets of gates, thus computing with travelling spikes is universal. However, in [3], we made a range of assumptions about origins, mechanisms of propagation and interactions of impulses of electrical activity. If the spikes of electrical potential do not actually propagate along the mycelium the model might be incorrect. The sandpile model designed in present paper is more relaxed because does not any auto-catalytic processes: avalanches can be physically simulated by applying constant currents, chemical stimulation to mycelium network. This is because the avalanches can be seen as movement of cytoplasm of products of fungal metabolism.

Whilst thinking about potential experimental implementation initiating avalanches is just one part of the problem. Selective control of the Woronin bodies might bring substantial challenges. As previous studies indicate the Woronin bodies can block the pores due to cytoplasmic flow [60] or mechanical stimulation of the cell wall, high temperatures, carbon and nitrogen starvation, high osmolarity and low pH [35, 59, 62]. We are unaware of experimental studies on controlling Woronin bodies with light but we believe this is not impossible.

**Acknowledgements** AA, MT, HABW have received funding from the European Union's Horizon 2020 research and innovation programme FET OPEN "Challenging current thinking" under grant agreement No 858132. EG residency in UWE has been supported by funding from the Leverhulme Trust under the Visiting Research Professorship grant VP2-2018-001 and from the project the project 1200006, FONDECYT-Chile.

## References

1. Andrew A (2018) On spiking behaviour of oyster fungi Pleurotus djamor. Sci Rep 7873
2. Adamatzky A (2018) Towards fungal computer. Interface focus 8(6):20180029
3. Andrew Adamatzky, Martin Tegelaar, Han AB Wosten, Anna L Powell, Alexander E Beasley, and Richard Mayne. On boolean gates in fungal colony. *arXiv preprint* arXiv:2002.09680, 2020
4. Alexander E Beasley, Anna L Powell, and Andrew Adamatzky. Memristive properties of mushrooms. *arXiv preprint* arXiv:2002.06413, 2020
5. Beck J, Ebel F (2013) Characterization of the major Woronin body protein hexa of the human pathogenic mold aspergillus fumigatus. Int J Med Microbiol 303(2):90–97

6. Michael W Berns, James R Aist, William H Wright, and Hong Liang. Optical trapping in animal and fungal cells using a tunable, near-infrared titanium-sapphire laser. *Experimental cell research*, 198(2):375–378, 1992
7. Bitar J, Goles E (1992) Parallel chip firing games on graphs. Theoret Comput Sci 92(2):291–300
8. Robert-Jan Bleichrodt, Marc Hulsman, Han AB Wösten, and Marcel JT Reinders. Switching from a unicellular to multicellular organization in an aspergillus niger hypha. *MBio*, 6(2):e00111–15, 2015
9. Robert-Jan Bleichrodt, G Jerre van Veluw, Brand Recter, Jun-ichi Maruyama, Katsuhiko Kitamoto, and Han AB Wösten. Hyphal heterogeneity in aspergillus oryzae is the result of dynamic closure of septa by Woronin bodies. *Molecular microbiology*, 86(6):1334–1344, 2012
10. Robert-Jan Bleichrodt, Arman Vinck, Nick D Read, and Han AB Wösten. Selective transport between heterogeneous hyphal compartments via the plasma membrane lining septal walls of aspergillus niger. *Fungal Genetics and Biology*, 82:193–200, 2015
11. Bonfante P, Anca I-A (2009) Plants, mycorrhizal fungi, and bacteria: a network of interactions. Ann Rev Microbiol 63:363–383
12. Carlile MJ, Watkinson SC, Gooday GW (2001) The fungi. Gulf Professional Publishing
13. Chessa A, Eugene Stanley H, Vespignani A, Zapperi S (1999) Universality in sandpiles. Phys Rev E 59(1):R12
14. Christensen K, Fogedby HC, Jensen HJ (1991) Dynamical and spatial aspects of sandpile cellular automata. J Stat Phys 63(3-4):653–684
15. Christensen M (1989) A view of fungal ecology. Mycologia 81(1):1–19
16. Collinge AJ, Markham P (1985) Woronin bodies rapidly plug septal pores of severedpenicillium chrysogenum hyphae. Exp Mycol 9(1):80–85
17. Cook M (2004) Universality in elementary cellular automata. Complex Syst 15(1):1–40
18. Cooke RC, Rayner ADM et al (1984) Ecology of saprotrophic fungi. Longma
19. Dai Y-C, Cui B-K (2011) Fomitiporia ellipsoidea has the largest fruiting body among the fungi. Fungal Biol 115(9):813–814
20. Deutsch P, Gailly J-L (1996) Zlib compressed data format specification version 3.3. Technical report
21. Dhar D (1990) Self-organized critical state of sandpile automaton models. Phys Rev Lett 64(14):1613
22. Durand-Lose J (2001) Representing reversible cellular automata with reversible block cellular automata. Discret Math Theor Comput Sci 145:154
23. Durand-Lose JO (2000) Reversible space–time simulation of cellular automata. Theor Comput Sci 246(1-2):117–129
24. Gajardo A, Goles E (2006) Crossing information in two-dimensional sandpiles. Theor Comput Sci 369(1-3):463–469
25. Goles E (1961) Sand piles, combinatorial games and cellular automata. In: Instabilities and nonequilibrium structures III. Springer, pp 101–121
26. Goles E (1992) Sand pile automata. In Annales de l'IHP Physique théorique 56:75–90
27. Goles E, Margenstern M (1996) Sand pile as a universal computer. Int J Mod Phys C 7(02):113–122
28. Goles E, Margenstern M (1997) Universality of the chip-firing game. Theor Comput Sci 172(1–2):121–134
29. Greenlaw R, James Hoover H, Ruzzo WL et al (1995) Limits to parallel computation: P-completeness theory. Oxford University Press on Demand
30. Griffin DM et al (1972) Ecology of soil fungi. Ecology of soil fungi.
31. Held M, Edwards C, Nicolau DV (2008) Examining the behaviour of fungal cells in microconfined mazelike structures. In: Imaging, manipulation, and analysis of biomolecules, cells, and tissues VI, vol 6859. International Society for Optics and Photonics, p 68590U
32. Held M, Edwards C, Nicolau DV (2009) Fungal intelligence; or on the behaviour of microorganisms in confined micro-environments. In: Journal of physics: conference series, vol 178. IOP Publishing, p 012005

33. Howard PG (1993) The design and analysis of efficient lossless data compression systems. PhD thesis, Citeseer
34. Imai K, Morita K (2000) A computation-universal two-dimensional 8-state triangular reversible cellular automaton. Theor Comput Sci 231(2):181–191
35. Jedd G (2011) Fungal evo-devo: organelles and multicellular complexity. Trends Cell Biol 21(1):12–19
36. Jedd G, Chua N-H (2000) A new self-assembled peroxisomal vesicle required for efficient resealing of the plasma membrane. Nat Cell Biol 2(4):226–231
37. Leonhardt Y, Kakoschke SC, Wagener J, Ebel F (2017) Lah is a transmembrane protein and requires spa10 for stable positioning of Woronin bodies at the septal pore of aspergillus fumigatus. Sci Rep 7:44179
38. Lew RR (2005) Mass flow and pressure-driven hyphal extension in neurospora crassa. Microbiology 151(8):2685–2692
39. Lindgren K, Nordahl MG (1990) Universal computation in simple one-dimensional cellular automata. Complex Syst 4(3):299–318
40. Margolus N (1984) Physics-like models of computation. Phys D 10(1):81–95
41. Margolus N (2002) Universal cellular automata based on the collisions of soft spheres. In: Adamatzky A (ed) Collision-based computing. Springer, pp 107–134
42. Martin B (2013) On goles' universal machines: a computational point of view. Theor Comput Sci 504:83–88
43. Martínez GJ, McIntosh HV, Seck-Tuoh Mora JC (2003) Production of gliders by collisions in rule 110. In: European conference on artificial life. Springer, pp 175–182
44. Martínez GJ, McIntosh HV, Seck-Tuoh Mora JC (2006) Gliders in rule 110. Int J Unconv Comput 2(1):1
45. Martínez GJ, Seck-Tuoh Mora JC, Vergara SVC (2007) Rule 110 objects and other collision-based constructions. J Cellular Autom 2:219–242
46. Maruyama J-i, Juvvadi PR, Ishi K, Kitamoto K (2005) Three-dimensional image analysis of plugging at the septal pore by woronin body during hypotonic shock inducing hyphal tip bursting in the filamentous fungus aspergillus oryzae. Biochem Biophys Res Commun 331(4):1081–1088
47. Momany M, Richardson EA, Van Sickle C, Jedd G (2002) Mapping Woronin body position in aspergillus nidulans. Mycologia 94(2):260–266
48. Moore C, Nilsson M (1999) The computational complexity of sandpiles. J Stat Phys 96(1–2):205–224
49. Moore RT, McAlear JH (1962) Fine structure of mycota. 7. observations on septa of ascomycetes and basidiomycetes. Am J Botany 49(1):86–94
50. Morita K, Harao M (1989) Computation universality of one-dimensional reversible (injective) cellular automata. IEICE Trans (1976-1990) 72(6):758–762
51. Neary T, Woods D (2006) P-completeness of cellular automaton rule 110. In: International colloquium on automata, languages, and programming. Springer, pp 132–143
52. Ng SK, Liu F, Lai J, Low W, Jedd G (2009) A tether for Woronin body inheritance is associated with evolutionary variation in organelle positioning. PLoS Genetics 5(6)
53. Olsson S, Hansson BS (1995) Action potential-like activity found in fungal mycelia is sensitive to stimulation. Naturwissenschaften 82(1):30–31
54. Rayner ADM, Boddy L et al (1988) Fungal decomposition of wood. Its biology and ecology. Wiley
55. Reichle RE, Alexander JV (1965) Multiperforate septations, Woronin bodies, and septal plugs in fusarium. J Cell Biol 24(3):489
56. Roelofs G, Koman R (1999) PNG: the definitive guide. O'Reilly & Associates, Inc.
57. Slayman CL, Scott Long W, Gradmann D (1976) "Action potentials" in Neurospora crassa, a mycelial fungus. Biochimica et Biophysica Acta (BBA)-Biomembranes 426(4):732–744
58. Smith ML, Bruhn JN, Anderson JB (1992) The fungus Armillaria bulbosa is among the largest and oldest living organisms. Nature 356(6368):428

59. Soundararajan S, Jedd G, Li X, Ramos-Pamploña M, Chua NH, Naqvi NI (2004) Woronin body function in magnaporthe grisea is essential for efficient pathogenesis and for survival during nitrogen starvation stress. The Plant Cell 16(6):1564–1574
60. Steinberg G, Harmer NJ, Schuster M, Kilaru S (2017) Woronin body-based sealing of septal pores. Fungal Genet Biol 109:53–55
61. Tegelaar M, Bleichrodt R-J, Nitsche B, Ram AFJ, Wösten HAB (2020) Subpopulations of hyphae secrete proteins or resist heat stress in aspergillus oryzae colonies. Environ Microbiol 22(1):447–455
62. Tegelaar M, Wösten HAB (2017) Functional distinction of hyphal compartments. Sci Rep 7(1):1–6
63. Tenney K, Hunt I, Sweigard J, Pounder JI, McClain C, Bowman EJ, Bowman BJ (2000) Hex-1, a gene unique to filamentous fungi, encodes the major protein of the woronin body and functions as a plug for septal pores. Fungal Genet Biol 31(3):205–217
64. Tey WK, North AJ, Reyes JL, Lu YF, Jedd G (2005) Polarized gene expression determines Woronin body formation at the leading edge of the fungal colony. Mol Biol cell 16(6):2651–2659
65. Trinci APJ, Collinge AJ (1974) Occlusion of the septal pores of damaged hyphae ofneurospora crassa by hexagonal crystals. Protoplasma 80(1-3):57–67
66. Wergin WP (1973) Development of Woronin bodies from microbodies infusarium oxysporum f. sp. lycopersici. Protoplasma 76(2):249–260
67. Wolfram S (1984) Universality and complexity in cellular automata. Phys D 10(1–2):1–35
68. Wolfram S (1994) Cellular automata and complexity: collected papers. Addison-Wesley Pub. Co
69. Ziv J, Lempel A (1977) A universal algorithm for sequential data compression. IEEE Trans Inf Theory 23(3):337–343

The manufacturer's authorised representative in the EU is Springer Nature Customer Service Centre GmbH, Europaplatz 3, 69115 Heidelberg, Germany. If you have any concerns regarding our products, please contact ProductSafety@springernature.com

Printed and bound by CPI Group (UK) Ltd, Croydon, CR0 4YY
06/10/2026
02225252-0015